Project Management of Large Software-Intensive Systems

Controlling the Software Development Process

Project Management of Large Software-Intensive Systems

Controlling the Software Development Process

By

Marvin Gechman

CRC Press
Taylor & Francis Group
Boca Raton London New York

CRC Press is an imprint of the
Taylor & Francis Group, an **informa** business

CRC Press
Taylor & Francis Group
6000 Broken Sound Parkway NW, Suite 300
Boca Raton, FL 33487-2742

© 2019 by Marvin Gechman
CRC Press is an imprint of Taylor & Francis Group, an Informa business

No claim to original U.S. Government works

Printed on acid-free paper

International Standard Book Number-13: 978-0-367-13671-0 (Paperback)

Visit the Taylor & Francis Web site at
http://www.taylorandfrancis.com

and the CRC Press Web site at
http://www.crcpress.com

Contents

SECTION 2 SOFTWARE MANAGEMENT AND QUALITY DOMAINS

SECTION 4 SYSTEMS AND SOFTWARE ENGINEERING DOMAINS

SECTION 5   CRITICAL SOFTWARE ELEMENTS DOMAIN

Foreword

Software-intensive systems have become the norm for everything from interpersonal communications to aircraft systems. In today's world, the development of successful products is highly dependent on a well thought out Systems Engineering process. In its absence, costs skyrocket, development timelines stretch out and many development programs fail completely.

Project Management of Large Software-Intensive Systems lays out a proven and extensive set of methods, processes and considerations that go into managing a large software development program. Whether you are a student or an experienced Program Manager, you will find a wealth of information in these pages.

Marvin Gechman brings to bear a lifetime of personal experience in managing large software projects, including the Space Station, information and communications systems and missile defense programs. His insights add immeasurably to the detailed descriptions of each stage and process throughout the book. Careful attention is paid to getting the requirements right and establishing clear, measurable indices for success. This book provides detailed guidance that will benefit all Project Managers in executing successful software development efforts.

Mica R. Endsley, PhD
U.S. Air Force Chief Scientist (2013–2015)

Preface

Software is both ubiquitous and critical for the operation of all major system developments.

This is a Software Project Manager's Guidebook and reference manual hereafter referred to as the "Guidebook." The purpose of this Preface is to provide introductory comments and describe the book's goal, how it is organized, who should read it, its objectives, tactics used to achieve those objectives and why tailoring the recommendations in this Guidebook to the needs of your project is essential.

Successfully managing the development of large complex software-intensive systems is difficult. This Guidebook is specifically directed to Software Project Managers (SPM); however, it is a valuable resource and reference for all members of the System Development Team, all stakeholders, and is a foundation of knowledge for all those who someday aspire to become a successful SPM.

Goals

The Guidebook is focused on medium-to-large systems with the intent that smaller systems will tailor out what does not apply to their needs. Large systems can include millions of lines of software code, produced by multiple developers geographically dispersed, where the software interfaces with a multitude of hardware elements as well as with human interfaces, and it all must work within the context of an environment of policies and regulations. Making sure you understand how to make all of that come together seamlessly, and perform the intended system functions, is the principal goal of this Guidebook.

Providing prescriptive methods of exactly how to develop medium-to-large software-intensive systems is *not* a goal or the intent. The principal goal is to essentially provide a *shopping list* of all the tasks and activities with which you *could* be involved with as the Project Manager of a complex software-intensive system. If you are managing a large system, you need to make sure that *all the critical activities* are performed. The other activities and tasks should be performed only if they are *applicable* to your project and *add value*.

Systematic Processes and Structure

Many will debate whether Software Engineering (SwE) has yet achieved the distinction of being a *real* recognized engineering discipline as opposed to an art. When those critics understand the *breadth and depth* of what is involved in producing and successfully delivering large, complex software-intensive systems, they should conclude that SwE is well on its way to *achieving the rigor, the structure and systematic processes* needed to qualify as a true engineering discipline.

The reality is that developing software-intensive systems is both an *art and a science*. The deep and insightful creativity needed to produce complex software-intensive systems will *always involve some elements of an art*. The intent of this Guidebook is to nudge SwE more toward a systematic science because you must have structure. The dictionary has five definitions of science and four of them include the word "systematized." The dictionary also defines "structure" as "interrelated parts ... put together systematically." Therefore, it can be concluded that *system and structure* are so interrelated that you *cannot have a workable system without systematic processes and structure*.

> **Lessons Learned.** Regarding structured processes, the simplistic rule I follow is: do not implement excessive software development *processes* to a point where you are trying to *kill a mouse with an elephant gun*; at the same time, never cut back the software development *process* so much that you are trying to *catch an elephant with a mouse trap*! Keep the process as simple as possible, do what needs to be done and don't cut corners. Even though common sense is not common, always use common sense.

The Phantom Silver Bullet

In his classic 1975 book *The Mythical Man Month* (Brooks, 1995 update), Frederick Brooks concluded that there was no "silver bullet" on the horizon to provide a tenfold improvement in software productivity. Today, there is still no single approach, or magical task, that can be undertaken to solve or mitigate *all* the problems faced by Project Managers in managing the development of software-intensive systems, and delivering them on time, within budget, and with the needed functionality.

Regardless, there have been very substantial improvements over the past four decades. Even if those improvements were a tenfold improvement in productivity, doing so over 40 plus years cannot realistically be considered a *magical* silver bullet. Following the management and process recommendations in this Guidebook *will not guarantee* success of your software-intensive project. However, if large complex software-intensive systems follow just a few, or none, of the applicable processes, *serious problems are guaranteed* and the potential for failure will be extremely high.

Organization of this Guidebook

Effective Software Project Managers must address, and successfully administer, *many activities* and the tasks that comprise those activities. A good way to do that is by breaking down the myriad of activities into smaller, more manageable entities, and organizing them so they can be addressed in a systematic and thorough manner. Following that approach, Software Project Management activities and tasks have been grouped into *six functional interactive domains* as depicted in Figure P.1.

The six domains that comprise the Software Project Manager's functional areas are shown as *six overlapping, synergistic and interactive ellipses of a Venn diagram*. Each ellipse in Figure P.1 represents a Software Project Management *Domain* and these six domains constitute the theme and basic framework of this Guidebook. There are 16 chapters,

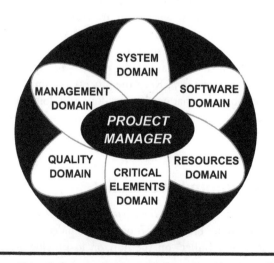

Figure P.1 Software Project Manager's Six Functional Interactive Domains

divided into five parts, plus ten appendices, as shown in Tables P.1 and P.2. *Part 1 should be reviewed by all readers of this Guidebook.*

The appendices are intended to supplement the 16 chapters with additional details. The Index in Appendix J is especially important as it is a major asset in locating specific topics of interest since many topics are discussed in multiple locations.

Who Should Read This Guidebook

This Guidebook will be a valuable resource to the following users, regardless of their industry:

■ *Software Project Managers*: The Guidebook is specifically directed to Software Project Managers at all levels—especially to Software Managers who have little or no experience at managing large, complex software-intensive *systems*. Its intent is to provide recommendations and describe proven practices that will help achieve successful project implementations. Experienced Project Managers will find this Guidebook valuable because it provides up-to-date insights into structured management processes and approaches that have been proven to work.

■ *Programmer Managers*: While in college, Programmers are taught the technical basics of their skill and then they expand that skill in the workforce. Eventually, they earn and achieve a position of leadership and management. While they have the technical skills, they may not have any, or very little, management skills when they are promoted to a Software Manager position. If you are in that position, stop whatever else you are doing and read this Guidebook.

■ *Project Teams*: Although this Guidebook is focused on project management, it is a valuable asset to *all team members* on any software-intensive system under development. It will help the teams understand what is (or what should be) going on during the development process. Its greatest value to the Project Team will be to help make sure that every member of your team *understands what they should be doing* at each stage of the development process. Other Software Team members with specific skill sets must read sections focused on their area of specialization such as:
  – Software Risk Management Team (Section 5.2)
  – Software Configuration Management Team (Section 6.3)
  – Software Quality Team (Chapter 6)
  – Software Measurements Team (Chapter 15)

Also, if you are a team member of a software project that is thrashing about and showing signs of severe management

Table P.1 Organization of the Software Project Management Guidebook

Part	Part Name	Chapter Name	Semester
One	Software Project Management Fundamentals	1. Software Project Management Introduction 2. Software Project Management Activities 3. Software Life Cycle Processes 4. Software Development Methodologies	One
Two	Software Project Management and Quality Domains	5. Software Management Domain 6. Software Quality Domain	
Three	Managing the Software Resources Domain	7. Managing the Software Project Team 8. Managing Software Costs and Schedules 9. Managing Software Facilities, Reuse and Tools	
Four	Managing the Systems and Software Engineering Domains	10. Systems Engineering Domain 11. Software Engineering Domain 12. Collaboration of Systems and Software Engineering	Two
Five	Managing the Critical Software Elements Domain	13. Software Documentation Work Products 14. Software Estimating Methods 15. Managing Software Performance with Measurements 16. Managing Software System Sustainment	

Table P.2 Organization of the Guidebook Appendices

Appendix	Appendix Name
A	Software Acronyms
B	Software Definitions
C	Software Roles and Responsibilities for Skill Groups
D	Criteria for Evaluating COTS and Reusable Software Products
E	Pre-Milestone A Plans and Strategies
F	Annotated Outline of the Software Development Plan
G	Exit Criteria for Software Reviews
H	Chapter Highlights
I	References by Category
J	Index

problems, be a "good Samaritan" and surreptitiously put a copy of this Guidebook on your manager's desk.

■ *College Students*: Regardless of their curriculum, many college students will wind up working in the high tech industry involving computers and software. All students, especially those in the *Computer Science* curriculum, can gain an extensive oversight from this Guidebook of what is really involved in

the development of medium-to-large-scale software-intensive systems. That insight may not be covered, or not well covered, in a typical college curriculum. Furthermore, this Guidebook provides a perspective of the tasks and responsibilities of a Software Manager to prepare students for the time when one day they might become a project or Program Manager. The Guidebook is organized so Parts 1–3 can be covered in the first semester and Parts 4–5 in the second semester.

■ *System Engineers*: Chapter 10 describes the Systems Engineering Domain, so it is a must read for all System Engineers working on your project. In addition, Chapter 12 is a detailed description of the critical need for the *collaboration* of System Engineers and Software Engineers—especially in large system developments. Chapter 12 provides System Engineers with an understanding of what support they need to provide *to* Software Engineers and what support they can expect (and need) *from* Software Engineers during the System Development Life Cycle. The Guidebook emphasizes this collaboration because it is often overlooked and is critical to the success of large software-intensive systems.

■ *Information Technology (IT) Departments:* All technical staff members working in IT departments, especially the Project Managers, will find this Guidebook a valuable asset.

■ *Senior Management*: Non-software Senior Managers can also gain a valuable understanding of the software development process from this Guidebook. That increase in understanding is likely to result in indispensable support senior management can provide to your project. Senior Managers do not have to read the entire Guidebook to gain that understanding. Chapters 2–3 can be considered an executive overview, possibly augmented with Chapter 1.

■ *Business School Students:* College students majoring in management, business, and business-related topics, can gain important management insights from chapters such as:

- Software Project Management Introduction (Chapter 1)
- Software Project Management Activities (Chapter 2)
- Software Project Management Domain (Chapter 5)
- Software Quality Management Domain (Chapter 6)
- Managing the Project Team (Chapter 7)
- Managing Software Costs and Schedules (Chapter 8)
- Managing Software Facilities, Reuse and Tools (Chapter 9)
- Software Estimating Methods (Chapter 14)
- Project Performance Oversight with Measurements (Chapter 15)
- Managing Software System Sustainment (Chapter 16)

■ *Program Management Office (PMO):* Customers frequently establish a PMO, also called the *Contract Acquisition Team*, directly responsible for management and control of the entire *program* and the *projects* that comprise the program. Many federal and military software-intensive systems have an Acquisition Team often called the *Acquisition Program Office* (APO). The APO is usually composed of military members, and government-funded contractors, responsible for *assuring mission success*. Also, your organization may have a corporate-wide PMO that has produced policies, procedures, tools, etc. for you to use and follow.

Government-funded contractors may be called *Systems Engineering Technical Advisors* (SETAs) who are commercial companies hired for technical support. The acquisition teams also typically include engineers from *Federally Funded Research and Development Centers* (FFRDCs) who work for independent nonprofit companies, maintaining centers of excellence in various technical disciplines providing technical support to government agencies.

Regardless of the team they are part of, every member of the PMO can gain from this Guidebook an invaluable understanding of the Software Project Management and development *processes*, as well as the specific *tasks they need to track* to make sure those processes are performed in a satisfactory and timely manner. Military members assigned to an APO who have the oversight responsibility, but have little or no previous experience managing large software-intensive systems, should find this Guidebook to be indispensable.

Guidebook Objectives

In support of the above goals, the objectives of this Guidebook were to make it *comprehensive, easy to understand and as brief, simple and interesting as possible*. Clearly, there are no simple solutions to complex problems, and the management of large software-intensive systems is certainly not simple. Another major objective was to plan the Guidebook so it would have a *wide audience*—by defining the fundamentals for students and new Software Project Managers in the early chapters and then drilling down into the details for experienced Managers. The details may be overwhelming for neophytes.

Managing large complex software-intensive systems involves many interactive elements. It can get complicated, so attempting to document the full process in a relatively simple and brief manner is difficult but achievable. The key strategy is to describe the project management approach within a *structured framework*.

Structured Framework. There are *36 principal activities* (discussed in Chapter 2) which a Software Project Manager must address. Of course, each activity is composed of several tasks that can collectively add up to approximately 250 tasks. I have been through the process of requirements analysis, design, development, test and implementation of both large and small software-intensive systems many times and that knowledge has been captured and documented. In general, the structured framework in this Guidebook includes:

■ *Systematic steps* needed to achieve successful deliveries of large, complex software-intensive systems including when and with whom *collaborations* need to take place.
■ *Work products* needed, when they are needed, and who should prepare them.
■ *Pitfalls* that can, and usually do, jeopardize the success of the system (Murphy's Law applies).
■ *Tools* that could or should be used for effective management and project control.
■ *Measurements* needed to provide useful metrics to *assess project status* and enable Software Project Managers to take Corrective Actions in a timely manner.

Lessons Learned. If you ask ten Software Engineers the same question, you will usually get 8–12 different answers, so it will not be a surprise if experienced Software Project Managers do not agree with *all* the guidelines and recommendations presented. The principles and guidelines described are based on my 57 years of practical experience covering a wide range of software application areas. Since I have *never* been part of a failed software-intensive project, there is a great track record to substantiate the recommendations and guidelines described. Why have I not been part of a failed software-intensive project? Because I always follow a well-structured, but tailorable framework, and that is what this Guidebook is all about. (Okay, perseverance, focus and hard work also played a big part.)

Tactics Used to Achieve Objectives

In order to make the Guidebook as useful, simple and understandable as possible, and to make sure it contains the relevant and essential information as succinctly as possible, the following tactics were used:

- *Graphics*: A graphical emphasis is heavily displayed in this Guidebook to more clearly describe the software development processes and the tasks involved. There are over *80 figures*; if "a picture is worth a thousand words" a lot of text is eliminated. Also, figures contain information in a format that is easier to understand when describing interrelationships.

- *Short Stories*: Short stories are a unique feature of this Guidebook. Stories are captivating; they engage the reader. Technical management books tend to be dry and often difficult to read. Short stories make them more interesting and easier to read. For those reasons, this Guidebook interlaces 85 short stories called *Lessons Learned* into related topic areas. All the Lessons Learned stories are true events experienced by the author and some are topic-related anecdotes.

- *Tables, Bulleted Lists and Acronyms*: Tables and bulleted lists are used to organize a wealth of information in a minimum amount of space. There are over 130 tables plus many bulleted lists. Common acronyms are used throughout the Guidebook. Acronyms are used in the real world, so newcomers to the Software Engineering industry need to start using this technique of reference.

- *Stay Focused on the Subject*: Since Software Engineering covers a wide range of subject areas, it is easy to understand how other authors veer off their main topic and incorporate subjects that they are interested in or know a lot about. Related topics are only included if they apply to *managing* the processes and ultimate delivery of a successful software-intensive system.

- *Simplicity*: A mandatory objective was to keep it as *simple as possible* and not get so deep into a subject that it becomes time-consuming and difficult to understand. Readers who need to delve deeply into a subject should review books written on that subject (see Appendix I). Many books have been written on essentially every subject covered in this Guidebook. Providing the readers with as much information as possible, clearly written, and in as few words as reasonable, was a primary goal. It is up to the reader to decide how close it comes to what is called "elegant simplicity." If I may quote Leonardo da Vinci: "*Simplicity is the ultimate sophistication.*"

- *Usefulness and Size*: The usefulness of any guidebook diminishes as its size and complexity increases since Software Managers typically do not have the time to dig into a huge esoteric textbook. The intent was to produce a *compendium* and not a 1000 page tome. The Webster definition of a compendium is a "concise but comprehensive treatise," and that is the intent.

- *Easy to Read*: Techniques such as repetition and italics are used for emphasis and to highlight important information; bold italics are used for additional emphasis.

- *Reference Handbook*: It is expected this Guidebook will be read thoroughly the first time and then used often by practitioners as a reference guide. An effort was made to streamline the search for specific topics. The detailed *Index* (Appendix J) is a valuable asset in searching for needed information. The beginning of every chapter, and many sections, describes its contents. Most chapters contain pointers to topics covered in more depth in other sections of the Guidebook. Topics are numbered down to a third level (e.g., 1.2.3). Figures and tables are consecutively numbered preceded by the chapter number (e.g., 2.1, 2.2, 2.3).

- *Roadmaps*: Parts 2–5 of the Guidebook contain the six domains depicted in Figure P.1. A "roadmap" to the contents of each of the six domains is shown in Figure 2.1. Also, Figure 4.4 is a flowchart of the Software Implementation Process with pointers to where each activity is described.

All of these features collectively add an expanded dimension to usability and readability that is helpful because, as you will soon discover, there is a great depth of content in this Guidebook.

Tailoring the Guidebook

The processes and activities described in this Guidebook must be tailored to specific requirements of your project or contractual structure to which it applies. Following are some tailoring guidelines.

Value Added. Tasks that add unnecessary costs, or create work products that *do not add value* to your project, commensurate with the effort they require, *must be* *eliminated*. Consider the value of the task from both a short-term as well as a long-term impact. If a task is on the borderline of providing value added, lean toward performing it. Also, always avoid gold plating (see 5.1.5).

Tailoring. Although process tailoring guidance may be provided by your customer (the organization that prepared your contract and is funding the effort), you are likely to be asked to provide suggested tailoring of the process. Generic tailoring guidance is provided in some software standards. Tailoring of the process by the customer may be specified in their *Statement of Work* (SOW), compliance documents or in the *Contract Data Requirements List* (CDRL) Section of your contract.

The figures and tables used in this Guidebook are *examples*, and they are *expected to be tailored* to the needs of each program as well as compliant with your corporate *Standard Software Process* (SSP) if you have one. Example figures throughout this Guidebook are intentionally made reasonably simple to convey the fundamental concepts.

Small Systems. A small software-intensive system will need to do a significant amount of tailoring to the contents of this Guidebook. A small system will *not* need to, and should not need to, perform many of the described activities. Regardless of the size of your project, always *tailor out everything that is not applicable or feasible or does not add value to your project*. Use this Guidebook as a framework for tailoring it down to meet the needs of your project.

Acknowledgments

After 57 years of work experience, you can imagine the huge list of professionals I have worked with and for, many of whom I am indebted to. Any names not mentioned do not minimize the value of their counsel. But there have been a few individuals who have been exceptionally instrumental in the development of my career. They are: Col F. F. Groseclose at Georgia Tech; Dr. Dimitri Chorafas and Dean Donald Marlow at The Catholic University of America; Ralph Wyscarver at NSA; Doug Climenson at RCA; Don Orkand and Allan Mann at ORI; David Waite at IDC; Dave Brown and Lew Goldish at TMA; Joseph Bennett at IGI; Ray Bellas, Norm Alperin and Allan Slusher at Lockheed Martin; and Suellen Eslinger at The Aerospace Corp. To these individuals, and to the many colleagues from whom I have gained knowledge, wisdom and capability, my *sincerest thank you*.

Also, my insightful wife Annie was an excellent sounding board for ideas, and I am grateful for her invaluable critique and patience. Finally, the quality of this Guidebook was enhanced with the help of the initial manuscript reviewers, especially Mica Jones and Steve Montgomery, initial ideas provided by Aimee Falkenbury, plus the invaluable assistance of John Wyzalek and the entire production teams at Deanta Global and CRC Press—Taylor & Francis Group.

Author

Marvin Gechman has 57 years of practical experience in all aspects of Software Engineering including the role of president or senior management in four Software Engineering consulting firms for 18 years, plus 5 years in the government and 34 years in the aerospace industry where he focused on space, communications and missile defense systems. After graduating from Georgia Tech with a BS in Industrial and Systems Engineering, he worked for the National Security Agency; during that time, he also obtained a MS degree in Systems Engineering and Computer Science providing his systems perspective focused on Software Engineering.

For the next two decades, he was an information systems consultant and software Engineer on large computer automation projects for multiple government agencies, businesses, municipalities, and research laboratories covering a wide range of software application areas. Marv was involved in early library automation, was the principal architect of an information service for businesses in New England, and was the subject of a TV documentary on the information explosion. He worked for two think tanks in the Washington, DC. area, was vice president of a custom programming firm, and program manager at a mass storage company where he managed data storage system installations at major computer centers.

Marv joined the Lockheed Martin Corporation where he worked on the management of large government and military software-intensive systems including the following roles: software manager on the NASA Space Station Program; manager of a missile defense Software Engineering Process group; chief architect for several software metrics tracking systems and tool environments; major contributor to 11 wins of 12 proposals at Lockheed Martin; and he was selected for the prestigious award as the person most responsible for Software Process Improvement in the corporation.

He retired from Lockheed Martin, and for the next 7 years Marv was president of a consulting firm providing Software Engineering support to aerospace companies. He then gained an inside government perspective for 8 years as a principal engineer at The Aerospace Corporation, an independent government-funded space research and development center that provides technical support to the Air Force and the intelligence communities on their space programs. His responsibilities included participation in RFP preparations; contractor source selections; author of a support plan for a worldwide space situational awareness program; contributor to an early Software Cost Estimation Model for the NRO; senior technical advisor to a Navy communications program; team lead for a software measurement information service; and a core member of the senior Air Force Program Management Assistance Group (PMAG) Team providing baseline reviews of large ongoing Air Force space programs.

During his career, Marv published 127 articles and reports including a book for NASA, several booklets, and he authored many plans, policies, procedures, standards, specifications and guidebooks. He is the principal author of a key *Software Development Plan Guidebook* and contributing author of the *Software Measurement Standard for Space Systems* both to be used in future Air Force procurements. He was also co-author of *Software Acquisition Management* guidebooks; responsible for review and analysis of major Air Force, NASA and contractor documentation; author of an important Guidebook describing the collaboration of systems and Software Engineering; and he taught courses at The Aerospace Institute.

Although semi-retired, Marv is currently president of Escon Software and System Consulting, Inc. where his current focus is on software documentation. He has five daughters, ten grandchildren, (at least) five great-grandchildren and two poodle daughters. Marv and his wife live in the Phoenix, Arizona, area. He can be contacted at *info@esconinc.net.*

SOFTWARE PROJECT MANAGEMENT FUNDAMENTALS

1

Chapter 1

Software Project Management Introduction

The beginning of wisdom is the definition of terms.

—Socrates (470–399 BC)

The essence of this introduction is indeed the beginning, or basic foundation, of what you need to know in order to better understand the rest of this Guidebook. Chapter 1 briefly covers the key Software Project Management fundamentals in the following ten sections:

- *Why You Should Read This Chapter First*: To understand the terminology and acronyms (1.1).
- *Business Operations, Programs and Projects*: Definition of a project and its relationship to programs and business operations (1.2).
- *What Is Project Management*: Why projects must be planned, executed and managed (1.3).
- *A System Perspective:* Why it is imperative that Software Project Managers have a system perspective, plus definitions of Systems Engineering, software-intensive systems, system boundaries, System of Systems, an overview of the System Life Cycle (SLC), and the value of a one-page big picture overview (1.4).
- *Software Project Planning*: An introduction to the importance of the two key software planning documents: the Software Development Plan (SDP) and Work Breakdown Structure (WBS) (1.5).
- *The Software Project Team*: Discussions of project organization, software roles, team organization, and the importance of the Chief Software Engineer (CSwE) and the Software Integrated Product Team (SwIPT) (1.6).
- *A Customer Focus*: Why you must figure out how to please your customer (1.7).
- *Software Classes and Categories*: Describes why having a mechanism to assign proper designations to

each software entity will result in major cost savings by eliminating unnecessary documents, reviews, measurements and testing (1.8).
- *Software Development Standards*: Needed to produce consistent software work products (1.9).
- *Why Software-Intensive Systems Fail or Overrun*: The core success factors needed for successful development and delivery of software-intensive projects (1.10).

1.1 Why You Should Read This Chapter First

All users of this Guidebook should read this introductory chapter first as it establishes a basic understanding of the terminology used and it briefly describes the Guidebook infrastructure that constitutes a foundation for understanding the principles of Software Project Management.

Lessons Learned. This Guidebook is not a standard! Therefore, there are no mandatory "shalls" as you will find in a typical technical standard. There are many variables involved with managing software projects. This Guidebook does *not* contain prescriptive directions as to *precisely how* to manage software-intensive system development. What it does provide is *approaches, techniques* and *recommendations* for managing software-intensive systems based on many years of proven practices and successful developments. Use new approaches where they fit, but please *don't abandon the wisdom gained from experience.*

Acronyms. Acronyms are used extensively in this Guidebook. Appendix A contains a list of all acronyms used in this Guidebook. Table 1.1 is a list of the most common acronyms used throughout the Guidebook; memorizing this set of acronyms will speed up and facilitate reading the Guidebook.

It is expected that, when this Guidebook is used a reference, readers will revisit individual sections for reminders when needed. Consequently, acronyms are typically redefined when first encountered in each chapter. Definitions of key terms used in this Guidebook are included in Appendix B.

Learning acronyms is almost like learning another language. Those who balk at learning a lot of acronyms must realize that, if they want to work in the Software Engineering profession, acronyms are an inherent part of the daily life of a software engineer. The pinnacle of acronym usage is in government and military work because, for example, it is much easier to say "DIACAP" than it is to say, "Department of Defense Information Assurance Certification and Accreditation Process."

Terminology. The Guidebook is focused on large software system developments with the intention that smaller sized systems will *tailor out what does not apply* to them. Essentially, the Guidebook is a "shopping list" of what *must* be done, what *should* be done, what *can* be done, and what *may* be done (if applicable to your project). In that context, the following terms—and what they mean—are used throughout this Guidebook:

- *Must:* Highly recommended for compliance with most software standards and as a best practice.
- *Should:* Recommended for accuracy and completeness.
- *Can:* Discretionary but should be seriously considered for inclusion.
- *May:* Discretionary or used to show examples; Using the term "may" implies that other good options exist—choosing between them is left up to the Project Manager.

1.2 Business Operations, Programs and Projects

Business Operations. A project should not be confused with general business operations that are ongoing, repetitive, or essentially permanent functional activities of your organization that produces products or services. A simplistic example: If a boat company has a speedboat division that manufactures and sells speedboats, the daily activities of that division is an ongoing business *operation*. When that division decides to design and build a *new* model speedboat, that effort is a *project*.

Table 1.1 Common Acronyms

CSwE	Chief Software Engineer	SDP	Software Development Plan
CCB	Configuration Control Board	SDR	Software Discrepancy Report
COTS	Commercial Off-the-Shelf	SEIT	Systems Engineering, Integration & Test
C/R	COTS/Reuse (a software class)	SEPG	Software Engineering Process Group
CUT	Coding and Unit Testing	SI-SU	Software Item–Software Unit
IT&V	Integration, Testing and Verification	SIP	Software Implementation Process
IMP	Integrated Master Plan	SLC	System Life Cycle
IMS	Integrated Master Schedule	SPM	Software Project Manager
JMR	Joint Management Review	SPR	Software Peer Review
JTR	Joint Technical Review	SQA	Software Quality Assurance
MSDL	Master SDL	SE	Systems Engineering
SCS	System Critical Software (a Sw class)	SS	Support Software (a software class)
SCM	Software Configuration Management	SwIPT	Software Integrated Product Team
SCR	Software Change Request (or Report)	SwE	Software Engineering
SDF	Software Development File (or Folder)	TIM	Technical Interchange Meeting
SDL	Software Development Library	UI&T	Unit Integration and Test
SDLC	Software Development Life Cycle	WBS	Work Breakdown Structure

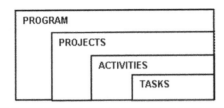

Figure 1.1 Functional hierarchy.

Programs. Likewise, projects should not be confused with programs since *programs typically consist* of *multiple projects*. In your company, there may be only a subtle difference between programs and business operations. As shown in Figure 1.1, you should view the overall hierarchy as the program consisting of projects, projects consisting of activities and activities consisting of tasks. If your program is a system, then its constituent projects could be called "subsystems" (or sometimes segments).

Program Managers. This Guidebook is focused on Software Project Manager's activities. However, it is also applicable to *Program Managers* who are managing multiple software-intensive projects. If you are a Program Manager, then you are at a higher level of management, and one more step removed from the trenches, so you will have to modify the recommendations in this Guidebook to match the needs of your responsibilities and position.

Projects. A project is a temporary endeavor to produce a specific product, service, upgrade or result within a defined period of time and is usually constrained by funding. Projects are:

■ Any size ranging from small and simple to very large and very complex
■ Short- or long-term with starting and planned completion dates plus intermediate milestones
■ Groups of planned activities and related tasks that collectively meet the project objectives

The Project Management Institute (PMI, 2013) describes project management as accomplished through the application and integration of the project management processes including initiating, planning, executing, monitoring, controlling and closing the project. For software projects, Chemuturi and Cagley (2010) classify typical software projects into the following types:

■ Full or partial Software Development Life Cycle projects
■ Reused software product customization and implementation
■ Porting, migration or conversion of existing software.
■ Web application projects
■ Agile development projects
■ Maintenance projects (defect repair, functional expansion, modification, etc.)

1.3 Software Project Management

Magic is not a participant in the delivery of large successful software-intensive systems; they must be planned, well-executed and managed.

A Software Project Manager (SPM) must be given the *responsibility*, *authority* and *support* from top management to *drive*, *direct* and *monitor* the project. According to a U.S. Air Force software acquisition handbook (USAF, 2003), project management is a discipline that performs the following functions:

■ Plans the project and establishes its life cycle goals
■ Employs skills and knowledge to achieve project goals through the performance of defined activities
■ Controls costs, schedules, risks, project scope and product quality through defined processes
■ Organizes and coordinates resources: people, equipment, materials, facilities and finances
■ Leads, organizes and motivates a Project Team to achieve the project goals
■ Tracks and evaluates the progress of the project and takes Corrective Actions when needed

Performing these project management functions, in an organized framework of activities and processes, is the job of the SPM. The actual role and authority of the SPM depends largely on the structure of the organization in which the SPM works. For example, Table 1.2 shows a strong preference for a *project-oriented* organization where the SPM has

Table 1.2 Project Manager's Authority Depends on the Organization Type

Organization Type	Project Team	SPM Level of Authority
Project-oriented	All (or most) team members are *assigned full-time* to the project and report directly to the SPM	High to very high
Matrix-oriented	Team members are *loaned* to the project by their home organization, to whom they report administratively, but work full-time for the SPM	Medium to high
Functional-oriented	Team members provide *support* to the project remotely through their functional Managers with whom the SPM must negotiate for resources	Medium to low

a high level of authority over a full-time dedicated team. In a functional-oriented organization, maintaining a constant level of staffing will be more difficult.

Software Project Manager Skills. Project Managers of software-intensive systems need both *management and technical skills*. Normally a SPM should not be doing the detailed technical work. However, SPMs will be directing the technical work performed by others, so the SPM must have, at a minimum, essential knowledge of the appropriate technical fields plus extensive knowledge of:

■ The *System Life Cycle* introduced in Subsection 1.4.4 and described in Section 3.2
■ The *Software Development Life Cycle* (SDLC) described in Section 3.4
■ The *Software Implementation Process* (SIP) described in Section 3.5 and Chapter 11

As the Software Project Manager on medium-to-large software-intensive projects, the essential level of technical expertise means your ability to understand *what* is being done by your team, not necessarily a detailed understanding as to exactly *how* it is being performed. If you become too deeply involved with the technical details while managing a large project, you are likely to fall into the trap of *micromanagement*, spending a disproportionate amount of your time doing technical work at the expense of performing your principal project management functions. Be directly *involved* but not *absorbed*.

Table 1.3 is a list and brief description of general skills needed by Project Managers (not in priority order). These skills are discussed in later chapters of the Guidebook. Also, the type of skill needed on the project over the life cycle varies considerably depending on the size of your project. On smaller projects, you may actually be acting as a *Programmer Lead* and will indeed be performing a lot of technical work. If you are directly managing a group of Programmers, you should know as much, if not more, than the members of your team. On medium-to-large projects, Software Project Management is *not a part-time job* so detailed technical work should be performed only when absolutely necessary.

Lessons Learned. SPMs need to make hard decisions; however, sometimes it is best to intentionally delay a decision or action even though some may call you a procrastinator. For example, "sticky" issues are brought to you for resolution where no viable solution is evident. Sometimes it is wise to hold off taking any immediate action because certain types of problems have a way of *resolving themselves*, so it is better to wait until the dust settles.

I am not recommending you become a procrastinator, but I do suggest you develop an ability to recognize the type of problems that have a *propensity for self-healing* and when you should not be prematurely over-reactive. In some cases, indecision may be a key to success as claimed in the old adage: *Flexibility is a key to tactical success; Indecision*

Table 1.3 Skills Needed by Software Project Managers

Project Manager Skills	Description
Requirements Management	Determine scope of the project at the start. Prepare plans and procedures to verify that the system is responsive to the requirements. Control scope changes.
Risk Management	Monitor project to detect and mitigate existing and potential risks.
Cost Management	Estimate project costs. Develop and control the project budget. Use earned value.
Schedule Management	Develop a Work Breakdown Structure, estimate the duration and sequence of each activity, and prepare and control the project schedule.
Performance Management	Measure, monitor and report performance of activities and take Corrective Actions.
Project Team Management	Build a quality staff and develop their skills as individuals and as team members. Plan an effective team organization. Demonstrate good leadership qualities.
Software Quality Management	Establish software quality plans, quality assurance and quality control procedures.
Integration Management	Develop and coordinate execution of the SDP. Establish a change control process to ensure all elements of the project are working efficiently and effectively.
Supplier Management	Plan, select, procure and administer services, equipment and materials needed.
Communications Management	Define reporting structure including information distribution, frequency and content.

is a key to flexibility. It's always a good idea to pause before you push the send button. However, there are definitely times when you *must* make an immediate decision where delay is not an option.

The road to life is paved with flat squirrels who couldn't make a decision.

—Anonymous

Lessons Learned. A good Project Manager, like a good Engineer, must be very *thorough.* That means omitting nothing and being careful (like looking both ways before crossing a one-way street). Good Engineers check their work twice; great Engineers check their work three times. Being thorough should not be confused with being precise because precision means "explicit and strictly defined" with no variation. A Software Project Manager must be flexible and open-minded because, when dealing with large complex systems, frequent changes are a certainty.

1.4 A System Perspective

The idea of a systems perspective evolved over time. In past decades, software-intensive systems continuously grew larger and early attempts to develop large-scale systems used approaches that worked successfully in small-scale developments. Those approaches often failed when scaled up because individual components that worked well by themselves did not work when they interfaced with many other hardware and software components in large-scale integrated systems.

1.4.1 What is a Software-Intensive System?

The basic concept of a *system* is a *bounded collection of interactive elements that interact with other elements.* From the perspective of a Software Project Manager, a software-intensive system is:

> *An organized collection of hardware, software, people, policies, methods, procedures, regulations, constraints and related resources interacting with each other, within the system's set boundaries, followed to accomplish common system objectives.*

A software-intensive system is where the Software Component of a system is *the predominant factor* in accomplishing the common system objectives.

Lessons Learned. Some hardware engineers I have known will object to this definition because

they typically believe the predominant element of every system is *always the hardware.* At one time that was true, but not now. Systems used to be 80% hardware and 20% software; that relationship is now reversed, and the software portion is growing in size and complexity.

System Characteristics. It is important to realize that there is a major difference between a software-intensive *system* and a software *product.* If the final deliverable product is *just the software code,* with some required documentation, then the delivered product is a software *package or module.* Software code by itself is obviously software-intensive, but it is *not a* software-intensive *system* as defined here because it does not have the characteristics of a system such as the following attributes:

- *System Performance*: System capabilities are greater than the capability of any component.
- *System Composition*: Structure and behavior subdivided into subsystems and elementary parts.
- *System Structure*: System components, hardware and software that directly or indirectly interface with each other.
- *System Behavior*: Each process fulfills an intended function or purpose, and the processes have structural and/or behavioral relationships.
- *System Boundary and Interface*: Limits its components, processes and interrelationships when it interfaces with another system and may be a subset of a super-system within which it operates.

1.4.2 What is Systems Engineering?

Systems Engineering (SE) is defined herein as a formal interdisciplinary field of engineering focused on how complex projects should be designed, implemented, tested and managed. The System Engineering process identifies mission requirements and translates them into design requirements at succeeding lower levels of detail to ensure operational performance. Control of the evolving development process is maintained by Systems Engineering through a continuing series of reviews, audits and testing. Complex systems are *progressively decomposed* until reaching the lowest elementary level. There are many aspects of SE not covered, or briefly discussed in this Guidebook, often referred to as the "ilities" including reliability, producibility, transportability, availability, maintainability, etc.

System Engineers take a *system point of view,* meaning they are not primarily concerned with the individual devices that make up the system but are focused on the system as a whole, including *all* the interrelationships and the *overall behavior of the entire system.* This approach comes from basic "systems theory" that views the world as a complex system of interconnected parts, scopes the system by defining its boundaries, and then makes simplified representations (or models) of the system structure in order to understand it and to predict the behavior of the system. At the highest level, the essence of Systems Engineering is *structure*. System Engineers understand the system by bringing structure to it and by defining the set of elements, or parts, so connected or related as to form the whole system.

Software Engineering can be considered a part of Systems Engineering. On many programs and projects, Software Engineering is located organizationally within the Systems Engineering group. That is because the two disciplines are not merely related, they are intimately interlaced. The important collaboration between system and Software Engineering is discussed below and in Chapter 12. Whereas System Engineers may manage the building of a dam or a bridge, the creation of new models of automobiles, the launching of satellites into space, or the development of new types of electronic devices, Software Engineers deal with the software aspects of those systems. Those aspects are never trivial.

It is important to understand what Systems Engineering is *not* in the context of this Guidebook. Information Technology (IT) departments in many organizations have an important function called "Systems Engineering." That function typically deals with the operation and performance of all the computers, corporate servers, networking, multicomputer systems, operating systems and other IT related issues. That function is *not* the Systems Engineering function referred to in this Guidebook. I view each project as *one system*, so I refer to the engineers working on a system as "System Engineers." "Systems Engineering" (plural) is used only because that is what everyone calls it.

1.4.3 The Collaboration of Systems and Software Engineering

Software Engineering (SwE) and Systems Engineering are tightly coupled in today's software-intensive systems—and that includes essentially all major systems.

In nearly all application areas, including aerospace, education, medicine, transportation, etc., software-intensive systems are providing unprecedented system functionality. The growth of software involvement is essentially exponential. Using jet fighter airplanes as an example: The F16 in 1976 relied on software for about 40% of its capability; the F-22s in 1997, relied on software for about 80% of its capability; and in 2006 the F-35 relies on software for 90% or more of its capability. The same growth of software for automobiles has occurred as current vehicles have millions of lines of code and the pinnacle is driverless cars, currently being tested, that are controlled entirely by software.

Software considerations need to be involved in *every system decision*, and SwE expertise should be utilized throughout the Development Life Cycle by System Engineers. At the same time, Software Engineers need to be continually looking at *the impact their decisions make* on the system, and they must convey this information to the System Engineers. The criticality of this collaboration and the exchange of information grows exponentially with the size of the program.

To ensure success, system and software requirements, architectural design, implementation, integration and testing must be a collaborative effort between the two disciplines and have the ability to evolve with changing customer or environmental needs. The "Systems Engineering Acquisition Team," or the customer's Program Office, is usually responsible for ensuring such participation exists. The Chief Software Engineer, and relevant Software Engineers, should be members of the Systems Engineering Acquisition Team.

Early involvement of Software Engineering in the System Life Cycle is a major help in mitigating the risks of excessive cost, effort and rework because poor system architecture choices can negatively impact the software development effort. Chapter 12 is devoted to the detailed identification and description of the collaborative tasks performed by System and Software Engineering. In Chapter 12 that collaboration is discussed from two perspectives:

■ There are system development activities that are the responsibility of Systems Engineering, but these activities are *supported by Software Engineering*.
■ There are software development activities that are the responsibility of Software Engineering, but these activities are *supported by Systems Engineering*.

1.4.4 System Life Cycle Overview

The System Life Cycle is a *cradle to grave* perspective starting with an initial conceptual analysis of the proposed system and ending with an ultimate replacement or retirement of the system. The SLC is sometimes referred to as the "System Acquisition Life Cycle." Figure 1.2 should be carefully reviewed as it is an important *conceptual overview* involving the hierarchical relationship of:

■ The *System Life Cycle* that includes *System Definition*, *Software Development Life Cycle* and the *System Sustainment* activities
■ The *Software Development Life Cycle* which incorporates part of *System Definition*, plus the *Software Implementation Process* and the *System Integration, Testing and Verification* (System IT&V)

(*NOTE*: Software Engineering literature sometimes defines the acronym "SDLC" as the *System* Development

Life Cycle; this Guidebook identifies the system-level life cycle as the SLC.)

Although Figure 1.2 is an important overview, it is intended to be a conceptual introduction to the architecture of the SLC process showing the five activities embedded within the SLC. Figure 1.2 will be expanded during the discussion of life cycle processes in Chapter 3, with additional figures that are elaborations of Figure 1.2. Hardware development is performed at the same time as the SDLC.

An easy way to understand the System Life Cycle can be depicted using the well-known engineering V-chart modified to show the software development and typical review process. As shown in Figure 1.3, the left side of the "V" contains the design activities and related reviews, and the right side contains the integration and test activities and related reviews.

Figure 1.3 does not contain all of the activities and reviews; the complete process will be discussed in later chapters and in Figures 3.3, 3.4, 4.4 and 12.2. Specific activities performed by Software Project Managers are covered in subsequent chapters of this Guidebook, especially Chapter 2 which is entirely focused on SPM *activities*. Additional system-related subjects discussed are:

- *Systems* Engineering activities are addressed in Chapter 10.
- *Software* Engineering activities and the software development process are discussed in depth in Chapter 11.
- Chapter 12 describes the collaborative activities of systems and Software Engineering along with more details of the system and software life cycles.
- Integration, Testing and Verification activities are covered in several sections.
- The System Life Cycle does not end at the release of a software product into production. A working product in the hands of the customer is just the beginning of the longest and most costly portion of the System Life Cycle—System Sustainment, often referred to as "System Maintenance," is covered in Chapter 16.

1.4.5 System Boundaries

You must provide the context for your project by defining the *boundaries and scope* of your contractual responsibilities. In other words, what your project includes and what it does not include. In addition to the starting and ending points, you need to know the *origin of the inputs* as well as the specific *destination of the outputs* of your interfaces.

Boundaries are important because most systems interact with other systems. These interdependencies between systems

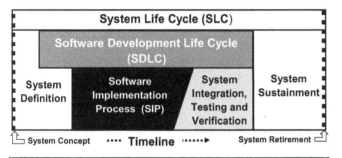

Figure 1.2   Conceptual overview of the System Life Cycle.

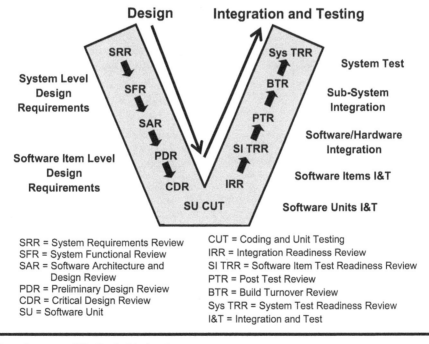

SRR = System Requirements Review
SFR = System Functional Review
SAR = Software Architecture and
         Design Review
PDR = Preliminary Design Review
CDR = Critical Design Review
SU = Software Unit

CUT = Coding and Unit Testing
IRR = Integration Readiness Review
SI TRR = Software Item Test Readiness Review
PTR = Post Test Review
BTR = Build Turnover Review
Sys TRR = System Test Readiness Review
I&T = Integration and Test

Figure 1.3   System Development Life Cycle V-chart.

can be complex and can take an inordinate amount of your resources. Even if they are not your responsibility, to make everything work together seamlessly, you and your team must define and describe these boundary interfaces because they can have a huge impact on your cost and schedule.

1.4.6 System of Systems

The software-intensive system you are developing can incorporate one or many systems within the boundaries of your project. Your project could then be considered a System of Systems (SoS). An SoS can become incredibly complex as illustrated by the Commercial Aircraft SoS in Figure 1.4.

A System of Systems can be viewed as a set, or arrangement of systems, that results when independent and useful systems are integrated into a larger system that delivers unique capabilities. A subsystem is a set of elements, which may be a system itself, but also is a *component of a larger system*. Subsystems are sometimes called *segments*.

During system development, integration and operational problems frequently arise due to inconsistencies, ambiguities, and omissions in addressing quality attributes between system and software architectures. These problems are compounded in a SoS. The *architecture framework* for the Department of Defense is called *DoDAF*. It provides a good set of architectural views for a SoS architecture. It is important to remember that identifying and addressing *quality attributes* early in the process, and evaluating the architecture to identify *risks*, is a *key to success*.

In addition, *end-to-end mission threads*, augmented with quality attribute considerations, are needed to help develop and later evaluate the SoS and the constituent system and software architectures. If you are the Project Manager of an SoS, you must track it carefully.

Each of the Commercial Aircraft systems identified in example Figure 1.4 are themselves considered a SoS as they can be further broken down into component systems as

shown for the Navigation Control System. The elements of a SoS must be broken down into manageable components. If your project is part of a System of Systems, the need to clearly *define the scope and boundaries* of each system element becomes even more critical.

1.4.7 The One-Page Big Picture Flowchart Overview

The following lesson learned is one example, of many in my career, where a *one-page big picture flowchart overview* was a key mechanism for *understanding and identifying required system functionality* with the help and concurrence of the customer.

> **Lessons Learned.** Early in my career, I was working for a Washington, DC, think tank, and we had a contract to automate the processes of the Small Business Administration (SBA). I was the Lead Software System Engineer. The SBA had previously hired two different software contractors, but nothing usable was produced. I asked them to show me what had been done and was taken to what was essentially a large closet filled with boxes of punched cards (yes it was a long time ago). Each of the two previous contractors had different sets of punched cards with different formats. So I then asked to see the documentation that went with the punched cards and the expected response was: we don't have any documentation.
>
> I realized the previous work had been performed by hackers who either had no idea what they were doing, or they were trying to automate small tasks. I had to start from scratch. I asked the SBA to set up an appointment for me to talk in depth with someone who has a good understanding of the *overall SBA process*.
>
> Being somewhat naïve at that time, I was surprised to learn that there was no such person! There were people who knew exactly what processes were being followed at the SBA Headquarters operation in Washington, DC, but in order to understand the process at a district office, I would have to visit one of them. I obtained all the information about the processes at the SBA Headquarters and then traveled to one of the largest SBA district centers in Dallas.
>
> After returning from Dallas with ample information, I prepared an implementation plan and report. Although I had a lot of data, I was able to identify the salient operations of the overall SBA process and converted it into a flow chart

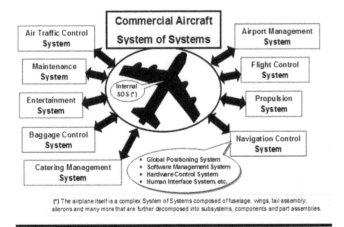

(*) The airplane itself is a complex System of Systems composed of fuselage, wings, tail assembly, ailerons and many more that are further decomposed into subsystems, components and part assemblies.

Figure 1.4 Context of a System of Systems.

where the *entire SBA process* was depicted on *one page* (a large 24 by 30 inch sheet that folded up nicely into a pouch in the back of the report). It was the first time SBA management had a *big picture* understanding of the overall SBA process—and on one page!

That one-page flow chart was instrumental in determining "what" to automate and it was also invaluable in obtaining understanding and approval by the customer. To keep it simple, the flowchart contained only the top-level processing tasks making it much easier to understand what needed to be automated and much easier to discuss the automation tasks with SBA management.

There are two reasons for telling this story:

■ The advantages of a one-page flowchart overview when discussing the project with the customer—especially top management.
■ Why the "big picture" view is important for every system and invaluable for large or complex systems. Keeping it as *simple as practical* was a critical element of success for this system as well as every other system I worked on.

For complex implementations, *all the functionality cannot be put on one page*. However, a very top-level overview of the process can be on one page, followed by *one-page expansions of each box shown on the top-level flowchart*. You can continue that process until you peel down to the precise details needed to describe the overall design for subsequent implementation by the Programmers.

Understanding the big picture is crucial.

Lessons Learned. Sometimes having a one-page big picture overview can get you in trouble. For example, I once went to work for a computer firm developing sophisticated solutions to data storage and retrieval systems. They had this one fellow who was brilliant and productive, so they let him get away with a lot. His approach was to make his system designs complex and confusing, and his briefings and documentation were the same, so no one really understood exactly what he was doing. I reviewed his report, combined the similar tasks, eliminated the side issues, and simplified it the point where I could graphically show essentially what he was doing on *one page*. When I presented this chart at the next update—well, let's say if he had a gun I would have been shot. He barely spoke to me again, but now everyone knew what he was doing.

1.5 Software Project Planning

Planning is critical; if you don't know where you are going—you will never get there!

Although software planning is performed throughout the software life cycle, *strategic planning up front usually makes the difference between success and failure* of a software development project. This Section describes the two fundamental pillars of software project planning, the *Software Development Plan* and the *Work Breakdown Structure*, plus the potential need for a Project Charter.

Project Charter. Preceding the formal planning process, an optional *Project Charter* may have to be produced. The Project Charter should be developed by the customer or organization requesting and funding the project prior to the formation of a Project Team. The Project Charter is a very high-level document, and it serves as the starting point for the SDP. It describes why the project has been initiated, what the project will accomplish, generally how and when the product will be developed or provided, who is responsible to perform the work, and who benefits from the project products or services.

Contents of the Project Charter, sometimes called the *Project Vision* or *Scope Plan*, is obviously dependent on the characteristics of the project. However, the following is a general outline of its contents:

■ *Title of the Project*: Descriptive but not too lengthy.
■ *Purpose of the Project*: Overall scope of the project including a brief statement of the problem(s) the project will solve, but not the solution.
■ *Expected Results*: Description of what will be accomplished and when the project is planned to be completed. Results must be measurable qualitatively or quantitatively.
■ *Assumptions*: Includes the conditions, environment or ground rules that will govern the execution of the project, and which must be acknowledged for its successful completion.
■ *Roles and Responsibilities*: Identifies key project participants, their roles and responsibilities, including the Project Manager and major supporting agencies.
■ *Authority Statement*: The level of authority and approval given to the Project Manager.
■ Signatures: Signoffs of key project stakeholders acknowledging their concurrence.

1.5.1 Software Development Plan

A poorly planned software development effort is likely to fail—that makes the SDP a critically important software management tool for essentially all software development efforts.

Every software development program *should* have a Software Development Plan, but for large programs it is a *critical element* for success. The SDP may also be called a *Project Plan, Project Work Plan, Software Project Management Plan, Software Development Management Plan* or something similar, but it is called an SDP in this Guidebook. The SDP *documents the processes by which the software will be designed, developed, integrated, tested and managed.*

An SDP, or similar document, is *required* by essentially all software development standards. It is prepared by the software organization performing the development. For government work, it is typically required and submitted with the developer's (contractor's) proposal. Whenever the SDP is referred to in this Guidebook, replace it with the equivalent document name used by your program or organization.

The SDP is focused on *what* will be done and should be backed up with detailed *operational procedures* that describe *how* to do it. It is not uncommon for the detailed standards of corporate operational procedures to fill a bookshelf. But even for large systems, the SDP itself is seldom larger than 200 pages. However, the *SDP package* may include *annexes or addenda* that could be bound with the SDP or bound separately. SDP packages on large programs can become quite large and may include annexes containing plans for management and control of quality, risks, configuration, metrics, testing, subcontractors, budgets, reviews, resources, reuse, integration and test, roles and responsibilities, etc.

The SDP describes the set of processes, methodologies, tools, development and testing environments, and life cycle models, appropriate for the scope and complexity of your project, that will be used consistently by *all* team members. Teammates and subcontractors may need to prepare *site-specific* versions of the project-level SDP to address issues of concern only to them, as long as their site-specific SDP does not conflict with the project-level SDP. Specific conflicts with the project-level SDP is acceptable as long as formal waivers are requested and approved. The SDP is discussed in more detail in Subsection 5.1.1, and Appendix F contains an example annotated outline of the SDP.

An incomplete, poorly written, unorganized or inadequate SDP is a clear red flag!

Lessons Learned. Do not underestimate the importance of the SDP. Bidders on government contracts with a deficient SDP are significantly decreasing their chances of winning contracts, and those with a deficient SDP who are awarded contracts have an historically high probability of cost and schedule overruns. A good SDP will significantly *increase the probability* of a successful software-intensive system. I was on Government Contract Source Selection Teams and could not

believe the quality of SDPs ranging from poor to excellent submitted by major aerospace corporations. Spend the time up front to produce a good SDP because it *will* have a big payoff in terms of winning contracts and especially during management of the project.

1.5.2 Work Breakdown Structure

The Work Breakdown Structure is a key software budgetary planning and control activity. The WBS organizes and decomposes, in a hierarchical structure, *all the project tasks* into smaller, more manageable and controllable components. It is *product-based* and includes the development of all hardware, software and documentation products plus requirements analysis, testing and anything else that must be accomplished to finish the project and deliver the system. The WBS provides a strategy for completing the project by dividing the entire project into a *hierarchical logical breakdown*. Graphically, the WBS is normally shown as a *tree structure*, as depicted in the example in Figure 1.5.

The WBS tree structure should be augmented with documentation in the form of an annotated *outline* that contains more details than shown in a figure similar to Figure 1.5. On smaller projects, it is not absolutely necessary to have both the tree structure and the outline. In its simplest form for small projects, the WBS can be portrayed with yellow sticky notes organized on a whiteboard.

The number of levels in a WBS must break down the planned work into *manageable* tasks and, as a general rule-of-thumb, they should not be *larger* than what can be accomplished in a short time period (varies by project but about 2–3 weeks). The WBS is also extremely useful in preparing network diagrams as described in Section 8.4. Updating the WBS, as required, should be performed by following the change control process described in Section 6.4.

There are many benefits you can realize from the WBS including:

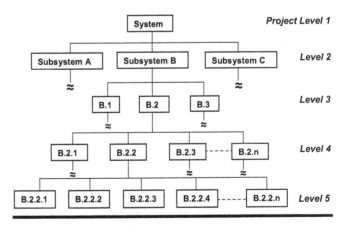

Figure 1.5   Example of a tree-structured WBS.

- Easier to estimate costs, schedule and resources in an organized manner
- Easier to assign work tasks and manage the project
- Focuses attention on project objectives allowing team members to visualize the entire project and where their responsibilities fit it
- Facilitates the identification of potential risk items
- A solid basis for monitoring project cost, schedule and performance

The WBS is discussed in more detail in Chapter 8. Section 8.2 contains an example WBS and describes the relationship between the WBS, the *Integrated Master Plan* (IMP), and *the Integrated Master Schedule* (IMS).

1.6 The Software Project Team

The progress of projects can proceed only as fast as the project team's skills will allow.

Cast of Characters. Chapter 7 is devoted entirely to managing your team. However, before you get into the details, you need to understand *who the key players are*. There are several individuals and teams involved in the development of software-intensive systems. Introducing these participants up front will help you understand their roles and functions when they appear during the development life cycle.

Not all of the participants listed alphabetically below in Table 1.4 report *directly* to Software Project Managers. From a managerial perspective, they are all part of *your team* and have the collective objective of producing a successful implementation.

Finding, recruiting, nurturing, and retaining a competent staff is the single most important function of a Software Project Manager.

1.6.1 Organizing Your Project within a Program

An organization chart showing the relationship of your software organizations to the overall program must be provided in the Software Development Plan along with an overview of your entire *project organization* structure. Figure 1.6 is an example of a *program organization* chart with an emphasis on software functions and subsystems.

For a software-intensive project, the organization chart should always be structured to facilitate software management visibility and software technical oversight. That can be accomplished by placing the software-related elements higher in the organizational structure rather than hidden as a sub-element within a lower organizational element. In addition to the overall program organization as shown in Figure 1.6, a detailed organization chart of your software group should

also appear in the SDP. Figure 1.7 is an example of a top-level organization chart for the Software Integrated Product Team (SwIPT) in a satellite ground subsystem.

Lessons Learned. I once went to work for a company that did not have an organization chart. Since I was a "newbie," I needed an organization chart to understand who was in charge of what—so I drew one and published it. Well, it caused quite an uproar. Some Managers were upset with me because their box was not on the same line as another Manager with whom they considered at the same level. My mistake was publishing it; I should have been more discreet and kept it to myself. These kinds of things are important to some people. Live and learn.

1.6.2 Software Development Roles

Development of very large and complex software-intensive systems may require hundreds of personnel involved in software management, development and support tasks that could include most of the software roles shown in Table 1.5. Some of these roles may be combined. For example, the Responsible Project Engineer (RPE) role may be assigned to the CSwE. For small and simple software-intensive systems, just *a few staff members* might assume all of these software roles. Appendix C, *Software Roles and Responsibilities for Skill Groups*, is a summary example of generic responsibilities for 11 Software Engineering skill groups.

1.6.3 Software Project Managers and Lead Programmers

Software Project Managers and Lead Programmers (LPs) have different responsibilities. To explain the differences between SPMs and LPs, two extreme examples will be described, and the readers should be able to visualize the variations that exist between those extremes. At one end of the spectrum is a small team composed of a very few co-located Programmers who are developing a simple software program for a single application. In this situation, the project's *Lead Programmer* is managing only the Programmers on the team.

The opposite scenario is a very large and complex software-intensive system involving millions of lines of code, being developed by multiple geographically dispersed organizations, where the software resides in many system components, interfaces with many hardware and other software components, and requires a significant amount of software documentation. In this latter example, the Software Project Manager must manage a full team that could include all, or most, of the roles listed in Table 1.5.

Table 1.4   Roles and Functions of Key Participants

Role	Function
Chief Software Architect	Large and complex software-intensive systems usually require a *Chief Software Architect* who has an understanding of the entire system architecture and can make sure all the elements are functionally compatible. On smaller, less complex systems, this role can be performed by the Chief Software Engineer or even a Lead Programmer on small systems.
Chief Software Engineer	The CSwE is responsible for technical oversight, process guidance, and the primary point of contact with the customer for all software technical matters in the system. Subsection 1.6.4 describes functions performed by the CSwE Team.
Customer	The organization that procures and manages the contract for a software-intensive system, software product or software service from a supplier for itself or another organization.
Developer/ Programmer	The technical staff that develops software products or performs development activities during the software life cycle process. It may include new development, modification, reuse, re-engineering, maintenance, or any other activity that results in software products needed to be responsive to the system requirements.
Lead Programmer	The Lead Programmer reports to the Software Project Manager and is a Senior Programmer who manages a group of less experienced Programmers. On large programs, there could be several Lead Programmers (See Subsection 1.6.3).
Program Manager	Typically, a program is composed of several projects and the SPMs report to the Program Manager. On programs that are essentially *all software*, the SPM could be performing the same functions as a Program Manager.
Sustainment Organization	The organization that is responsible for modifying and otherwise maintaining the software after transition from the development organization to the deployed operational environment. (see Section 16.4).
Software Configuration Management	Software Configuration Management is a critical activity—especially on large programs. The SCM Team: establishes formal baselines of products; maintains their integrity; tracks and controls Change Requests; and controls the changes (see Section 6.3).
Software Engineering Process Group (SEPG)	The SEPG ensures consistent implementation of the software development processes across the project and facilitates production of compatible software products. The SEPG is the heart of the Software Process Improvement (SPI) effort and is often chaired by either the CSwE or a Software Process Engineer (see Subsections 2.4.1 and 6.1.1).
Software Integrated Product Team (SwIPT)	This is your cross-disciplinary Development Team responsible for the Software Engineering effort and for *producing* the software products. The SwIPT is managed by the SPM and is typically a *matrix organization* assembled from members of the applicable engineering disciplines. On very small projects, the SwIPT may consist of only a single Software Developer (see Subsection 1.6.5).
Software Process Lead (SPL)	The SPL is on the CSwE Team and, on large programs, is a trained change agent who is responsible for facilitating all software process tasks for the program (see Subsection 6.1.4).
Software Project Manager	This Guidebook is specifically focused on the SPM; the individual with *responsibility and authority* for directing the software project, managing development of the software-intensive system, and managing the Project Team (see Chapter 7).
Subcontractors/ Suppliers	Subcontractor organizations enter into a contract with the customer or the project for the development and implementation of a portion of a system, software product or software service under the terms of a formal contract or an internal company transaction.
Software Quality Engineer (SQE)	The SQE is administratively part of the software quality organization and is assigned (either full-time or part-time) to your project (see Subsection 6.2.1).

(Continued)

Table 1.4 (Continued)   Roles and Functions of Key Participants

Role	Function
Software System Engineers	Software System Engineers are System Engineers who have a heavy background in software systems. They provide important functions during the planning and requirements analysis activities, during testing and integration, and as a vital communications interface between the customer and the Developers/Testers.
Stakeholders	Those persons and organizations that have an interest in the performance and completion of the project. The customer or end-user of a system is usually the primary stakeholder.
Systems Engineering Integration and Test	The SEIT Team is part of the Systems Engineering organization, and its members do not report administratively to the SPM. The SEIT organization has primary responsibility, with support from your team, for all of the system-level tasks described in Chapter 10.

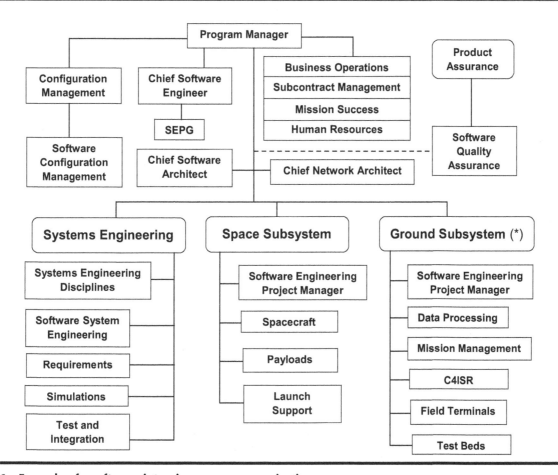

Figure 1.6   Example of a software-intensive program organization.

The need for a full-time SPM is dependent on the size of the team and the complexity of the system. Use your best judgment to decide, however, Table 1.6 can be used as a rule-of-thumb guideline.

1.6.4 Chief Software Engineer Team

The quality of a software system is directly related to the process used to create it—and the Chief Software Engineer is the core of the software process.

A key element of success, especially for a large software-intensive project, is the establishment of an effective and proactive Chief Software Engineer Team. Table 1.7 is an example list of the responsibilities a CSwE Team typically has for software oversight and process guidance as well as support the CSwE provides to the *Systems Engineering Integration and Test* (SEIT) Team.

On small projects, the CSwE would not normally have a team, and on even smaller projects the responsibilities of the CSwE could be handled by either the Project Manager, a

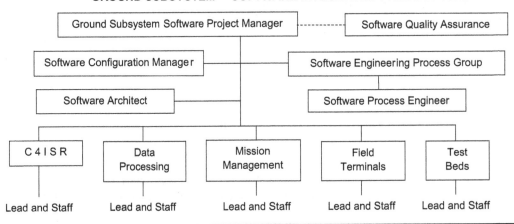

GROUND SUBSYSTEM —SOFTWARE INTEGRATED PRODUCT TEAM

Figure 1.7 Example of a ground subsystem SwIPT organization.

Table 1.5 Typical Software Roles in Large Projects

SOFTWARE MANAGEMENT	SOFTWARE DEVELOPMENT
Software Project Manager	Chief Software Architect
Responsible Project Engineer	Software Designers/Architects
Chief Software Engineer	System Programmers
Software Team Leads/Managers:	Application Programmers
■ Programming Team Leads	Software Coders
■ Integrated Product Team (SwIPT)	Software System Integrators
■ Configuration Management	Database Management Specialists
■ Subcontract Management	Modeling and Simulation Engineers
■ Software Quality Assurance	Software Test Engineers
■ Software Integration and Testing	Algorithm Specialists
■ Software Systems Engineering	Subject Matter Specialists (SMS)
SOFTWARE SUPPORT	SOFTWARE SYSTEM ENGINEERING
Configuration Management Team	Software System Engineers
Software Quality Engineers	Software Process Lead
Stakeholders	
Information Assurance Engineers	CONTRACT ASSIGNMENTS
Human System Integration (HSI) Engineers	Program Manager
Software Librarian(s)	Financial Services/Cost Accounting
Network Support Engineers	Customer and Customer Representatives
Software Technical Writers	

Table 1.6 When is a Software Project Manager Needed?

Team Size	Need for SPM or LP
2–4	LP only
5–10	SPM and 1 LP
11–15	SPM and about 2 LPs
Over 15	SPM and several LPs

SPM = Software Project Manager; LP = Lead Programmer.

highly experienced member of the staff or a combination of divided responsibilities.

On large programs, the CSwE should be accountable to, and *report directly* to the Program Manager, or to the Systems Engineering Manager. The CSwE is the primary point of contact with the customer's Acquisition Program Office (APO) for all software *technical* matters across the program. The CSwE Team also provides technical support to the SEIT Team for all software-related matters. In some complex programs, this responsibility may be shared with a Chief Software Architect. The CSwE should be a voting member

Table 1.7 Chief Software Engineer Team Responsibilities

Chief Software Engineer—Software Development Oversight Responsibilities
Reviewing the definition of software elements, or subsystems needed to satisfy requirements
Reviewing of high-level software architecture guidance to be used in the development of software implementing details through design reviews and technical exchange meetings
Reviewing software baseline implementation and subsequent changes
Assessing and guiding Configuration Management requirements, build approach, and implementation
Assessing the methods used to transition new builds into the operational baseline
Selecting program CASE tools and developing program-level training
Oversight of architecture development and data architecture, including the database management system, file system implementations, and support tools to ensure successful execution
Assessment of validation and verification approaches and subsystem operability
Reviewing subsystem timeline performance analysis
In concert with the SEPG, collecting and consolidating metrics from the IPTs
Analyzing metrics across the program and providing metric summaries and recommendations
Reviewing and approving software plans, specifications, test procedures and test results
Definition and oversight of software IT&V activities
Technology insertion planning and review
Chief Software Engineer—Software Process Oversight Responsibilities
Developing the Program-wide SDP, and monitoring for compliance by the SwIPTs and teammates
Reviewing software implementation processes to ensure compatibility with project goals
Defining and implementing the program's Quantitative Management Plan.
Assessing the cost, schedule, tools and technical and management risks of all software IPTs/teammates
Chairing the SEPG and the Software Configuration Control Board [Sw/CCB]
Chief Software Engineer—Software Support to SEIT
Participating, assessing and approving requirements allocations and decomposition of system functionality to software elements, domains or subsystems
Assisting the SEIT in trade studies and margin assessments related to software
Reviewing and approving subcontracts requiring software development or purchase of software

CASE = Computer Aided Software Engineering; SEPG = Software Engineering Process Group; IPT = Integrated Product Team; IT&V = Integration, Test & Verification; SEIT = Systems Engineering Integration & Test.

of the appropriate control boards and the Risk Management Board.

1.6.5 Software Integrated Product Team

The development of software products is performed by one or more Software Integrated Product Teams (SwIPT).

On your project, the SwIPT is your *cross-disciplinary team of Developers* who are a critical component of an effective and successful software development program. Although some members of your SwIPT may be full-time members of your staff, they are typically people from *matrix organizations*, assembled from the applicable engineering disciplines, to design, develop, produce and support an *entire* product or system. Members of the SwIPT should provide the following:

- Share a common understanding of the SwIPTs tasks, objectives and responsibilities.
- Provide the technical skills and expertise needed to accomplish the tasks and objectives.
- Collaborate internally and externally with other SwIPTs and relevant stakeholders.
- Provide the advocacy and representation to address all phases of the life cycle.

This Guidebook assumes SwIPTs are the basic *performing organization* and that they have the resources and authority to produce and deliver software products and execute processes. On very small projects the SwIPT may consist of just one person who is assigned all the software tasks; on very large systems the SwIPT could consist of hundreds of team members as discussed in Subsection 1.6.2. The software project you are managing may be composed of *a* single SwIPT or multiple SwIPTs.

The *Integrated Master Plan*, as described in Section 8.1, identifies responsibilities for the SwIPTs. The project's contract should identify key personnel and their responsibilities during software planning, development, test and transition to deployment. The IMP should designate software responsibilities through identification of program events, planned accomplishments as well as acceptance criteria for each milestone.

The organization of SwIPTs is normally consistent with the product hierarchy as defined in the Contract Work Breakdown Structure. *Primary responsibility for each WBS element should be assigned to a single SwIPT.* This allows SwIPTs to identify clear and measurable outputs and interfaces; it also facilitates the flow down of requirements to the SwIPTs. On large projects, the SwIPT would include a CSwE, probably a Chief Software Architect, plus a large team of Software Developers, Test Engineers and Subject Matter Experts.

At the program level, there may be an over-arching SwIPT led by the CSwE who reports directly to the Program Manager. If a program does not have a CSwE, the functions are performed by the person(s) having those responsibilities regardless of their job title. If you have a government contract, the Government Acquisition Team may have an oversight organizational structure similar to yours, and the SwIPT tasks may involve participation by the government's SwIPT members.

1.6.6 Interpersonal Communication

No methodology or Software Process Improvement strategy can overcome serious problems involving ineffective communication or the mismanagement of interpersonal conflicts.

Frequent and honest communication with your staff is a critical factor in increasing the likelihood of project success and the mitigation of problems. As the SPM, you should seek customer involvement and encourage end-user input in the development process to avoid misinterpretation of requirements, misunderstanding of changing customer needs, and unrealistic expectations. Project Managers, Software Developers, end-users, customers and project sponsors need to communicate frequently.

Bad news resulting from these conversations may be to your advantage if the problems are communicated early enough for you to take Corrective Actions. Also, casual conversations with the customer, team members and other stakeholders may surface a potential problem sooner than its appearance at formal meetings when the problem may be much more severe.

You have the right and obligation to criticize the work underway as long as it is provided in a way that is constructive, respectful and non-accusatory. Effective interpersonal communication, in an intellectually honest fashion, along with effective team conflict management and resolution, are the keys to successful Software Project Management. These issues are discussed in more depth in Section 7.5.

1.7 Customer Focus

Without the customer, you don't have a project!

The "customer" as referenced in this Guidebook refers to the organization *responsible for the program's contract including funding and approval of the final products.* It implicitly includes their representatives—such as consultants hired by the customer. Also, you may interface daily with the customer's Project Management Office also called the "Acquisition Program Office."

Even if the software-intensive project being developed is for internal use by another division or group in your company, they are still your "customer" and are very likely funding the project (through inter-departmental transfer) and they deserve exactly the same treatment as an external customer. The importance of a customer focus is also discussed in Subsection 7.7.2. As the Project Manager, it is self-evident that *you* must figure out *how to please your customer—even if they are wrong.* The following short story is a personal example to make that point.

> **Lessons Learned.** Many years ago, when automation was at its infancy, I was the Project Manager of a pioneering effort to automate library processes. We had a well-funded project with a large government technical library, and I needed to interface with the customer's library

director, who was more than a bit autocratic and knew *nothing* about software development.

At the kickoff meeting, I described the development plan which included *requirements analysis* and production of the *Software Requirements Specification* (SRS). The library director immediately objected and said there is no need for requirements analysis, that it is a complete waste of time, as all we needed to do was to "automate the current processes."

I did not object since it is not a good idea to argue with your customer (especially this one). So, this is what we did. We identified the relatively simple tasks where what needed to be done was clearly needed, and I assigned Programmers to produce the design and code for those tasks. In subsequent status reports that was the output shown to the library director so she would see *physical progress*.

Meanwhile, quietly in the "back room," a group of us worked on the system and software requirements resulting in an abridged version of the SRS. This was necessary because we needed to understand exactly what we were building since no one had prior experience with library automation at that time.

The bottom line is that if you have a technically unsophisticated customer, who impedes your ability to follow the required processes, use your ingenuity to figure out ways to workaround these obstacles to get the job done. We did that in this example, and it was a very successful pioneering library automation implementation, delivered on time and within the budget.

1.8 Software Classes and Categories

Project managers must make sure there is a mechanism in place to *assign the proper designation to each software entity*.

There are typically *three generic classes* of software in a software-intensive system:

■ *System Critical Software* (SCS) described in Subsection 1.8.1
■ *Support Software* (SS) described in Subsection 1.8.2
■ *Commercial Off-the-Shelf or Reuse* software (C/R) described in Subsection 1.8.3

Each software generic *class* can be further subdivided into *categories* as needed for your program, resulting in the identification of 4–8 categories of software for a typical program. The number of software classes, the number of categories within those classes, and the names of each are *not critical*.

What is important is that there must be a definition of the category assigned to each software entity because not every software entity needs to have the full set of documentation, the full set of reviews, the full set of metrics, and the same level of testing.

Assigning categories to software entities, such as the example categories shown in the following Tables 1.8–1.10, can result in *major time and cost savings by eliminating unnecessary documents, reviews, metrics and testing.* However, the simplicity of this approach is deceiving since obtaining agreements from all stakeholders on the appropriate category to assign each software entity is not always simple or obvious. As the Software Project Manager, you must work it out and obtain concurrence.

1.8.1 System Critical Software

System Critical Software, often called *Mission Critical* software, is physically part of, dedicated to, and/or *essential to full performance* of the system. SCS can be defined as:

System Critical Software consists of software functions that if not performed, performed out-of-sequence, or performed incorrectly, may directly or indirectly cause the system to fail.

System Critical Software plays a critical role in the overall dependability, reliability, maintainability and availability of every system. SCS may be expanded to two software

Table 1.8 System Critical Software Class and Category Definitions

Class Definition	Category	Category Definition
SCS SYSTEM CRITICAL SOFTWARE All applications software used to perform real-time operations and non-real-time functions implicitly required to implement the system functions allocated to software.	SCS-1	Deliverable applications software *that has a direct role in system operation* required for full system functionality.
	SCS-2	Firmware is software embedded in the deliverable hardware. Firmware is treated in the same way as software that executes in general purpose computers.

Table 1.9   Support Software Class and Category Definitions

Class Definition	Category	Category Definition
SS SUPPORT SOFTWARE Software that aids in system hardware and software development, integration, qualification, operations, test and maintenance	SS-1	Software Items that play a direct role in program and system development including software and system requirements qualification and acceptance testing for final "sell-off."
	SS-2	Support software that is typically prototype software, simulation software, or performance analysis and modeling tools (although some of this type of software may be selected to be in Category SS-1).
	SS-3	Non-deliverable and non-critical tools or test drivers that indirectly aid in the development of the other categories of software.

Table 1.10   COTS/Reuse Software Class and Category Definitions

Class Definition	Category	Category Definition
C/R COTS/REUSE SOFTWARE Non-developmental Software Items including Commercial Off-the-Shelf software. All C/R products must be treated and controlled as defined for the category targeted for its end use.	C/R-1	Non-developmental software that is *unmodified* COTS or Reused software.
	C/R-2	Non-developmental software that is *modified* COTS or Reused software.(*) (A distinction between vendor-provided and internally reused software can be made if meaningful to your program).

(*) Modifying vendor-provided COTS is a high-risk approach and is not recommended; however, modifying internally reused software can be effective since you have access to the full source code.

categories to specifically identify *firmware* as shown by the example in Table 1.8. If it is meaningful to your program, the number of SCS categories can be further expanded. The SCS implementation *process* is described later in Subsection 3.5.1.

1.8.2 Support Software

Support Software (SS) aids in system hardware and software development, test, integration, qualification and maintenance. The SS class may be composed of three Software Item (SI) categories, for example, SS-1, SS-2 and SS-3 as defined in Table 1.9.

SCS-1, SCS-2 and SS-1 software categories (*but not SS-2 or SS-3*) are usually *deliverable and contractually obligated*. These three categories must pass through *all* of the developmental phases, including all of the relevant software documentation, reviews, metrics, and testing, and are subject to external *Software Discrepancy Reports* (SDRs), also called problem reports, (see Subsection 6.4.1).

SS-2 software is used in non-operational environments and is normally not deliverable. Both SS-2 and SS-3 software categories do *not* go through the full software life cycle or receive external SDRs.

In some cases, important *Support Software* packages may be contractually deliverable. For example, deliverable Support Software may include training software, database-related software, software used in automatic test equipment, and simulation software used for diagnostic purposes during the sustainment activity. The Developers must decide the appropriate category for all software entities in compliance with contractual requirements. The Software Support *process* is described in Subsection 3.8.2.

1.8.3 Commercial Off-the-Shelf and Reused Software

COTS/Reused (C/R) software is *non-developmental* Software Items and is often referred to as *third-party software*. It includes Commercial and Government off-the-shelf (COTS or GOTS) software as well as reused software obtained from vendors or internal libraries, previously developed, or developed by other programs, set up specifically for reuse. "COTS" is a generic term and does not realistically mean you can go to a store and buy it off-the-shelf. The C/R class may be composed of two categories as described by the example in Table 1.10 or additional categories if meaningful to your project.

1.8.4 Software Category Features

A single SI may consist of different classes and/or categories. In that event, each part of the SI must be compliant with the documentation, review and testing requirements of the category assigned to it. All software releases must be configuration controlled in a Software Development Library (SDL) at the *subsystem level* or by the Master Software Development Library (MSDL) at the *program level* as described in Section 9.3.

If we view the categories as hierarchical levels, SCS is the top level, SS is the middle level and C/R the lowest level. Software cannot be moved up or "promoted" to a higher category level without additional development and testing. To achieve a higher category level, the software must be "re-engineered" and conform to the documentation, review and testing requirements imposed on the *higher* category level. All COTS and reused products must be treated and controlled as defined for the category targeted for its *end use*. COTS software is also discussed with Software Sustainment in Chapter 16.

1.9 Software Development Standards and Regulations

Applicable software standards and practices must be addressed in the SDP, or other documents such as a *Software Standards and Practices Manual*. That documentation should identify the programming language standards to be used including specific versions, a list of applicable standards dictated by the customer, toolsets to be used, operational details of the program's defined software process, use of COTS/Reuse software, the process for waivers or deviations, and portions of the applicable standards that need to be modified for your project.

These standards ensure that Developers produce *consistent* software development products. Standards also help ensure the similarity of the structure of all code/design units so that lines of code counts and software measurements can be applied consistently. Standards must provide *value added* to your project.

Software standards are discussed or referred to in many sections of this Guidebook, they are addressed in Section 4.5, and some common standards are listed in Appendix I. Standards are adjustable to your needs, but regulations are not; they must be followed. Regulations are imposed by government bodies, and if you don't comply with them you could be fined or even face imprisonment. For example:

- Health Insurance Portability and Accountability Act (HIPPA).
- Occupational Safety and Health Administration (OSHA) requires all organizations to provide for the protection and safety of all of your employees.

- Sarbanes-Oxley Compliance (SOX) Act of 2002 affects all publicly held corporations.

In addition to government regulations, there are all kinds of industry-specific regulations and laws that your project, and your company, must adhere to. As the Software Project Manager, you need to be aware of the mandatory regulations, and those that can or may affect your project, and incorporate provisions for these concerns into your SDP and other planning documents.

1.10 Why Software-Intensive Systems Fail and Overrun

The Standish Group has been periodically publishing their well-known CHAOS Report (Standish Group, 2016) containing software project *success/failure/challenged* rates starting in 1994. They have also attempted to analyze *success factors* at each publication. The rating of the top success factors change with each report. However the perennial major factors needed for project success appear to be:

- Executive management support (sponsorship)
- Customer (user) involvement
- Clear Objectives and requirements
- Optimizing scope with shorter project milestone durations (part of planning)
- Skilled staff
- Project Manager expertise

This means the absence of those factors are reasons for failure. The past few CHAOS reports have found a decrease in IT project success rates and an increase in IT project failure rates where:

- 32% to 35% were considered *successful projects* as they were completed on time, on budget and with the required features and functions (the Standish definition of success).
- 19% to 24% of the projects were considered *failures* because they were cancelled before they were completed, or they were delivered but never used.
- 44% to 46% were considered *challenged* meaning they were either finished late, over budget, or with fewer than the required features and functions.

If the data is accurate, based on their definition of success, 65% to 68% of the IT software projects were not successful. Software tools provided by the project management industry is estimated to be over $3 billion annually and predicted to boom to over $5 billion in a few years, but these new methodologies and software management packages don't seem to have any impact. Maybe the real solution is to create better Software Project Managers (achievable if they read this Guidebook!). Some serious researchers point out problems with the CHAOS reports, most notably:

■ Unlike published academic research, the data needed to evaluate the claims is kept private so their data or methods cannot be independently verified.

■ Their definition of success is narrow; success means cost, time and content were accurately *estimated up front*; if the project did not meet those estimates, it was categorized as *challenged*. In the real world, successful projects usually undergo changes as they are developed, and they may take longer to incorporate new functionality not initially planned for, and the resulting project may indeed be successful.

Popular, and often quoted, reports that contain questionable data is not a new phenomenon as described by a personal experience below.

Lessons Learned. The cost department at a company I worked for decided to find out what it cost to develop software. They sent out a questionnaire to a large number of completed software-intensive programs, and one key question asked for the *cost of the software* portion of their program. Their final report was very professional and thorough, and it was referenced widely in the industry.

The big mistake made by the cost department was the Cost Analysis Team that prepared the report did not have any Software Engineers either on the team or used as a consultant. As a result, the key question about software cost did not identify the scope of *what to include in the cost figures*. Some programs surveyed considered all the requirements analysis and integration tasks to be part of Systems Engineering and did not count it as a software cost. Some programs charge Software Quality Assurance (SQA) and Software Configuration Management (SCM) to those departments. Other programs consider all of those tasks *part of the software effort*. Since the data collected had no consistency in scope, the well documented cost results were meaningless.

Table 1.11 Core Success Factors for Software-Intensive System Implementations

1. Obtain Senior Management Commitment to Your Project
2. Define and Document Your Customer's System and Software Requirements
3. Prepare and Follow a Software Development Plan and a Defined Process
4. Prepare and Follow a Work Breakdown Structure
5. Hire the Best Programmers and Support Staff Available and Provide Training
6. Select and Document a Clear Software Development Methodology
7. Take a Big Picture Systems Approach
8. Measure and Track Quality, Performance and Progress
9. Build Quality Into Software from the Start of the Life Cycle
10. Identify, Manage and Mitigate Software Risks
11. If You Change *Anything* When Developing a Complex System: Retest
12. Prepare Software Documentation as Needed by Your Project
13. Provide Configuration Management of the Work Products
14. Don't Make Decisions Based Only on Budget or Schedule Issues
15. Collaborate With Hardware and System Engineers
16. Use Software Standards Where Applicable
17. Document Lessons Learned to Facilitate Process Improvement
18. Lead, Motivate and Manage Effectively and Ethically
19. Keep Solutions Simple and Concise and Avoid Gold Plating
20. Do Not Concur With Unrealistic Schedules or Budgets
21. Don't Take Risky Shortcuts
22. Always Use Common Sense

Core Success Factors. Why do software-intensive projects fail, or result in severe cost overruns, or are delivered well beyond the planned completion date? The principal answer is: they do not follow all of *the core success factors* needed for successful software-intensive system implementations. My list of the core success factors in Table 1.11 are not listed in order of importance. They are all important. I am sure you can add more to the list, but these are the *essential elements* for successful development and deployment of large software-intensive systems. Memorize this list of 22 success factors or tape it to the wall near your desk.

Chapter 2

Software Project Management Activities

Activity is the only road to knowledge.

—George Bernard Shaw

Chapter 2 is focused on the key activities performed by a Software Project Manager (SPM). These activities are described in this chapter within the context of the six overlapping domains as depicted in Figure P.1, the areas of control normally addressed by SPMs, the SPM activities focused on Software Process Improvement (SPI) and System Sustainment activities, and a list of the common SPM pitfalls. The key SPM activities are discussed in the following six sections:

- *Overview of the Software Project Management Domains*: Describes how the multitude of project management tasks are organized into six manageable interactive domains, the Software Project Management activities within each domain, and the chapter or section where each activity is described (2.1).
- *Software Project Management Objectives*: Provides the overall objectives of Software Project Management including balancing scope and resources in order to achieve project control (2.2).
- *Software Project Management Control Areas*: Contains a brief introduction to the key Software Project Management control areas including management of software: requirements and requirements traceability; risk; performance measurement tracking and control; product quality; costs and schedules; resources (your budget, team, facilities, and tools); work products; suppliers; estimation; and integration and testing (2.3).
- *Software Process Improvement*: Covers highlights of the SPI management control areas including software: process group; quality assurance; Configuration Management; Corrective Action; Peer Reviews; an

introduction to the Capability Maturity Management Integrated (CMMI) processes; and the application of Six Sigma to software (2.4).

- *System Sustainment*: An introduction to the longest and most expensive phase of the System Life Cycle—System Sustainment (2.5). Sustainment is discussed in depth in Chapter 16.
- *Software Project Management Pitfalls*: There are of course many potential pitfalls a Software Project Manager may encounter but most of the major ones are listed (2.6).

2.1 Overview of the Software Project Management Domain

The overlapping ellipses of the Venn diagram in Figure P.1 is an introduction to the concept of six interactive Software Project Management *functional domains* working together synergistically to produce a successful result. Collectively, the six domains cover *all* of the software development, management and support activities necessary to assure a successful software project implementation. Figure 2.1, however, is a more useful figure because it can be used as a *roadmap* to this Guidebook as it indicates where each activity is described.

Figure 2.1 shows the contents of each domain and the chapter or section in the Guidebook where each activity is addressed. To complement Figure 2.1, the following is a brief description of what is covered in each of the six SPM domains and the chapters in the Guidebook where they are discussed:

- *Chapter 5—Project Management Domain*: Covers essential software management activities including project planning, risk management, Technical

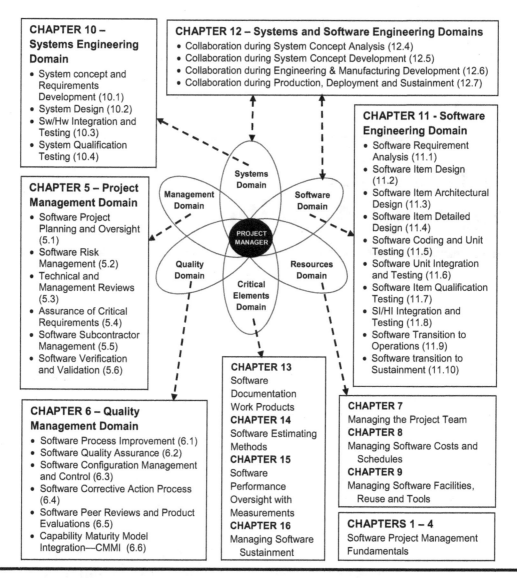

CHAPTER 10 – Systems Engineering Domain
- System concept and Requirements Development (10.1)
- System Design (10.2)
- Sw/Hw Integration and Testing (10.3)
- System Qualification Testing (10.4)

CHAPTER 12 – Systems and Software Engineering Domains
- Collaboration during System Concept Analysis (12.4)
- Collaboration during System Concept Development (12.5)
- Collaboration during Engineering & Manufacturing Development (12.6)
- Collaboration during Production, Deployment and Sustainment (12.7)

CHAPTER 11 - Software Engineering Domain
- Software Requirement Analysis (11.1)
- Software Item Design (11.2)
- Software Item Architectural Design (11.3)
- Software Item Detailed Design (11.4)
- Software Coding and Unit Testing (11.5)
- Software Unit Integration and Testing (11.6)
- Software Item Qualification Testing (11.7)
- SI/HI Integration and Testing (11.8)
- Software Transition to Operations (11.9)
- Software transition to Sustainment (11.10)

CHAPTER 5 – Project Management Domain
- Software Project Planning and Oversight (5.1)
- Software Risk Management (5.2)
- Technical and Management Reviews (5.3)
- Assurance of Critical Requirements (5.4)
- Software Subcontractor Management (5.5)
- Software Verification and Validation (5.6)

CHAPTER 6 – Quality Management Domain
- Software Process Improvement (6.1)
- Software Quality Assurance (6.2)
- Software Configuration Management and Control (6.3)
- Software Corrective Action Process (6.4)
- Software Peer Reviews and Product Evaluations (6.5)
- Capability Maturity Model Integration—CMMI (6.6)

CHAPTER 13 Software Documentation Work Products
CHAPTER 14 Software Estimating Methods
CHAPTER 15 Software Performance Oversight with Measurements
CHAPTER 16 Managing Software Sustainment

CHAPTER 7 Managing the Project Team
CHAPTER 8 Managing Software Costs and Schedules
CHAPTER 9 Managing Software Facilities, Reuse and Tools

CHAPTERS 1 – 4 Software Project Management Fundamentals

Center diagram labels: Systems Domain, Management Domain, Software Domain, PROJECT MANAGER, Quality Domain, Resources Domain, Critical Elements Domain

Figure 2.1   Software project management activities mapped to the six domains.

and Management Reviews, Critical Requirements Assurance, Subcontractor Management and Software Verification and Validation.

■ *Chapter 6—Quality Management Domain*: Covers Software Quality Assurance, Software Configuration Management and Control, Corrective Action Process, Peer Reviews, Product Evaluations and SPI.

■ *Chapters 7, 8 and 9—Resources Domain*: Covers management of your Software Team, budgets, schedules, facilities, reused software, and Software Engineering tools.

■ *Chapter 10—Systems Domain*: Covers software-related activities that are under the purview of Systems Engineering including system requirements analysis, system design, software/hardware integration and testing, and system acceptance testing.

■ *Chapter 11—Software Domain*: Covers Software Engineering development activities including requirements analysis, architectural and detailed design,

Software Unit coding, testing and integration testing, Software Item qualification and acceptance testing and transition to operations and sustainment. Chapter 12 covers the collaboration of systems and Software Engineering.

■ *Chapters 13, 14, 15 and 16—Critical Elements Domain*: Covers software documentation, software work products, project performance oversight with measurements, software estimation methods and managing long-term sustainment often referred to as Software Maintenance.

2.2 Software Project Management Objectives

There are so many issues a Software Project Manager must address that it is difficult to adequately state the Software Project Management (SPM) objectives in a single sentence.

One generic definition of SPM objectives is to *produce software products according to a plan while simultaneously improving the project's capability to produce better products.* However, this does not cover all the objectives. The overall SPM functional objectives can be more completely described as the ability to efficiently and effectively:

■ Perform all the activities and management tasks embodied in the six SPM domains as presented in Section 2.1 above.
■ Manage the principal SPM control areas as described in Section 2.3.
■ Implement and manage the Software Process Improvement tasks as described in Section 2.4.
■ Complete the sustainment tasks as described in Section 2.5 and Chapter 16.
■ Avoid the SPM pitfalls as listed in Section 2.6.

How Much Software Management Control is Needed? It may be stating the obvious, but large and complex projects require more management control than small, simple projects. There is a huge difference in the potential for problems spiraling out of control between these two extremes. There are no hard rules for how much control is needed as size and complexity increase, but you can use Figure 2.2 as a rule-of-thumb. As the SPM you must determine how much control is needed and, an even more difficult task, make sure you are not over-controlling.

There is almost nothing that can spiral so far out of control, or so quickly, as a mismanaged large software-intensive system.

Balancing Project Scope and Resources. A major objective facing all SPMs is the task of *balancing scope and resources in order to achieve project control.* Figure 2.3 is a depiction of this "balancing act." If either scope or resources change, it will have an impact on the balance. The scope side of the balance ball tells you what you need to do and the environment within which you must do it including the required system functionality and project constraints (schedule constraints

and external regulations). The resources side includes your staff, allocated budget, facilities and tools available.

Once balance is achieved, you can exercise effective control of the project, and you can take Corrective Action when the scope or resources change beyond acceptable limits.

2.3 Software Project Management Control Areas

Figure 2.3 also graphical depicts ten principal Software Project Management areas that need to be integrated and controlled. Each of these areas are briefly described in Subsections 2.3.1 through 2.3.10. These descriptions are intended to be an introduction to each topic with pointers to the location(s) where the topic is covered in more depth in this Guidebook. Communications management is not listed, but that important area is discussed in Section 7.5.

2.3.1 Requirements and Traceability Management

A software-intensive system, coded to perfection by world-class Programmers, will fail if it does not address the needs of your customer. There is nothing, except for a capable staff, that will contribute more to the success of your project than having a valid set of requirements.

System and software requirements, sometimes called the *scope of the system*, define the objectives of your project and exactly what must be produced or achieved to be responsive to the needs of your customer's system. If the requirements are not complete and correct, you may deliver a system responsive to the *identified requirements*, but there is a high probability the project will fail because the users will abandon it if it does not solve their important problems and meet their

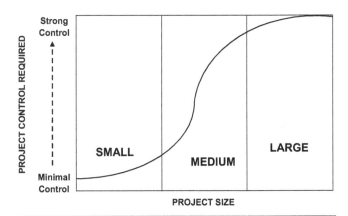

Figure 2.2 Project control mapped to project size.

Figure 2.3 Balancing project scope and resources to achieve project control.

important needs. *System* requirements tasks are described in Section 10.1 and *software* requirements tasks in Section 11.1.

Requirements are often complex and usually not fully defined up front; in that case, a life cycle methodology that allows for *frequent refinement* of requirements during development, such as evolutionary or spiral development, must be used as described in Section 4.1. If a function is *not* asked for it will *not* be produced, and what *is* asked for *will* be produced even if it's not needed to be responsive to your customer's needs. To use an old cliché, *defining the problem is half the solution*.

Requirements Management and Traceability Database. An automated requirements management and traceability database should be used by every large software-intensive project. Examples of such tools are *Dynamic Object-Oriented Requirements System* (DOORS), *System-Level Automation Tool for Engineers* (SLATE), *Requirements and Traceability Management* (RTM) and *Requisite-Pro*. In this Guidebook, the requirements management and traceability tool is called the *Requirements Database*.

Table 2.1 is an example of recommended traceability products that should be produced for each software class and category (SCS, SS and C/R are described in Section 1.8).

The SDP must describe the approach to be followed for establishing and maintaining bi-directional traceability such as between:

- Levels of requirements (System/Subsystem/Element/ Software Item/Software Unit)
- Requirements and design and the software that implements the design
- Requirements and qualification test information
- Required versus measured computer hardware resource utilization

Scope Creep. An important aspect of your SPM requirements management responsibilities is to control what is called requirements *scope creep*. That means allowing the volume of changes to become so large and so invasive as to severely impact the project's cost and schedule. Of course, if the customer is paying for these changes, and agrees to the impact of those changes on cost and schedule, you are probably okay. If not, you must "draw a line in the sand" and actively and aggressively campaign to restructure the program's cost and schedule when scope creep becomes an issue. Aside from its impact on cost and schedule, scope creep could be so severe that it affects your system architecture design and that could cause a catastrophic impact requiring a restart of the entire project. There are many pitfalls to Project Management, but excessive scope creep is more like quicksand.

2.3.2 *Software Risk Management*

Software risk management is the process of identifying, measuring and assessing the risk and then developing strategies to manage its mitigation or elimination. Risk mitigation strategies include:

- *Eliminating* the risk by taking positive action to find and fix the root cause(s) of the risk to eliminate the risk or reduce its actual or potential negative impact
- *Transferring* the risk to another element of the project where there is a better opportunity to mitigate that risk
- *Avoiding* the risk through implementation of a work-around task (but the risk is still there)
- *Accepting* some or all of the consequences of a particular risk (and hope it doesn't occur)

Risk management in Software Project Management begins with the business case for starting the project, which includes a cost-benefit analysis as well as a list of fall-back options for project failure, often called a "contingency plan." Software risk management and mitigation are discussed in more depth in Section 5.2 as part of the Software Management Domain.

Another interesting view of risk management is referred to as *Opportunity Management* where the potential risk could have a positive, rather than a negative impact. Theoretically,

Table 2.1   Traceability Requirements by Software Category

Software Requirements Traced To:	SCS-1 and SCS-2	SS-1	SS-2 & SS-3	C/R-1 and C/R-2
Parent Requirements	Required	Required	Required	Required
Software Builds	Required	Required	Required	Required
Use Cases	Required	Required	Not Required	Not Required
Software Units	Required	Required	Required	Not Required
Software Test Cases	Required	Required	Required	Required
Software Test Procedures	Required	Required	Not Required	Required

SCS = System Critical Software; SS = Support Software C/R = COTS/Reused software.

it is handled exactly as risk management, but using the term "opportunity" rather than the negative term "risk" helps to focus a team's mindset about possible positive outcomes of any given risk in their project.

2.3.3 *Project Performance Measurement, Tracking and Control*

Nothing is more terrible than activity without insight.

—**Thomas Carlyle (1795–1881)**

If you cannot measure it, you cannot control it.

—**Anonymous (He really gets around a lot)**

A fundamental requirement for effective project management is the ability to *measure and track performance in a timely manner to determine the status of your project and allow you time to take Corrective Actions.* You need *information* in order to do that and much of that information comes from measurements. When planning your project, you must make sure there are ample periodic checkpoints to facilitate project tracking (see Subsection 5.1.4).

In his 1990 classic (Humphrey, 1990), Watts Humphrey captured the importance of measurement with: "If you don't know where you are, a map won't help." The IEEE Software Engineering Body of Knowledge (SWEBOK) (Bourque and Fairley, 2014) states: "The importance of measurement and its role in better management is widely acknowledged … effective measurement has become one of the cornerstones of organizational maturity."

Software measurement (often called software metrics) is discussed from a management perspective in Chapter 15. That entire chapter is devoted to performance oversight with measurements. Typical objectives of a software measurement initiative are to provide:

- Tracking information for project management to facilitate the control and reduction of the project's software cost, schedule and technical risks
- A practical, efficient and up-to-date management methodology and basis for quantitative software development control, status determination, timely Corrective Action and activity re-planning
- Relevant and timely information to help Software Leads and Software Engineers perform their responsibilities correctly, on-time, within budgets and to produce a higher quality product
- Historical records of performance for trend analysis and other value-added information, to support continuous SPI

Chapter 15 also includes: a typical software measurement set of Management Indicators; the base measurements needed to define the status of each indicator; an example of a software measurement construct that defines the data that will be collected; the computations that will be performed on that data; how and when the data will be reported and analyzed; management indicator thresholds to identify non-nominal conditions; and red flags from the measurements indicating potential problems.

2.3.4 *Software Cost and Schedule Management*

You need accurate insight of your project's status, and you must perform continuous monitoring of its progress in order to control the project's costs and schedules. To be effective and reliable, project cost management requires the use of standardized processes and tools to support the process. Furthermore, the management of cost and schedules for software-intensive projects cannot be viewed as a *singular* activity because it involves multiple activities that *interact, overlap and support* each other. For example, successful management of software costs and schedules depends, at a minimum, on effectiveness of the:

- Project planning process including the cost and schedule estimation process
- Formal program-level master plan directly mapped to a program master schedule
- Collection and analysis of cost and status measurements
- Identification and mitigation of software risks
- Management of project costs with the earned value management control system
- Corrective Action Process
- Work Breakdown Structure (WBS)

2.3.5 *Software Product Quality*

Essentially everything a Software Project Manager does is directly or indirectly related to quality.

Textbooks have been written on every topic in this Guidebook and software quality is no exception. A study of 250 large software projects by (Jones, 2017) concluded that *poor quality control* was the largest contributor to cost and schedule overruns. It also concluded that *poor Software Project Management* was the most likely cause. Chapter 6 is entirely devoted to software quality. In addition, software quality activities and processes are introduced later in this chapter in the following process improvement Subsections:

- Software Quality Assurance (2.4.2)
- Software Configuration Management and Control—including change management (2.4.3)

- Software Corrective Action Process (2.4.4)
- Software Peer Reviews and Product Evaluations (2.4.5)
- Capability Maturity Model Integrated—CMMI (2.4.6)

2.3.6 Software Resource Management

Part 3 of the Guidebook contains three chapters describing management and control of your project's principal resources which includes your software staff, project funding, schedules, facilities, software reuse and tools:

- *Chapter 7—Managing the Software Project Team*: Software is labor-intensive, and the success of your project is highly dependent on the capability of your team, so this chapter might be the most important chapter in the Guidebook.
- *Chapter 8—Managing Software Costs and Schedules*: Involves determining the policies, procedures, and documentation that will be used for planning, executing, and controlling project costs and schedules (see also 2.3.4).
- *Chapter 9—Managing Software Facilities, Reuse and Tools*: Includes hardware, software, procedures, and documentation necessary to support the life cycle software development effort. Software reuse includes any existing software product (i.e., specifications, designs, test documentation, executable code, and source code) that can be effectively reused to develop the software system. Computer Assisted Software Engineering (CASE) tools are intended to enhance the efficiency of software development.

Stakeholders. As the Software Project Manager, you are responsible for managing stakeholders. You must first determine *exactly who* the stakeholders are. A stakeholder is anyone who has a vested interest in, or is affected by, the system being developed. Besides yourself and your team, the major stakeholders normally include:

- The customer or sponsor funding the project; this could include your finance department.
- Operational entities who are the eventual *users* of the system such as manufacturing, sales, marketing, shipping, engineering, accounting, system operations, control centers, and various levels of Managers.
- Other related systems that your system interfaces with including *sources* of information you use and *recipients* of services or information you provide to them.

Figure 2.4 is a rule-of-thumb matrix regarding stakeholder's "influence" versus stakeholders "involvement" in the successful completion of your project. Figure 2.4 also

Figure 2.4 Stakeholder management matrix.

provides general guidelines on how to manage the four categories of stakeholders.

Expectation Management. Managing stakeholder's "expectations" of what your system will do must be one of your highest priority management control activities. Make sure your customers are not expecting a Ferrari when they are paying for a bicycle. Excellent communication with your stakeholders is an essential foundation of controlling expectation management. Keep stakeholders informed on what is going on through timely progress reporting (at least weekly) or techniques such as daily stand-up meetings, if that is feasible for your project. See Chemuturi (2010) and Boehm (2000) for detailed discussions of stakeholder expectation management.

2.3.7 Software Work Products

Software work products are essential artifacts of the software development process.

Documents are the only visible and permanent products during several phases of the development process. After your software-intensive system is delivered, it will be very expensive (maybe impossible) to support the software during the sustainment activity without adequate documentation. When insufficient, poor, or no documentation is provided, and changes need to be made during sustainment, it is usually necessary to *reverse engineer* the software, or abandon it altogether, and start all over again resulting in a very expensive, wasteful and time-consuming task.

In addition to the continuity provided during System Sustainment, the following is a list of the most obvious benefits and value of good software documentation *during* development—not sometime later:

- *Staff Turnover Training*: There is always a need to provide project information to new staff members, and good software documentation is an invaluable asset for training them.

- *Customer Insight*: Customer involvement in the development process is necessary to provide control, redirection if necessary, frequent oversight, an effective interface with the Developers, and review and approval of selected software development documentation.
- *Evidence of Status and Progress*: Software specifications, plans and reports, along with periodic software reviews, will provide you and the customer with a measure of the project's status and progress.
- *Clarification of Complexity*: A complex software-intensive system may be so complex that there is *no one person on the project who fully understands the details about the entire system*. Good documentation is an important asset to *all* members of the Development Team, and it serves as an important archive of the rationale for key decisions and the technical architecture.

Unfortunately, Programmers have a general aversion to documentation. Maybe that is because documents are associated with structure and bureaucracy or a belief that documents are "unnecessary paperwork" since most Programmers prefer to sit in front of the computer and pound out code rather than write or read documents. Maybe it is because no one likes to read poorly written documents and that is frequently what is produced.

To overcome this hurdle, try convincing the authors to strive to produce documents that are as interesting to read as possible. You must also convince your management on the importance of good documentation and, on very large projects, to hire good technical document writers and pay them consistent with their importance. If your documentation sparkles with quality and content, you can't imagine what a *positive perception it will have by your customer* on your software product. Chapter 13 is devoted to software documentation work products.

2.3.8 Software Procurement Management

Requiring software subcontractors to provide deliverable products against a set of confusing requirements will result in serious issues. Management of Software Subcontractor Teams is *almost always a significant challenge* for Software Project Managers. On large contracts, there may be numerous subcontractors contributing software products, and they are typically geographically dispersed (sometimes offshore).

Section 5.5 of the Guidebook discusses the Subcontractor Management Team, and compliance with the program-level SDP, performance reviews, and the Statement of Work. In addition, Section 8.3 covers management of software budgets including procurement of contractors, consultants, capital equipment, travel and training.

2.3.9 Software Integration and Testing Management

Managing software integration and testing is a continuing task.

Software integration and testing activities occur throughout the software development process at various stages of the system buildup, culminating in System Qualification Testing (SQT) and acceptance by your customer. The stages, terminology and acronyms used for software integration and testing buildup, with an implicit Verification and Validation (V&V) buildup, is illustrated in Figure 2.5.

Another instructive graphical example of a *typical software integration and test buildup process* during software development is shown in Figure 2.6. It depicts the buildup of an example software-intensive system with two subsystems (called Adam and Eve in the figure) and a decomposition of subsystem Eve into three Software Items (called SI-1, SI-2 and SI-3 in the figure).

The example in Figure 2.6 also shows the breakdown of the Software Items (SI) into Software Units (SU), Coding and Unit Testing (CUT) of the SUs, integrating those SUs back into SIs, and then passing them through the SI Qualification Testing (SIQT). The three SIs, the hardware, and the two subsystems are then progressively integrated and tested until the buildup is complete and the system is ready for System Qualification Testing.

Software Integration, Testing and Verification (IT&V) involves activities where individual software modules are combined and then tested and verified as a group. The

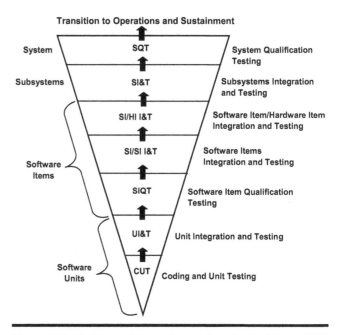

Figure 2.5 Software integration and testing buildup—process and terminology.

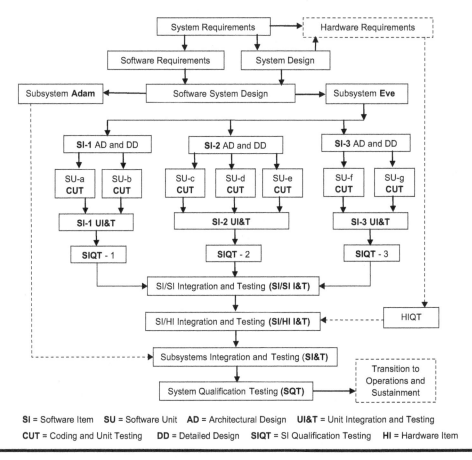

SI = Software Item   **SU** = Software Unit   **AD** = Architectural Design   **UI&T** = Unit Integration and Testing

CUT = Coding and Unit Testing   **DD** = Detailed Design   **SIQT** = SI Qualification Testing   **HI** = Hardware Item

Figure 2.6   Software integration and testing buildup process—example.

software IT&V process is described in Section 3.6; Software V&V is discussed in Section 5.6. Verification confirms that the next step in the software development process is responsive to its previous step. Validation confirms that the final product is responsive to the original system requirements.

Regression Testing is performed to make sure that recent changes have not impacted the software that was previously tested and caused unintended and undesirable consequences. Regression Testing (discussed in Subsection 3.9.3) can occur at any level of testing. To complement Figures 2.5 and 2.6, the following briefly describes each of the progressive stages in the software integration and testing buildup and where each stage is discussed in this Guidebook:

■ *Software Coding and Unit Testing*: Converts detailed design of the Software Units into computer code and databases that have been inspected, and unit tested (11.5).

■ *Software Unit Integration and Testing (UI&T)*: Performs systematic and iterative integration builds on Software Units that have individually completed Code and Unit Test, and builds them up to a higher level (11.6).

■ *Software Item Qualification Testing (SIQT)*: Demonstrates that a Software Item meets the software and interface requirements allocated to it (11.7).

■ *Software Items Integration and Testing (SI/SI I&T)*: Integrates Software Items with other interfacing SIs, testing the resulting groupings to determine if they function as intended, and continuing this process until all interfacing SIs are integrated and tested (10.3) and (11.8).

■ *Software/Hardware Item Integration and Testing (SI/HI I&T)*: Involves integrating SI with interfacing Hardware Items, testing the resulting groupings to determine if they function as intended, and continuing this process until all interfacing HIs in the system are integrated (10.3) and tested (11.8).

■ *Subsystems Integration and Testing (SI&T)*: This activity includes the integration of all subsystems and the elements within each subsystem. SI&T must precede SQT (10.4).

■ *System Qualification Test*: A formal test demonstrating that software functional and interface requirements have been met for a release of the system (10.4).

■ *Software Transition to Operations and Sustainment*: Involves testing the executable software on the target hardware at the customer's site and updating all code, files, documents, customer manuals, and turnover of the system to the users (11.9).

2.3.10 Software Estimation Management

Early estimates of cost and time are important for a variety of reasons, and accurate estimates are essential to many people in your organization. For example, Pre-Sales Teams need cost estimates to *price* custom software, Senior Managers need estimates to decide *if and how* they will proceed with the Software Project, Finance and Program Managers need estimates for resource allocation, planners need time estimates to make sure the Software Project Plan is consistent with the larger program plan and to inform customers about deadlines, milestones and deliveries. Chapter 14 is devoted to software estimation.

The estimated cost and time you give at the start of a project *may haunt you for the whole project*, so it is very important to have *as good an estimate as you can make from the beginning*. You might make it perfectly clear that the early estimate you are giving is a ROM (Rough Order of Magnitude) but, somehow, this ROM has a life of its own, and you are asked well into the project why you were so far off.

> **Lessons Learned.** While conducting software estimates for a large government proposal, we were told to *assume the best case* to minimize cost. My team did that, and I presented it at a briefing to the Program Manager. At the end of the briefing, he asked me if I would stake my reputation on whether the cost could not be 20% higher or lower than the estimate. I said that it could be higher but unlikely to be lower. He said, okay, make it 20% lower!
>
> I had no choice, did as instructed and we won the contract. However, I avoided serious implementation problems because the final requirements were so different than the original requirements we had the opportunity to prepare a re-estimate with realistic numbers. If this happens to you, make sure you go on record as clearly stating that you do not concur with the final numbers but will carry out the Manager's edict. You may not be as lucky as I was. The estimation task is serious business and is not trivial.

2.4 Software Process Improvement

Process improvement assumes you have a process that can be improved!

There are many software organizations that develop successful software systems without a formal process and without following most of the approaches recommended in this Guidebook.

The key word here is "formal" because everyone follows *some process* even though he or she may not call it that. Such team members tend to have "jack-of-all-trades" roles by performing and assuming (or trying to assume) all of the roles.

It is clear that teams without a formal process can work effectively but only on small projects; such teams cannot realistically be scaled up to large projects and, if they try to do that (and they do), the common result is not good. An excellent process improvement section on "Life Without a Software Process" is provided in (Stellman, 2006).

Volumes have been written on the subject of Software Process Improvement. Section 6.1 of this Guidebook describes the key elements of SPI and is focused on its management implications. Highlights of the SPI activity are defined and summarized here in the following Subsections:

- Software Engineering Process Group (2.4.1)
- Software Quality Assurance (2.4.2)
- Software Configuration Management and Control (2.4.3)
- Software Corrective Action Process—including change management (2.4.4)
- Software Peer Reviews and Product Evaluations (2.4.5)
- Capability Maturity Management Integrated—CMMI (2.4.6)
- Six Sigma Applied to Software (2.4.7)

2.4.1 Software Engineering Process Group

The Software Engineering Process Group (SEPG) plays a key role—especially when software development activities span organizational, administrative, geographic, or functional boundaries.

The SEPG is an important organizational element because it ensures consistent implementation of the software development processes across the project and facilitates production of compatible software products. A key responsibility of the SEPG is to coordinate, develop and maintain internal software work instructions and procedures. The SEPG is not a software implementer like the SwIPT but may include similar membership.

> **Lessons Learned.** From front-line experience, I can say that you cannot overstate the value and importance of a SEPG. I was the SEPG Lead on a large missile defense program that involved multiple companies and the customer's project office scattered across the USA. Collectively, the SEPG representatives included over 50 people from 12 organizations. We had monthly meetings at different sites with telephone conferences as needed in between. I set up six active working groups, and one provided a monthly metrics report that compiled the status from all companies, briefed

Program Managers and the customer, and we did that every month for 5 years. We did not write code, but we were the "glue" that held the whole program together. Smaller programs or projects also need a SEPG but on a reduced scale consistent with their needs.

SEPG responsibilities also include *planning, managing* and *coordination* with stakeholders and other groups. The principal groups that interface with the SEPG are Software Quality Assurance (SQA), Software Configuration Management (SCM), and Software Integrated Product Teams (SwIPTs) including SwIPTs at the corporate level. Subsection 6.1.1 describes the responsibilities, membership and functions of the SEPG.

2.4.2 Software Quality Assurance

> *Software quality is a ubiquitous concern in all Software Engineering developments.*

There are many aspects of software quality, and many ways to achieve it. Software quality must be formally defined and discussed with your team, so they will clearly understand what quality means on your project. SQA, sometimes called *Software Quality Management* or *Software Product Assurance*, performs the planned and systematic pattern of actions necessary to help assure that software, and software-related products, satisfy system requirements and support delivery of a successful product.

SQA is a support organization and is separate from the software group because it must be an independent group in order to conduct its business in a fully objective manner. The SQA organization has a responsibility to provide project management with visibility into the software development process and products by performing independent audits and assessments. A Software Quality Engineer (SQE) should be assigned to your project as discussed in Subsection 6.2.1.

The job of SQA is to *help you* determine if your quality objectives have been met. It must be understood that:

> *The SQA group is not responsible for software quality; as the SPM, you and your software Development Team (SwIPT) are responsible for building quality into the product, and you are accountable for delivering a quality product.*

2.4.3 Software Configuration Management and Control

A *system* involves the combination of all hardware, software, firmware, people, and other constraints working synergistically together to produce a desired operational result. Software-intensive systems interact with all system elements. The *configuration* of a system is the functional and/or physical characteristics of the elements comprising the system. The *configuration* can also be a *version* of the hardware, software and firmware *for a specific increment* during development of the full system.

As defined by SWEBOK (Bourque and Fairley, 2014) "Configuration Management is the discipline of identifying the configuration of a system at distinct points in time for the purpose of *systematically controlling changes* to the configuration, and *maintaining the integrity and traceability* of the configuration throughout the System Life Cycle."

The SCM process is an *essential development control activity* that begins during requirements definition. SCM is responsible for the tasks necessary to *control baselined software products* and to *maintain the status of those baselines* throughout the development life cycle. Section 6.3 describes management aspects of the SCM functions and baselined products are discussed in Subsection 6.3.2.

2.4.4 Software Corrective Action Process

As the Software Project Manager, you are responsible for taking timely Corrective Actions or at least making sure this process takes place. Corrective Actions are triggered when:

- Performance deviates significantly from the plan (a definition of "significant deviation" must be determined by mutual agreement between you and the customer)
- Defects are identified in the software work products
- Enhancements and improvements are proposed and approved

Most software developments automatically involve *constant change*, and if you try to eliminate change you would be removing the opportunity to take advantage of lessons learned and it would scuttle the Corrective Action Process. If you don't incorporate change, it will result in limitations to the system and early obsolescence, which is essentially a system's death certificate before it is born. However, making changes too often brings with it *unintended consequences* so all proposed changes to the software under development must be carefully evaluated as well as controlled. The opportunity to measure progress and identify issues that need Corrective Action can come from the reviews and evaluations, test results, and other quantitative management data.

A discussion of the Corrective Action Process (CAP), also called *change management*, is covered in Section 6.4. Section 6.4 also covers the drivers of change management, Software Discrepancy Reports (SDRs), Software Change Requests (SCRs) and the Configuration Control Boards (CCB).

2.4.5 Software Peer Reviews and Product Evaluations

Software Peer Reviews (SPRs) are *structured methodical examinations* of software work products by the Developers, with their peers, to identify existing defects, to recommend needed changes, and to identify potential problems. The performance of SPRs can provide immeasurable value to your project, and they are an established best practice. Regardless of the size of your project, if you intend to produce high quality software work products, Peer Reviews must be performed. Peer Reviews must be an integral part of the development process and focus on the identification and removal of defects as early and efficiently as possible because it is significantly more expensive to fix the defect later in the life cycle.

The SPR process, preparing for it, conducting Peer Reviews, and analyzing Peer Review results is discussed in Section 6.5. Some software standards require Software *Product Evaluations* to be performed in addition to the SPRs. However, Peer Reviews are an acceptable alternative to conducting evaluations as long as the specific requirements for the product evaluations are satisfied.

Technical and Management Reviews are structured to support the evolution of natural products during the Software Development Life Cycle. Section 5.3 describes three types of reviews: Joint Technical Reviews (JTR); Joint Management Reviews (JMR); and Technical Interchange Meetings (TIM).

2.4.6 Software Capability Maturity Model Integration

The Capability Maturity Model Integration is a *process improvement, training and process maturity appraisal* program that was developed by the Software Engineering Institute (SEI) at Carnegie Mellon University. Many books and reports have been written on utilization of the CMMI for process improvement in a number of industries (Chrissis et al., 2010; Ahern et al., 2001).

Key elements of the CMMI are discussed in Section 6.6 but only as a summary to familiarize Project Managers with the conceptual framework of the CMMI, how CMMI can be used to assess the maturity of your project and, most important, how it helps to identify *where the shortcomings are* that need attention.

2.4.7 Six Sigma Applied to Software

Six Sigma is a set of techniques and tools to improve quality. It originated with statistical quality control of manufacturing processes to improve the quality of the output of a process by identifying and removing the *causes of defects*. A Six Sigma process will statistically expect 3.4 defective features per million parts (Five Sigma at 3.4 defects per 100,000 opportunities; Four Sigma at 3.4 defects per 10,000 opportunities, etc.).

Some users have combined Six Sigma ideas with *Lean Management* to create a methodology called *Lean Six Sigma*. Lean management is focused on eliminating waste while Six Sigma's focus is on eliminating defects and reducing variability. The use of Black Belts, Green Belts and other implementation roles as itinerant change agents has fostered an industry of training and certification.

There is no intent here to evaluate the values versus criticisms of Six Sigma, but the question is whether or not it has application to the management and quality improvement of software development. The Six Sigma approach has indeed been successfully used to improve software quality at many software organizations. However, there are some who believe that Six Sigma does not apply to software.

For example, Binder (1991) believes the hardware analogies do not apply and points out that the software process contains many uncontrollable variations from one "part" to another. In addition, many defects may contribute to a single software failure, so we often do not know how many faults a system contains. Software is clearly not a mass-production process.

> **Lessons Learned.** I tend to agree with Binder. However, it does appear that Six Sigma has been successful at many software organizations because it invokes a formal, structured process to improve software quality compared to the haphazard approach those organizations used in the past. The bottom line is, if Six Sigma works for you—use it.

2.5 System Sustainment

System Sustainment (often referred to as "System Maintenance") typically *consumes about two-thirds (or more) of the total System Life Cycle cost*. Some might argue it is also the most important activity of the System Acquisition Life Cycle—it clearly is the *longest* in duration and the *most expensive*. The assertion that it is *the* most important activity is true only if the system you delivered meets the customer's needs and if it is actually used for a long period of time.

Maintaining a deployed system, responsive to changing customer needs over its lifetime, can be just as challenging, as the original system development task. From a customer and user point of view, if the system you delivered served them well for years, if not decades, then the System Sustainment period may be much more important to them than the original development tasks.

Chapter 16 is devoted to System Sustainment. It includes the processes, procedures, people, materiel, and information required to fully support, maintain, and operate the software portions of a system over long time periods until the system is eventually replaced or retired.

2.6 Software Project Management Pitfalls

Murphy's Law says that anything that can go wrong will go wrong. When it comes to software development, Murphy was an optimist!

There are many red flags that SPMs must be aware of to produce a successful software-intensive system. The pitfalls listed below are organized in terms of your *failure* to do something you should be doing, *allowing* something to happen that should not happen, and the *lack* of not having something that you should have but don't have. These pitfalls, not in priority order, should be used in conjunction with the list of 22 success criteria at the end of Chapter 1.

- Red flag if failure to:
  - Compile a complete understanding of your customer's system requirements
  - Properly allocate your resources
  - Hold effective and timely reviews, audits, code walk-throughs and inspections
  - Record, track and resolve errors and action items
  - Collect and use appropriate metrics for timely Corrective Action
  - Pre-plan and define needed support very early in the life cycle
  - Build-in quality from the start of development
  - Be aware of the dependencies between projects
  - Define tool usage guidelines before using them or provide adequate training

- Red flag if allowing:
  - Software documentation to become out-of-date making it irrelevant
  - System and software requirements to rapidly change and impact the scope of the system
  - Software Developers to test their own software beyond Unit Testing
  - Cost and schedule constraints to determine when testing is complete
  - Critical documentation to be deferred to latter stages of the life cycle
  - Incomplete requirements traceability between documents
  - An incomplete definition of the needed SI for your system
  - Testing to be compressed to meet a scheduled milestone
  - SQA, SCM and IV&V findings to be ignored
  - Applicable standards to be ignored or not effectively tailored to your project
  - A breakdown in communications with the team, the customer or stakeholders
  - Too much process making the team inflexible and hamstrung
  - Incomplete or unrealistic project schedules and unrealistic deadlines to exist
- Red flag if the project lacks:
  - Comprehensive critical planning documents such as the SDP and WBS
  - A clear and effective change management process
  - Clear, frequent and meaningful completion milestones
  - Documented plans, procedures and reports for software testing
  - A structured software development process supported by CASE tools

Chapter 3

System and Software Life Cycle Processes

The processes followed by a project are the secret to its success.

Chapter 3 describes the *set of structured processes* that are implicit components of the System and Software Development Life Cycle Process. Its purpose is to describe the processes since Software Project Managers (SPMs) must *manage* them, and in order to do that, they must *understand* them and the activities that comprise each process. These processes apply to both large and small software-intensive developments, and they are basically the same general processes, but they *differ significantly in scope* depending on the size and complexity of the system being developed.

Effective management and control of these processes is a major theme of this Guidebook, however, following them will *not* guarantee success. Following defined and structured processes, applicable to your project, will dramatically increase the probability of a successful software-intensive system implementation. Not following defined and structured processes deprives the Project Manager of the mechanism that is a fundamental element in managing the project and guiding it toward success.

This chapter addresses the following key processes, directly involved with software development that SPMs must manage, and the sections where they are discussed:

- *Software Project Manager Processes*: Processes that SPMs must manage (3.1)
- *System Life Cycle (SLC) Process*: Covers the entire system from cradle to grave (3.2)
- *System Definition*: The first element of the SLC (3.3)
- *Software Development Life Cycle (SDLC) Process*: The complete Software Development Life Cycle from System Definition to the start of System Sustainment (3.4)

- *Software Implementation Process (SIP)*: The full cycle of software implementation tasks from initial software build planning to Software Item Qualification Testing (3.5)
- *System Integrated Test and Verification (IT&V) Process*: Activities where individual software modules are combined and then tested and verified as a group (3.6)
- *System Sustainment*: The processes, procedures, people, equipment, and information required to fully support, maintain, and operate the system over a long time period (3.7)
- *System Critical and Support Software Processes*: The significant advantages of assigning a class and category to every software entity (3.8)
- *Software Testing Process*: Strategies, alternatives and approaches to managing the testing process (3.9)

The processes noted above that are at the "system" level are usually the responsibility of the Systems Engineering Integration and Test (SEIT) organization but, as described in Chapters 10 and 12, Software Engineering plays an important supporting role. In addition to the processes described in Chapter 3, Table 3.1 lists software management and software quality-related processes that are described in other locations in this Guidebook and are not repeated in this chapter.

3.1 Software Project Management Processes

The processes that SPMs must manage overlaps the SPM control areas described in Section 2.3. All of these processes are important. They must support each other since they are equivalent to links in a chain—if one is defective, or ineffective, the probability of failure will increase.

With the possible exception of "Cowboy Coders," *most Programmers follow some or all of a process*—even though they may not realize it or may not refer to what they do as a "process." Why? Because there appears to be a strong cosmic force of some kind that prevents many Programmers from realizing that what they do is a *process*! (my apologies for this

bit of sarcasm; see Chapter 7 and especially Section 7.6 for a thoughtful discussion of managing Programmers).

3.1.1 What is a Software Process?

A process is a series of activities, involving tasks, procedures, constraints and applied resources, that produces one or more planned outputs. Another way to help answer this question is to define the process-related terms. The following definitions were partly derived from (Humphrey, 1990):

- *Software Engineering Process*: A complete set of Software Engineering activities needed to transform user's requirements into operational software
- *Software Development Process*: A defined set of activities, composed of procedures and practices, used to develop and evaluate software products
- *Software Process Architecture*: A framework within which project-specific software processes are defined including the relationships of those processes

Figure 3.1 is a graphical example of a sequential software development process depicted in a circular fashion. It is another perspective of the linear sequential (Waterfall) Software Implementation Process discussed in Subsection 4.1.7. The sequential process starts with software requirements analysis, followed by the design, coding, integration and testing, and then transition to operations and sustainment (maintenance). Details of each activity comprising the software development process are covered in Chapter 11.

When the cycle is complete, and the software becomes operational at the sustainment site(s), the subsequent

Table 3.1 Software Management and Quality Processes External to Chapter 3

Process	Location
Performance Tracking and Control	5.1 and Chapter 15
Risk Management	5.2
Integrated Test and Verification	5.6, 10.3, 11.6, 11.7
Software Process Improvement (SPI)	2.4, 6.1 and 6.6
Configuration Management Process (CMP)	6.3
Corrective Action Process (CAP)	6.4
Peer Reviews and Product Evaluations	6.5
Requirements and Traceability Management	2.3.1, 10.1 and 11.1
Software Acquisition Life Cycle	Chapter 11 and 12.2
Earned Value Management	8.5

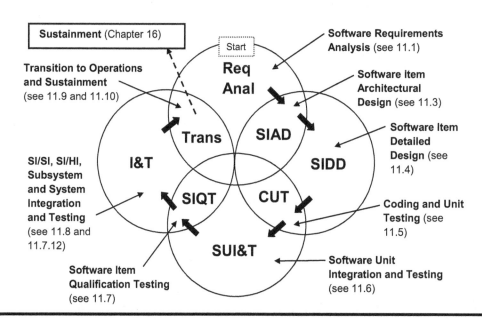

Figure 3.1 Overview of the software development process.

required system modifications and enhancements can restart the process.

3.1.2 The Cost of No Process

The relative cost of not following a software development process is depicted in Figure 3.2. The figure indicates that, during early software life cycle activities, the extra cost for following a structured process is greater than the cost of not following a structured process. However, during the latter activities of the Software Development Life Cycle, the cost of not following the process is significantly greater.

Comparing the areas under the curves clearly indicates that, collectively at the conclusion of the project, it is *much less expensive to follow a process*. The extra upfront effort has a large payback.

Figure 3.2 is a notional chart. It is not the result of an extensive statistical study and is not intended to be exact. Its purpose is to show the relative overall value of following a process from a cost perspective. There are a lot more additional advantages of following a defined process including a higher quality product that you will produce along with a happier customer.

3.2 System Life Cycle Process

The SLC *covers the entire system from cradle to grave* starting with an initial conceptual analysis of the proposed system and ending with eventual replacement or retirement of the system. As introduced earlier in Subsection 1.4.3, the SLC process is envisioned as being composed of five elements, and they are described in the following sections and chapters:

- System Definition (Section 3.3)
- Software Development Life Cycle (Section 3.4)

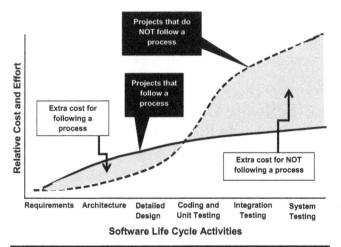

Figure 3.2 Relative cost of not following a defined software development process.

- Software Implementation Process (Section 3.5 and Chapter 11)
- System Integration, Testing and Verification (Section 3.6)
- System Sustainment (Section 3.7 and Chapter 16)

Figure 1.2 in Subsection 1.4.4 was a graphical introduction to the System Life Cycle. Figure 3.3 is an *important elaboration* of that figure as it provides more detail of all the SLC activities starting with the system concept all the way to ultimate system retirement. Figure 3.3 identifies the elements between those extremes and these elements are discussed below. Study Figure 3.3 until you fully understand it.

To help clarify some terms in Figure 3.3: A software *build* is a portion of a system that satisfies, in part or completely, an identifiable subset of the total end-item or system (see Subsection 5.1.4); A software *spiral* refers to an evolutionary software development model (see Subsection 4.1.5).

3.3 System Definition

System Definition (or description) is the first element of the SLC as shown in Figure 3.3. It starts with the objective of analyzing, specifying and documenting top-level system requirements. The birth of a software-intensive system is generally not a point in time but typically takes place over a period of time. For example, here is one scenario. Someone or a group identifies a problem area or a needed service. They report it to management who may decide to look into the matter by assigning a Tiger Team (a small, short-term special purpose group) to investigate it. If the Tiger Team confirms the need, they may propose an approach thus leading to the birth of the preliminary system concept.

After management approves moving forward with the proposed system and provides initial funding, system concept analysis, and concept development tasks are performed providing more details of the proposed system. These conceptual ideas are followed by initial system planning, development of fundamental system requirements and sometimes a very high-level overall system design.

The system concept and system requirements analysis is discussed in Chapter 10 as part of the Systems Engineering Domain. Section 10.1 covers the objectives and approach and definition of the operational concept. Section 10.2 is focused on the system and subsystem design. See those sections for a more complete discussion of system concept, system requirements analysis, and system design.

Project Charter. A Project Charter (as discussed in Section 1.5) may need to be prepared by, or under the direction of, the Upper Management Planning Team. The Charter documents the results of the initial system's planning task.

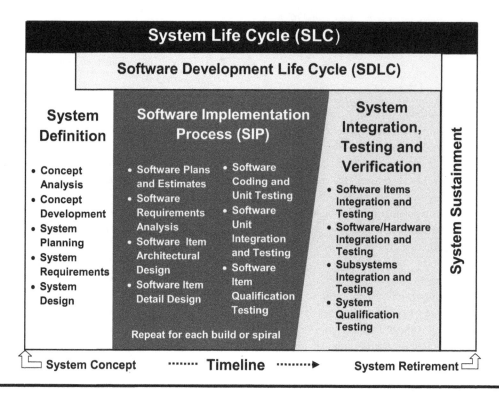

Figure 3.3 Overview of the System Life Cycle.

The focus of the Project Charter is on all aspects of the overall *system solution* being considered including:

■ Required functionality, operational concepts and capabilities
■ Technology development strategies
■ Alternative architectures and candidate solutions
■ Risk reduction activities and control schemes
■ Initial software development planning
■ Initial estimated system costs and benefits

Software Subject Matter Experts. During System Definition, the entire focus is at the *system level*. Essentially all modern day systems have a major Software Component, so Software Subject Matter Experts (SSMEs) are needed as technical support for all software-related issues. Their challenges are:

■ Early recognition of system capabilities that will or may require software contributions during system development, and
■ Assessment of the feasibility of the proposed software solutions. System and Software Designers evaluate actual alternatives in later milestones. However, *infeasible or unviable* capabilities or implementation options should be identified and modified or eliminated as early as possible.

In addition, SSMEs are needed for estimating software development costs. The initial cost estimates include an analysis of the amount of software for new code to be developed as well as the potential for reused code. Once the software size is estimated for each type of software, an independent analysis translates software size estimates into more accurate cost estimates using software cost estimation models. The cost estimates are used to help decide if the proposed project should proceed.

System Capabilities Documents. For large systems, *initial capabilities documents*, often called *Concept of Operations* (CONOPS), are often prepared. Although they may have limited software content, Software Engineers must identify software implications of the required user capabilities. The initial and preliminary System Definition tasks should not be confused with the detailed system requirements and System Design tasks covered in Chapter 10. Section 12.5 discusses software involvement during system concept analysis.

3.4 Software Development Life Cycle Process

The Software Development Life Cycle, as shown in Figure 3.3, starts during System Definition and ends at the start of System Sustainment. Embedded within the SDLC are: part of the System Definition task; the entire SIP; and the System Integration, Testing and Verification process.

Figure 3.4 is an expansion of Figure 3.3 as it is a more detailed graphical representation of the relationship between the SLC, SDLC, SIP and the System Integration and testing

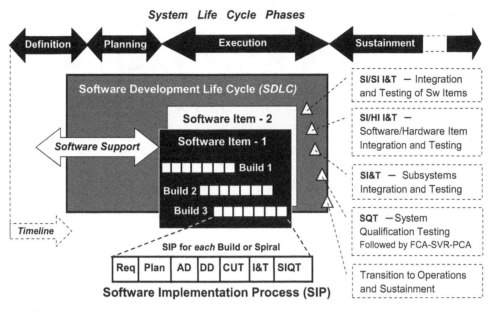

Req = Software Build Requirements Plan = Software Build Planning and Estimating
AD = Software Architectural Design   DD = Software Detailed Design   CUT = Coding and Unit Testing
I&T = Software Unit Integration and Testing   SIQT = Software Item Qualification Testing
FCA = Functional Configuration Audit   SVR = System Verification Review PCA = Physical Configuration Audit

Figure 3.4 Overview of the System and Software Development Life Cycles.

buildup prior to transition to operations and sustainment. There is a lot going on in Figure 3.4; take time to understand it as that would be a giant step forward in understanding the overall processes.

Figure 3.4 shows the SIP embedded within the Software Development Life Cycle. The SIP for each build consists of seven sequential activities (with appropriate iterations). The example in Figure 3.4 involves two Software Items (SIs) under development including three builds in SI-1. The SIP covers *the full cycle of software implementation tasks* performed for each build (or spiral) from initial software build planning to Software Item Qualification Testing (SIQT).

3.5 Software Implementation Process

Figure 3.5 depicts the full software integration and testing process. The SLC phases shown at the top of Figure 3.5 are a top-level system acquisition perspective. The seven SIP activities are part of the *Software Engineering Domain,* and those activities are described in depth in Chapter 11. The SIP is repeated for each build or spiral and may be repeated multiple times during System Sustainment to implement new functions and make needed modifications. The seven SIP activities are:

- *Software Build Requirements*: Software Systems Engineering should define the level of requirements satisfaction needed by each build or increment to implement a specified level of system functionality.

Within a subsystem, additional influences may dictate when capabilities are needed. This may include such factors as developing required software infrastructure or addressing areas of high-complexity. Also, naming conventions for each build must be established up front by assigning unique alphanumeric designations.

- *Software Build Planning and Estimating*: Software build planning is covered in Subsection 5.1.4 and software estimating has its own chapter (14). Software project planning and estimating are considered (in this Guidebook) as part of the Project Management Domain. Regardless of where these activities appear in the Guidebook, software planning and estimating are essential components of the development process. The re-planning activity is a critical ongoing task because changes almost always need to be made during development.

- *Software Architectural Design*: Architectural (or preliminary) Design precedes Detailed Design. The objective of SI Architectural Design is to describe the high-level organization of the SIs in terms of the planned functionality and relationships of the Software Units (SU).

- *Software Detailed Design*: The objective of SI Detailed Design is to determine and define the implementation details for each SU. It involves decomposing the SIs from the SI Architectural Design into the lowest level SUs in sufficient detail to map the design to the features of the selected programming language, the target hardware, operating system, and network architecture.

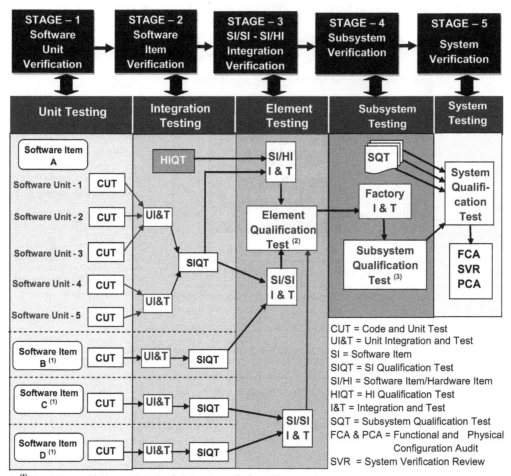

Figure 3.5 Software integration and testing process.

■ *Coding and Unit Testing*: Converts the SU Detailed Design into computer code and databases that are inspected, unit tested, and confirmed as responsive to the design.

■ *Software Unit Integration and Testing*: A systematic and iterative series of integration builds of SUs that have successfully completed Code and Unit Test, and building them up to a higher level SU, or SI, for the current build (see Sections 3.6 and 11.5).

■ *Software Item Qualification Testing*: Demonstrates that the Software Item meets the system and interface requirements allocated to the SI being tested. (see Sections 3.6 and 11.6).

3.6 System Integration, Test and Verification Process

The System Integration, Testing and Verification process is embedded within the SDLC as shown in Figures 3.3 and 3.4. The system IT&V process involves activities

where individual software modules are combined and then tested and verified as a group. It occurs after unit testing of the code but before System Qualification Testing. Integration testing takes, as its input, software modules that have been unit tested, it groups them in larger aggregates, applies tests defined in the *Software Test Plan* (STP) to those aggregates, and delivers as its output the integrated system ready for system testing. There may be a separate *Integration Test Plan,* or it may be incorporated in the STP.

Discussion of the system IT&V, presented in Section 3.6 below, includes the IT&V *stages* as well as the IT&V *process*. IT&V is also described as part of the Systems and Software Engineering Domains in Chapters 10 and 11 including:

■ Software Unit Integration and Testing (Section 11.5)
■ Software and Hardware Item Integration and Testing (sections 10.3 and 11.7)
■ System Qualification Testing (Section 10.4)

The system IT&V approach must be consistent and compliant with the system-level integration and verification test plan often called the system *Master Test Plan* (MTP). The rationale for software testing is based on an incremental buildup of tested requirements with a simultaneous incremental verification buildup. An example of the software integration and testing process is graphically depicted in Figure 3.5 showing a systematic and incremental buildup of testing software requirements. The generic software IT&V process normally involves *five testing stages* as shown in Figure 3.5.

3.7 System Sustainment

The last activity of the System Life Cycle is System Sustainment. This activity is not covered in Chapter 3 because Chapter 16 is devoted to the sustainment activities (often called maintenance), and that chapter is focused on the management issues encountered during sustainment. Software Sustainment includes the processes, procedures, people, materiel and information required to fully support, maintain and operate the software portions of a system over a long period of time (sometimes decades). Most Software Engineering literature is focused only on software development activities, but sustainment is actually *the longest and most expensive activity of the System Life Cycle.* From the user's standpoint, it may also be considered the most important activity.

3.8 System Critical and Support Software Processes

As discussed in Section 1.8, every software entity must be assigned a *class and category* because:

> *Not every software entity needs to have the full set of documentation, the full set of reviews, the full set of metrics, and the same level of testing.*

Significant savings can be realized if software entities can be placed into a class that does not require the same level of attention required by software that is critical to full functionality and performance of the system. Section 1.8 described *three generic classes of software* in a software-intensive system:

- *System Critical Software* (SCS) discussed in Subsection 3.8.1
- *Support Software* (SS) discussed in Subsection 3.8.2
- *Commercial Off-the-Shelf or Reuse Software* (C/R) discussed in Section 9.7

3.8.1 Development Process for System Critical Software

SCS was defined in Subsection 1.8.1 as consisting of all applications software used to perform real-time operations and non-real-time *functions implicitly required* to implement system functionality allocated to software. Figure 3.6 is an example graphical overview of a software development process for the SCS class of software. Tailor it for your project.

The SCS process is a comprehensive process since it provides the *critical* application software functionality. SCS development must support the development of as many SIs and builds as needed to meet customer requirements, contract milestones and system objectives. As shown in Figure 3.6, the SCS development process begins with requirements analysis and definition for each SI using system-level documents, such as the *Technical Requirements Document* (TRD) and the subsystem-to-subsystem *Interface Specifications.*

Requirements from these specifications are allocated to software and hardware, and the allocated software requirements are functionally decomposed, elaborated and documented in the *Software Requirements Specification* (SRS) and the *Interface Requirements Specification* (IRS). For the SCS class, Detailed Design, coding, integration, and testing activities are performed for each SI within each build. Once the SIs are integrated and tested for a build, the build is delivered, along with the *Software Version Description* (SVD), to the cognizant Software Development Library for Configuration Management control as discussed in Section 6.3.

3.8.2 Development Process for Support Software (SS)

Support Software (SS) was defined in Subsection 1.8.2 as software that *aids* in the development of hardware and software, integration, qualification, operations, test and maintenance. Figure 3.7 is an example graphical overview of a software development process for the SS class of software.

Although Support Software operates only in non-operational environments, the SS-1 category normally requires the same level of documentation as SCS software. However, reviews for the SS-1 category may not be as formal or as frequent as it is for SCS. The principal differences between the examples for SCS software (Figure 3.6), and SS software (Figure 3.7) processes, is that for (SS-1):

- There is no formal Software Specification Review (SSR), PDR, CDR, IRR, PTR and BTR as shown for System Critical Software; they are replaced by Technical Interchange Meetings (TIMs).
- The Architecture and Detailed Design phases are merged and followed by a TIM.

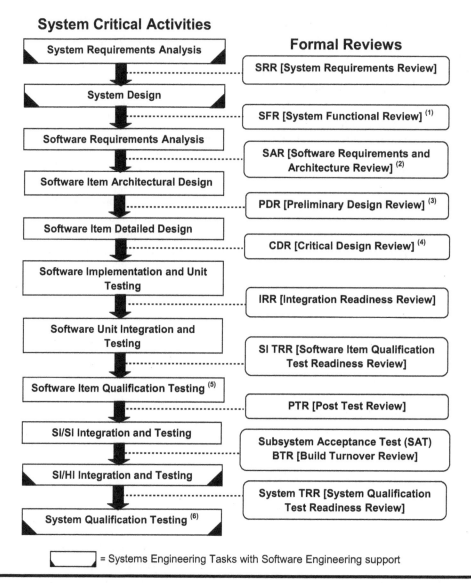

Figure 3.6 System Critical Software development process. (1) The SFR was formerly called the System Design Review (SDR). (2) SAR and PDR may be combined for Object-Oriented development because requirements definition and Architectural Design are usually iterative. The SAR was formerly called the Software Specification Review. (3) An optional SBRAR [Software Build Requirements & Architecture Review] may be held in addition to the PDR. (4) An optional SBDR [Software Build Design Review] may also be held in addition to the CDR. (5) Software Item Qualification Testing may be performed within each build. (6) SQT can be followed by FCA, SVR and PCA prior to transition to operations.

The SS-2 development process should be expected to be similar to the SS-1 process, but much less stringent. The principal differences between SS-1 and SS-2 software are:

■ SS-2 requirements information is normally maintained in a Requirements Database and referenced in SDFs, just as it is for SS-1, but formal preparation of the SRS, SDD and IDD is usually not required for the SS-2 class.

■ Informal SI and software build test descriptions and test results are maintained in SDFs. However, formal STP, STD and STR documents are usually not required for the SS-2 class.

■ TIMs performed for SS-1 may be replaced by Peer Reviews for the inspections and verifications of work products developed for the SS-2 class.

■ Applicable software metrics data should be collected for SS-2 software, but the metrics data set and the reporting frequencies should be significantly reduced for the SS-2 class.

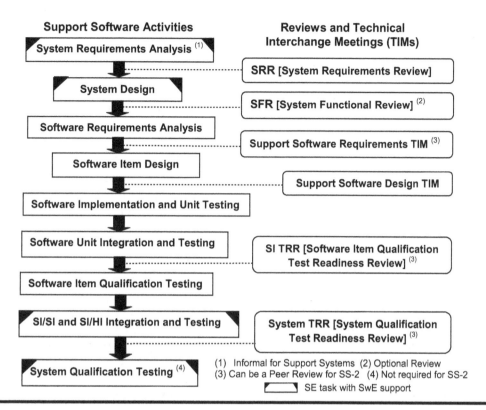

Figure 3.7 Support Software (SS) development process.

3.9 Software Testing Processes

Testing is not optional; it is an inherent and critical element of every software-intensive system.

Managing the test process basically involves: *estimating* the cost of testing; tracking and monitoring the *actual cost* of testing; and *making corrections* to stay on track. I could write a book on testing, so to avoid producing a 1000 page Guidebook, seven test topics have been selected to briefly discuss how they relate to managing the testing process:

- Testing is Ubiquitous (3.9.1)
- Testing Strategies (3.9.2)
- Regression Testing and Refactoring (3.9.3)
- Test Risk Analysis (3.9.4)
- Requirements Testability (3.9.5)
- Test-Like-You-Operate (3.9.6)
- When to Stop Testing (3.9.7)

3.9.1 Testing is Ubiquitous

The system testing and software testing functions take place at different times, with different levels of fidelity, throughout the Software Development Life Cycle. In that context, *software testing is covered several times* as part of

the following activities, at the indicated locations in this Guidebook:

- Software Item Test Planning (5.1.6)
- System Test Planning (5.1.7)
- Software Testing, Verification and Validation (5.6)
- Software Coding and Unit Testing (11.5)
- Software Unit Integration and Testing (11.6)
- Software Item Qualification Testing (11.7)
- Software Item and Hardware Item Integration Testing (10.3 and 11.8)
- Subsystems Integration and Testing (10.4)
- System Qualification Testing (10.4)

3.9.2 Testing Strategies

As the SPM, especially on large programs, you should not be directly performing the actual testing. However, you should be directly involved with deciding the appropriate *testing strategy planned for your project*. Addressing strategy means a general approach rather than specific methods and techniques devised for component testing. The major *testing strategies* include:

- *Alpha and Beta Testing*: Often used for commercial software products before formal release to identify defects. The pre-release is given to a small representative group of users for trial testing. Alpha Testing is usually in-house, and Beta Testing is usually performed by external users.

■ *Back-to-Back Testing*: A testing strategy used when versions of a system are available. The versions are tested together, and their outputs are compared for consistency.

■ *Bottom-Up Testing*: Testing starts with the basic components and works upwards.

■ *Configuration Testing*: Analyzes software under various specified configurations when the software is built to serve more than one user.

■ *Installation Testing*: Verifies software performance in the target environment (see 11.8).

■ *Performance Testing*: Verifies that the developed software meets specified performance requirements.

■ *Recovery Testing*: Forces software to fail in a variety of ways and verifies that recovery from the forced failure is properly and gracefully performed.

■ *Regression Testing*: Verifies that modifications have not caused unintended effects (see 3.9.3).

■ *Security Testing*: Verifies that built-in protection mechanisms will prevent improper access.

■ *Stress Testing*: Relies on stressing the system by going beyond its specified design loads and design limits, i.e., how well the system can cope with overload situations.

■ *Thread Testing*: Used for testing systems with multiple processes where the processing of a transaction threads its way through these processes.

■ *Top-Down Testing*: Testing starts with the most abstract component and works downwards.

Larger systems are usually tested using a mixture of these strategies rather than any single approach. Different strategies may be needed for different parts of the system and at different stages in the testing process. Developing *Software Test Plans*, especially on large projects, is not a trivial task so you must provide ample time and resources to produce this important planning document.

> **Lessons Learned.** Whatever testing strategy is adopted, it is always sensible to adopt an incremental approach to subsystem and system testing. Do not integrate all components and then start testing; always *test the system incrementally*. Each increment should be tested before the next increment is added to the system. This process should continue until all modules have been incorporated into the system.

When a *new module* is introduced in the testing process, previous tests that did not find defects may now detect defects. These defects are probably due to interactions with the new module. The source of the problem may be localized so that should simplify defect location and repair.

3.9.3 Regression Testing and Refactoring

Regression Testing. This is a retesting approach used to confirm that the code changes fixed the identified problems, but its real value is finding *unintended impacts of code changes*. Good Regression Testing should be able to identify both localized and global impacts of code changes. It is not unusual to find *defects injected at a rate of 5%–10% of defect fixes*. The rate of unintentional errors, resulting from making code changes, depends heavily on the complexity of the design since complexity can make software defects more difficult to fix and often their side effects are less apparent.

As the SPM, you may become embroiled in a disagreement as to the need to perform Regression Testing because of its impact on cost and schedule. You may have to prepare a strong argument especially if your senior manager is the principal objector. Keep statistics to prove that your Developers are injecting enough errors to justify Regression Testing.

> **Lessons Learned.** The bottom line is that whenever there is a code change, *previous testing must be repeated* to confirm that those changes did not change the behavior of the previously tested software. In my judgment, if *any* system element changes, including hardware changes, *software testing needs to be repeated*.

A famous, and expensive, example of not following this advice was the 1966 Ariane V missile launch vehicle failure. The flight software worked flawlessly in the earlier Ariane IV launch. It was *not retested* for Ariane V because *no changes were made to the software*. However, changes were made to the hardware, and the control software reused from Ariane IV encountered conditions during the Ariane V flight test it did not understand and issued faulty commands, causing the vehicle to go out of control. Ariane V had to be destroyed.

The amount of breakage from unintended impacts of code changes can be reduced by *mitigating the sources* that could cause defects to be accidentally injected. Make sure the entire software Development Team is involved in the process of identifying those sources. Even with very good software development processes, some defects will still be injected so a *mandatory Regression Testing protocol* must be imposed on the project—especially for larger projects.

Refactoring. Adding, removing or changing source code in order to make the code easier to read, understand and maintain is called *refactoring*. Refactoring makes improvements by changing the design and approach, but it does *not alter the behavior or functionality* of the software. Many Programmers find opportunities to improve the design of the system, but these improvements usually do not become apparent to the Programmers until well into development of the code. There are many different improvement techniques that can be called refactoring (they will not be described here).

Making the system perform more efficiently, and making the code easier to understand and maintain, is a good thing. However, spending far too much time trying to make it progressively better and better can be detrimental to the project's cost and schedule targets. As the Software Project Manager, you must decide when to draw the line and decide, *after the contractual requirements are met, when good enough is good enough.* See Fowler (2000) for a full discussion of refactoring.

You can always make the product "better," so if this effort is allowed to continue for a long time it will destroy your budget. See the discussion on *gold plating* in Subsection 5.1.5 to help track the impact of prolonged refactoring. Of course, if lives are dependent on consistent performance of the software, perfection, or as close to it as possible, is the goal.

3.9.4 Test Risk Analysis

Since it is impossible to test every possible aspect of an application, every possible combination of events, every dependency interface and everything else that could go wrong, test *risk analysis* is appropriate to most software development projects. Use risk analysis to determine *where* testing should be focused. This requires judgment, common sense and experience. Use a checklist, similar to the example in Table 3.2, to obtain answers to these types of questions *tailored to your program.*

Typical testing performed for consumer software is insufficient for medical devices; the requirements for rigorous testing of life-critical devices will not be necessary for many kinds of business systems. Each project must come to terms with the level of testing they really need as well as the level and kind of test documentation that fits their risk tolerance and available test resources.

3.9.5 Requirements Testability and Verification

A critical characteristic of a *good* requirement is that it must be *testable* (also called verifiable). Testability (or verifiability) asks the question: "Is there a way the stated requirement can be verified?" If the answer is "no" then the specified requirement is invalid and must be reworked and rewritten.

When each software requirement is defined, the Test Team must specify the *verification methods* to be used to confirm how each requirement is to be satisfied. The four typical verification methods include *Inspection, Analysis, Demonstration* and *Test.* The verification methods, and verification levels, are usually documented in the SRS and the

Table 3.2 Software Testing Risk Analysis Check List

Which aspects of the application:	■ Are most important to the customer? ■ Can be tested early in the development cycle? ■ Are most complex and are likely responsible for most of the errors? ■ Were developed in a rush or panic mode? ■ Caused problems in similar/related previous project? ■ Had large maintenance expenses in similar previous projects? ■ Contain unclear or poorly thought out requirements or design?
Which functional capabilities:	■ Are most important to the project's intended operational purpose? ■ Have the largest safety impact? ■ Have the largest financial impact on users?
What specific types of tests will:	■ Address applications judged to be the highest-risk aspects? ■ Create the most customer service complaints? ■ Cover multiple functionalities? ■ Have the best high-risk-coverage?

IRS. The verification methods and levels should also be specified in the automated requirements management tool. *You should require test personnel to be involved with requirements definition from the beginning of the software life cycle, or shortly thereafter, to establish from the start a common understanding as to how each requirement will be verified.*

3.9.6 Test-Like-You-Operate/ Test-Like-You-Fly

If your software-intensive system is fully compliant with your customer's requirements, and it passes all of the qualification testing, it still might fail if you did not test it in its intended real operational (flight-like) conditions and expected environment.

Test-Like-You-Operate, also called "Test-Like-You-Fly," is an approach to testing complex software systems. This approach to testing is *not a replacement* for any testing function described in this Guidebook; it is an additional and optional testing approach where *all* the elements, or components of the system that interface with each other, must *all* be tested together *before* the system goes operational. Subsection 10.4.9 describes some interesting examples of total system failures, or degradation in system functionality that would have been avoided if the Test-Like-You-Operate approach was utilized. Some of those examples of failure seem to me to be the epitome of a lack of plain common sense.

3.9.7 The Value of Automated Testing

General approaches to test automation, such as code-driven testing or Graphical User Interface testing, is not covered here but you should be familiar with these approaches. As the SPM, you should be directly involved with decisions related to *what* to automate, *when* to automate, or even whether your project really *needs* to automate since these are crucial decisions. Selecting the correct features of your product as candidates for automation often determines the success of the automation initiative.

For large projects, or ongoing long-term projects, automated testing can be highly cost-effective. But for smaller projects, the time needed to learn and implement the automated testing tools is usually not worthwhile. Automated testing tools may not make testing easier. One problem with automated testing tools is that *if* there are continual changes to your product or system being tested, the scripts have to be updated so often, that it becomes a very time-consuming task. Another problem with such tools is that the interpretation of the results (screens, data, logs, etc.) can also be a time-consuming task. Do not interpret this as a vote against automated testing; just make sure it is cost-effective for your project.

3.9.8 When to Stop Testing

There are many variables involved in deciding *how much testing is needed* to produce a dependable software product fully responsive to your customer's needs. The key variables in this decision-making process are the size and complexity of the code, the ability of your testers, plus the required system criticality and dependability. The obvious reasons to stop testing are:

■ When all *test cases* (or a customer approved percentage) are successfully completed
■ When project testing *deadlines* are reached
■ When the *budget* for testing is fully consumed
■ When the testing *bug rate* detection falls below a preset acceptance level
■ When *functionality* testing achieves a desired level of confirmed performance

In addition, there are other techniques that can be used to approximate when the Test Team has performed an *acceptable level of defect detection*. Three example techniques are described below: using measurements, capture-recapture, and error seeding.

Defect Tracking Using Measurements. Measurements and tracking of defects can be used to determine when to stop testing. For example, the SPM and the customer may agree before testing starts that when the *defect detection rate* falls below a certain level, the testing effort will be considered completed. Some of the measurements that can be used for continual defect tracking include:

■ Defect Density (number of defects per thousand lines of code)
■ Defect Discovery Rate (Number of new defects and number found in defect insertion phases)
■ Defect Resolution Rate (Number defects that have been fixed)
■ Number of Test Cases Developed, Dry Run, Performed and Passed

Keeping track of the total number of errors detected does not help you very much because you don't know how many total errors there are, so you cannot calculate your percentage of success.

Capture-Recapture. An interesting approach to estimating the number of defects in a software product was proposed by Humphrey (2000). He was looking for a simple method to predict remaining defects based on the results from two or more independent Reviewers who conducted either Peer Reviews, inspections or testing. He found there was such a method being used in estimating animal populations called capture-recapture. An example application of this approach is calculating how many fish there are in a pond. For example, you randomly catch 30 fish, tag them, and then release them. A few days later you catch 25 fish randomly from the pond and find that five of them have your tags. The simple mathematical equation is:

$$\frac{30 \text{ tagged fish in the pond}}{X\left(\text{Total fish in the pond}\right)} \approx \frac{5 \text{ tagged fish in the sample}}{25 \text{ fish caught in the sample}}$$

$$\approx 150 \text{ fish in the pond}$$

This approach can be utilized to estimate defects in a software product where the pond becomes the software product being tested and the fish becomes the defects. For example, a software engineer inspects a software product and identifies and records the defects found. These defects are "tagged." A second software engineer inspects the same product and finds some of the tagged defects as well as other defects. Using the same formula as above for the number of fish in the pond you can estimate the number of remaining defects in the software product.

When you find the estimated number of defects according to the calculation, you can stop testing. (Davis, 2005) reported good results using this method but only if you have good Reviewers and the number of defects are not too small.

Error Seeding. Another interesting approach to determine when to stop testing is to plant some errors into the code (without telling the Testers). When using this error seeding approach, keep track of the ratio of the seeded errors found divided by the total number of planted errors. This ratio can be used as a measure of test progress; when all the seeded errors are found, you can stop testing because, statistically, the Test Team demonstrated that they did a good job of finding defects. This can be a reasonable approach as long as the seeded errors are relatively hard to find.

Chapter 4

Software Development Methodologies

Programmers subscribe to their chosen methodology of worship!

Software development methodologies are frameworks used to plan and control the process of developing software-intensive systems. Over the past several decades, a wide variety of frameworks have evolved each having specific strengths and weaknesses. There is no single *best* system development strategy; good options exist, and no one methodology is good for all types of projects. Forget about silver bullets and one size fits all.

As the Software Project Manager, you should have an in-depth understanding of the alternative methodologies and their strengths and weaknesses; then you can proceed with the methodology that *best fits your project*. Even better, you can select specific features from the methodologies that are applicable to your project and create your own hybrid methodology tailored to the needs of your project.

Chapter 4 provides a top-level overview of the most popular software development approaches, methodologies, models and standards and is discussed in the following sections:

- *Software Development Process Models*: Covers the most commonly used Software Development Process Models including a brief description of those models and their differences (4.1).
- *Software Analysis and Design Models*: Covers the most commonly used Software Analysis and Design Methodologies including a brief description of each (4.2).
- *Managing Agile Software Development Projects*: Agile is an umbrella term for multiple project management methodologies. This section is focused on Agile/Scrum which is an approach to software development that focuses on iterative goals set by the Product Owner through a backlog developed by the Scrum Team, facilitated by the Scrum Master (4.3).

- *Schedules and Activity Networks*: Describes the importance of a formal program *Integrated Master Plan* (IMP) coupled with a related *Integrated Master Schedule* (IMS) (4.4).
- *Software Standards:* Covers the need for and advantages of software standards to provide consistency during system development plus typical software product and testing levels (4.5).

As the Software Project Manager you must be involved with the selection of the development methodology because, once a decision is made, it will have a long-term impact to your project. The objective of Chapter 4 is to provide an introduction to that basic understanding providing you with a foundation for that decision.

4.1 Software Development Process Models

A Software Development Life Cycle Model should almost always be used to describe, organize, monitor and control software development activities. There is considerable confusion regarding the differences between some of the development models. Table 4.1 contains, in alphabetical order, a brief description of the most commonly used *Software Development Process Models*. To clarify the intent of each model, they are discussed in more detail at the indicated locations in the Guidebook.

Managing the *Agile* methodology is covered in a separate Section 4.3 since it is important enough to warrant its own section (actually, it is worth more than a section, but this Guidebook is intended to be a compendium and not a huge tome). Your project must select the strategy (or strategies) appropriate to the system you are developing with emphasis on the availability of requirements. The process selected should be defined in your Software Development Plan (SDP) as described in Subsections 1.5.1 and 5.1.1. More than one

Table 4.1 Software Development Process Models

Process Model	Brief Description
Agile Model	The term *Agile* covers a number of methods and approaches for software development. It is focused on frequent delivery and testing of small portions of the full system. Agile methods are based on the principles of human interaction, where solutions *evolve* through *collaboration of small self-organizing, cross-functional teams*. Agile is conceived to be flexible and more quickly able to respond to changes than traditional models (see 4.3).
Evolutionary Model	With this model, the software product is developed in a series of builds (blocks or stages) with increasing functionality. The requirements are defined for each evolutionary build as that build is developed. This is a "build-a-little, test-a-little" development process model that can provide an early operational capability for a portion of the entire system, and it is highly amenable to systems with evolving requirements (see 4.1.1).
Incremental Model	This model requires that all of the requirements are defined up front; the software product is then developed in a series of builds with cumulative increasing functionality. A portion of the software product is built and tested—one small increment at a time. This is also a "build-a-little, test-a-little" approach that can provide an early operational capability for a portion of the entire system (see 4.1.2).
Iterative Model	Not really a software development model but more of a *quality improvement approach* where a fully developed and delivered systems is periodically updated and improved with each new release of the product. After a system is developed using one of the other development models, the iterative approach is used to improve its quality (see 4.1.3).
Prototyping	This development approach involves building an early experimental portion of a system to better understand the requirements and interfaces, to test throughput speeds, develop environment testing, etc. Normally the product produced is built fast, without sufficient documentation, and not designed to be maintainable, so it normally cannot be used as the final product, but there are exceptions (see 4.1.4).
Spiral Model	The Spiral Model is a risk-driven software development process that has two main features: (1) A cyclic approach that grows a system's functionality and implementation incrementally while focusing on decreasing its degree of risk; and (2) A set of anchor point milestones for insuring stakeholder commitment to acceptable system solutions. Implementations using this model are often done in conjunction with the Evolutionary Model (see 4.1.5).
Unified Model	A variation of the Spiral Model is the Unified Process exemplified by the IBM Rational Unified Process® (RUP®). RUP is an iterative software development framework. However, it is not a single prescriptive process but an *adaptable process framework* intended to be tailored by selecting elements of the process applicable to each user. It involves an underlying Object-Oriented Model using the Unified Modeling Language (see 4.1.6).
Waterfall Model	A linear sequential software development model that requires all functionality and design requirements to be defined up front and each development activity to be completed before the next activity begins, although some overlap is allowed. The entire software product is not available until the last testing activities are completed (see 4.1.7).

Software Development Life Cycle Model may be needed for different types and applications of software used.

4.1.1 Evolutionary Software Acquisition Strategy

As defined in the dictionary, the word "evolution" refers to a "process of continuous change from a lower to a higher, more complex, or better state." In that context, evolutionary software development could be considered a *generic strategy*, applicable to other development methodologies where the functionality of the software product "evolves" or is refined during the course of each increment or iteration. Key to the success of evolutionary acquisition is *continuous customer involvement* in the articulation, validation and prioritization of system requirements over the life of the development process. An example development scenario following the Evolutionary Model is shown in Figure 4.1.

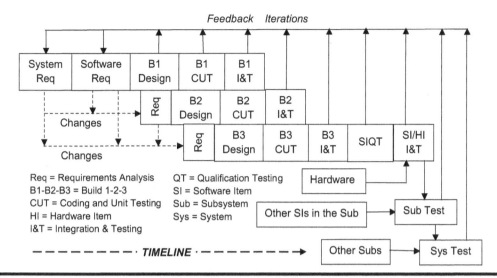

Figure 4.1   Evolutionary Software Development Model—example.

The figure shows three builds (B1-B2-B3) where each build adds more functionality. The requirements are defined or updated for each build. In this example, the requirements analysis for Build-2 starts during the Build-1 design, and the Build-3 requirements analysis starts during the Build-2 design. Software Item Qualification Testing (SIQT) takes place at the end of Build-3 followed by progressive integrations with Hardware Items (HI), other Software Items (SI) and then with the other subsystems. The Evolutionary Model is time-tested; consider it seriously for medium to large software-intensive system developments.

NOTE: Increment versus Iterate. Before discussing the Incremental Model and the Iterative Model, their differences should be noted because they are easily confused, and sometimes the two terms are used interchangeably. According to (Cockburn, 2008), *increment* fundamentally means *add onto* whereas *iterate* generally means *re-do* or *rework*. They are fundamentally different models, each serves a different purpose, and they need to be managed differently. Cockburn states: "The development process, feature set, and product quality all need constant improvement. Use an incremental strategy, with reflection, to improve the first two. Use an iterative strategy, with reflection, to improve the third." Please note that the word "iterate," "iterative" or "iteration" with a small case "i" refers to the common meaning of the word: "repetition." A capital "I" is used when referencing the model name.

4.1.2 The Incremental Model

The Incremental Model is a method of developing software where the product is developed in a series of *builds* with increasing functionality. A portion of the software product is built, tested and integrated—one small increment at a time. This is a *build-a-little, test-a-little* approach that can provide an early operational capability for a portion of the entire

system. The product is defined as finished when it satisfies all of the requirements that were defined up front.

The Incremental Model combines the elements of the Sequential Model (Waterfall; see 4.1.7) with the iterative philosophy of prototyping (see 4.1.4). The biggest problem with this approach is that it presumes you can define all of the software requirements up front. Assuming there will be no changes during development is unrealistic unless there is an edict to freeze the requirements for the current version. A variation of this model could include the provision for some feedback and changes during the build process. Regression Testing is an important element of the incremental build model.

A big advantage of the Incremental Model is that there is a working model of a portion of the system at an early stage of development making it easier to find functional or design flaws. Finding issues at an early stage of development is extremely cost-effective (see Subsection 6.5.1). Aside from needing requirements to be defined up front, a disadvantage of the Incremental Model is that it is typically applicable only to large software development projects because it may be hard to break a small software system into even smaller serviceable increments.

4.1.3 The Iterative Model

The Iterative Model can be defined as delivering an entire system and then changing the functionality of the system as needed during each subsequent release of the product. This model involves a cyclical process. A cornerstone of the Iterative Model is the fact that it is often very difficult to know upfront *exactly* what the customer really wants, so the process builds in as many *learning opportunities* as possible. The learning you gain from the iterations provides constructive information to the next iteration. This learning can come from end-users, Testers or the Developers themselves.

When the Iterative Model is used, the first delivery is likely to be a "core product." The subsequent iterations are the supporting functionalities or the add-on features that the customers want. The product is designed, implemented and tested as a series of incremental deliveries until it has all the functionality required by the users.

For example, a shrink wrap company may want to get a product to market within a prescribed time frame, regardless of its inferior quality, and then rely on future releases to fix the bugs and improve its quality. They may call this approach a "development model" because that is the iterative approach they follow. It can be argued that this is *not* a development model because the subsequent upgraded improvements could be considered the maintenance phase.

Another perspective of the Iterative Model can be explained with the analogy of a portrait painter who provides the customer initially with a very simple sketch; the customer decides the head is facing in the wrong direction so at this point it is easy for the painter to make that adjustment. The painter then adds a little more detail and presents it again to the customer for review. The customer makes recommended changes and the "iteration" starts over again and continues until the customer approves what will be the final version. For a software project where requirements are unknown, this approach may work, however, for a large or mega system this analogy does not generally apply.

The key to successful use of an iterative Software Development Life Cycle is rigorous validation of the new requirements, and verification and testing of each version of the software product against those requirements. As the software evolves through successive release cycles, tests have to be repeated and extended to verify each version of the software. Software project managers often consider each iteration a separate project.

4.1.4 Prototyping

Prototyping is the process of building an early experimental model of a defined system that provides partial functionality of the final product. Prototyping allows customers to evaluate development approaches and "try them out" before implementation and to better understand the requirements. Developing prototypes is an extra cost element but provides useful benefits especially for systems having a high level of user interactions such as online systems where users need to fill out forms or go through various screens before data is processed. A prototype can be used very effectively to give the exact look and feel before the actual software is developed. The major software prototyping types used widely are:

■ *Throwaway/Rapid Prototyping*: Throwaway prototyping is also called rapid or close-ended prototyping. This type of prototyping needs minimal requirement

analysis to build a prototype. Once the actual requirements are understood, the prototype is discarded, and the actual system is developed with a clearer understanding of user requirements.

■ *Evolutionary Prototyping*: Evolutionary prototyping, also called *breadboard prototyping*, is based on building actual functional prototypes with minimal functionality in the beginning but built up incrementally to eventually deliver the entire system with full functionality.

■ *Integration Prototyping*: Refers to building independent multiple functional prototypes of the various subsystems and then integrating all the available prototype subsystems to form the complete system.

The trap with the last two types is the tendency to *delay or avoid developing the documentation* needed by the final product and the difficulty, and cost, of recreating the documentation later.

Prototyping addresses the inability of many customers to specify their exact information needs and the difficulty of system analysts to understand the user's environment, by providing the customer with a tentative system for experimental purposes at the earliest possible time. The major problem with prototyping is that too often a prototype becomes a quick but inefficient approach to meeting customer requirements. The code will most often need major reconstruction, and most Programmers do not want to throw away their code so you may end up with patchwork code that is difficult to maintain.

Another problem with prototyping is the frequent interactive interface with the customer during prototyping may lead to requirements scope creep. Every time you believe you are finished there can be new improvements or new functionality proposed to make it better (see Gold Plating 5.1.5). You can avoid this if you plan a specific number of iterations and issue a cut-off date for adding new functionality to the system.

4.1.5 The Spiral Model

The Spiral Model is a risk-driven Evolutionary Software Development Model that couples the iterative nature of prototyping with the controlled and systematic aspects of the Linear Sequential Model where the software is developed in a series of incremental releases. Early releases typically produce a simple prototype, however, during later iterations more complex functionalities are added. The Spiral Model is a *risk-driven Software Development Process* Model that has two main features:

■ A cyclic approach that grows a system's functionality and implementation incrementally while focusing on decreasing its degree of risk

■ A set of "anchor point milestones" for insuring customer and/or stakeholder commitment to acceptable system solutions

Implementations using this model are often done in conjunction with either the Incremental or the Evolutionary Models. A simplified view of the Spiral Model is depicted in Figure 4.2. The Spiral Life Cycle is shown as a spiral that begins in the center of the spiral, eventually working its way outward, over and over again, until completion of the project. Figure 4.2 shows the four basic elements of the process: *Analysis* to determine the goals, objectives, constraints and alternatives; *Evaluation* of the risks, alternatives and mitigation approaches; *Development* and testing of the current iteration of the product; and *Planning and Review* for the next iteration of the development.

A more detailed view of the Spiral Software Development Process Model is shown in Figure 4.3. Various versions of the Spiral figure have been created and used, but they all contain the basic principles of the Spiral Model which can be summarized as:

■ The focus is on risk assessment and on minimizing project risk by breaking a project into smaller segments and providing more ease-of-change during the development process, as well as providing the opportunity to evaluate risks throughout the life cycle.

■ Each cycle of the spiral involves a progression through the same sequence of steps, for each portion of the product and for each of its levels of elaboration, from an overall concept-of-operation document down to the coding of each individual program.

Figure 4.2 Overview of the Spiral Software Development Process Model.

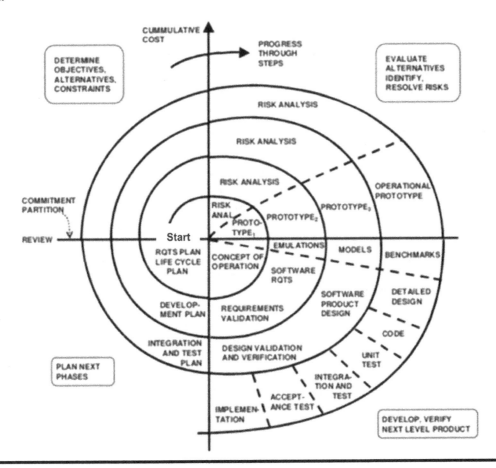

Figure 4.3 Spiral Software Development Process Model.

■ Each trip around the spiral traverses four basic quadrants: (1) determine objectives, alternatives, and constraints of the iteration; (2) evaluate alternatives; identify and resolve risks; (3) develop and verify deliverables from the iteration; and (4) plan the next iteration.

■ Each cycle starts with an identification of stakeholders and their objectives, and each cycle ends with a review and commitment.

■ Spiral development reduces technical risk by allowing implementers to drive the "bugs" out of new technologies through evaluation of the iterative prototypes.

The principal advantages of the Spiral Model include:

■ A realistic approach to the development activities because the software evolves as the process progresses. In addition, the Developers and the customer better understand and react to risks at each evolutionary level.

■ The model uses prototyping as a risk reduction mechanism and allows for the development of prototypes at any stage of the evolutionary development.

■ It maintains a systematic stepwise approach, like the classic Waterfall Model, and also incorporates into it an iterative framework that more realistically reflects the real world.

The principal disadvantage of the Spiral Model is that is it more complex and harder to manage than other models and it often increases development costs and schedule.

4.1.6 *The Unified Process Model*

A variation of the Spiral Model is the *Unified Process Model* exemplified by the IBM's Rational Unified Process® (RUP®). Other examples include "OpenUP" and the Agile Unified Process. RUP is an iterative software development framework. However, it is not a single prescriptive process but an adaptable process framework intended to be tailored by selecting elements of the process applicable to each user. It has an underlying Object-Oriented Model using the Unified Modeling Language (UML) which is a visual modeling system for graphically representing the use cases, class models, object interactions and components of the system.

UML provides a very robust notation, which grows from analysis to design. It is a language used to specify, visualize, and document the artifacts of a system under development. You can model just about any type of application running on any type of hardware, operating system, programming language, and network with UML. It is a natural fit for Object-Oriented languages and environments, but you can also use it to model non-Object-Oriented applications. For details on the use of the Unified Process Model refer to

(Kruchten, 2000). RUP is somewhat unique in that it is both a process model and a product set because it incorporates several optional supporting tools including:

■ *Rational Rose*: A UML editing tool that captures and models the software architecture
■ *Clear Case*: A software product Configuration Management tool
■ *Clear Quest*: A tool for tracking problem reports and changes
■ *Test Manager*: A tool for managing test cases

4.1.7 *The Linear Sequential Model*

Software development using the linear *Sequential* Model has been called the *traditional* approach, or *plan-driven development*, and is often referred to as a "*waterfall*," after a model described by Winston W. Royce in 1970 (Royce, 1970) (a former colleague of mine at Lockheed Missiles & Space Co). Royce did not describe the model as a waterfall, and he noted that a sequential implementation is risky. Despite Royce's warning, many commercial and government organizations developed and used several types of sequential waterfall-like development models. The effects of these Waterfall Models persist to this day, and they significantly impact the management approach to software development.

The linear Sequential Model requires each activity to be completed before the next activity begins, although a small overlap may be allowed to start some activities in a timely manner. The requirements and design activities are defined up front, and the *entire functional software product* is not available until the last testing activities are completed. If the first full working version of the program is not available until the project is finished, there can be disastrous results.

Figure 4.4 is a graphical example of an *Iterative Waterfall* version of the software development process showing iterative feedback loops and prototyping. The numbers in parenthesis indicate the sections in the Guidebook where each activity is described. Those development activities must be performed regardless of the software model followed.

A *pure* Waterfall approach is risky. It is highly unlikely for any *large system* to use a *pure* Waterfall Model, as its primary model for development, because requirements changes during the development process are essentially a certainty. A *pure* Waterfall Model can be used successfully for *small systems* when the product is *well-defined*, and the *requirements are known and stable*. That is never the case in medium-to-large developments.

Introduction of the Waterfall Model in the 1970s provided a major contribution to the understanding and management of the software development *process* because it was monumentally better than the haphazard software development approach typically followed at that time. The Iterative Waterfall with feedback loops and prototyping, as shown in Figure 4.4, is a more effective approach. However, other good options exist as described in this chapter. The Waterfall Model, or a version of it, is often used as an integral part of other methodologies and currently many information systems and projects are developed using versions of the Waterfall Model.

Lessons Learned. For those who balk at the thought of following a sequential process, I ask them to be realistic. You crawl before you walk and you walk before you run. *That is the way it is.* You cannot realistically do any meaningful coding until you understand what the code is supposed to do—you need a *design*. But before you can create a realistic design, you need to understand the *requirements* for what you are building. And you cannot test the final product until you have the requirements for the system you are testing to make sure the developed product is responsive to those requirements. *That is the way it is.* Live with it!

4.1.8 Other Process Models

Service Oriented Architecture (SOA). SOA is concerned with modeling business and software systems, for designing and specifying *service oriented* business systems within a variety of architectural styles, including enterprise, application and service oriented architectures and cloud computing. The SOA methodology typically includes a modeling language that can be used by the customer and the Developers.

Service oriented modelers typically strive to create models that provide *clear and comprehensive views* of the analysis, design and architecture of *all software entities* in an organization so they can be understood by individuals with diverse levels of business and technical understanding.

The first SOA-related methodology, *Service Oriented Modeling and Architecture* (SOMA), was announced by IBM in 2004. SOMA refers to the more general domain of service modeling necessary to design and create SOA. SOMA includes an Analysis And Design Method that extends traditional Object-Oriented and component-based Analysis and Design Methods which includes concerns relevant to and supporting the SOA philosophy.

Microsoft Solutions Framework (MSF). MSF, developed by Microsoft, requires the design of a series of models,

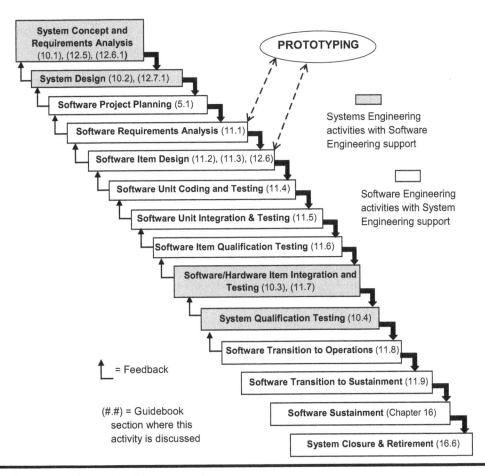

Figure 4.4   Iterative Waterfall Model with prototyping.

Table 4.2 Overview of the Software Analysis and Design Models

Analysis and Design Models	Brief Description
Object-Oriented Analysis and Design (OOA/OOD)	OOA/OOD analyzes and designs software applications by applying an *Object-Oriented* paradigm and visual modeling throughout the Development Life Cycle resulting in better communication between Developers, customers and stakeholders plus improved product quality (see 4.2.1).
Structured Analysis and Structured Design (SA/SD)	SA/SD is an age-old systems development technique that is time-tested and relatively easy to understand. In addition to modeling the processes, Structured Analysis includes data organization and structure, relational database design, and user interface issues. SA/SD is used during the software requirements analysis, Preliminary Design, and Detailed Design activities. SA/SD views a system from the perspective of the data flowing through the system (see 4.2.2).

including a risk management model and a team model. Each model has a specific purpose and outputs that contribute to the overall design of the system. MSF uses OOA/OOD concepts (described in 4.2 below), but it also examines a broader business and organizational context that surrounds the adoption and utilization of information systems.

4.2 Software Analysis and Design Models

The Software Development Process Models described above covered the different approaches to managing the development throughout the full life cycle. There is a second category of models that covers just the *front-end of the life cycle*: requirements analysis and design.

Table 4.2 briefly describes two of the two most commonly used Analysis and Design Models—OOA/OOD and SA/SD—followed by a more detailed discussion in the subsections indicated.

4.2.1 Object-Oriented Methodology

Object-Oriented Analysis and Design (OOA/OOD) is a popular approach for analyzing and designing business and engineering applications by applying an *Object*-Oriented paradigm and visual modeling throughout the Development Life Cycles to foster better stakeholder communication and improve product quality. Iteration by iteration, analysis models for OOA, and design models for OOD, are refined and evolve continuously driven by key factors like risks and its value to your business or the system you are creating. Many books have been written on OOA/OOD.

The OOA approach combines data, and the processes that act on that data, into entities called *objects*. OOA defines the various types of objects that are doing the work, that interact with one another in the system, by showing user interactions called *use cases*. Systems analysts use Object-Oriented

Methods to prepare models of real-world business processes and operations. The result is a set of software objects that represent people, things, transactions, and events.

Using objects (data) and methods (processes) needed to support a System Design, a System Developer can design reusable components that allow faster system implementation and decreased development cost. The Object-Oriented (OO) approach provides "naturalness" and reuse. It is natural because people tend to think about things in terms of tangible objects and, because many systems within an organization use the same objects (i.e., windows, dialog boxes and menus), the classes can be used repeatedly. Object-oriented analysis provides an easy transition to popular OO programming languages, such as Java, C++, C+, Perl, PHP, Smalltalk, Common Lisp, Python, Ruby and Delphi.

4.2.2 Structured Analysis/Structured Design Methodology

Structured Analysis and Structured Design (SA/SD) is a traditional systems development technique that is time-tested and easy to understand. Because it describes the processes that transform data into useful information, Structured Analysis is called a process-centered technique. In addition to modeling the processes, Structured Analysis includes data organization and structure, relational database design, and user interface issues. SA/SD is used during the software requirements, Preliminary Design, and Detailed Design activities. Structured Analysis views a system from the perspective of the data flowing through it. The function of the system being designed is described by processes that transform the data flows.

Structured Analysis takes advantage of information hiding through successive decomposition (or top-down) analysis. This allows attention to be focused on the most important details. The result of Structured Analysis is a set of related graphical diagrams, process descriptions and data definitions. There have been many variations of these models proposed by luminaries such as (DeMarco, 1979; Ward 1985) and many others, but they generally include:

- *Data Flow Diagrams (DFD)*: The DFD is a graphical representation of the *flow of data* through the system; it shows the top-level functionality and interfaces of the SI.
- *Database Dictionary*: Defines the basic organization of each database including a list of all files in the database, the number of records in each file, and the names and types of each data field. It does not contain any actual data, just bookkeeping information for managing the databases.
- *Context Diagram:* Shows external data interfaces with the SIs similar to a block diagram.
- *Entity Relationship Diagram*: Shows the relationship between the data elements. An entity can be viewed as a data store or terminator on the DFD.
- *State Transition Diagram*: Shows the transition between states (e.g., normal, impaired, critical) of the DFD as a result of changes in the incoming control flows.
- *Structure Chart*: Shows the breakdown of the system to the lowest manageable levels. It is used in structured programming to arrange the program modules into a hierarchical tree structure that recursively breaks the problem down into parts that are small enough to be understood.

Even though this front-end Analysis and Design Model is "older than dirt," you should consider using the SA/SD technique especially if you are not planning to use the OOA/OOD methodologies. Criticisms of the SA/SD approach include: difficulties in agreeing on the content of the DFDs; the large size of the documentation needed to understand the data flows; the burden of making changes to the DFDs as the functionality changes; difficulty in understanding the subject matter; and the difficulties designers encounter when converting the DFDs into an implementable format. Some users love it, and some users hate it. In any case, do what is needed and meaningful for your project and do not produce artifacts, just because the SA/SD technique says you should do so if it does not add value to your project.

4.3 Managing Agile Software Development Projects

Agile software development management, hereafter referred to as *Agile*, is an "iterative and incremental method of managing the design and build activities for engineering, information technology, and new product or service development projects in a highly flexible and interactive manner." (Wikipedia definition). Agile is highly iterative whereby deliverables are submitted in stages. Agile is most effective with small co-located teams.

The Agile movement gained momentum in the mid-1990s as a collection of *lightweight* software development methods in reaction to the more *heavyweight process-oriented* software development methods (such as the Waterfall Model), perceived by the critics to be too structured, too heavily regulated, too much documentation, and often micro-managed.

4.3.1 Types of Agile Methodologies

There is no intention in this Guidebook of providing a comprehensive discussion of the Agile methodologies. Books have been written about the Agile methodology, like many other topics covered, so Section 4.3 will primarily address issues that can be important to Software Project Managers. As the Software Project Manager (SPM), you should become familiar with at least the following most popular Agile lightweight project management methodologies to determine what is suitable for your project. Most of the Agile terms used are described in Subsection 4.3.10.

- *Scrum*: A holistic approach to software development that focuses on iterative goals set by the *Product Owner* through a *backlog*, which is developed by the *Delivery Team* through the facilitation of the *Scrum Master*. (Scrum is a short version of scrummage—a rugby term for two sets of forwards kicking the ball between them and back to their teammates).
- *Extreme Programming (XP)*: Extreme Programming involves practices based on a set of principles with a development goal that provides value by implementing tight feedback loops at all levels of the development process and using them to steer development. XP popularized Test Driven Development (TDD) and Pair Programming.
- *Crystal Clear*: An Agile type of methodology that focuses on co-location and osmotic communication (osmotic means apparent effortless absorption of ideas, feelings, attitudes, etc. That sounds to me like some kind of magic).
- *Kanban*: A lean framework for process improvement that is frequently used to *manage work in progress* within Agile projects; it has been specifically applied to software development.
- *Scrum ban:* A mix of Scrum and the Kanban approaches to project management that takes the flexibility of Kanban and adds the structure of Scrum to create another way to manage projects.

What does Agile Mean? As described by (Stuart, 2010) there are many pitfalls involved with an Agile implementation, but a common pitfall for *enterprise-wide Agile adoption* is a lack of defining exactly *what does Agile mean* in your organization. Agile is an umbrella term since it may include more than one methodology (as described above), and more than one technique or practice (e.g., continuous integration, pair programming or test driven development). Without a clear understanding of what Agile means, each Project Team may be doing something very different even though they all call it an Agile development.

A white paper by (Stuart, 2012) referenced a 2008 survey reporting that over 71% of teams using Agile were using some variation of Scrum, about 8% were using XP, and none of the other methods accounted for more than 5% each. As a result, this discussion of Agile will focus on Agile/Scrum. In order for Agile/Scrum to work effectively, there are some important environmental factors that must be in place. They include:

■ Self-managing teams that are mature—meaning the team must work cohesively, be accountable to commitments and able to maintain respectful and productive relationships

■ A software product where your customer has agreed to an incremental delivery schedule—the key here is to determine how the software product can be broken down into small segments that can be designed, built and tested in about two weeks each

■ Stakeholders and a parent organization that are committed to the Agile philosophy

4.3.2 Core Principles of the Agile/ Scrum Methodology

Scrum is an Agile management *framework* for software development. The term "agile" implies flexibility as well as rapidity and the core principles exemplify those goals. In summary, they are:

■ *Self-Organizing Teams*: A powerful aspect of Agile is the self-organization of its multidisciplinary teams where the team members continually interact while working on specific short-term goals. *Note:* The multidisciplinary teams are similar to the Software Integrated Product Teams (SwIPT) described in this Guidebook, but the SwIPTs are not self-organizing.

■ *Self-Directed Teams*: Agile teams interact like self-correcting organisms based on a constant feedback mechanism. (Beware of an adage: If everyone is in charge, no one is in charge!)

■ *Daily Stand-Up Meetings*: Short daily early morning meetings provide self-directed teams with timely feedback information the team uses to monitor and adjust their daily tasks.

■ *Information Sharing*: Information is shared by *all* team members; communication is facilitated by co-location

of the team members sometimes augmented with "team rooms" containing informative charts such as burn-down charts showing the work remaining.

■ *Minimal Processes and Principles*: The Agile approach is to down-size processes to the bare minimum set needed to address only the current work. Processes, procedures, documentation, checklists, etc. are replaced in Agile by *guiding principles*. The Scrum Master, or the Project Leader, decides which principles to employ and how they are implemented.

■ *Frequent Releases*: Most Agile approaches release software more frequently than the traditional methods. By delivering code early and often, Agile Teams get frequent checkpoints to validate their understanding of the requirements. The basic unit of development in Agile/Scrum is called a *Sprint* (an iteration of a specific duration fixed in advance for each Sprint). The Sprint duration is usually between one week and one month, with two weeks being common.

Sprints normally start with a planning meeting where the Product Owner and the Development Team agree upon exactly what work will be accomplished during the Sprint. The Development Team decides how much work can realistically be accomplished during the Sprint. The Product Owner determines the approval and acceptance criteria. Acceptance normally requires the software comprising the Sprint to be fully integrated, fully tested, documented as required, and potentially shippable to the customer.

■ *Continuous Testing*: Agile Teams test at the end of each Sprint or iteration to ensure that all of the planned features are performed. This frequent testing helps to reduce risk because more issues are identified early in the life cycle.

■ *Supporting Infrastructure*: In order to produce working software, especially for large implementations, that is sufficiently stable, and at the prescribed level of quality, a *supporting infrastructure* is needed that may include:
  – Continuous builds that are validated by practices such as automated smoke tests.
  – An automated unit-level testing framework.
  – Automated system-level Regression Testing.
  – An automated software release and status process.

■ *Customer Collaboration*: Communication is a hallmark of the Agile/Scrum methodology, and face-to-face communication maximizes its effectiveness. Co-location of the customer, or the customer's representative, is necessary in order to interface with the Agile Team daily. However, as the complexity and size of the system grows, it can be very difficult to find a single person who is a good communicator, available full-time and understands the entire system.

4.3.3 Is Agile a Silver Bullet?

Lessons Learned. I believe there are never two sides to every story—there are always *three* sides! You can call the three sides *the positive, the negative* and *the neutral* or call them *the optimistic, the pessimistic* and *the realistic*. Whatever you call them, this Guidebook takes the third side: *neutral and realistic* and that is how Agile is discussed.

There are people who believe that the Agile methodology is the *only* way to go. Treatment of the Agile methodology in this section is neutral and realistic because both the pros and the cons are discussed, and other options are described, so it is left up to the Project Manager to decide what is the best fit for his or her project. Hopefully, as the Software Project Manager you will have the opportunity to decide what methodology is best for your project rather than being forced to adopt a mandatory Agile methodology by your home organization regardless of its fit.

Most Agile methodologies are *team-level* workflow approaches. These approaches can be highly effective at a local small team level, but they do not easily scale up to geographically dispersed large systems, and they fail to address the entire system architecture, system requirements, system planning needs, and effective management of large systems. Successful implementations show that very large geographically dispersed software system developments require extensive planning, vigorous control, formalized communication abilities, considerable (applicable) documentation and comprehensive integrated testing procedures; all of that is essentially inconsistent with the Agile core principles.

Since there are no silver bullets, I do not believe that Agile is the greatest thing since sliced bread—but many advocates believe that to be the case. Regardless, if applied properly in the right environment, even on large systems, Agile can be an effective methodology (as described in Subsection 4.3.5).

If a self-directed team means no one is in charge, it can have inherent risks. Large complex software-intensive systems are typically one-of-a-kind, first time, expensive and risky developments; you need to embark on such a journey with as much *clarity and direction* as you can get. There may be a software development standard process produced by your organization that can be followed, but someone must decide if the standard process can be followed for your project, if it needs to be tailored, or if you need to develop a hybrid process for your project. Also, self-directed teams often cannot make the quick decisions needed during development because of differences of opinion. It may be possible for the process to be managed by a self-directed team, but a large project must be managed by someone who has clear *responsibility and authority* to make those decisions: the Software Project Manager.

4.3.4 Agile Operational Considerations

Working Software. The goal of every software project, regardless of size, is to produce *working software*, but what does that mean? Some strong proponents of the Agile methodology believe that *delivery of working software code is the only goal*. Even if the software does not interact with hardware, that opinion is not valid because you must successfully *test and verify* that the software is responsive to your customer's requirements, it must include the documented information needed to run and maintain the system, it may involve user training, and it usually involves much more. If you are developing a complex *mega system*, the goals are much more than just "code that works."

Definition of Done. Defining "done" is important for every software development, but for Agile developments it is of paramount importance to the team. The definition of done is specific to each project. For Agile developments the fundamental definition of "done" at a minimum should include:

■ The code for each Sprint or story is complete, and it compiles without error
■ Required tests have been written, executed and passed
■ Functionality is fully compliant with the user story and test results have been reviewed and approved by the customer
■ Applicable documentation was produced, reviewed, approved and checked in

Control of Agile Projects. In Chapter 2, Figure 2.2 displayed the amount of project control that is needed compared to the size of the project. That figure clearly shows that large and complex projects require more management control than small, simple projects. Figure 4.5 is a modification of that figure with the additional mapping of the principal area of the Agile focus versus the focus of this Guidebook. Agile is best suited for and most effective on small systems. The recommended processes and procedures in this Guidebook

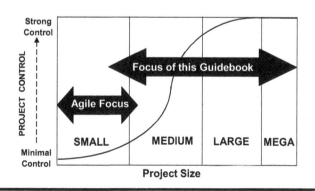

Figure 4.5 Project control mapped to project size and focus areas.

are best suited for and highly effective on medium through large and mega systems.

It is almost certain you have never seen a diagram of an Agile group with a Software Project Manager box at the top. The very nature of the Agile/Scrum methodology, and its fundamental building blocks, seem to be the antithesis of the systematic structure and defined processes needed to successfully produce, control and deploy large complex software-intensive systems as described in this Guidebook. Furthermore, if Agile is misapplied, and force-fitted into situations that it is not suited for, it would be a good example of trying to put square pegs in round holes.

4.3.5 Application of Agile to Medium and Large Systems

Despite the Agile focus issue discussed in Subsection 4.3.4, Agile is important because, if properly applied in the correct environment, it can be and has been an effective approach. When properly applied, Agile can be used in a large project by integrating the development activities of multiple Scrum Teams. There have been many attempts to scale up the Agile philosophy to large projects with varying levels of success. Some of the Agile scaling technique in current use include:

■ *Scrum of Scrums (SoS)*: The Scrum of Scrums technique is the oldest and likely the most commonly used. It is used for integrating the work of multiple Scrum Teams (normally 5–9 members each) working on the same

project. The SoS approach is described below when used in combination with a traditional approach on large systems.

■ *Large-Scale Scrum (LeSS)*: An expanded version of a single Scrum Team focused on producing the entire product with one Product Owner and a single Product Backlog.

■ *Scaled Agile Framework (SAFe)*: A knowledge-based technique used to scale lean-Agile developments in large software enterprises.

■ *Disciplined Agile Delivery (DAD)*: The DAD approach is to build and test the most risky or difficult parts first to confirm feasibility of meeting the project goals.

■ Nexus Scrum of Scrums (Nexus SoS): Best used for 3–9 Scrum Development Teams with a common Product Backlog.

Since Agile is ideal for small applications, it is reasonable to apply the Agile methodology to *small pieces of a large system*. As displayed in Figure 4.6, Agile can be performed in *combination* with one of the non-Agile approaches described in this chapter. Figure 4.6 also identifies the equivalent Agile terminology. This combination of both methodologies will work if you can identify Software Units (SUs) that are small enough and consistent with the Agile criteria. Then the Agile developed SUs can be combined to produce higher-level software components (also called higher-level SUs). From that point on, one of the traditional methods takes over as the full system is integrated and incrementally built up.

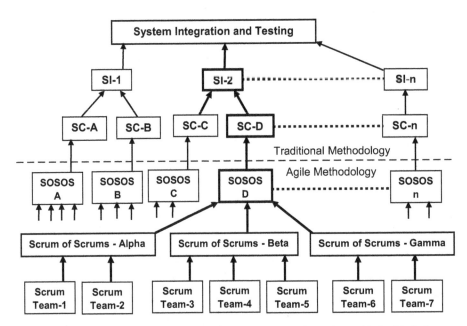

Figure 4.6 Combination of Agile and traditional methodologies in large systems.

When using the Agile methodology combined with a traditional approach, the work products produced must be reviewed and approved, usually by the Chief Software Engineer (CSwE), to make certain the developed products are compliant with system requirements and to confirm that a *self-directed Agile Team* did not decide to go in another direction.

Attempting to apply a *fully Agile approach* to the entirety of a very large software-intensive system is extremely difficult if not impossible. Successful developments of large complex systems require processes that are generally inconsistent with the core principles of the Agile methodology. If you plan to follow some type of a scaled Agile framework, consider hiring an Agile consultant who has experience with Agile scaling. You can then decide what scaling option is best for your project or, even better, customize an option and then have a proactive plan to improve it incrementally during the developmental process.

4.3.6 Agile/Scrum Roles

The principal roles performed may be called by different names depending on the specific Agile methodology being used but the basic concepts are similar. There is no formal role for a Software Project Manager or Chief Software Architect as these roles are incompatible with the Agile concept. *If* a Project Manager is involved with an Agile project, they are typically focused only on resolving issues that block progress originating outside the team's boundaries (i.e., hands off the team!).

Scrum Master. In the Agile parlance the *Scrum Master*, also called the *Team Leader* or *Coach*, *essentially* performs some functions similar to the Software Project Manager including acting as the interface between the Agile Team and the outside world. The Scrum Master is the *keeper of the Agile process*, responsible for making the process run smoothly, removing obstacles that impact productivity, facilitating the critical meetings, understanding Scrum well enough to train and mentor the other roles, and educating and assisting other stakeholders. The Scrum Master maintains a constant awareness of the status of the project but does *not manage the team*, and does *not assign tasks* to team members since task assignments are the internal responsibility of the team. An important function of the Scrum Master is to prevent new requirement changes from being introduced during a Sprint. This allows the team to focus, without distraction, on performing their tasks and benefit from the resulting stabilizing environment.

The Scrum Team. The Agile/Scrum Project Team is a *self-organizing cross-functional group* of people who perform the hands-on work of developing and testing the product. Since the team is responsible for producing the product, they also have the authority to make decisions about how to perform the work. The team is therefore *self- directing*: Team members decide how to break work into tasks and how to allocate tasks to individuals for each Sprint. The team size is normally kept in the range of 5–9 people. A *Scrum Team* refers to the team plus the Scrum Master and Product Owner.

Product Owner. The Product Owner is the keeper of the customer requirements and the *single point of contact regarding all requirements* and their planned order of implementation. The Product Owner is the interface between the organization, the customer, their product related needs, and the team. He or she buffers the team from feature and bug-fix requests that come from multiple sources. Each project has only one Product Owner. In other Agile frameworks (such as Extreme Programming) the Product Owner is usually the customer or the customer's representative. Communication is a core responsibility of the Product Owner who also serves as the team's representative to the overall stakeholder community.

When several teams work on one product, they should generally use a single Product Owner (who has the authority to make business decisions) and a single Product Backlog with customer-centric requirements. Each Scrum Team should strive to become a *feature team*, able to build a complete slice of a deliverable product.

The Customer. Your project's customer, or the customer's representative, plays a substantial role in an Agile project. During the Agile planning process, the customer is the leader as they specify what they want and the order they want to receive it. The ideal Agile concept has the customer, or a proxy, *co-located* with the Agile Team in order to maximize the benefits of daily interaction during the sprints. When co-location is not practical or possible, planned daily telephone or video conferences can be held. The modern office can be viewed as a "virtual office" since team members, and the customer can communicate all day by teleconferencing with VoIP headsets.

For larger software-intensive systems, the customer typically prepares detailed specifications such as the *Request for Proposal* (RFP), *Statement of Work* (SOW), a *Technical Requirements Document* (TRD) and sometimes the *Work Breakdown Structure* (WBS). All the specifications are expected to be updated as specific requirements change during the development process. In many (or most) Agile developments the customer is not involved with (or cares) *how* the software is built; the customer is deeply concerned with *what* the software produces. Consequently, it would appear that involving the customer in meetings about the detailed structure of the software may be a waste of their time and yours and may slow you down.

4.3.7 Agile/Scrum Terminology

Many new terms have been "invented" for the Agile methodology, and many of them are metaphors to help the team develop a *vision* of what they are trying to accomplish. It does appear that some Scrum names are *new names for existing terms.* My attempt to define "essential equivalents" of the Agile terminology is shown in the example Rosetta Stone in Table 4.3 listed alphabetically.

4.3.8 Adopting the Agile Philosophy

The Agile approach could be considered more of a modern *philosophy* than a development methodology. This philosophy is focused on value to the customer, effective collaboration, rapid responsiveness to changes, and efficiency in the approach to product delivery. As described by (Kennaley, 2010), "The goal is not to just "do agile" but to "be agile." Simply utilizing an agile process will yield some benefit, however, if being agile is the goal a *culture of agility* needs to be created."

Some companies who decide to introduce Agile practices do so without thoroughly thinking through *why* they are doing it and without confidence that switching to Agile will bring long-term benefits. Often, executive management decides to make the switch because their peers are going the Agile route. For such companies, Agile adoption is like a fashion statement. Such companies care more about keeping up with the mainstream than making a well thought-through and carefully planned step forward.

Table 4.3   Agile/Scrum Nomenclature Rosetta Stone

Scrum Name	Essential Equivalents(*)
Backlog or Product Backlog Item	Requirements not yet implemented
Cadence	Time between sprints or releases
Feature Team	Systems Engineering Integration and Test (SEIT) Team
Information Radiator	Status Chart
Poker Planning	Estimating
Product Level Testing	System Testing or Release Testing
Product Owner or Chief Product Owner	Responsible Product Engineer or Chief Software Engineer or Software Project Manager
Product Owner Team (or Council)	Integrated Product Team
Product Roadmap	Software Development Plan
Product Vision	Concept of Operations(**)
Release	Release or Version
Retrospectives	Project Review (Lessons Learned) Meetings
Scrum Master (or Coach)	Software Project Manager/Team Leader
Scrum of Scrums	Meeting of two or more Project Teams
Spikes	Milestones
Sprint	Build (or Software Unit) Iteration
Sprint Backlog	SU Requirements Prioritized
Story Points	Requirements Estimation
Team-Level Testing	SI Qualification Testing (SIQT)
Time Boxes	Time Durations/Sprint Iterations
User Stories	Requirements or features needed
Velocity	Throughput

(*)   There are subtle differences for some of these equivalents to the Scrum name.
(**)   Could also be the Capabilities Development Document (CDD), the Technical Requirement Document, or the Project Charter.

There are many common sense Agile transformation readiness prerequisites that these companies either overlook or intentionally ignore. They go after what may seem to be a very good cause, but in the majority of cases, *poorly substantiated motives for Agile adoption* brings about failure. There is only one chance to make a first impression, any future attempts to reintroduce Agile at a later point in time usually has very little support. As the Software Project Manager, you may, or may not, have any influence on whether or not your organization has adopted Agile. Hopefully, you do have the option of deciding if Agile, or some hybrid version of it, is the right methodology for your project.

Physical versus Intellectual Philosophies. There is a major conceptual difference between a software development project that produces a *software package*, comprised of only the code plus required documentation, compared to a software-intensive *system* (SIS) involving millions of lines of code that interfaces with many Hardware Items, plus lots of documentation and human interactions to make it all work. The SIS can be considered *dynamic* in nature (lots of moving parts) whereas the software package can be viewed fundamentally as *intellectual or cognitive* in nature. The Agile methodology is primarily the intellectual environment, but most of this Guidebook is focused conceptually on *managing the dynamic nature of a large SIS*. You must apply the philosophy that fits your project.

> **Lessons Learned.** Over the years of my professional and personal experiences, I have noticed a mindset that I call the *Fixed Belief System*, and it is prevalent in all walks of life. For example, there are people who always buy cars from the same company because they firmly believe that a specific company makes better cars. This may or may not be true, but it is an example of a Fixed Belief System—the opposite of taking the time to impartially *evaluate merits of the alternatives.*

So what does that have to do with Agile? In many organizations, the Agile movement is like a vigorous religious cult and its strong proponents display the epitome of a Fixed Belief System. Agile has definite value and applicability to specific situations, but it *should not be forced* into environments where it does not seamlessly meet the tenants of the Agile methodology. It is almost impossible to modify a person's Fixed Belief System. Successful Software Project Managers should be *open-minded* and have only one professional Fixed Belief System: *do only what is applicable, feasible and needed to make your project a success.*

4.3.9 Agile Implementation Considerations

Transition to Agile. Many organizations have successfully made the transition from traditional software development

methodologies to an Agile methodology. Organizations that develop large systems typically have more difficulty adopting Agile across their enterprise. Whether or not Agile should be adopted by your organization is beyond the scope of this Guidebook and may be beyond your control as a Project Manager. Even if your organization is currently using an Agile methodology, it may not be the best approach for your project. If senior management dictates the *mandatory use of Agile*, they are doing a disservice to your project because it deprives you of the right to select the best fit methodology.

Agile Individualism. Discussions of historical backgrounds and the philosophical issues related to Software Engineering and Software Project Management have been intentionally avoided because those issues may not make you a better manager. However, over the past decade, there has been a growing "individualistic orientation" in the USA described by (Copper, 2016). It is apparent that this orientation for individualism is occurring in several areas including politically, religiously and technologically. Therefore, it is not surprising that millennials—especially young Programmers—want to do things their way. Agile development is a perfect environment for them to exercise that individualism.

I have no problem with this individualistic orientation whatsoever. In fact, if the Agile approach is the *only* way to get Developers *involved and excited* about the software development process, then I encourage using the Agile approach *where it fits*. Agile advocates should be commended for their creativity and novel approach of introducing a new modern software development methodology that is more "exciting" than traditional approaches. Despite Agile's shortcomings, evident when efforts are made to apply it beyond small, simple systems with co-located Developers, *Agile is orders-of-magnitude better than the all too common chaotic ad-hoc approach which is completely void of any process.*

The Need for Agile Training. Adopting Agile processes, such as Scrum, may appear to be simple but it is actually difficult to implement well. If your staff is new to Agile, they need new skills. Old roles need to be refined and revised. Existing processes need to be refined, and the organization's culture must change. Organizations will find that initial successes with Agile for small projects may not scale up very well to larger projects or throughout the entire organization. This is partly because the initial teams were likely made up of Agile enthusiasts, whereas larger teams joining later include people who are lukewarm or even resistant to Agile. The value of training for new tools and processes cannot be overemphasized.

Scrum of Scrums. As discussed above, this is a technique for operating Agile/Scrum at a larger scale where multiple teams work on the same product. It provides a forum to discuss progress on their interdependencies, focusing on how to coordinate and successfully deliver software, especially in areas of overlap.

Typically, each Scrum Team designates one member as an *ambassador* to participate in the Scrum of Scrums with ambassadors from the other teams. The ambassadors may be technical contributors or each team's Scrum Master. Rather than simply a progress update, the Scrum of Scrums meeting is focused on how teams are collectively working to resolve, mitigate, or accept any risks, impediments, dependencies, overlaps, and assumptions that have been identified. The Scrum of Scrums is responsible for delivering the working software of all teams at the end of the Sprint or for releases.

Management by Peer Pressure. Sprint *planning* meetings commit teams to what they will do; Sprint *review* meetings publicly announces what was actually performed. One of the most powerful attributes of Agile is the *power of peer and organizational pressure* because when the deliverable product does not meet the definition of "done," the team cannot hide from their failure to produce. As described by (Chemuturi et al., 2010) "The process of public commitment and public demonstration provides a level of control that a Project Manager cannot." The positive performance resulting from this built-in peer pressure is apparently one of the reasons why the Agile approach has been so successful.

Agile Compatibility with Capability Maturity Model Integrated (CMMI). The CMMI, framework for building process improvement systems, is discussed in Section 6.6. There has been debate as to whether Agile and CMMI can co-exist on a software-intensive system. It is clear (to me) that the answer is yes. Although it is obvious, there are large differences between CMMI and Agile methods, (Boehm and Turner, 2004) claim that both approaches have much in common. They believe neither way is the "right" way, and that the two approaches are complementary where different fragments of each should be combined into a hybrid version. I agree that a combination of Scrum and the performance improvement framework of CMMI will provide more adaptability and predictability than either one alone.

4.4 Schedules and Activity Networks

Every large system, and most mid-sized projects, must have, *and follow*, a formal program management methodology by maintaining an approved *Integrated Master Plan* coupled with a related *Integrated Master Schedule*, or equivalent, to provide a complete schedule and activity network for all program and system activities. The IMP and IMS must be maintained electronically and available through an accessible electronic data management system (see Sections 5.1.9 and 9.5.1).

If the IMS is at a relatively high level, it must be augmented with lower-level detailed subsystem schedules for software planning, design, development, integration and test. These detailed subsystem schedules must be maintained and monitored at the subsystem level with oversight by Project Management and the CSwE. Subsystem schedules must be integrated with the IMS. If any conflicts between the IMS and subsystem schedules occur, the system-level IMS always prevails. The SDP should identify the group responsible for software compliance with the IMS delivery schedule.

As discussed in Subsections 1.5.2 and 8.2, the IMP and IMS must be organized by a systematic *Contract Work Breakdown Structure* (CWBS or WBS) to provide a complete schedule and activity network for *all program activities*. The IMS must include software activities showing the time-phased interrelationships of events and accomplishments for software builds, and the SwIPTs must manage and control their respective schedules within the IMS structure.

Software Schedules. The summary and detailed schedules for the software activities can be updated weekly, monthly or as needed to be consistent with overall program schedules. The software development schedules must include the details of the proposed builds and how they relate to overall program milestones. Eventually, the software schedules may get all the way down to the "inch-stones" with tasks identified at the level of individual engineers. Typically, the schedules are prepared by "rolling waves" which can be for 6-month periods or developed build-by-build.

An overall master schedule may be included with the Software Development Plan. However, once the contract starts, the schedules, especially the detailed software schedules, are typically updated so frequently that the SDP should only have pointers to where current software schedules can be found.

Software Activity Network. The SDP should include, or reference, an activity network depicting sequential relationships and dependencies among all software activities, and identify those activities that impose the greatest restrictions on the project. The activity network can identify the *Critical Path*. If any part of a software development activity is on the Critical Path, then any slippage of that activity will impact the completion date. Additional management attention to that activity is needed to make sure it is performed as planned or determine how to remove it from the Critical Path. Section 8.4 has more detail.

4.5 Software Standards

Software standards ensure that Developers produce *consistent* software development products. You must identify standards to be followed for your project including product levels and terminology.

4.5.1 Standards for Software Products

Applicable standards for software requirements, architecture, design, code and test must be documented. The recommended location for these standards is in a *Software*

Standards and Practices Manual that can be a stand-alone document or an addendum to the SDP. Software development standards should include the programming language standards to be used, a list of applicable standards required by the customer, operational details of the program's defined software process, use of COTS/Reuse software, the process for waivers or deviations, and portions of the applicable standards that need to be modified.

Standards also help ensure the similarity of the structure of all code/design units so that lines of code counts and software measurements can be applied consistently. *Standards must provide value added to your project*; if that is not the case, and irrelevant standards are being imposed on your project, mount a campaign to reverse that edict. As the Project Manager, it is your responsibility to make sure the relevant standards are being followed. The SDP should also document the process for waivers or deviations to portions of the applicable standards. Changes to standards must be:

- Justified, documented, and submitted by the cognizant Software Item Lead
- Approved by the SwIPT Lead and SQA and submitted to the SEPG for concurrence
- Entered in the *Software Development File* or the *Software Engineering Notebook* (SEN)

The SEPG should review software standards and tools usage and provide the means for sharing knowledge and lessons learned with other SEPGs across the project and the organization.

4.5.2 Software Product and Testing Levels

The hierarchy of software-related specifications must be produced in accordance with the program's *Specification Tree*. Typical hierarchical software product levels, and their directly related testing levels as used in this Guidebook are depicted in Figure 4.7. Whatever hierarchy is chosen for your program you must ensure it is followed to avoid confusion and to significantly enhance the probability that all the pieces of the software puzzle will fit together properly.

The following is a brief description of each Product Level:

- *System*: It is the *entire* system covering all hardware, software and system-related entities. It is "Product Level One" of the Work Breakdown Structure (see 1.5.2).
- *Subsystem*: A subsystem (may also be called a segment) is the next level down from the entire system; subsystems are typically a physical or functional subset of the entire system.
- *Element*: An element (or module) is a larger and more complex group of integrated Software Items. Use of the Element Product Level is optional and dependent on the size and complexity of your software subsystem. The relationship between an element and a SI may be reversed.

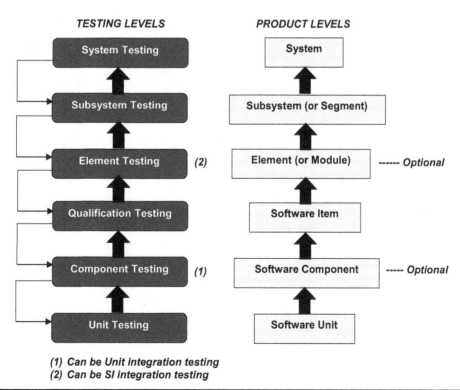

(1) Can be Unit integration testing
(2) Can be SI integration testing

Figure 4.7   Software product and testing levels.

◼ *Software Item*: An SI is a an aggregation of Software Units and/or Software Components that satisfies an end-use function and is used for purposes of specification, interfacing, qualification testing, Configuration Management, etc. Software Items are selected based on trade-offs among software function, size, host or target computers, support strategies, plans for reuse, criticality, interface considerations, need for separate controlled documents and other factors. In the past, SIs were called *Computer Software Configuration Items* (CSCI).

◼ *Software Component (SC)*: Can be a group of integrated Software Units. The use of the SC Product Level is optional and depends on the size and complexity of the software. In the past, the use of the SC as a Product Level of the software hierarchy was popular, but it is not now commonly used. The SU in this Guidebook can be interpreted as "SU/SC" if applicable.

◼ *Software Unit*: The SU is the smallest module of software, but it is more complicated than that. Software Units may be a subdivision of a Software Item, a component of that subdivision, a class, object, module, function, routine, or database. Software Units may occur at different levels of a hierarchy and may contain other Software Units. Software Units in the design may or may not have a one-to-one relationship with the code and data entities (routines, procedures, databases, data files, etc.) that implement them.

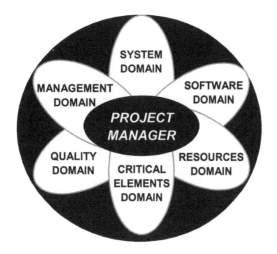

SOFTWARE MANAGEMENT AND QUALITY DOMAINS

NOTE:
There are 6 domains: two are covered in Part 2, two in Part 4, and one each in Parts 3 and 5

Chapter 5

Software Management Domain

A goal is a destination with a planned path and a deadline.

The *Software Management Domain* is focused on the specific management tasks a Software Project Manager must address. However, Software Project Management is involved with *all six* of the interleaved domains described in Section 2.1. All the domains are important, they all involve project management, they all must work together to achieve success, and they all have various goals to attain. The Software Management Domain is addressed in this chapter by the following subject areas:

- *Software Project Planning and Oversight*: Involves detailed upfront planning for the entire Software Development Life Cycle including updates as the project progresses (5.1).
- *Software Risk Management and Mitigation*: Addresses the management process for identification, mitigation, and tracking of software-related risks (5.2).
- *Technical and Management Reviews*: Joint reviews demonstrate progress-to-date on project work products and provides a forum for discussing programmatic issues and risks (5.3).
- *Assurance of Critical Requirements*: Involves oversight of key software requirements that are critical cornerstones for safety, security, privacy protection, reliability, maintainability, availability, performance, etc. (5.4).
- *Software Subcontractor Management*: Addresses management of the subcontractors on your project and the need for a Subcontractor Management Team (5.5).
- *Software Verification and Validation*: Verification is the process of making sure that each function works correctly. Validation is the procedure to make sure the final product is fully responsive to the original system requirements (5.6).

- *System Retirement or Disposal*: Introduced here but described in Section 16.6 (5.7).
- *Definition of Project Success*: The elusiveness of what success means and the importance of defining the word "done" (5.8).

5.1 Software Project Planning and Oversight

If you failed, you did not plan to fail; you just failed to plan.

—Anonymous (Here he is again)

The major objective of this software management activity is to complete and document detailed planning for the software development task. The planning activity is an *ongoing task*. It may be initially performed with the project proposal, and usually is repeated several times with changing requirements of the program and with a better understanding of the project by the planners while it is being performed. The planning task is critical at the start of the Development Life Cycle as it is the foundation for initially producing the software plans required to implement and perform the software development process and for the identification and formation of the Software Teams required to execute those plans.

Software management has cognizance over the *Software Development Plan* (SDP) including the software management and quality control plans. The preparation of the *Software Configuration Management Plan* (SCMP), and the *Software Quality Program Plan* (SQPP) should be assigned to your Software Configuration Management (SCM) and the Software Quality Assurance (SQA) groups.

Readiness Criteria. A summary example of the readiness criteria for the Project Planning and Oversight activity is shown in Table 5.1; it includes the entry and exit criteria,

Table 5.1 Readiness Criteria: Project Planning and Oversight

Entry Criteria	Exit Criteria
■ A management decision to initiate planning has been issued. ■ Customer system requirements are available. ■ The system-level Integrated Management Plan and Integrated Management Schedule are available. ■ The software Work Breakdown Structure has been defined down to the Software Item level. ■ The Program Risk Management Plan has been baselined. ■ Software, Systems, and Project Teams are sufficiently staffed to support the software planning activity.	■ Software plans are placed in the electronic database. ■ Software size estimates are established; budgets and schedules are baselined. ■ The SDP is reviewed and approved by all Software Team members and the customer.

Verification Criteria
■ Program software plans are reviewed and approved. ■ Program and senior management are provided with the status of ongoing product engineering activities (including requirements definition) on a periodic and event-driven basis. ■ SQA performs process and product audits for the software planning activities.

Measurements
■ Program schedule showing planning activities—estimated and actual. ■ Staffing levels planned versus actual. ■ Effort hours budgeted and actual. ■ Milestone due dates—contractual, estimated and actual (see Chapter 15).

verification criteria to ensure completion of required tasks, and the project measurements usually collected.

Software project planning and oversight is covered in ten subsections:

- Software Development Plan (5.1.1)
- Software Development Planning Tasks (5.1.2)
- Software Resource Estimating (5.1.3)
- Software Build Planning (5.1.4)
- Software Development Tracking and Oversight (5.1.5)
- Avoiding Gold Plating (5.1.6)
- Software Item Test Planning (5.1.7)
- System Test Planning (5.1.8)
- Planning for Software Transition to Operations and Maintenance (5.1.9)
- Tracking and Updating Project Plans (5.1.10)

Three extremely important elements of Software Project Management, software cost, schedules and the Work Breakdown Structure (WBS), are missing from Chapter 5 because they are covered in Chapter 8 and elsewhere. The WBS was introduced in Subsection 1.5.2 and is described in Section 8.2.

5.1.1 Software Development Plan

The Software Development Plan, or an equivalent document whatever you may call it, is the *key* software planning document. The SDP describes the mechanism for documenting and tracking the software development effort and activities

required by the software-related provisions in your contract. There may be multiple SDPs at different levels of the program, but the top program-level SDP must define software activities *common to all development sites.*

The SDP is considered a "living" document that must be updated periodically throughout the Software Development Life Cycle. Updates are usually planned to occur at program milestones, and a schedule or table should be included in the SDP to identify the planned update timeline. Also, the SDP must be kept consistent with updated versions of other program-level plans, and it should also contain a figure depicting the relationship between the project SDP and other key program plans.

Content of the SDP. The SDP must address the following *applicable* software development issues:

- Control Issues
  - How will software development be *managed* and with what *controls*?
  - What *standards*, *practices* and *guidelines* will be followed and how will they be enforced?
  - What *reviews* will take place, who are the attendees and when will the reviews take place?
  - How are software development responsibilities managed and flowed down to *subcontractors*?
  - What software products will be subject to formal *Configuration Management* and when?
  - How will access to *classified* data and products be controlled?

- Process Issues
  - What *processes* will be followed to conduct software requirements analysis, design, coding, testing, integration and qualification?
  - Who is *responsible* for each software development task and what is their *reporting* chain?
  - What software *documentation* will be produced, in what format, and when?
  - How will *compliance* with the SDP be assured and what are the plans for updating the SDP?
- Measurement Issues
  - What software management *measurements* (metrics) are planned and what is the process for collection, reporting, analysis and *Corrective Actions*?
  - What is the top-level software development *schedule* and what are the reportable *milestones* (or, are pointers included as to where that information can be found)?
  - What mechanism is in place to inform management if the current software project is consistent with planned *schedules*?
- Quality Issues
  - What methods will be employed to identify and mitigate *software risks*?
  - What process and methods will be used to ensure the *quality* of the software product?
  - How will *errors* be detected, documented and corrected?
- Testing Issues
  - What is the process for ensuring *systematic testing* of the developed software?
  - What development and *testing Support* Software, environment, and tools are required?
  - What software is *deliverable* to the customer and what are the transition plans?

Format Options for the SDP. A comprehensive SDP for large programs is almost always composed of multiple parts. Typically, there are two basic approaches to SDP formats: projects with a single SDP and projects with a system-level SDP plus SDPs prepared at each development site:

- *The Single SDP Approach*: A project may elect to have a single SDP and mandate that it be followed by all subsystems. That approach works very well when all Developers, including subcontractors, are co-located and using the prime's infrastructure.
- *The Site-Specific SDP Approach*: On large programs and projects, typically involving numerous organizations that are geographically dispersed, site-specific SDPs are usually needed because of significant corporate differences in software organizations, management policies, development environments, and unique operational processes, procedures and products.

Subsystem site-specific SDPs (also called SDP Annexes) can be produced containing specific and/or unique policies and procedures applicable to them only that expands on, but *does not conflict with*, the policies and procedures defined in the program-level SDP—except for approved waivers. The SDP contents should be submitted to the Developer's (or contractor's) Configuration Control Board (CCB) for approval before it is released for implementation.

Site-specific SDPs are written and maintained by their development sites, and they provide additional standards and procedures specific to their site. Table 5.2 is an example organization of a typical complete "SDP package" containing three parts including site-specific annexes. Programs using the single SDP approach would not have Part 3.

Importance of the SDP. The quality and attention to detail in the SDP is a major *source selection evaluation criteria* when bidding for government work. Regardless, a good SDP at the start of *any* project is like a good flight plan that gives you a clear idea as to where you are going even though the route may be modified more than once during the flight.

The SDP also builds the foundation for the teamwork and disciplined trust between the customer and the Developer that is vital to Software Life Cycle cooperation and success.

Table 5.2 Components of a Typical SDP Package

PART—1	Program-Level Software Development Plan ■ Appendices
PART—2 SDP Addenda	SDP Management Plans ■ Software Quantitative Management Plan ■ Software Measurement Plan ■ Software Subcontract Management Plan ■ Software Risk Management Plan ■ Software Reviews Plan ■ Software Data Management Plan ■ Software COTS/Reuse Plan ■ Software Resource Estimation Plan ■ Software Integration and Test Plan ■ Software Maintenance Plan
	SDP Quality Control Plans ■ Software Configuration Management Plan ■ Software Quality Assurance Plan ■ Software Process Improvement Plan ■ Software Corrective Action Plan ■ Software Product Inspection Plan ■ Software Standards [Coding, Design, etc.]
PART—3 Site-Specific SDPs	Annex A: Site 1 Specific SDP Annex B: Site 2 Specific SDP Annex C: Site 3 Specific SDP

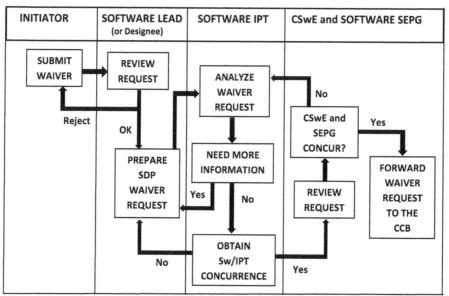

SDP = Software Development Plan   Software IPT = Integrated Product Team   CSwE = Chief Software Engineer SEPG = Software Engineering Process Group   CCB = Configuration Control Board

Figure 5.1 SDP waiver approval process—example.

Preparing a comprehensive SDP does not guarantee project success. However, *a poor SDP* at the start of a medium to large size program *is essentially a guarantee of serious problems ahead.*

Because of the importance of the SDP, an example annotated outline of the SDP is included in Appendix F. Since I authored seven SDPs during my career, I produced a comprehensive description of *The Elements of an Effective SDP* that you can obtain as listed in Appendix I (Gechman, 2011).

Waiver Processing. If a software development site has a justifiable reason for not complying with a required procedure in the SDP, they must submit a request for a waiver in accordance with the waiver process described in the SDP. All waivers should require the pre-approval of the Chief Software Engineer (CSwE) and the Software Engineering Process Group (SEPG) prior to being sent to the CCB for approval. Figure 5.1 is an example of a waiver processing process.

Waivers can have a significant impact so the approval process should be formal. When a waiver is initiated, the justification for it should be presented to the Software Lead. If the Software Lead (or designee) agrees with the need for a waiver, a Waiver Request should be prepared requesting authority to deviate from the requirements in the SDP. The Software Lead should forward the Waiver Request to the subsystem SwIPT for approval. The SPM and the affected SwIPT should review the Waiver Request, request more information or clarification if necessary, and either approve or deny the Waiver Request.

If the Waiver Request is approved, the SwIPT should forward it to the CSwE. If not approved or found to be a duplicate, it should be returned to the Software Lead with reason(s) for disapproval. If the CSwE and SEPG concur with the Waiver Request, it should be forwarded to the CCB for approval.

5.1.2 *Software Development Planning Tasks*

Software planning is iterative and should start when the assignment of planning roles has been made. Software planning responsibilities normally reside with the SwIPT leads for development implementation planning and the *project-level* SEPG for process planning. The first step in planning is to review software requirements since the scope of the software task is established by identifying system requirements to be satisfied by the software products. Table 5.3 is an example list of typical planning tasks including the location in this Guidebook where each task is discussed. The principal locations are bold highlights.

Integrated Management Plan and Integrated Management Schedule. Large system developments have (and need) a master program-level plan and schedule. They are often called the system-level *Integrated Master Plan* (IMP) and the *Integrated Master Schedule* (IMS). Software development planning information should be prepared by the subsystem SwIPTs by augmenting the IMP and IMS with more detailed development schedules. Once these detailed schedules are complete, development oversight begins by monitoring the activities, products and processes and taking Corrective Actions when necessary.

Lessons Learned. Detailed schedules could be in the SDP, but it is recommended to reference their location in the SDP since schedules typically change more often than the SDP is updated.

Table 5.3 Software Planning Tasks

Software Planning Tasks	Locations
Methods for developing and maintaining the SDP	**1.5.1**, **5.1.1**, 5.3.2, 6.6.3, 10.1, 13.3
Software Data Management (DM)	4.4, **5.1.10**, 10.2, 9.5, **12.5**, 13.4
Software size and resource estimation	3.2.2, 3.4, 3.7, 5.1.3, 9.5.1, 13.4, **Chapter 14**
Software build planning	3.4, **5.1.4**, 8.2, 8.7
Software integration and test (I&T) planning	**2.3.9**, 3.6, **3.7**, **10.3**, 8.5, 8.7
Software Development Environment and support tools	6.3, **12.1**, 12.3, **12.11**
Software acceptance, delivery, installation, transition, operations, sustainment and retirement planning	**5.1.9**, 8.8, 8.9, 9.8.2, 16.2, 16.3
Software Configuration Management	**2.4.3**, 5.1, **6.3**, **6.4**, 8.6, 12.3, 16.1
Software evaluations with formal and informal reviews	**2.4.5**, 6.1.2, 6.2, **6.5**, 13.4
Software Quality Assurance (SQA)	2.3.5, **2.4.2**, 5.1, **6.2**, 10.8.2, 13.4
Problem resolution methods and Preventive Action	5.1, 5.2.3, 6.2, **6.4**, 10.3.4, 10.4.5, 8.5.8, 9.10.5, 12.4, 15.2
Software risk management	1.6.3, **2.3.2**, **5.2**, 6.4.3, 6.6.3, 9.5.2, 9.6.3, 13.4, 16.5
Software metrics covering products and processes	1.5.1, **2.3.3**, 3.10.8, **5.1.4**, 6.1.2, 13.4, **Chapter 15**
Security and Privacy issues	3.10.2, 5.2.5, **5.4.1**, **10.4.3**, 9.5, 9.8.2
Oversight of software subcontracts	5.1.2, **5.5**, 11.2, 11.3.2
Software schedules with critical interdependencies	2.3.4, 2.6, **4.4**, 5.1, 8.6, 8.8, 9.8.2, **11.4**, Appendix D
Software organization, roles and responsibilities	1.5.1, **1.6**, 8.1.4, 8.2.3, 8.4.3, 8.5.3, 8.6.3, **10.1**, **Appendix C**
Required resources, skills and staffing plan	**10.1**, **10.2**, **10.3**, 11.3
Training plans and training requirements	2.3.7, 4.3.9, 6.1.3, 8.9.9, 10.5, **10.8.4**, **10.9**, **11.3.5**
Software Operations and Maintenance	**Chapter 16**

Corrective Action Process (CAP). Unplanned updates to the SDP must be handled through the CAP as described in the SDP or the *Corrective Action Plan* if you have one. All changes to the SDP should require approval by the *CSwE* and the program *SEPG*. This Guidebook assumes the project has a CSwE, or an equivalent position, as described in Subsection 1.6.4. Changes to the SDP also should require *CCB* approval as discussed in Subsection 6.4.3. The SEPG is described in Subsection 6.1.1.

Software Entity Database (SWED). A database, that can be called the *Software Entity Database*, should be produced and periodically updated to provide a mechanism for identifying, profiling and tracking all *Software Items* (SIs) on the project. Each subsystem should be responsible for their data input to the SWED, but the CSwE should be responsible for compiling this information into a single centralized and controlled database for the project or the program. This database may include for each SI in the project: a functional description of the SI, class, category, size, the percent of new versus reused code, responsible Developer(s) and their contact information, and language(s) used.

5.1.3 Software Resource Estimating

Software resources, including physical, personnel, cost and computer resources, should be estimated before software development begins. These estimates are used to establish software development schedules, risk mitigation plans, and commitments and should be documented in a *Software Resource Estimation Plan*. Details of managing the software resources are provided in Part 3 of the Guidebook.

Software personnel should participate with other affected groups (Systems Engineering, SQA, SCM, test, etc.) in the overall program planning throughout the program as members of Software Integrated Product Teams (SwIPTs). The SwIPTs are discussed in Subsection 1.6.5. Commitments, or changes to commitments, made to individuals and external groups must be reviewed with management regularly.

Staffing Estimation. In order to determine the level of staffing required, the planning function should consider program constraints including milestones, reviews, document and product deliveries, internal milestones, incremental builds, technical constraints, and any changes in scope. Estimates of Source Lines Of Code and software development productivity play an important role in staffing estimates.

Re-planning. The software groups should participate, when required, in re-planning activities to address contract changes, process improvements, or when measured performance varies from planned performance. The related data that is generated must be maintained and placed in the applicable *Software Development Files* (SDF) or *Software Engineering Notebooks*. Software personnel also should participate in contract/subcontract modification activities (such as engineering change proposals). Chapter 14 covers software estimating fundamentals, methods, size growth and estimation tools.

5.1.4 Software Build Planning

A software *build* is a portion of a system that satisfies, in part or completely, an identifiable subset of the total end-item or system requirements and functions. There are often multiple internal builds leading to a deliverable build for an increment in the life cycle. *Requirements met in one incremental build are also met in all successive increments.* The final build is the complete software system. A *release* is a build version that is delivered for acceptance testing and subsequently may be released or delivered for operational use. Incremental builds can be planned for each SI, or group of SIs.

Software Build Plan. A table, similar to the example in Table 5.4, must be included in the SDP, or a separate document referenced by the SDP, to show the intended software delivery plan. The table must include a unique number, often called the *Program Unique Identifier* (PUI), for each Software Item and its name, the responsible developing organization, and *Equivalent Source Lines of Code* (ESLOC) planned for each build. Subsection 14.1.3 provides an explanation of how ESLOC is derived. As shown in Table 5.4, the version (preliminary, initial, update, fixes) can be identified for each delivery.

Software Master Build Plan (SMBP). A comprehensive SMBP must be provided to map the incremental functionality, capabilities and requirements allocated to each build. The CSwE, or Build Manager, usually maintains the SMBP with the approval of the Software CCB. Once approved, the SMBP should be controlled by SCM. In a large program, this is *not a trivial issue*.

As the Project Manager, you, the CSwE, or the Software Build Manager, should routinely report the status and

Table 5.4 SI Build Delivery Plan—Example

PUI	Software Item Name	Developer	Build-1	Build-2	Build-3	Total
	Total for 18 SIs:		102,000	65,000	130,000	297,000
1.0	Decision Support		45,000	28,000	48,000	121,000
1.1	Decision Analysis	Able Corp	I	UWDR	U	
1.2	Analytical Algorithms	Able Corp	–	I	U	
1.3	Scenario Analysis	Able Corp	–	I	U	
1.4	Test Bed Controls	Baker Co.	P	I	U	
1.5	Traffic Control	Baker Co.	I	UWDR	U	
1.6	Simulation Analysis	Baker Co.	P	I	U	
2.0	Services Support		30,000	18,000	12,000	60,000
2.1	Routing Analysis	Charlie Co.	P	I	U	
2.2	User Support	Charlie Co.	–	I	UWDR	
2.3	XYZ Services	Charlie Co.	I	UWDR	UWDR	
	etc.	etc.				

P = Preliminary Version; **I** = Initial Delivery; **U** = Updated Delivery; **–** = No Delivery; **UWDR**= Updates for Work-Offs of Discrepancy Reports; "##,###" = Number of ESLOC per Build; **PUI** = Program Unique Identifier.

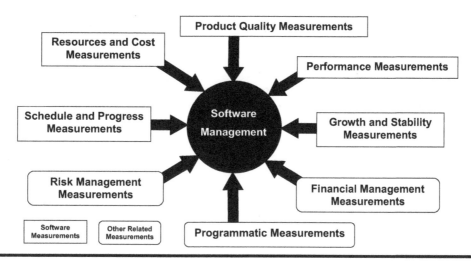

Figure 5.2 **Software management from a measurement perspective.**

changes to the involved upper-level Program Manager(s). The SMBP may also be called the "Master Software Integration and Verification Plan" or may be referred to as a "Build Functionality Matrix." The SMBP is a software document, and it must be compliant with the Master Test Plan (MTP) which is a system-level document covering hardware and software.

Build Planning Updates. Software build planning should occur for each program increment and each deliverable build and be updated continuously throughout the program. Build plans are typically updated only when the plan contents change significantly as determined by the SwIPT Lead.

When planning the builds, the project schedule, ESLOC, and functional content estimates must be taken into consideration. As the program matures, additional design, requirements, technical content, and testing approaches should be added. The build activities should be documented in detailed schedules and then incorporated into the IMS along with staffing and budget-plan information.

5.1.5 Software Development Tracking and Oversight

The software tracking and oversight effort begins once software planning is complete. Subsystem SwIPTs, Software Leads, the Chief Software Engineer, and Software Quality Assurance, can monitor software development status by:

- Collecting and evaluating software measurement data (see Chapter 15)
- Performing product quality and process audits (see 6.1.2 and 6.2)
- Supporting software reviews (see 5.3)
- Performing risk management activities (see 2.3.2, 3.9.4 and 5.2)

Software Measurement Oversight. Throughout the development process, software measurement data must be used to compare actual software size, cost, schedule and progress against the established plan so that you can take timely Corrective Actions (see Section 6.4). If those metrics indicate out-of-tolerance conditions, you or Subsystem SwIPT members must perform an analysis to identify the problem. At that point, you must take the appropriate Corrective Action and identify the potential risks including cost and schedule impacts. The *Software Measurement (or Metrics) Plan* is an important addendum to the SDP. Chapter 15 is entirely devoted to software measurements.

The status of software should be reviewed (usually weekly) at subsystem-level meetings and at monthly program status meetings. In addition, software status should be provided to the customer monthly and also at quarterly reviews. Software management and control must be integrated into your overall program management scheme. Figure 5.2 is a depiction of software management from a measurement perspective.

As shown in Figure 5.2, software measurement oversight involves:

Software-specific measurements:

- Schedule and Progress Measurements
- Resources and Cost Measurements
- Product Quality Measurements
- Performance Measurements
- Growth and Stability Measurements

Non-software-specific measurements:

- Risk Management Measurements
- Financial Management Measurements
- Programmatic Measurements

Cost Account Oversight. Software work packages are typically cost accounts within the *Earned Value Management System* (EVMS) used by many contractors (EVMS is described in Section 8.5). The cost account must be at a level of detail sufficient to maintain control of the associated software development activities. Ongoing metrics collected on the cost accounts and work packages must be reported to program management and available to the customer.

Schedule Oversight. Schedule review meetings are normally conducted weekly or more often if the schedule is consistently changing. Schedule metrics (using weekly milestone accomplishments, including subcontractor data) should be reported along with the status of Corrective Action/recovery plans. IMS and detailed schedules should also be reviewed at lower levels within the SwIPTs.

Headcount Oversight. SwIPTs should monitor headcount on a weekly basis and strive to identify potential problems early. Updates of accomplishments, actual budgeted and forecasted headcounts should be conducted monthly. Forecasts should be updated and reported in internal cost performance reports that include cost/schedule variances and changes in the latest revised estimate. Costs and schedules should be controlled by monitoring headcount, expenditures, and assessing progress.

Product and Process Oversight. Product evaluations, software reviews, process audits and assessments are used by Subsystem SwIPTs, CSwE, and SQA as a means to determine compliance with the standards established by the SDP. Noncompliance of baselined products is handled via the *Corrective Action Process* (see Section 6.4). Process audits must be performed by SQA, with support from the CSwE, to determine compliance with the processes specified in the SDP. SQA must be responsible for documenting and verifying closure of a non-compliance issue. Subsection 6.5.6 describes the software product evaluation process. You or your senior management must implement and maintain the mechanisms for interfacing to and communicating with the customer.

5.1.6 Avoid Gold Plating

A passion for excellence and efficiency is much less expensive to attain than a passion for perfection.

There is a tendency for all Engineers to overachieve. That is often called "gold plating." Tasks performed that are *not* called out in the contract as a system requirement can be considered gold plating unless the task is a derived requirement essential to system functionality. Rework to make it "better" and then more rework to make it "even better," and continuing this "polishing the apple" approach always has a severe impact on project cost and schedule. It is understood that Engineers typically want to continue working on their task to make it as good as possible. As the Software Project

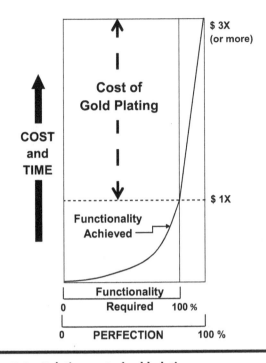

Figure 5.3   Relative cost of gold plating.

Manager, you must be careful to *avoid this perfectionist pitfall*. For example, Figure 5.3 is a notional curve showing the relative cost of perfection versus system functionality required. The cost and time to achieve perfection is very high.

Where human lives and safety are involved, perfection can be justified, and the customer must pay for it; otherwise, *perfection is too expensive*. When you reach 100% of the functionality required by the contract, you have fulfilled the contract requirements. If the time and cost at that point is $1X, as shown in Figure 5.3, perfection can be *at least* three times more expensive and time-consuming.

Lessons Learned. The last 5% is costly. It may be human nature to delay submitting your product until it is perfect, but it is a Software Manager's responsibility to see to it that gold plating does not happen. I have observed that, if human lives and safety or *not* involved, 95% of the planned functionality often satisfies 95% of the users; that last 5% is a very steep part of the curve so, if you can avoid it, you could realize a big cost savings. This is a tricky game to play. Most "shrink wrapped" software products probably do not come close to 95% of full functionality when the product is initially delivered to the marketplace.

5.1.7 Software Item Test Planning

The testing of Software Items being developed by the subsystems should be performed by the respective subsystem Software Test Engineers. They are responsible for documenting their *Software Test Plan* (STP), *Software Test Description* (STD), and *Software Test Report* (STR) to verify that the SIs meet their allocated subsystem requirements.

A preliminary version of the STP is usually produced during Software Design activities, but the full STP is the result of the SI test planning activity. Production of the STD and STR is performed during Software Item Qualification Testing (SIQT) and is discussed in Section 11.7. Test activities for the critical (SCS) and support (SS) software classes must also be documented in the *Software Development Files* as described in Section 9.4.

Software Test Plan. The STP describes plans for qualification testing of SIs and is an important software document. It describes the test environment to be used, identifies the tests to be performed, provides schedules for the testing tasks, defines the resources needed, and addresses the planning tasks required to conduct the SIQT. Table 5.5 is an example summary of the readiness criteria in terms of entry, exit and verification criteria to ensure completeness of the STP.

5.1.8 System Test Planning

The SDP should contain the approach for providing support to System Test Planning. If there is a *System Verification and Test Plan* (it may also be referred to as an *Integration, Test and Evaluation Plan*), it should be prepared by the SEIT. It is the key planning document for system testing. System Integration and System Test activities are described in sections 10.3 and 10.4.

The System Test Team should be responsible for performing the actual system testing. Software Developers and/or Software Test Engineers have a *support role* in System Test Planning that may include reviewing test preparation materials and providing software test support items, such as reusable software test documentation, simulators, drivers and analysis tools. Software Engineers also support anomaly analysis to determine if the problem is due to software only, hardware only, or a combination. If Regression Testing on the software builds is needed, SCM must provide the software builds against which the tests are conducted. Regression Testing is discussed in Subsection 3.9.3.

5.1.9 Planning for Software Transition to Operations and Maintenance

Planning for Software Transition to Operations. The SDP must describe the approach for performing the software installation planning. This activity involves the preparation for, and the installation and checkout, of the executable software at a user site. Planning and preparation should start relatively early in the life cycle to ensure a smooth transition

Table 5.5   Readiness Criteria: Software Test Plan

Entry Criteria	Exit Criteria
■ The appropriate STP data items and other reference materials are obtained. ■ Software requirements are established in the Software Requirements Specification and Interface Requirements Specification (IRS) and are traceable to a parent requirement. ■ The top-level software architecture is established. ■ The Requirements Test Verification Matrix (RTVM) specifies the test verification method and level for each requirement in the SRS and IRS. (See 11.1.2).	The following tasks have been defined and documented in the STP: ■ Test environment (sites, hardware, software, test tools, test facilities, test data, etc.) needed to conduct the life cycle tests. ■ Test scenarios to be performed including the schedule for executing the test activities. ■ Traceability between SI requirements and the related tests and test phases where the requirements are verified. ■ Personnel, organizations, responsibilities and management activities needed to develop and implement the planned testing. ■ The objectives for each test including test level, type, conditions, data to be recorded, qualification method(s), data analysis, assumptions and constraints, safety, security and privacy considerations. ■ Approach to related issues such as data rights, training, Regression Testing, delayed functionality and deliverable documentation. ■ Criteria for evaluating the test results.
Verification Criteria	
■ The tests identified fully validate the system requirements being tested. ■ The occurrence and timing of the test phases in the life cycle, plus the entrance and exit criteria for each test phase, has been identified and documented. ■ The terminology and format is consistent between the SRS, RTVM, IRS and STP. ■ The STP has successfully passed its Peer Review.	
Measurements	
Statistics from the STP Peer Review. (See Chapter 15)	

to the user's site. It should include the preparation of documentation and software products required by the user to perform operational tasks. This includes the code for each SI and supporting documentation including the preparation of user manuals and user training materials as the pertinent information becomes available (See Section 11.9).

Planning for Software Transition to Sustainment. The SDP must contain the approach for performing software transition planning. Transition planning involves advance planning and preparation that should start early in the life-cycle to ensure a smooth transition to the maintenance organization. It must include the installation and checkout of the software at the maintenance site. Section 11.10 and Chapter 16 discuss Software Maintenance and sustainment issues.

5.1.10 Tracking and Updating Project Plans

Electronic Data Interchange Network (EDIN). The plans identified in Subsections 5.1.1–5.1.9 must be made available via an EDIN *accessible to all stakeholders and the customer.* The EDIN is described in Subsection 9.5.1 including the potential for access via the "cloud." Once baselined, unplanned modifications to these plans should be initiated and tracked using the Corrective Action Process described in Section 6.4. Modifications that are planned, such as scheduled updates to baselined documents at major milestones, must also be electronically available. Unplanned modifications, may also be captured as lessons learned. The SDP should also cover the approach to enforcement of planned updates to the plans.

Software Engineering Process Group. The SEPG should review the software development process at SEPG meetings usually held monthly. These meetings can determine the effectiveness of the process through analysis of software metrics, requests from subsystem SwIPTs, recommendations from SEPG members, the customer and their representatives, process audit information from SQA, and program directives.

If other software or program-level plans are affected by the approved change to the SDP, the CSwE must ensure that responsible parties are notified of the SDP update and ensure that all inter-group commitment changes are coordinated. The SEPG, described in subsections 2.4.1 and 6.1.1, should coordinate this activity.

5.2 Software Risk Management and Mitigation

> *Risk management is one of the most important responsibilities of a Software Project Manager: it must be a proactive continuous process employed throughout the Software Development Life Cycle.*

All software development projects have risks. Some are minor, and your project can live with them. Other risks have a severe impact and must be addressed and mitigated—either eliminated or made less severe. Software risk management addresses the management process for *identification*, *tracking* and *mitigation* of software development risks throughout the software development process that involves potential adverse technical, programmatic, schedule, cost, or supportability impacts.

As the Software Project Manager, you must provide proactive risk management direction to ensure that your project makes an early and continuing identification of its software risk items, develop a strategy for handling known risk items, identify and document an agenda to handle new risk items as they surface, and track progress versus plans for risk item mitigation.

5.2.1 Program-Level Risk Management

Risk Management Board (RMB). If your project is a part of a large program composed of several projects, there should be a program-level RMB. Alternatively, each project in the program could have its own RMB. In any case, this high-level RMB is the primary entity for evaluating risks, watch list items, and unfavorable concerns. The Chief Software Engineer should be a member of the RMB. Usually, there is no Risk Management Board at the software level, but there is nothing to preclude it if you need a *Software Risk Management Board* to supplement the program-level RMB.

Risk Management Plan (RMP). If your RMB is doing their job, they should have produced a program-level *RMP* that must be followed by all projects in the program. The RMP may also be called a *Risk Mitigation Plan* or a *Risk Response Action Plan*. The program-level RMP should identify the process for risk planning, identification, assessment, prioritization, handling, and monitoring.

5.2.2 Software Risk Handling Plan

The *Software Risk Handling Plan* (SRHP), typically an addendum to the *Software Development Plan*, is the principal plan for identifying and mitigating software risks on your project. The SRHP should assign risk severity levels, define specific risk handling plans where appropriate, and describe the process for ensuring implementation of the SRHP. The software risk handling process must be consistent with the program-level RMP and with the program's process for elevation and resolution of all software risk items of sufficient concern to the program.

There will likely be some software risks that are *not big enough* to make it to the program-level RMB. Those known risks should be described in the SRHP. Measurements used to control and track risks are addressed in Chapter 15. Your SDP needs to be clear as to what type of risks are handled at the program level and what is handled at the project level (assuming your program involves multiple projects). Handling software risks are an important integral function of the software development process. Risks must be managed and coordinated across subsystems and development tasks. The overall software *risk mitigation approach* can be summarized as follows:

■ Software build planning must be consistent with the *Software Risk Handling Plan.*
■ The risk handling approach must encourage cross-support from SEPGs, particularly for process improvement and risk reduction groups.
■ Software risk assessment is an integral part of each formal review.
■ Oversight of software tasks across the program by the CSwE helps to reduce software risk by early identification and resolution of issues or potential issues.
■ Incremental software development helps to mitigate risks. Each increment, technical, programmatic, schedule, cost, or supportability issue and requirement change should be assessed, baselined, prioritized, tracked and resolved early in the Development Life Cycle.
■ Prototypes can evaluate hardware and software integration, and integration of multi-Developers. These tools can be used as an integration and demonstration facility to evaluate risks early in System Design and development.

■ Measurements and Critical Path analysis (see Section 8.4) are integral to risk management and provide guidance for reducing, preventing or eliminating adverse impacts.

Development of each SRHP is the responsibility of the affected SwIPT. SRHPs must be approved by the Risk Management Board. SwIPT leads are accountable for implementing the SRHP and reporting risk status within the scope of their assigned work elements.

5.2.3 Software Risk Management Process

During software development, software risk management involves identifying, assessing, mitigating, monitoring, documenting and closing risks. Individual risk plans define mitigation tracking measures and Corrective Action when thresholds exceed the limits as defined in the RMP and SRHPs. Figure 5.4 is an example overview of a risk management process.

5.2.4 Software Risk Handling Alternatives

All risks, regardless of severity, must be handled in some manner. It is a good idea for the SRHP to include a risk assessment and response guideline similar to the graphical example in Figure 5.5.

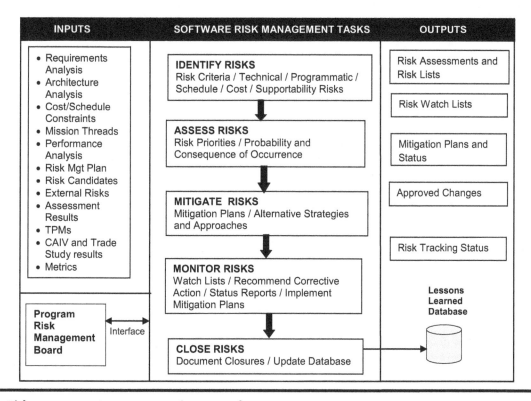

Figure 5.4   Risk management process overview example.

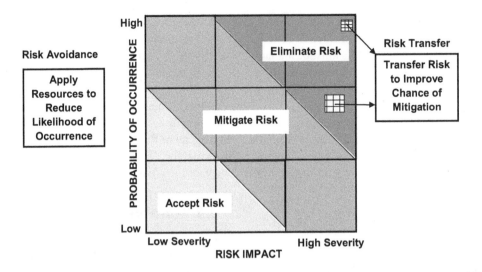

Figure 5.5   Risk assessment guidelines.

Handling software risks typically includes the following options:

- *Risk Avoidance*: If potential risks are identified early enough during the planning process, you can plan to simply avoid them by taking alternative actions.
- *Risk Acceptance*: The impact of some risks may be so low that you *can take the risk* they will not happen, and if they do, the project can absorb the impact without a serious outcome. Some risks are never resolved. The risk might "fade away" over time such as the risk one of your COTS vendors will go out of business. Depending on your acceptable *risk tolerance*, you may be able to accept some risks that have a very low impact if they occur. You must determine *how much-calculated risk is acceptable*. Such risks might be acceptable in order to meet cost and schedule constraints as long as lives are not involved.
- *Risk Mitigation*: For known potential risks to your project, you can prepare plans in advance to take positive action to reduce the impact and severity of the risk so the project can cope with it when and if it does happen. Risks that are not candidates for mitigation must still be watched via some tracking metrics so that changes to a risk status is flagged for you in a timely manner.
- *Risk Elimination*: If you identify risks with potentially catastrophic consequences, they *must be eliminated*. If they cannot be *entirely* eliminated, they must be addressed in order to convert them to one of the above options. If neither is possible, and the probability of the risk occurring is high, you must change your approach or consider cancelling the project.
- *Risk Transference*: If your staff does not have the expertise to handle specific complex areas of your contract, or if they do not have the experience with a required

programming language, you might consider using a subcontractor or a consultant with the needed expertise. *The risk does not disappear*; you simply transferred it to someone else. If that risk taker does not deliver, the risk is back in your lap, so transference has its own set of risks.

5.2.5 *Software-Related Risks*

Software-related risks can be so prevalent that it is difficult to manage them. In order to manage them, however, you must identify them. The following red flags cannot be considered an exhaustive list, but it is a guideline of typical software risks you may encounter:

- Poorly defined, incomplete, complex or unstable system and/or software requirements
- Inability to properly estimate the software size, complexity and account for growth
- Incompatibilities between the software development effort, schedule and performance baselines
- Too rapid buildup of your staff at the start of your project
- Inadequate interaction and coordination between system and Software Engineering
- Availability, suitability, integration and sustainment of COTS software
- Extensive security or Safety Critical requirements driving use of new immature technology
- Uncertainty that the developed software will meet its requirements and be maintainable
- Uncertainty that the planned percent of software reuse will be realized (this can be a big impact)
- Using tools, architectures, technologies and methods outside your staff's expertise

- Concern over the complex System Integration and control of hardware and software components
- Using customer-furnished equipment or acquired software with unknown performance abilities
- The risk of planned technology advances and program plans for technology insertion
- Availability and suitability of software test beds
- Consistency of the Software Test Environment with the target operational environment/platforms
- Limited or unstable funding allocations restricting staffing, training, tools and facilities

5.3 Technical and Management Reviews

Joint Technical and Management Reviews demonstrate progress-to-date on project work products and provide a forum for discussing programmatic issues and risks. Periodic Technical and Management Reviews of software products and status must be conducted at the following levels:

- *System level*: to review system-wide project status, to identify program cost and schedule issues, and to address system technical issues (e.g., inter-subsystem interface problems)
- *Subsystem level*: to review subsystem-wide project status and to identify subsystem-specific cost, schedule, and technical issues (subsystems may be called segments)
- *Software Item level*: to review development progress and to identify Software Item-specific cost, schedule, and technical issues
- *Software Unit level*: for feedback on in-progress technical tasks and issues

Joint Technical and Management Reviews must be conducted in concert with the *Integrated Master Plan and Schedule* (IMP/IMS) events and milestones. There could be a *Software Review Plan* addendum to the SDP defining the objectives of each type of review, the entry and exit criteria for each review, when the reviews occurs, what products are reviewed, and for what software categories these reviews must be conducted. Appendix G contains recommended exit criteria for software reviews.

The software review process must be structured to support the evolution of natural products during the Software Development Life Cycle. This can be accomplished by utilizing the following types of reviews:

- *Joint Technical Reviews (JTR)*: conducted to review the status, correctness, and completeness of in-progress and final software products and to discuss technical issues (see 5.3.1)

- *Joint Management Reviews (JMR)*: conducted to demonstrate the current status of products and as a forum for discussion of status, schedules, programmatic issues and risks (see 5.3.2)
- *Technical Interchange Meetings (TIM)*: similar to JTRs but are conducted in a less formal manner and may be focused on support software or specific software requirements, architecture or design issues

5.3.1 Joint Technical Reviews

JTRs must be conducted to ensure product correctness and completeness and to elevate management and your customer's visibility into the status of evolving products. JTRs focus on evaluating the adequacy and completeness of in-process or final software products. Table 5.6 presents an example of the software product reviews by activity and the type of technical or management review to be utilized for five software categories.

The Joint Technical Reviews must be attended by persons with technical knowledge of the specific software products. JTRs have the following objectives:

- Review and evaluation of evolving software products using the example software product evaluation criteria in Appendix D
- Review and demonstrate proposed technical solutions; identify and resolve technical issues
- Review project status and obtain feedback on in-progress technical tasks
- Identify near and long-term technical, cost and schedule risks
- Identify mitigation strategies for identified risks, within the authority of those present, and identify risks to be raised at Joint Management Reviews and the Risk Management Board
- Ensure ongoing communication between Software Management, Developers and the customer

Following the appropriate inspections and document reviews, the overall technical assessment of the product is typically presented at the JTRs, which ultimately provide software program and project status information presented at the JMRs (discussed below). The frequency of the reviews at each level can vary depending on the objectives of the specific reviews. When a Software Item includes a mixture of software categories, the JTR requirement must be at the most stringent category included in that SI. JTRs should be performed for each build.

Lessons Learned. SPMs often have to participate in JTRs and JMRs. If the JTR is for your project you are likely to be the sponsor;

Table 5.6 Software Product Reviews by Activity and Category—Example

Subsystem Reviews by Activity	SCS-1	SS-1	SS-2	SS-3	C/R	Sw Level	Frequency
Software Requirements Definition Activity							
Subsystem Software Requirements and Architecture Review (SAR)	JTR	NA	NA	NA	JTR	Sub	Once
Subsystem Support Software Requirements TIM	NA	TIM	NA	NA	NA	Sub	Once
Software Design Activity							
Software Preliminary Design Review (Sw PDR)	JTR	NA	NA	NA	NA	SI	Once
Software Critical Design Review (Sw CDR)	JTR	NA	NA	NA	NA	SI	SSB
Support Software Design TIM (SSD TIM)	NA	TIM	NA	NA	NA	SI	SSB
SI Qualification Testing Activity							
SI Qualification Test Readiness Review (SIQ TRR)	JTR	JTR	NA	NA	JTR	SI	SSB
Subsystem Qualification Testing Activity							
Subsystem Qualification Test Readiness Review	JTR	JTR	NA	NA	JTR	SI	SSB
Management Reviews							
Reviews of Process Compliance Audits	JMR	JMR	JMR	JMR	JMR	Sys/Sub	NA
Monthly Status Reviews	JMR	JMR	JMR	JMR	JMR	Sys/Sub	M
Program Status Reviews	JMR	JMR	JMR	JMR	JMR	Sys/Sub	Q

Type of Review: JMR = Joint Management Review; JTR = Joint Technical Review; TIM = Technical Interchange Meeting; NA = Not Applicable; **Level:** Sys = System; Sub = Subsystem; SI = Software Item; Sw = Software; **Frequency:** M = Monthly; Q = Quarterly; SSB = Subsystem, Spiral or SI Build; **Software Categories:** SCS = System Critical Software; SS = Support Software; C/R = COTS/Reuse Software.

if your project is part of a large program, you will certainly be a participant. Prior to a JMR in front of the customer, I was once instructed to prepare a *20-minute* software status report which I did. When I was introduced, the Vice President leading the JMR said: "Marv is now going to give a *10-minute* update on the status of software." As I walked to the podium, I tried to figure out how to cut my briefing in half but could not do so in 6 seconds. All I could do is give the briefing as planned and see how long they would let me go. Because my briefing was so informative, I got through about 90% of it before they cut me off. After that incident, every briefing I gave thereafter included a pre-planned identification of the charts I could cut if, for any reason, my presentation time was cut short. Pre-planning of this sort is highly recommended.

5.3.2 Joint Management Reviews

JMRs must be periodically conducted to ensure product completeness and to elevate both management and customer *visibility* into the development process and evolving products. JMRs are used to review the current state of deliverable products, as well as project costs and schedules. Attendees must be persons with the authority to make cost and schedule decisions (with supporting staff) as needed. The objectives for software JMRs include the following:

- Keep management informed about project status, directions being taken, technical agreements reached, and overall status of evolving software products.
- Resolve issues that could not be resolved at JTRs.
- Mitigation strategies for near and long-term risks that could not be resolved at the JTRs.
- Identify and resolve management level issues and risks not raised at JTRs.

- Obtain commitments and customer approvals needed for the timely accomplishment of the project.
- Joint Management Reviews that normally apply to software are:
  - Program Quarterly Status Reviews.
  - Monthly Status Reviews.
  - Process Compliance Audits.

As the SPM, you will be an intimately involved with Joint Management Reviews along with your CSwE and appropriate members of the SwIPTs. All participants support JMRs by providing progress-to-date and overviews of the software products. Identification of the JMRs that apply, the schedule for each, the process to be followed for each, the documents to be reviewed, and the personnel involved should be defined in the SDP or in a *Software Reviews Plan* as an addendum to the SDP.

A guideline to the specific software documentation product reviews can be described in a table in the SDP similar to Table 5.7. It is an example of the *evolution of document maturity* from draft versions to preliminary versions and then to baselined versions followed by updates as needed. The example in Table 5.7 maps software documentation to formal software reviews and is complementary to Table 13.1 that relates the development of software documents to software development activities.

> **Lessons Learned.** At one time I was the Software Lead on the Space Station Freedom program at Lockheed. Since we were a subcontractor, every month the VP in charge of our program participated in a JMR at the prime contractor's site. Since his presentation was to their Senior Managers it had a time constraint; each area of our responsibility was restricted to a *one chart* update. There was so much software activity going on that it was difficult to get a meaningful update on one chart. So, I came up with a technique of multiplying the content density of that one chart. I did that by using the four corners of each rectangular box on the chart to contain, or represent, some data. It was a simple but effective way of conveying more of the key information in the same limited space.
>
> However, I used only three of the four corners. After our VP finished my chart, the prime contractor's VP asked, "Is that a Gechman chart?" The answer was "yes," and the prime's VP said: "Tell Marv that we are upset he did not use all four corners!" This humorous reaction was an example of the rapport I was able to establish with our customer; it made my job much easier and much more enjoyable. I highly recommend

that you always try to do the same with your customer. Also, be creative in preparing briefing materials and try to get the top level on one page as discussed in Subsection 1.4.7.

5.4 Assurance of Critical Requirements

Critical strategies must be identified in the SDP to ensure that software groups provide additional oversight and focus on incorporating critical requirements into the SIs. There are almost always some key software requirements that are critical cornerstones such as for safety, security, human factors, privacy protection, reliability, maintainability, availability, performance, etc. Software Project Managers must develop and employ strategies to ensure that these critical requirements are satisfied.

The critical strategies should be documented in the SDP. It should include both test and analyses activities to ensure that the requirements, design, implementation and operating procedures, for the identified computer hardware and/or software, *minimize or eliminate the potential* for violating established risk mitigation strategies. The SDP should also indicate how evidence is to be collected to prove that the assurance strategies have been successful.

5.4.1 Software Safety

Software safety requirements involve Software Items or Software Units (SUs) whose failure may result in a system condition that can cause death, injury, occupational illness, damage to or loss of equipment or property, or damage to the environment. Each software-related Safety Critical requirement identified must be documented in the safety requirements Section of the Software Requirements Specification (SRS) and identified by a unique product identifier.

If aviation safety standards are specified in the contract as compliance documents, your SDP or an addendum to it must describe the approach for complying with those standards and regulations. There are other software safety standards that you may be required to adhere to in specific industries such as transportation, aerospace, nuclear power plants and national defense. Safety and reliability issues also occur in normal every day safety-related scenarios such as the software that controls elevators, medical databases, microwave ovens, and many more.

The activities required for ensuring that Safety Critical software requirements are met for the program must be shared between the System Safety group, at the program level, and the Subsystem Software Team. Each SwIPT should assign responsibilities for safety issues and for coordination with System Safety. The Software Team is responsible for developing system software that is safe to operate and compliant with all appropriate safety standards and requirements.

Table 5.7 Software Documentation Maturity Mapped to Software Reviews

Software Document	SFR	SAR	PDR	CDR	IRR	SI TRR	PTR or BTR	FAT TRR	SQT TRR
Software Development Plan (SDP)		B	U	U	U				
Software Metrics Report	P	B	U	U	U	U	U	U	
Software Master Build Plan (SMBP)		D	P	B					
Software Requirements Specification (SRS)		P	B	U	U				
Interface Requirements Specification (IRS)		P	B	U	U				
Interface Control Document (IFCD)		B	U						
Software Design Description (SDD)			P	B					
Software Architecture Description (SAD)			P	B					
Software Test Plan (STP)			P	B	U	U	U		
Interface Design Document (IDD)			D/P	B	U				
Database Design Description (DBDD)			D/P	B	U				
Software Installation Plan (SIP)				D			P		
Software Transition Plan (STrP)				D			P		
Software User Manual (SUM)				D			P		
Firmware Support Manual (FSW)					D		P		B
Computer Programming Manual (CPM)					D		P		B
Software Test Description (STD)						D/P	B	U	
Software Test Report (STR)							B		
Software Version Description (SVD)						D	P	B	U
Software Product Specification (SPS)						D	P	B	U

D = Draft; **P** = Preliminary; **B** = Baselined; **U** = Updated (As Required); **SFR** = System Functional Review; **SQT TRR** = System Qualification Test—Test Readiness Review; **SAR** = Software Requirement and Architecture Review; **IRR** = Integration Readiness Review; **PDR** = Preliminary Design Review; **SI TRR** = Software Item—Test Readiness Review; **CDR** = Critical Design Review; **PTR or BTR** = Post-Test Review or Build Turnover Review; **FAT TRR** = Factory (or Element) Qualification (or Acceptance) Test—Test Readiness Review.

The general approach to managing software Safety Critical development activities for the program should be to *integrate safety management into the Software Life Cycle activities*. System Safety should play an integrated role in the software development process providing the System Safety group with visibility into the software development activities that are critical to program safety issues, as well as providing the SwIPTs with the input required to ensure that safety issues are addressed effectively. Details regarding software safety should be included in a *System Safety Program Plan* or equivalent.

The SDP should require Software Safety Engineers to *define classifications for Safety Critical SIs and SUs*. All SIs and SUs should be categorized according to these Safety Critical classifications. To prepare these classification levels, consideration should be given to: the severity and probability of hazards the SIs or SUs may contribute to (as determined by a hazards analysis); the potential for the SIs or SUs to provide Safety Critical monitoring or mitigation actions; and how the SIs or SUs handle and protect Safety Critical data. System and Software Safety Engineers should:

- Participate in system and software requirements analysis to generate additional functional or performance requirements to assure safe operations and safety contingency actions.
- Monitor these additional software requirements to assure they are properly specified and traced to documented Safety Critical hazards.
- Assure that unsafe operations are not specified by existing requirements.
- Participate in design reviews to prevent unsafe approaches from being applied.
- Track internal and external safety-related interfaces to assure they are fully documented and unambiguous.
- Participate in the review of test procedures to assure Safety Critical requirements are properly interpreted and tested.
- Participate in the evaluation of Safety Critical code changes and review regression tests.
- Document Safety Critical criteria used in selecting COTS, GOTS and reused code.

5.4.2 Software Security

Software security (renamed Information Assurance) involves SIs and SUs whose failure may lead to a breach of system security or a compromise of proprietary or government classified data. Each software-related Security Critical requirement identified should be documented in the security and privacy protection requirements section of the SRS. Security requirements can be derived from the System Specification.

Security concerns can have a significant impact on your software architecture.

If applicable, security services provided by the program for all projects should be documented in an *Information Assurance* (IA) Plan and should provide "layers" of structured defense from commercial packages (such as anti-virus software and firewalls) to elaborate National Security Agency (NSA) approved Type-1 encryption algorithms. The SDP should state that software portions subject to security product certification and accreditation must be developed in accordance with the Security Plan.

Software security requirements should be flowed down to subcontractors with the normal requirements analysis process. The Software Design activity must conform to the security architecture as described in the IA Plan. Also, when developing the software schedules, and the build plan, the security certification and accreditation need dates must be accounted for.

5.4.3 Human Systems Integration

A system (or product) is ultimately judged good or bad by its end-users.

Human System Integration (HSI), also called the Human-Computer Interface, is a disciplined and interactive Systems Engineering approach for integrating human considerations into System Development, Design and Life Cycle management. Applying HSI techniques improves total system performance and can reduce human errors and the cost of operations across the system's life cycle. Significant improvements in performance can be made with design features that often cost almost nothing to implement. *How people perform with technology is a critical component of total system performance.* Human performance can be substantially improved, and the likelihood of errors reduced, simply by designing a system that is compatible with the characteristics of the people who operate and maintain it.

If good requirements are generated, the system will support users in performing their tasks effectively and efficiently. This can be achieved with the support of trained HSI Engineers who provide the applied technical expertise. One of the first Systems Engineering objectives of a project is to define the "who, what, when, where, and why" of the system being designed and built. That normally takes place during the Concept Development phase. At that early phase of the life cycle, inputs from potential users may not be available, or as useful, as inputs from the HSI Engineers because the system analysis they can conduct enables them to understand the role of the human in the system and the specific tasks people perform during the required human system interaction.

The Department of Defense (DoD) has mandated inclusion of HSI in the development of military systems. DoD Instruction (DoDI) 5000.02, Enclosure 7 addresses HSI, stating that the Program Manager should plan for and effect HSI, beginning early in the acquisition process and throughout the product life cycle. It charges the Program Manager with the responsibility for ensuring HSI is considered at each program milestone. The design aspects of HSI are discussed in Subsection 8.4.2.

Some Program/Project Managers do not fully appreciate the ways in which HSI can improve system performance, or they remain confused about how to effectively incorporate HSI into their programs. This is due to a number of fundamental gaps in understanding what HSI can do. As the SPM, you should *not* subscribe to the following seven HSI myths identified by (Endsley, 2016):

- *Myth 1*: HSI means asking users what they want: User input is an important element of good System Design, but relying solely on their input is insufficient evidence of good HSI since users often miss the subtle features of technology that can negatively impact human performance.

■ *Myth 2*: HSI means including the newest technology and equipment: It may be really cool, but applying the latest advancements may actually lead to slower performance and higher error rates on critical tasks.

■ *Myth 3*: HSI should be performed at the end of a program: That is the *worst time* to address HSI. The cost of incorporating changes recommended by Human Factors Engineers late in the design phase or even later during implementation is far more costly than incorporating those changes during the System Definition and analysis phases.

■ *Myth 4*: Anyone can do HSI: Most systems, especially large systems, should require the participation of experienced professionals in the areas of human factors, Industrial Engineering, psychology, Software Design, and other related skills. HSI is a multidisciplinary profession and, like most areas of engineering, there is a significant body of knowledge that must be acquired. Their expertise should be applied to help conduct System Definition, task analysis, hardware and software user interface design, prototyping, test and evaluation, and development of user documentation and training materials.

■ *Myth 5*: We can just train around HSI problems: There is a long history of trying to use training to compensate for poorly designed user interfaces such as a green light meaning "Stop" or pushing a lever down when you want to go up. Good training is important, but it is not a substitute for good System Design.

■ *Myth 6*: With automation, we don't need to worry about HSI: Actually, the opposite is usually true. Experience with automated systems over the past 30 years has shown that, in many systems, automation can make the user's job more complicated and prone to errors.

■ *Myth 7*: HSI costs too much: Applying HSI techniques in a timely manner will save the project money by avoiding extensive and expensive rework later—especially during sustainment.

Lessons Learned. I once spent a year as the Software Safety Engineer on the launcher subsystem of a major missile defense program. Because of the variety of environmental and situational variables, the launching required a soldier technician to push a sequential series of buttons and controls and then leave the immediate area prior to the actual launch to avoid being roasted. Figuring out the proper sequence was not that difficult. However, making absolutely 100% sure that hitting the *wrong sequence* of buttons would *not cause a misfire*, resulting in the death of the technician, was very difficult. It took the team of System Engineers that I worked with much longer than originally planned to check out every possible combination of a wrong sequence. The lesson here is to make sure you do not underestimate the importance of the Human Systems Integration function. Also, it is always a good idea to have a cash *management reserve for unexpected tasks* that must be performed and likely to take longer than expected.

One example of an extreme violation of HSI principles was a principal cause of John Denver's deadly crash of his experimental airplane in 1997. The builder of the plane failed to follow standard location protocols for the fuel tank change switch. Rather than placement where a pilot could easily reach the switch, it was located behind the pilot's left shoulder, requiring a turning motion that could have, and did have, fatal results. Ignoring HSI considerations can have catastrophic consequences.

5.4.4 Privacy Protection

Privacy Critical requirements are those requirements on SIs and SUs whose failure may lead to a compromise of private personal data such as training scores or personnel evaluations. Each software-related *Privacy Critical requirement* identified should be documented in the security and privacy protection requirements section of the SRS and identified by a unique product identifier.

5.4.5 Dependability, Reliability, Maintainability and Availability

System Critical Software (SCS) was described in Subsection 1.8.1 as a *class* of software *required* for implementation of a portion of the system functionality allocated to software.

Failure Modes, Effects and Critically Analysis (FMECA). System Critical Software functions can be more easily managed and tracked if a *FMECA*, or a similar process by any other name, is performed for all new or modified SCS functions. FMECA is a comprehensive systematic analytical method to identify potential failure conditions and to contribute to the improvement of designs for products and systems to avoid those conditions. It has been used for hardware since the 1940s by the FAA, the auto industry, military and space systems. FMECA can, and should, be expanded to include software and human interaction considerations. Also, a list of SIs and SUs that are System Critical should be created and maintained.

Software Dependability. Dependability is the sum result of effective strategies for *reliability, maintainability* and *availability* (RMA) and the SDP should describe the overall approach to develop these strategies. Software RMA practices must be incorporated throughout all software development activities; they provide the building blocks for dependability. Effective strategies for RMA also helps to ensure that software meets system requirements with minimum risks, maintains the integrity of the Software Design, and minimizes life cycle costs.

Software Reliability. Software reliability models should be used to assist in making predictions about the software system expected failure rates. The reliability tasks must be integrated with Quality Assurance, Product Evaluations, Maintainability, and other Engineering activities to avoid duplication and provide a cost-effective program. Software reliability should involve detection, reporting, quantification, and correction of software deficiencies throughout all design, development and testing activities.

Software Maintainability. There are two major aspects of software maintainability: software *restorability* and software *reparability*:

- Software restorability is defined as the process of restoring the software to an operational state after the occurrence of hardware or software failures. Ineffective software restorability can be a large contributor to downtime and thus can significantly affect system availability. The need for rapid software restorability is a major driver of the software architecture and design task.

- Development of maintainable software, from a software reparability perspective, involves planning and establishing the software development methodology, environment, standards, and processes with an objective of making Software Maintenance changes efficiently and effectively (e.g., not requiring an engine disassembly to change the oil).

Some methodologies, such as Object-Oriented Design, development and programming, may produce software-related products that are more maintainable than other approaches. The design must be captured and retained in the Software Engineering tools and subject to Configuration Management (CM) processes. Similarly, the software CM tools provide support to Software Maintenance needs. Other tactics that can be described in the SDP to improve maintainability may include:

- The Software Development Environment (SDE), covered in Section 9.1, must be sized to include sufficient capacity to support post-deployment Software Support requirements, thus promoting long-term maintainability.

- Software standards must be established for each programming language to ensure that consistent programming styles are applied by all Developers and that the software and supporting documentation are complete and understandable.

- The Software Product Evaluations should assess compliance with the standards to ensure that they are consistently applied.

System Availability. A high availability rate for access to the system is the by-product of effective RMA practices as well as accurate estimation of user needs. By performing modeling and trend analysis, based on historical trend data and collected metrics, software reliability and availability can be predicted and the necessary Corrective Actions taken to achieve RMA and system requirements.

5.4.6 Assurance of Other System Critical Requirements

Critical software requirements should be tracked and monitored throughout all the software development activities just like other software requirements. However, in addition to the standard testing and quality assurance procedures for other software requirements, the SwIPT should follow an assurance strategy designed to ensure that hazardous or compromised conditions are eliminated or minimized for each development activity. This strategy should be to:

- Identify and document critical requirements in the appropriate SRS sections.
- Document the specific Software Units that contribute to these critical requirement.
- Define specific SI testing procedures that execute all affected SUs to determine compliance.
- Execute the security and privacy testing procedures at each SI build when affected Security Critical and Privacy Critical SUs have changed.
- Execute the safety-related test cases at each SI build for SIs with Safety Critical SUs, even if the units have not changed.
- Update safety analysis, models and modeling results at any time as required.

The CSwE and SQA should review the procedures followed by the SwIPT and the products produced for critical requirements compliance as part of the normal reviews of each development activity. The CSwE should focus on identifying evidence that the general strategy stated above is being implemented. SQA should evaluate the process of performing the critical requirements testing, the successful completion of the testing, and the proper documentation of the results.

5.5 Software Subcontractor Management

The management of Software Subcontractor Teams is almost always a significant challenge.

On large contracts, there are usually numerous subcontractors contributing software products, and they are typically geographically dispersed (sometimes offshore). A *Lessons Learned Database* should be maintained on past experience with each teammate to tailor preparation of their Statement of Word (SOW) so as to avoid problems you, or your organization, encountered in previous software subcontracting management efforts. Subcontract management is discussed in four subsections:

■ *Subcontractor Management Team (SCMT)*: Why the SCMT must be established (5.5.1).
■ *Subcontractors Compliance with the SDP*: The Chief Software Engineer should be responsible for monitoring software subcontractor compliance with the program-level SDP including compliance of subcontractor site-specific SDPs with the program-level SDP (5.5.2).
■ *Subcontractor Performance Reviews*: Subcontractor progress performance should be regularly reviewed and assessed by the SCMT and the CSwE (5.5.3).
■ *Subcontractors Statement of Work*: The SOW must be focused on a clear, unambiguous definition of all Developer responsibilities including product deliveries, software metrics reporting, interfaces with the SCMT, and authorization for CSwE oversight (5.5.4).

5.5.1 Subcontractor Management Team

On large projects, a *SCMT* must be established. It should be led by your Subcontract Program Manager who has overall responsibility for monitoring technical, schedule and cost performance of the software subcontractors. On smaller projects, the SPM could take the dual role of Project Subcontractor Manager. Also, a *Software Subcontract Management Guidebook* should be prepared. The Software Quality Engineer (SQE) and the Chief Software Engineer must support subcontract management by technically monitoring the software portion of the subcontract. Table 5.8 is an example of typical SCMT membership and their responsibilities.

5.5.2 Subcontractor Compliance with the SDP

The CSwE should be responsible for technical monitoring of software subcontractors through attendance at the software subcontractor's formal reviews and status reviews, as applicable, and by regular metrics reviews. The CSwE must evaluate the performance of the software subcontractors and prepare subcontract evaluation reports as required. The CSwE can delegate these responsibilities.

Program-Level SDP. *The program-level SDP must apply to all software subcontractors.* The CSwE has the authority to enforce the processes described in the program-level SDP. Software subcontractors must follow the processes, procedures, and documentation defined in program-level SDP unless a formal waiver has been approved.

Site-specific SDPs. If needed, site-specific SDPs can be written and maintained by subcontractors at their sites; they provide additional standards and procedures specific to each site. *Site-specific SDPs expand upon, but must not conflict with, the processes and procedures defined in the program-level SDP unless a waiver has been approved.* The CSwE and SQE must perform software subcontractor product and process audits to determine compliance with program-level SDP and with the contract.

5.5.3 Subcontractor Performance Reviews

The progress and performance of the subcontractors should be regularly reviewed and assessed by the SCMT. These reviews address the total performance of each team toward meeting its objectives. The SEPG oversees the software development process and provides approval of all subcontractor specific appendices to the program-level SDP to ensure that software development methods, standards, practices and procedures are consistent with the contract.

5.5.4 Subcontractor Statement of Work

The subcontractor's Statement of Work must be focused on a clear definition of responsibilities, software metrics reporting requirements, interfaces with the SCMT for oversight, and identification of your CSwE as the single point of contact for software. Each team member performing software development, algorithm analysis, simulation development, or data set development should have its own SOW. The SOW should delineate development software products, scope, required reviews, schedule milestones, status reporting, performance evaluation criteria and acceptance criteria.

5.6 Software Verification and Validation

Software Verification and Validation (V&V) are two related but separate procedures. Because they are so related, there can be confusion as to how they differ.

■ Verification: Ensures that each developed software function works correctly. Verification techniques include Managerial Reviews (see 5.3), Peer Reviews (see 6.5) and code walk-throughs (see 11.4.4).

Table 5.8 Subcontractor Management Team Members and Responsibilities

Member	SCMT Responsibilities
Subcontract Program Manager	■ Management of software subcontract technical, cost, and schedule performance ■ Ensures all software technical, cost, and schedule requirements are satisfied ■ Facilitates subcontractor's ability to plan and perform software more efficiently ■ Works program-level issues and manages software award fee program ■ Ensures subcontractor compliance with program plans and procedures
Subcontract Administrator	■ Single point of contact for contractual matters and administers the software subcontract ■ Negotiates and awards software subcontracts and approves vouchers/invoices ■ Ensures proper flow down of software technical requirements and software subcontract terms and conditions ■ Maintains configuration control of contractual documentation sent to the subcontractor ■ Receives, logs, and distributes incoming correspondence from subcontractors
Responsible Engineer or CSwE	■ Develops software specifications and associated technical documentation ■ Ensures subcontractor understanding of the software technical requirements ■ Coordinates approval of subcontract software deliverables and documentation ■ Conducts a technical evaluation of subcontractor's software proposals ■ Develops and/or approves software test plans, procedures and acceptance plans ■ Participates in or witnesses subcontractor software acceptance testing ■ Provides independent evaluation of subcontractor's technical progress and performance
Business and Financial Operations	■ Analyzes cost/schedule performance data including key indices and variance analyses ■ Helps subcontractor develop cost account plans and incremental planning packages ■ Integrates subcontractor budgets, costs, IMP and IMS into the centralized database system
Mission Assurance (SQA)	■ Monitors data, configuration, and quality processes used by the subcontractor ■ Ensures the software subcontract quality implementation program is consistent with the program's Software Quality Program Plan ■ Ensures that the configurations of all deliverable Software Items are identified with a clear audit trail ■ Chairs the Subcontractor Functional and Physical Configuration Audits (FCA/PCA)

■ Validation: Ensures that the final product incorporates all of the system requirements. System testing also validates the inclusion of all system requirements. Validation techniques include White Box and Black Box testing:

- *White Box Testing*: An *internal* perspective of the code functionality often assisted by automated testing tools.
- *Black Box Testing*: An *external* perspective with little concern or knowledge of the internal structure of the software. Inputs are provided, and the outputs are compared to the expected outputs.

Figure 5.6 is a simple way to understand the difference between them. Verification is shown as making sure that each step in the software development process is responsive to its previous step. Validation is the procedure to make sure the final product is responsive to the original system requirements. Incremental validations could take place during the activities preceding System Test.

A common phrase used to explain the difference between validation and verification is: validation determines if you built the *right product*; verification determines if you *built it right*. The confusion between them is compounded depending on whether you are talking about V&V of the entire system or V&V of the software work products. In the latter case, for example, even though a software document may be complete, and it followed the standard and format for that document (you built it right), the contents of the document may not address or satisfy the customer's needs (you built the wrong product).

5.6.1 Software Verification and Validation with Associate Developers

On large contracts, the lead prime contractor organization may enter into what may be called "Associate Contractor Agreements" with other prime contractors whose performance will impact your program and project. Such arrangements facilitate *joint participation and collaboration* in meeting program requirements. The objective of these agreements is to create *ground rules* and an environment for *safeguarding* each other's technical and/or proprietary information and

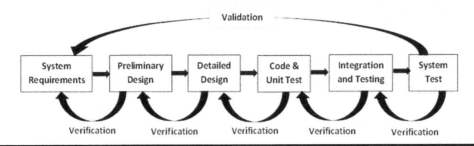

Figure 5.6 Software verification versus software validation.

resolving issues, to the maximum extent possible, without customer involvement or intervention.

As the need for these agreements become known, the lead prime contractor must advise the other prime contractors, in addition to other organizations that may be defined in the contract, where such agreements are necessary and then proceed with establishing those relationships. From a software perspective, Associate Developers are one type of software stakeholders.

5.6.2 Independent Software Verification and Validation Coordination

If required on the contract, Independent Verification and Validation (IV&V) contractors will interface with subsystem software development as a member of their respective Integrated Product Teams (SwIPTs). In addition to supporting software-related SwIPT tasks, the IV&V contractors may also perform audits of Software Development Files and software processes.

If audits are performed, the IV&V contractor must coordinate in advance with the SwIPT Lead and identify required Software Developer and Test Support. Additional details for interfacing with external software IV&V subcontractors may be defined in the SDP Annexes.

SwIPT personnel, including Software Developers and Test Engineers, should interface with the IV&V representatives to allow identification and resolution of software issues

and problems at the lowest practical level and as early in the development process as possible. The IV&V representatives must interface with the SwIPTs and Developers, both formally and informally, through the following mechanisms:

- At scheduled software development status meetings, reviews, TIMs and SwIPT meetings
- Through the test process, either when the IV&V representative is a test witness or when the IV&V representative conducts independent testing
- Through periodic inspections and audits of Software Development Files, other software products, as well as the software development process

Table 5.9 is an example of the type of software products that must be provided to the IV&V agents and how problems are reported. Problems identified by the IV&V agents must be resolved through the Corrective Action Process (see Section 6.4).

As members of appropriate SwIPTs, the IV&V representatives should offer advice, assistance, and subject matter expertise in the development of program documentation, but must not co-author such documentation. This allows the IV&V representatives to preserve the degree of objectivity required to effectively discharge the IV&V role of verifying and validating both adherence to the software development process, and the adequacy, sufficiency, and performance of software products.

Table 5.9 Software IV&V Evaluations

Tasks	Software Products Reviewed	Problem Reporting Mechanism
Technical Interchange Meetings	Development phase documents	Action items
Formal software reviews	Development phase documents	Action items
Audits	SDFs, SDP	Audit report
Independent testing	Software test documents	SCR/SDRs

SDF = Software Development File (or Folder); SDP = Software Development Plan; SCR = Software Change Request; SDR = Software Discrepancy Report.

5.7 System Retirement or Disposal

Project Managers need to address the strategy and requirements for system retirement or disposal and ensure that sufficient information exists so that it can be carried out in a way that is in accordance with all legal and regulatory requirements relating to environmental, safety, security and occupational health issues. It is also an excellent idea to be proactive in planning for the *disposition* of government property that was used during the development of your project.

System disposal is one of those contract tasks that will often get little or no attention until contract completion, but it should be prepared for in advance to help the process run smoothly. For the retirement of government contracts, work with your local *Defense Contract Management Agency* (DCMA) for the procedure and options to return, purchase, reuse, sell, or scrap the government-owned property.

Planning for the disposal of the system while you are working so hard to develop it may seem to be counterproductive. However, system disposal can be a big headache if not properly planned for in advance. See Section 16.6 for more details on system retirement and disposal.

5.8 What is a Successful Project?

Demonstrating your product complies with the test cases, and successfully completing testing at various levels, is an obvious measure of success. Customer acceptance of your software product or system is a clear successful achievement. However, the definition of a successful product can be elusive.

The customer could be elated if the delivered system performed even better than expected even though the project was over budget and delivered late. The customer could be very unhappy if the delivered system did not provide the level of benefits planned for and expected even though you delivered the required system on time and under budget. Also, if there was disharmony between your Developers and the customer during development, the customer might be very unhappy with your team regardless of the quality, timeliness and functionality of the delivered system.

However, the definition of "done" should not be elusive; if it is, you are responsible for not making sure that every member of your staff precisely understands what is being built and what is meant by "*done*." You must be precise as to what reviews, testing, documentation, performance, scripting, etc. will be required for the code. The definition of *done* may change during development. The project kickoff is a good time to get an agreement on *when* a task, feature, Software Unit, or Sprint is finished.

If you don't define "done" up front, development costs and schedule could spiral out of control.

Chapter 6

Software Quality Domain

The quality of a software product stems, in large part, from the quality of the process used to create it.

—Watts Humphrey

The *Software Quality Domain* is basically focused on *Software Quality Control* (SQC) which is sometimes called Software Quality Management (SQM). There is no Section in this chapter titled SQC or SQM because *all six of the sections in this chapter collectively provide a synergistic ability to control software quality*. Although these activities are discussed individually, they heavily interact and support each other. The Software Quality Domain is described in the following six sections of Chapter 6:

- *Software Process Improvement (SPI)*: Describes the Software Engineering Process Group (SEPG) and procedures for software process improvements (6.1)
- *Software Quality Assurance*: A key to the delivery of successful software-intensive systems (6.2)
- *Software Configuration Management and Control*: An essential management and product control activity (6.3)
- *Software Corrective Action Process*: The change management mechanism (6.4)
- *Software Peer Reviews and Product Evaluations*: A structured, methodical examination of software work product quality during development (6.5)
- *Capability Maturity Model Integrated*: An effective framework for building Software Process Improvement systems (6.6)

Continuous Quality Improvement. A consistent improvement of quality is achieved through a pervasive culture, an attitude, and a philosophy for analyzing capabilities and processes and improving them repeatedly to achieve the major objective of improving the quality of your software. In order to achieve customer satisfaction, you must meet or exceed the functional requirements of the desired system as well as their reasonable expectations for quality, schedule and cost. Continuous quality improvement is focused on work processes, that is, "how" things get done. The basic philosophy is (or should be):

Empowering everyone on your team, to act on the basis of understanding and following the processes in which they are involved, resulting in better software products and services.

Achieving that goal depends on the systematic analysis of what is being done, for whom, and why, and the systematic application of process improvement methodologies. But the process begins with the involvement of management at all levels. As a Manager of a software-intensive system, you must use your own expertise and judgment to coach your team members in a way that brings out their individual talents and knowledge and encourage them to inform you about potential problems before they occur so that you can take timely Corrective Action.

In that context, you must remove intimidation as a management style in order to avoid any hesitation by your team members to question policies and procedures and to bring up new ideas (no matter how bizarre) in the search for better performance. Indeed, eliminating "fear" in an organization is an enormously powerful means of releasing energy to bring about real operational improvements.

Process and Quality. *All work is part of some type of process.* Simply stated, "a process is the transformation of input into output through work that adds value" (ISO 9000:2005; Lockheed, 1994). The relationship between quality and cost is "added value," and *Continuous Quality Improvement* is a proven approach to increase value to your organization and to your customers.

A process can be very broad or very narrow in scope. In past decades, quality management was focused on a purely product-oriented, inspection-based approach. The contemporary approach to continuous improvement now deals with *building quality into every process and product.* It deals not only with the specific software product being made but, equally important, *the process* used to produce the product. Since the process is the basic foundation of quality and productivity, real improvements in quality and productivity should be made at the *process level*.

Effective process management involves the constant monitoring and control of the process for continuous improvement in the quality of your software end products. It also involves the application of process-control methods to facilitate predictable high-quality product outcomes. Visible signs of an *effectively managed process* would include:

■ The processes are documented, routinely evaluated and improved.
■ Products consistently meet customer requirements.
■ Rework is low to none.
■ Process performance and capabilities are analyzed and documented.
■ Statistical methods for process control are routinely applied.

6.1 Software Process Improvement

The *Software Process Improvement* activity covers the tasks for identifying Software Process Improvement areas and developing new process policies and procedures to implement those improvements. SPI is an important function and Software Project Managers are responsible for assuring SPI performance. Software Process Improvement is described in the following Subsections:

■ *Software Engineering Process Group*: The SEPG is the heart of the SPI effort (6.1.1).
■ *Software Process Audits and Assessments*: To verify compliance with the processes (6.1.2).
■ *Infrastructure of the SEPG*: The operational organization focused on SPI (6.1.3).
■ *Software Process Lead:* A trained change agent and SPI facilitator (6.1.4).
■ *Sharing Lessons Learned*: To avoid repeating software problems in future programs (6.1.5).

The *Capability Maturity Model Integrated* (CMMI), an effective and widely used Software Process Improvement Model, is described in Section 6.6.

Software Process Improvement Model. Figure 6.1 is my concept of a simple Software Process Improvement Model to which I have given the acronym *DAPPER* (Define; Approve; Prioritize; Plan; Execute; Results). The DAPPER Model is similar in type, but more detailed, than the model called IDEAL^sm (Initiating; Diagnosing; Establishing;

Acting; Learning), used in the past by the Software Engineering Institute (SEI), or DMAIC (Define; Measure; Analyze; Improve; Control) used by Six Sigma for improving an existing business process. You can use both of them along with DAPPER as a basis to develop your own version of a Software Process Improvement Model specific to your needs.

6.1.1 Software Engineering Process Group

A Software Engineering Process Group, or an equivalent organization, must be established as it is the heart of your SPI effort. The SEPG is focused on: evaluating the implementation and progress of the defined software development process; identifying areas for potential process improvement; and performing evaluations to determine if the observed deficiencies are the result of non-compliance with, or inadequacy of, your current policies and procedures.

If the deficiency is *non-compliant* with a process step, the SEPG must determine the nature of the non-compliance and either identify an improvement or recommend to the Chief Software Engineer (CSwE) how to correct the compliance deficiency. If the deficiency is an *inadequacy*, the SEPG must analyze the affected process area and develop proposed process improvements with corresponding changes to the Software Development Plan (SDP). Table 6.1 contains the typical SEPG membership and their responsibilities.

SEPG Focus. The focus of the SEPG is to: assess the current process status; define, document, maintain, monitor and improve the program software process; establish software training requirements; establish a program-specific archive (usually a physical library often called the SPAR—Software Process Assets Repository); and work to improve the program's software maturity rating (see Subsection 6.6.5). An example of how to focus Software Process Improvement initiatives, in terms of goals, approach and the measure of success, is shown in Table 6.2.

SEPG Mechanisms. The SDP should define organizational and procedural mechanisms that support the SEPG in its role of identifying potential process improvement areas. For example, these mechanisms can include:

■ Assigning a program-level Software Process Lead, who is responsible for facilitating software process tasks for the program (see 6.1.4)
■ Conducting Software Quality Assurance (SQA) and Software Configuration Management (SCM) tasks (see Sections 6.2 and 6.3)
■ Conducting process and product audits by the CSwE and SQA
■ Analyzing software Management Indicators (measurements—see Chapter 15)
■ Performing the problem reporting and the Corrective Action Process (see Section 6.4).

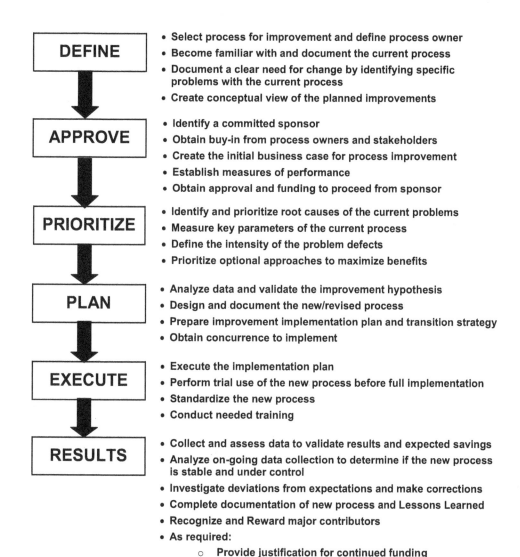

DEFINE
- Select process for improvement and define process owner
- Become familiar with and document the current process
- Document a clear need for change by identifying specific problems with the current process
- Create conceptual view of the planned improvements

APPROVE
- Identify a committed sponsor
- Obtain buy-in from process owners and stakeholders
- Create the initial business case for process improvement
- Establish measures of performance
- Obtain approval and funding to proceed from sponsor

PRIORITIZE
- Identify and prioritize root causes of the current problems
- Measure key parameters of the current process
- Define the intensity of the problem defects
- Prioritize optional approaches to maximize benefits

PLAN
- Analyze data and validate the improvement hypothesis
- Design and document the new/revised process
- Prepare improvement implementation plan and transition strategy
- Obtain concurrence to implement

EXECUTE
- Execute the implementation plan
- Perform trial use of the new process before full implementation
- Standardize the new process
- Conduct needed training

RESULTS
- Collect and assess data to validate results and expected savings
- Analyze on-going data collection to determine if the new process is stable and under control
- Investigate deviations from expectations and make corrections
- Complete documentation of new process and Lessons Learned
- Recognize and Reward major contributors
- As required:
  - Provide justification for continued funding
  - Prepare long range improvement plan
  - Continuously improve the process

Figure 6.1   A Software Process Improvement Model called DAPPER.

■ Leading the joint technical and management reviews (see Section 5.3)

The SEPG is also responsible for producing the program- and project-level SDPs. Site-specific SDPs can be produced by your subcontractors to address local issues as long as they do not conflict with the program-level SDP. If there is a conflict with the program-level SDP, it must be approved through the waiver approval process (see Subsection 5.1.1).

6.1.2 Software Process Audits and Assessments

The initial objective of an audit is to verify compliance with the documented procedures. Where non-compliance is found, the cause must be identified and evaluated for potential process problems (e.g., lack of training, inadequate documentation, difficulty in implementing the procedure, etc.).

Audits. The Chief Software Engineer, the Software Process Lead, and SQA should witness project and product audits to verify process compliance as well as to evaluate product quality. They should audit Subsystem SwIPT tasks for compliance with software development policies and procedures as defined in the project's SDP. These audits typically consist of reviewing Software Development Files (SDFs), and other software products, and interviewing the Software Developers performing the particular process being audited. Audit results should be presented to the SEPG.

Measurements. The SEPG must also monitor the monthly software measurement (metrics) reports to identify problem trends. Where such trends are found, an analysis should be performed to determine if the trend is caused by a process deficiency. An examination of software measurement for the other subsystems should be performed to determine if similar results are found. Based on these examinations, recommendations for process improvement should be developed and presented to the CSwE for further action. Software measurements are covered in Chapter 15.

Process Improvement Proposals. Proposals for process improvements may also originate from joint technical and Management Reviews or any Subsystem SwIPT.

Table 6.1 SEPG Membership and Responsibilities

Chief Software Engineer (CSwE). The CSwE is the SEPG Chair, ensures SEPG tasks are assigned and performed in a timely manner, allocates resources for the SEPG (including people, time and equipment), and encourages and actively supports the SEPG tasks. This function may be performed by the SPM
Software Process Lead (SPL). The SPL, typically the SEPG Administrative Lead, ensures the SEPG meetings are well organized (agenda, facilities, etc.), meeting results are recorded (meeting minutes), tracked (action items, working group status), participates as the expert on software processes and Software Process Improvement methods, and is the project point of contact for Software Process Improvement issues and SEPG tasks
Software Configuration Management (SCM). The SCM Lead participates as the expert on SCM processes and procedures and is the project point of contact for SCM process improvement issues
Software Quality Assurance (SQA). The SQA Lead is the point of contact for SQA issues and audits SEPG tasks to assure conformance with contractual requirements, plans, standards and procedures
SEPG Members. Each organization responsible for a software product assigns a member and alternate to represent the organization as their SEPG Representative who is *their* organization's point of contact for Software Process Improvement and SEPG tasks

In either case, these proposed process improvements should be submitted as Software Change Requests (SCRs) against the applicable work product and forwarded to the Software Configuration Control Board (Sw/CCB). The Sw/CCB can then assign it to the SEPG, or a development site, for evaluation and implementation.

Change Implementation. Recommendations for improvement to the development process must be presented to the SEPG, for review and evaluation of the cost and schedule impacts, and then to the CSwE for approval. Approved changes at the program level should be implemented by the CSwE. Approved changes concerning a subsystem's SDP Annex (produced by a subcontractor) should be implemented by that development site.

Whenever a process change is approved for implementation, the SEPG should recommend the criteria by which they plan to measure the success or failure of the change. Process change implementations should include adequate notification to all affected development and management staff, within a reasonable period of time for resolution of comments, concerns and questions.

Once the process improvement is implemented, the SEPG should monitor the effect of the change to determine its impact on the software development process. This monitoring not only determines if the desired effect is achieved, but also whether any positive or negative side effects are generated. You should always keep in mind that there are often *unintended consequences to every decision.*

6.1.3 Infrastructure of the Software Engineering Process Group

The SDP must describe the infrastructure within which the process improvement initiative operates. The Project Manager, or designee, must support the SEPGs at the subsystem and program levels. The Program SEPG provides direction to the subsystem SEPGs and may receive services from the corporate SEPG. The Subsystem SwIPT Lead is normally a member of the Program SEPG. These relationships, and

Table 6.2 Focus of the Software Process Improvement Initiative

Area	Subsystem SwIPT	Program or Project SEPG
Goals	Produce and maintain software products that satisfy the system requirements	Perform continual process improvement through quantitative feedback from the process and from innovative process improvement tools
Approach	Use the most effective Software Engineering tools and techniques including metrics collection	Perform measurement analysis, assess processes, maintain a library of experiences, create and update software standards
Measures of Success	Delivery of quality software products on time, within budget and fully responsive to the requirements	Improve processes to result in improved products, reuse growth, and efficient collection, storage and retrieval of Lessons Learned

the general functions provided, are shown in the example Figure 6.2.

Software Process Training. An important cornerstone of an aggressive process improvement program is training. The SEPG (at the subsystem, program and corporate levels) must support training activities in the techniques of process improvement as well as training on the software development processes. This includes support in constructing training courses and materials for Software Engineers, Leads and Managers in implementing and improving the software development process.

Training also includes orientation sessions and procedural training in the techniques and skills of the software development process for the Software Developers. Training plans should be developed, maintained and monitored by training coordinators and/or software process leads in coordination with SwIPT Software Leads. Software Training Plans are discussed in Section 7.8 and budgeting for training is covered in Subsection 8.3.5.

6.1.4 Software Process Lead

The Software Process Lead (SPL), sometimes called the Software Process Engineer, is a trained change agent who is responsible for facilitating all software process tasks for the program. The SPL is critical to an effective Software Process Improvement program. The SPL may report to the CSwE (the SEPG Chair), or the SwIPT Lead. The SPL is usually the Administrative Lead of the SEPG. Typical functions performed by the SPL are listed in Table 6.3.

NOTE: Do not confuse the Software Process Lead with a "Process Engineer" in the semiconductor industry responsible for production process operations such as laminating, etching, polishing, plating, etc.

6.1.5 Sharing the Lessons Learned Database

In order to avoid repeating software problems from past programs, each Program Manager should contribute to a software Lessons Learned repository that is (hopefully) maintained at the organization level. During the development activities Program Managers should:

■ Identify processes that do work well, processes that do not work well, processes that add no value, areas where new processes need to be developed, and processes that could not be addressed due to resource constraints and possible opportunities for workarounds.

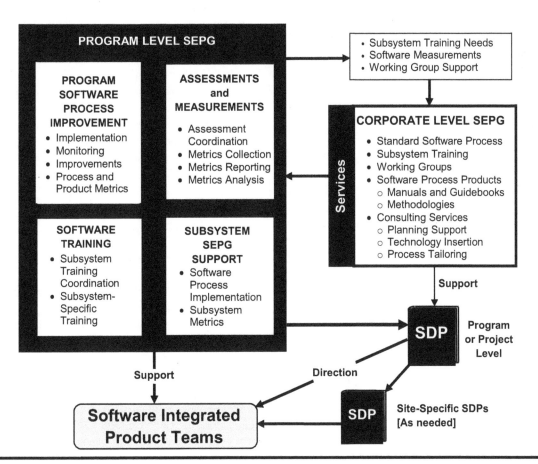

Figure 6.2   Software Engineering Process Group infrastructure.

Table 6.3 Typical Software Process Lead Functions

SPL Function	Function Description
SEPG Administrative Chair	Prepares for SEPG meetings (reserve room, agenda, etc.). Prepares and distributes SEPG meeting minutes. Ensures the SEPG follows established applicable program software processes
Working Group Oversight	Ensures that working groups are effective; ensures that each working group understands roles, responsibilities and specific task assigned to the working group; and monitors progress of working groups, etc
Software Process Improvement Monitoring and Appraisals	Ensures that the appropriate measurements on process tasks are collected, analyzed, and distributed. Focal point for software process appraisals
Software Process Improvement Recommendation Coordination	Reviews, supports and presents process improvement recommendations from program personnel, internal and external groups and customer, industry data on current best practices, company and industry standards to the SEPG
Software Process Improvement Reporting	Reports status of SEPG and process improvement activities to the Software Team, program management, company management and SEPG as applicable

■ Record the reasons for any changes to required performance, increases in required resources, schedule extensions, changes in required or actual manpower, or any other factors or events that affect the outcome of the program (especially cost, schedule, and performance).

The collection and sharing of Lessons Learned between programs and projects can, in the long run, provide substantial benefits to the organization.

6.2 Software Quality Assurance

Software Quality Assurance is sometimes called "Software Quality Management" or "Software Product Assurance." SQA performs the planned and systematic actions necessary to help assure that software, and software-related products, satisfy system requirements and support delivery of a successful product. SQA is a support organization; it is separate from the software group and its members do not report to the Software Project Managers (SPM). The SQA organization has a responsibility to provide project management with visibility into the software development process and products by performing independent audits and assessments. It must be understood that:

> *The SQA group is not responsible for Software Quality; the Software Development Team is responsible for building quality into the product and the Software Project Manager is accountable for delivering a quality software product.*

The SQA assessments provide assurance that products and processes conform to contractual requirements and established plans, standards, and procedures. SQA is a member of the Software Engineering Process Group, as discussed in Subsection 6.1.1, and they participate in performing the process improvement tasks. Software Quality Assurance activities are described in the following Subsections:

■ *Software Quality Engineer and Program Plan*: Includes principal roles of Software Quality Engineers and the Software Quality Program Plan (6.2.1).
■ *Software Process Audits and Product Reviews*: Covers software quality evaluations (6.2.2).
■ *Software Quality Factors*: Definition of key software quality factors (6.2.3)
■ *SQA Records:* Provides objective evidence of performance (6.2.4).
■ *SQA Independence*: SQA must report directly to its SQA parent organization (6.2.5).
■ *SQA Non-compliance Issues*: Candidates for Corrective Action (6.2.6).

6.2.1 Software Quality Engineer and Program Plan

Software Quality Engineer (SQE). Depending on the size of your project, at least one SQE should be assigned to your project by the SQA organization. On small projects, the SQE would not likely devote full-time to your project. The SQE does not directly report to the Program Manager, the Software Project Manager, or the Chief Software Engineer, but works very closely with them on a dotted line relationship as shown in the organization charts in Figures 1.5 and 1.6.

SQEs work directly with their assigned Development Teams to resolve identified problems at the lowest level before elevating them for resolution at a higher management level. SQEs also identify SQA tasks and responsibilities to be implemented on the program. A detailed SQA audit schedule should be provided in the Software Quality Program Plan

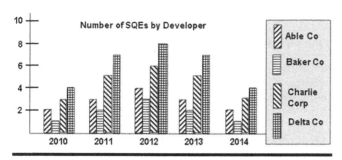

Figure 6.3 SQE staffing projection.

(SQPP) (see next paragraph) in terms of the SQE staffing projections over the life of the program at each Software Developer site as illustrated in example Figure 6.3.

Software Quality Program Plan. Identification of the evaluations to be performed and the criteria to be used for those evaluations should be defined and described in a SQPP or similar document. This plan is normally prepared by SQA to direct the SQE, and the SQE Team if there is more than one SQE, in performing the evaluations. SQE tools, techniques, and methodologies to be employed by each software Development Team member should also be defined in the SQPP and can be augmented by the subsystems in their SDP Annexes if necessary.

The SQPP is maintained and updated as needed and is usually an addendum to the Software Development Plan. In addition to a complete description of the evaluations to be performed and the criteria to be used for those evaluations in the SQPP, SQA planning should also be addressed from an overview perspective in the SDP.

6.2.2 Software Process Audits and Process Reviews

Both the subsystem level and the program-level SQA organizations perform two major types of evaluations, process audits and product reviews:

- Process audits are conducted by SQA to assure effective implementation of the software development process as defined in the SDP.
- Product reviews are performed by program and subsystem level SQAs as a participant in formal reviews. SQA product reviews verify that the software products conform to system requirements. Software product evaluations are discussed in more depth in Subsection 6.5.6. The SQA organization should conduct ongoing evaluations, in accordance with the contract and the SDP, to ensure:
  - Adherence of the software development processes, software work products, and software services to their applicable process descriptions, standards and procedures.
  - That each required software product does exist, and that it has undergone Software Peer Reviews and product evaluations, testing (when applicable), and Corrective Actions (for identified problems).

6.2.3 Software Quality Factors

The subject of software quality is much broader than the tendency to focus on software discrepancies and failure statistics. Evaluating the quality of developmental software products requires a "big picture" perspective. There are a number of quality models that present ways to tie together the different quality attributes. Table 6.4 is my succinct list of key software quality factors plus a brief definition of each quality factor and its related attributes.

6.2.4 Software Quality Assurance Records

The SQA group must maintain records for all evaluations performed in order to provide objective *evidence* that the evaluations were conducted. The records should consist of observations or formal findings along with their resultant Corrective Actions, disposition, metrics and closure. These records and reports must be retained in a repository for the duration of the contract and made available to the customer and management as required by the contract.

Technical and management performance data should be collected and organized into a process database. These data should be analyzed by either SQA or the SEPG, or both, to determine performance trends, and areas for potential process improvement. An essential element of technical performance data is the collection and analysis of measurements (metrics) as described in Chapter 15.

6.2.5 Software Quality Assurance Independence

Software Quality Engineers support software development as an active member of the subsystem they are supporting. However, SQEs must maintain a direct reporting line to their SQA organization and not be in a direct reporting line to the program they are supporting. The SDP must make it clear that the SQE responsible for conducting the Software Quality Assurance evaluation must not be the person who developed, or is responsible for, the software work product. The SQE responsible for assuring compliance with the contract must have the resources, authority and organizational freedom to permit objective SQA evaluations and to initiate and verify the Corrective Actions needed.

Independence in SQA is obtained by having a separate reporting chain to Product Assurance management. If SQA findings cannot be resolved at the lowest level possible, it must be evaluated by the SQE to the next higher level of

Table 6.4 Key Software Quality Factors

Quality Factors	Definitions of the Software Quality Factors
Correctness	A key attribute indicating if the software conforms to the user requirements and system performance. Related attributes include: Accuracy, Completeness; Interoperability, Suitability, Functionality, Verifiability; Security; Traceability and Consistency.
Efficiency	How well the software utilizes resources in terms of its performance versus the amount of planned resources to be used. Related attributes include: Accessibility, Accountability, and efficiency of interfaces with peripheral devices.
Interoperability	How well the software interfaces with other required operational systems. Related attributes include: Modularity, Simplicity, and Data and Communications Commonality.
Maintainability	How easy is it to modify and upgrade the capabilities of the software. Related attributes include: Expandability, Flexibility, Testability, Stability, Consistency, Simplicity, Conciseness, Modularity, and Self-Descriptiveness.
Portability	Ability of the software to be transported from one environment to another. Related attributes include: Adaptability, Installability, Replaceability, Reusability, Conformance, Modularity, and System Independence.
Reliability	Capability of the software to maintain its performance level under stated conditions for the required length of time. Related attributes include: Integrity, Functionality, Maturity, Accuracy, Completeness, Consistency, Simplicity, and Error Tolerance.
Survivability	How well the software performs under all environmental conditions. Related attributes include: Reliability, Portability, Interoperability and Maintainability.
Testability	Ease and ability of the software to effectively test its system requirements. Related attributes include: Accessibility, Communicativeness, Simplicity of the Structure, Self-Descriptiveness, Instrumentation, Portability, Modularity and Stress Testability.
Usability	From a human engineering perspective, how easy the software is to use and understand. Related attributes include: Learnability, Operability, Robustness, Understandability, Accessibility, Training, Communicativeness, and Input/Output Efficiency.

management. Figure 6.4 is an example of a program's independent reporting structure for SQA and the typical problem resolution interfaces as the resolution are elevated to a higher management level. The objective is always to resolve the problem at the lowest possible level.

6.2.6 Software Quality Assurance Non-Compliance Issues

All non-compliance issues, identified through audits, reviews, normal SQA monitoring, ad-hoc findings, etc., are candidates for Corrective Action or Preventive Action as described in Section 6.4. Correction of non-compliance issues is typically handled with an automated tool, an audit database, and an established escalation mechanism to ensure that the appropriate level of management can resolve the issues. In using tools, selected by the program, non-compliance issues should be documented, tracked to resolution, and resolved within a given time frame. The Software Quality Assurance organization also provides metrics data to support management decision-making as detailed in either the Quantitative

Management Plan, or the Software Measurement Plan, or both, as discussed in Chapter 15.

6.3 Software Configuration Management and Control

Software Configuration Management is an essential development control activity that begins during requirements definition. Formal software control starts with the establishment of the *Allocated Baseline*, which identifies the Software Items (SIs) that must be formally managed in coordination with the Configuration Control Boards described in Subsection 6.4.3. A *baseline* is the initial standard or measure against which future status, progress, and changes are compared and measured.

> *SCM is responsible for all tasks necessary to control baselined software products and to maintain the current status of the baselined products throughout the Development Life Cycle.*

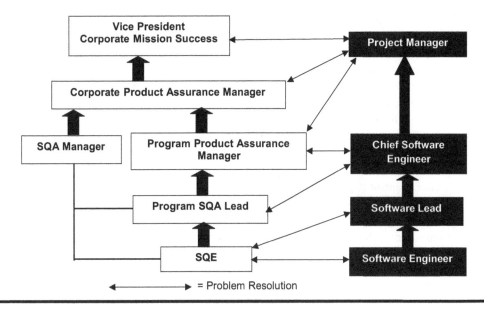

Figure 6.4 SQA independent reporting and resolution structure.

The basic Software Configuration Management responsibilities are to:

- Establish formal baselines of identified products and maintain the integrity of baselines.
  - Identify software configuration items (CI), components, and related products that will be placed under Configuration Management and when that takes place.
  - Create and release baselines for internal use and for delivery to the customer.
- Track and control Change Requests for the configuration items and control the changes.
  - Establish and maintain a Configuration Management and change management system for controlling software product changes.
  - Establish and maintain records describing configuration items and perform configuration audits to maintain the integrity of the configuration baselines.

The SCM activities may be performed by both the Development Team as well as the customer's technical representatives. A general description of the division of SCM responsibilities is shown in Table 6.5. On government contracts, the customer's technical representatives may include Systems Engineering Technical Advisors (SETAs) or Federally Funded Research and Development Centers (FFRDC). On non-government contracts, the customer technical representatives may include customer employees, customer consultants, end-users of the system, or possibly some stakeholders.

Table 6.5 Division of SCM Responsibilities

Development Team	Customer Technical Representative
Defines and documents Software Configuration Management processes	Reviews software processes for quality and ensures process compliance
Implements Software Configuration Management tools and environments	Reviews Software Configuration Management tools and environments for quality and process compliance
Conducts SCM boards	Participates in Software Configuration Management boards
Ensures the integrity of all software configuration items	Audits delivered software products for baseline integrity
	Incentivizes the Developer to control software baselines

Details of the SCM activity must be covered either in the SDP, in an addendum to the SDP, or in a *Software Configuration Management Plan* (SCMP). A good option is to include an SCM overview in the SDP with the details documented in the SCMP. The SCMP is described in Subsection 6.3.4. The importance of an effective Configuration Management process cannot be overemphasized—especially on large projects. If program or project software is being developed at multiple worldwide sites, the importance of good Configuration Management is even more critical to success.

Software Configuration Management activities are described in the following Subsections:

- Configuration Management Control Levels: The three-tiered SCM scheme (6.3.1)
- Formal Baselines: Tracking versions of controlled software source code (6.3.2)
- Software Configuration Identification: Identifying software products under control (6.3.3)
- Software Configuration Control: Controlling modifications to baselined products (6.3.4)
- Software Configuration Status Accounting: Configuration status records (6.3.5)
- Software Configuration Audits: Configuration audits to verify changes (6.3.6)
- Packaging, Storage, Handling and Delivery: Preparation for system delivery (6.3.7)

6.3.1 Software Configuration Management Control Levels

Typically, SCM has a *three-tiered* Configuration Management control scheme as shown in Figure 6.5. It consists of a program-level, subsystem level, and development site control of software libraries.

SCM tasks should be performed at each software development site utilizing the site's Software Development Library (SDL) for configuration control of the developed products. On large programs, the SDLs are typically subdivided into two levels—at the Developer level and at the element or subsystem level, as shown in Figure 6.5. The software products *move up the levels* to until they reach the program's Master Software Development Library (MSDL). The development and subsystem sites may be co-located.

SCM establishes and maintains the integrity of specified software work products, controls and promotes stable baselines, maintains status accounting of the baselines throughout the life of a project, and controls the build process through product delivery. Responsibility for SCM should reside with the Software Group Lead. The principal SCM performers within subsystem and development sites are the SCM Lead and the SCM Librarian.

Library Levels. The specific organization of the SDL and MSDL must be tailored to each program; however, Table 6.6 is an example of the library levels, names, and who controls them.

Specific guidance regarding structure and control of the software libraries should be provided in the SDP and/or the SCMP. A discussion of Software Development Libraries, source code version control, and electronic SDL file partitioning is addressed in Section 9.3.

6.3.2 Formal Baselines

Baselines are initial standards or product versions against which future status, progress, and changes are compared and measured. The budget and schedule can serve as baselines. Software baselines change and become specific versions of controlled requirements, source code, build files, user documentation or data for an increment, build, or release. A brief comparison of the differences between the various software baselines is shown in Table 6.7. A requirements baseline involves an approved Software Requirements Specification (SRS) under SCM control.

Software Version Control. Version control, also called revision or source control, involves the *management of changes to source code, documentation* and *other collections of information*. It is an important component of Software Configuration Management because it is common for multiple versions of the same software, or documentation, to be deployed in different sites and for Software Developers to be

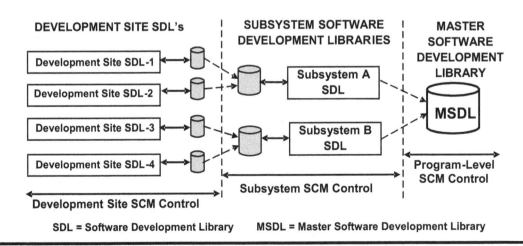

Figure 6.5   Relationship of the SDLs to the MSDL.

Table 6.6   Software Development Library Levels and Controls—Example

Library Level	Library Name	Controlled By	SCCB	SCM
1	Software Work Area	Software Developer		
2	SCM Control, Integration and Test	Site SCM	Controls Promotion between levels	Controls Access to each Level
3	Software Build and Qualification Testing	SCM at Site or Subsystem		
4	Subsystem Qualification	Subsystem SCM		
5	System Verification and Validation	System SCM		

SCM = Software Configuration Management; SCCB = Software Configuration Control Board.

Table 6.7   Comparison of Formal Software Baselines

Baseline Name	Baseline Contains	Where Baseline Exists
Functional	System requirements	Documents and databases
Allocated	Architectural Design	Set of documents (SRS, IRS, etc.)
Design	Detailed Design	Set of documents (SAD, SDD, etc.)
Development	Code under development	Source code
Product	Tested and approved system modules	Documentation and executable programs
Operational	Full system in use	Customer site(s)

working simultaneously on multiple version updates. Errors or features of the software may only be present in certain versions, so it is important to be able to retrieve and run different versions of the software to determine in which version(s) the problems occur.

Software Developers could retain multiple copies of the different versions of a program, and label them appropriately, but this simple approach is risky as many near-identical copies of the program have to be maintained and it often leads to mistakes. Granting read–write–execute permission to a group of Developers is also necessary; this adds the pressure of someone managing permissions so that the code base is not compromised. Consequently, systems to automate some or all of the revision control process are available to ensure that the majority of effort in managing version control is efficient and hidden.

6.3.3 Software Configuration Identification

Configuration Identification consists of identifying the software products to be placed under configuration control. These software products are the development products identified in the SDP, the hardware and software in the Software Development Environment, plus other documents and data as required by the contract.

Program Unique Identifiers. The output of Configuration Identification is a configuration controlled list of configuration items. Each software requirement, or product, must be *uniquely identified* with a *Program Unique Identifier* (PUI) numbering system. Massive confusion will engulf a large software-intensive program if it lacks a carefully controlled identification system. This is not a trivial issue.

The Chief Software Engineer, or designee, should be responsible for ensuring that a common, and unique, software PUI scheme is used across the entire program or project. The identification scheme must be at the level at which the software entities will be controlled, for example, computer files, electronic media, documents, SIs, SUs and hardware elements.

Computer Assisted Software Engineering (CASE) Tools. CASE tools, used to support Configuration Identification, should be identified as well as how their features (e.g., versioning, branching, and labeling) are used to track and control promotion and delivery of deliverable items. CASE tools are discussed in Chapter 9.

6.3.4 Software Configuration Control

Software configuration control is the systematic control of modifications to baselined products throughout the product's life cycle. The SDP or the SCMP (see below) must describe the SCM process for controlling baselined products and establishing common SCM change procedures. Subsystem SCM procedures may be provided in the Subsystem SDP Annexes. The policies and process for approving and implementing changes to baselined software products must be defined in the SDP or SCMP. If needed, more detailed *operational procedures* can augment the direction provided in the SCMP.

Software Configuration Management Plan. The SCMP documents the policies and procedures for conducting required Software Configuration Management for all SIs. The SCMP establishes the plan for creating and maintaining a uniform system of configuration identification, control, status accounting, and audit for the software work products throughout the software development process. The SCMP can be organized into five sections:

- *Section 1 Introduction*: Presents and defines the scope and purpose of the SCMP.
- *Section 2 Applicable Documents*: Lists the compliance document(s) and other documents that are referenced, or related to, the SCMP.
- *Section 3 Organization and Resources*: Describes the overall structure of the Software CM Organization, personnel and resources to be employed.
- *Section 4 Software Configuration Management Activities*: Covers details of the major Software CM functions and activities (a major portion of the SCMP).
- *Section 5 Glossary:* Lists the abbreviations and acronyms.

Objectives of the SCMP. The objective of the SCMP is to define the process to be used by SCM personnel in managing the configuration control of the software work products. Specifically, the SCMP should provide the following guidelines, direction and/or procedures in order to:

- Identify the software development items to be baselined.
- Provide change control and visibility of the changes made to software work products through the configuration control procedures.
- Control incorporation of all approved software changes, and related documentation to the Master Software Development Library or the Software Development Libraries, and the subsequent release of the SI to integration and test, and system test.
- Provide status accounting of software work product changes submitted to the SDL or MSDL.
- Ensure that only approved changes are incorporated into the baselined software work products.
- Maintain a configuration audit system to ensure that records, which are provided to the MSDL or SDL, are consistent with documentation and software work product identification.

Three levels of configuration control are depicted in Figure 6.5 and five levels listed in example Table 6.6. Regardless of the number of CM control levels, overall responsibilities at each level should be defined in the SDP or SCMP including roles and procedures. In addition, the approach to CM related tasks should also be addressed (such as support to multiple baselines, a distributed development environment, data integrity and data restoration).

Scope Creep. There are many problems affecting (maybe infecting) the development of software but what makes it worse is that many Managers have a common misbelieve those are "not really problems" but simply characteristics of how software is developed. A lot of software projects fail or are canceled due to excessive time and cost overruns resulting from a non-existent, or poorly controlled, change management system.

For example, in order to be accommodating and a team player, a Manager may agree to what appears to be a "simple" change that seems like "a good idea," even though the change is not a documented requirement. But, surprise, it turns out to be difficult to implement, and similar changes start to pile up, progress starts to slow down, and the entire project is thrown out of control. This is sometimes called *scope creep*. It is clear that changes are almost always needed, however, changes must be controlled through the change control process to avoid massive disruption and potential project cancelation or failure. (How many "little" projects have you ever started that turned out to be anything but little?)

6.3.5 Software Configuration Status Accounting

SCM must prepare and maintain *records of the configuration status* for all baselined software products and includes maintenance of the records required to support configuration auditing. Configuration status accounting data includes the current version of each baselined product, a record of changes to the software product since being placed under configuration control, and the recording and reporting of:

- When each baseline was created and when each SI completed the initial build placing the software or database under CM control.
- Descriptive information about each SI.
- Description and status of each Software Discrepancy Report (SDR). The status items could be: approved, disapproved, awaiting action, incorporated, or closed.
- Change status and traceability of changes to controlled software products.
- The status of the technical and administrative documentation associated with each product baseline and/or update to a product baseline.
- Closure and archive status.

6.3.6 Software Configuration Audits

SCM must *perform periodic configuration audits to verify that changes were made* in accordance with the Corrective Action Process as described in your SDP or SCMP. SQA can witness and support these audits. Configuration Audits should be used to ensure that submitted software is accompanied by appropriate documentation and approvals, is correctly delivered and merged, and is correctly included in the software builds. The audits must ensure that each software entity incorporates only the approved changes scheduled for inclusion at the time of the audit. The degree of formality of the configuration audits may differ at the different levels of configuration control.

The Software Development Libraries (SDL) and the Master Software Development Library should be audited at least quarterly. A sampling of software releases must be checked against the *Software Version Description* (SVD) for correctness and completeness. You also must ensure that there is a SVD for each release.

Functional and Physical Configuration Audits. Software engineers may be requested to support the *Functional Configuration Audits* (FCA), *Physical Configuration Audits* (PCA) and in some cases a *System Verification Review* (SVR). Both software and system FCAs and PCAs may be conducted independently or concurrently. The SVR is often conducted concurrently with the System FCA.

Your contract may require a Software FCA to be conducted as part of the System Qualification Test (SQT) or following the SQT. The purpose of a Software FCA is to demonstrate that each SI was successfully tested and complies with the software and interface requirements of its functional requirements and design documentation. To complete the Software FCA, Software and System Engineers must reach a technical understanding of the validity and degree of completeness of the *Software Test Reports* (STR) and the applicable software user documentation. If software FCAs are contractually required, they should be conducted on every SI in the system.

The Software FCA, discussed in Subsection 12.6.5, is a prerequisite to the software Physical Configuration Audit discussed in Subsection 12.7.1. The purpose of the Software PCA is to conduct a formal examination of the *as-built*, and *as-coded SI*, against its design documentation in order to establish the product baseline. The Software PCA includes a review of the *Software Product Specification* (SPS), the *Interface Design Description* (IDD), the *Software Version Description* and all the required operational and support documentation.

If there are differences between the physical configuration of the SI and the configuration used for the Software FCA, they must be identified at the Software PCA. Approved and outstanding changes against the SI must also be provided along with approved deviations and waivers to the requirements specifications. FCAs and PCAs may be conducted on a single SI, a related group of SIs, or incrementally such as blocks. Results of the Software PCA can become an entrance criterion for the System Verification Review discussed in Subsection 12.6.6.

6.3.7 Packaging, Storage, Handling and Delivery

The SCM procedures for packaging, storage, handling, and delivery of software products must be provided in the SCMP for both the SDLs and the MSDL. Master copies of delivered software products must be maintained in the MSDL for the duration of the contract.

Packaging. The package for a software delivery, at a minimum, consists of the *Software Version Description* and the electronic media. The SwIPT responsible for the delivery prepares the SVD in collaboration with SCM. The SVD typically requires Engineering Review Board (ERB) and CCB approvals. SCM is responsible for providing the appropriate identification labels. SCM and SQA should perform the package content verification review, using a *Verification Checklist*. A hardcopy listing of the files should be attached to a signed hardcopy checklist.

Upon successful completion of the review and approval process, a formal transmittal contracts letter should be generated and submitted. Copies of the completed Verification Checklist and contracts letter must be maintained in the CM Library.

Storage. The storage requirement can be satisfied through the implementation of supporting library systems. There should be at least three basic components of the library system: a Software Library Management System; a Documentation Library; and Data Storage Backup:

- The Software Library Management System allows the software to be maintained in a central location, yet each host or client has access. A CASE tool can track the baseline changes, marking the transition throughout the activities of the Software Life Cycle. SCM must control the software libraries to provide a disciplined structure for progressive development, integration, test, and implementation of software within a controlled, well-defined environment.
- The Documentation Library is a repository for the most current approved and controlled documentation.
- The Data Storage Backup component provides daily incremental backups and system backups (performed daily or weekly) to ensure recovery from an uncontrollable situation.

Handling. A SCM CASE tool is ideally suited to administratively manage the handling of software through the version database directory structure. All elements in the database must be read-only—at all times. They only become Read/Write when they are checked out for updating by an authorized user.

Delivery. Release packages for each increment must be delivered to the Program SCM organization. All deliveries must receive SwIPT approval prior to shipping. Software deliveries should be on removable media. Installation and checkout of the delivered products at customer-designated facilities should be performed if applicable.

If problems arise during the installation, checkout, or test, these problems must be documented in a *Software Discrepancy Report* and then resolved. SDRs are discussed in Section 6.4.

Delivery Preparation. SCM is responsible for accumulating the SIs for milestone deliveries. This responsibility includes overseeing the scheduling, storage, handling, and delivery of the project media. All SIs to be delivered must be examined by SQA for specification compliance, SOW compliance, open items unresolved, SDR closure, and test verification status. No delivery should leave the facility without proper authorization from the Program Director or designate. CM should retain records of all deliveries

Software Version Description. An SVD document must also be prepared or updated for each release to provide a history of version changes, subsystem/element data, references to related documentation, and references to known problems. In preparation for delivery of the system to the customer, the CM organization must create the formal SVD document with detailed information on all software components, including COTS and reuse software, and their associated version numbers, plus special instructions necessary for system installation.

6.4 Managing the Software Corrective Action Process

The *Corrective Action Process* (CAP) is often called *change management*. Corrective action is triggered when performance deviates significantly from the plan, when defects that must be corrected are identified in the software work products, or when enhancements and improvements are proposed. A definition of "significant deviation" must be determined by mutual agreement between you and the customer. The opportunity to measure progress and identify issues that need Corrective Action can come from the reviews and evaluations, test results, and other quantitative management data. The CAP is described in the following three Subsections:

- Discrepancy Reports and Change Requests: Reporting problems and changes (6.4.1)
- Corrective Action System: Handling problems and issues for software work products (6.4.2)

- Configuration Control Boards: Hierarchical responsibilities of the change control boards (6.4.3)

6.4.1 Software Discrepancy Reports and Software Change Requests

To report problems or changes with baselined software products, *Software Discrepancy Reports* and *Software Change Requests*—or similar names—must be used as part of the Corrective Action Process. The SDR may also be called a "Software Deficiency Report" or a "Problem Report." The problem with calling everything a "problem" is that *every issue is not a problem*.

> **Lessons Learned.** At one time I presented a report to program management that showed there were many Software Problem Reports being processed. I was asked to find out why we had so many problems. My investigation revealed that most of the problems were not really "problems." Many were requests for improvement, some were simple mistakes easily corrected (like a typo), and some were queries regarding observed discrepancies that needed to be reviewed to determine if they were potential problems. That was when we got rid of calling everything a "problem" and instituted the SCR and SDR.

The SCRs/SDRs are inputs to the Corrective Action Process discussed in Subsection 6.4.2. The following describes the basic differences between SCRs and SDRs:

- *Software Change Request*: An SCR is typically used to enhance or improve the software product or change commitments, plans or a baseline. It is inappropriate to classify recommended improvements or enhancements as a "problem." Also, if an anomaly is found early in the process, the change requested will prevent the problem from occurring later in the process when correcting it will be much more costly.
- *Software Discrepancy Report*: SDRs document unexpected error conditions or anomalies that occur and is deemed as an incorrect action (or reaction) of the software product.

> **Lessons Learned.** At Lockheed Martin, when a change was made early in the development process that avoided a future problem, we called it a "save," and statistics were kept on the number of saves to show the cost saving advantages of early Peer Reviews. Most of the time, the cost savings of early Peer Reviews were impressive.

At the subsystem level, SCRs/SDRs are under the control of the Software Lead but may have to be passed to a higher-level board for approval. The Chief Software Engineer should be responsible for SCRs/SDRs at the program level. SCRs/SDRs must be used to report a known or suspected problem or discrepancy or change with software products under any level of configuration control above the Developer of the product. The originator of the SCR/SDR should be responsible for completion of the *issue description* but not necessarily the person who fixes the issue.

Candidate items for inclusion into the CAP Tracking System are: project name, issue originator, item number, item name, software element or document affected, origination date, category, severity, issue description, analyst assigned to the problem, date assigned, date completed, analysis time, recommended solution, impacts, problem status, approval of solution, follow-up actions, name of corrector, correction date, version where corrected, correction time and description of the solution implemented.

6.4.2 Software Corrective Action Process

If you expect the unexpected and prepare for it, you can manage it when it happens!

A software *Corrective Action Process* must be implemented for your project in order to handle problems or issues detected in software work products under configuration control, and other problems or issues related to the contract. The CAP must use SCRs/SDRs as inputs to the process and the process should include mechanisms to ensure that:

- All detected problems and issues are promptly reported and entered into the CAP Tracking System, organized and checked for duplication.
- Corrective Actions are promptly initiated when the problems or issues are identified.
- Status is tracked and resolutions are achieved.

- Records of the problems and issues are maintained for the duration of the contract.
- Software problems are classified by category and severity.
- Analysis is performed to detect trends.
- Corrective Actions are evaluated to determine if changes are correctly implemented without introducing additional problems.

Details of the Corrective Action Process can vary from location to location. These details should be documented in each Developer's SDP Annex or in a *Corrective Action Plan* addendum to the SDP. The SCR/SDR process can also be detailed in the *Software Configuration Management Plan* and/or in lower level procedures.

The division of responsibilities for Corrective Action tasks between the Development Team and the customer's Program Office should be clear but is often confusing. A table should be included in the SDP to specifically clarify the division of responsibility. It can also be expanded to identify specific organizations performing each task.

Figure 6.6 is an example overview of a Corrective Action Process for Software Discrepancy Reports. As shown in Figure 6.6, once the SDR has been generated and logged in at the program level, the SDR is assigned to a Responsible Software Engineer for investigation. The investigator must recommend the Corrective Action needed and record the actions taken to either correct the problem or provide a workaround solution. When this is accomplished, the SDR is returned to the responsible Configuration Control Board for disposition.

Once a process problem is defined, it must be assigned a *priority* and *severity level*. Table 6.8 is an example of definitions for software problem severity levels. The status of problems must be reported and tracked. The CSwE, the SEPG, or both should perform trend analysis on process problems and report adverse trends to the appropriate level of management. Process issues are closed out when SQA verifies that the Corrective Action is in place and there exists objective

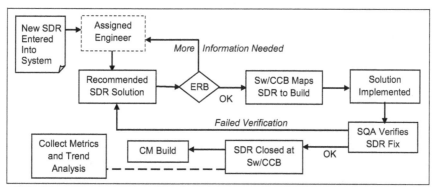

SDR = Software Discrepancy Report   CCB = Configuration Control Board   SQA = Software Quality Assurance
ERB = Engineering Review Board   CM = Configuration Management

Figure 6.6   Corrective Action Process overview.

Table 6.8 Software Problem Severity Levels

Severity	Severity Level If the Problem
1	Can seriously impact any requirement identified as "critical" —or— Prevents the accomplishment of an "essential" system capability
2	Adversely affects the accomplishment of an "essential" system capability and there is *no* feasible workaround solution. —or— Adversely affects the technical, cost, or schedule risks to the project, or to its life cycle support system, and there is *no* feasible workaround solution available
3	Adversely affects the accomplishment of an "essential" system capability but there *is* a feasible workaround solution. —or— Adversely affects the technical, cost, or schedule risks to the project, or to its life cycle support system, but there *is* a feasible workaround solution available
4	Is an inconvenience or annoyance to the system users/operators but does not affect a required operational or system essential capability. —or— Is an inconvenience or annoyance for development or maintenance personnel, but does not interfere with accomplishment of their responsibilities
5	Has little or no impact and does not match description of severity levels 1–4

evidence that the process is being followed. Process audit Corrective Action requests should be retained by SQA.

Corrective Action measurements must also be collected including SCRs/SDRs opened, closed and deferred; and aging metrics for SCRs/SDRs open for 30, 60, and 90 days, plus root cause. SCM is usually responsible for control and status updating of the SCR/SDR databases and overseeing the change management process. All SCRs/SDRs should be retained by SCM through the end of the contract.

6.4.3 Configuration Control Boards

On large programs, there is typically a hierarchy of *Configuration Control Boards* with different levels of control and responsibilities. On smaller programs, there is less formality; however, some group(s) must exist to review and approve changes and resolve conflicts like an *Engineering Review Board* or a "Tiger Team." Figure 6.7 is an example of the typical relationship of CCBs.

Software CM supports all of the change boards depicted in the example shown in Figure 6.7. At the very top of

the hierarchy is the customer's technical support CCB (or Program Office CCB). They participate in, and monitor, the activities of the lower level CCBs as needed.

The *program level* boards include the Configuration Control Board (Program CCB) and the Software Configuration Control Board (Sw/CCB). At the *subsystem level*, there may be three lower level boards: the Subsystem Configuration Control Board (Subsystem CCB), the Software CCB, and the Engineering Review Board. The suppliers (subcontractors) may also have similar levels of control boards and, as the Software Project Manager, you should have cognizance of the effectiveness of their boards.

Problems or issues for Software Units (SUs) with only subsystem *internal* changes should be handled with SCRs/SDRs by the *subsystem's* control board. Baselined software with *external* subsystem interface changes must be handled with SCRs/SDRs by the *program-level* Software Configuration Control Board (Sw/CCB) or the *program-level* CCB. Problems or issues for SIs with no interfaces should be handled with SCRs/SDRs by the respective subsystem's control board. The program-level Sw/CCB also has cognizance over the Product Development Baseline.

The Responsible Software Engineer (RSE), usually one of the Software Leads, provides support to the subsystem boards. All of these boards must be described in the SDP or the SCMP. The functions and relationships of the five typical CCBs are briefly described below:

- *Program Configuration Control Board*: The Program CCB operates under the authority of the customer's CCB and approves changes to baselined documents that affect cost, schedule, program constraints or scope issues. There may also be a high-level Program Configuration Evaluation Board (CEB). The CCB has cognizance over Allocated and Product Baselines.
- *Software Configuration Control Board*: The Sw/CCB operates under the authority of the Program CCB; it has control over software changes at the system and subsystem-to-subsystem software integration and test level. The Sw/CCB has cognizance over the Product Development Baseline for software, the Master Software Development Library, and software risk management.
- *Subsystem Change Control Board*: The subsystem CCB must have control over changes found in baselined SIs. It reviews SCRs/SDRs, provides impact assessments, assigns appropriate personnel, and oversees the resolution and verification. The subsystem CCB must have cognizance over the subsystem's Allocated and Product Development Baselines as well as the subsystem's Software Development Library.
- *Subsystem Software Configuration Control Board*: This subsystem board reviews changes to baselined

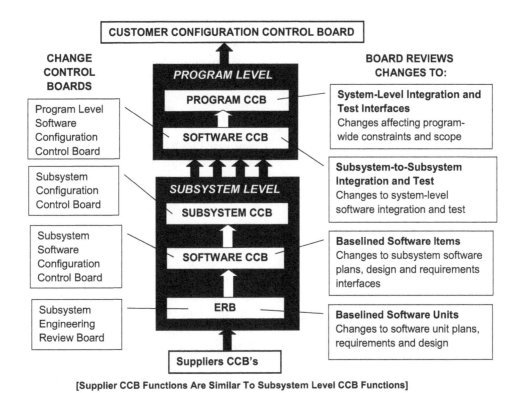

Figure 6.7   Relationship of the Configuration Control Boards—example.

SIs including changes to subsystem software plans, designs, and requirements interfaces.

- *Engineering Review Board*: This lower level board performs the same type of functions as the Subsystem Software CCB but only controls changes found in baselined *SUs*. This function can take place at the subsystem or at a lower level.

Integration of the Corrective Action Process. The Corrective Action Process must be integrated across disciplines (Software, Hardware, Systems Engineering), all SwIPTs, the Developer organizations (prime and suppliers), and the System Development Life Cycle activities (from requirements definition through System Test). In addition, the Corrective Action Process must be integrated with the risk management, Configuration Management, and the process improvement processes. For example, risks, and risk mitigation actions, can become problems that need Corrective Action.

6.4.4 When You Need a Paradigm Shift

If you collect all of the relevant information available, you thoroughly analyze it, and you make the best-informed judgments and decisions for your project as to the best course of action, you did your job. However, sometimes the decisions you make are wrong, and the problems they create do not become apparent until well into the project. *Do not be afraid to admit you made a mistake.*

Instead, take Corrective Action as soon as you can to revise the action plan and redirect the team. It will do great harm to your project if you fail to change course by being rigid and bull-headed. In this circumstance, you must implement a *paradigm shift*—a change to a new set of rules.

The characteristics and evolution of paradigms are well covered by Barker (1998). As a futurist, Barker makes sense of what is happening in the business world. Understanding the paradigm shift concept is a first step of trying to influence or control the future—of both your organization and your project. If you are not getting the expected results, find out why and make the needed changes to your project, regardless of how drastic or painful it may seem, before it is too late to affect the outcome.

These comments about the need to make paradigm shifts are applicable to decisions made by your Board of Directors down to the workbench level. For example, IBM had a good business model with computer mainframes until Apple came along with laptops that forced IBM to apply the new paradigm in order to play the new game. Kodak was a leader in their industry until digital cameras came along and, since Kodak was apparently guilty of "paradigm paralysis," they had to declare bankruptcy.

Contingency Planning. Contingency planning is an element of change management but is often not included in the normal planning process. That is because it really is part of the post-planning process. In other words, after you and the team create the best possible plan, and associated planning documents, you should spend some time focused on contingency planning. This is often called "what-if" analysis. Pre-planning of this type can be very useful when things just don't go the way you expected.

> **Lessons Learned.** If you don't already know this, you will find out that "things" have a way of not going the way you expect them to. Most of the time, contingency planning does not take place until unfolding events requires you to do it, but having alternate contingency plans in your back pocket can be very helpful in time of need. I recommend a *big* back pocket.

> *Consider the field mouse, how sagacious an animal it is, who never entrusts his life to one hole only.* —*Plautus (254–184 BC)* A perfect example of contingency planning.

6.5 Software Peer Reviews and Product Evaluations

A *Software Peer Review* (SPR) is a structured, methodical examination of software work products by the Developers, with their peers, to identify defects, to recommend needed changes, and to identify potential problems. Software Peer Reviews are an important software quality activity. To some, SPRs are a pain in the neck (and other places). I have medical advice for those people: get medication because Peer Reviews are not going away.

> **Lessons Learned.** I have witnessed the following scenario many times. A busy Programmer is sent to a Software Peer Review for the first time. He or she arrives, with a grumpy and nasty look on their face, and they make a statement such as "let's get this over with quickly so I can go back to real work." When the Peer Review is over, a high percentage of these same people usually say something like: "I thought this was going to be a waste of time, but I cannot believe what we just did; those errors may not have been detected for a long time."

The performance of Software Peer Reviews can provide immeasurable value to your program, and they are an established best practice. Regardless of the size of your program, if you intend to produce high-quality software work products, *Peer Reviews must be performed.* The Software Peer Review and Product Evaluation activities are described in the following Subsections:

- Software Peer Review Objectives (6.5.1)
- Software Peer Review Process (6.5.2)
- Preparing for Software Peer Reviews (6.5.3)
- Conducting Software Peer Reviews (6.5.4)
- Analyzing Software Peer Review Data (6.5.5)
- Software Product Evaluations (6.5.6)

6.5.1 Software Peer Review Objectives

Software Peer Reviews are a critical element to the development of high-quality software work products. Peer Reviews are an integral part of the development process and focus on the identification and removal of defects as early and efficiently as possible. The reason the defects should be removed as early as possible is because the later in the life cycle the defect is found, the more expensive it is to fix.

Figure 6.8 is a "notional view" of the relative cost of making software changes during the Software Development Life Cycle. The figure highlights the significant difference in cost between finding and fixing an error during Coding and Unit Testing versus finding and fixing it after release to the customer. There are some estimates that the cost difference in fixing errors in the field versus fixing them during the requirements activity can be as much and 100:1. Your total *project cost can go out-of-sight* if serious problems are allowed to occur late in the Software Life Cycle.

Peer Reviews evaluate deliverable (and System Critical non-deliverable) work products for correctness, completeness, accuracy, and consistency. They help to ensure adequacy of these quality attributes prior to transition from one activity to the next. Peer reviews also identify areas of needed change and improvement in the product, assure compliance to standards, and ensure satisfaction of functional, performance and interface requirements. Also, action items, defects, technical decisions, and measurements resulting from these reviews are documented and tracked.

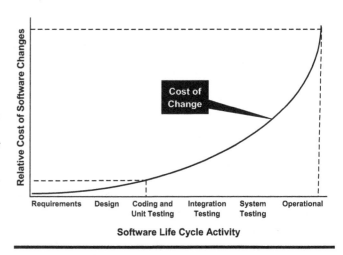

Figure 6.8 Relative cost of fixing errors during the life cycle.

6.5.2 Software Peer Review Process

The SDP, or the *Peer Review Plan* if there is one, must define the processes to be followed for each type of Peer Review to be used on each of the software work products and for each software category (categories are described in Section 1.8). The process should include preparing for the Peer Review, conducting the Peer Reviews, and analyzing the resulting Peer Review data. The planning for Peer Reviews is part of the software development planning process. The *evaluation criteria* for each software product must be supplied to the Peer Review participants. Figure 6.9 is an example overview of a typical Software Peer Review Process.

System Critical software (whether deliverable or non-deliverable) must undergo the most formal, robust Peer Review process for their initial development and for significant changes. Support software and minor changes to software work products may undergo a less formal Peer Review.

6.5.3 Preparing for Software Peer Reviews

A software project typically adopts the standard Peer Review process of its parent organization for conducting Software Peer Reviews. SPR preparation should consist of entry/exit criteria, definition of the steps, and roles with appropriate participants identified. Peer reviews must be scheduled, planned, and tracked by the subsystem and/or SwIPT leads in order to:

- Determine what type of Peer Review will be conducted on each work product.
- Identify key reviewers who must participate in each Peer Review plus invited reviewers.
- Ensure that each software work product satisfies the Peer Review entry criteria.
- Ensure all reviewers understand their roles in the Peer Review.
- Confirm that the participants have reviewed each work product *prior to the Peer Review*, using the predetermined review checklist for that product.

A subsystem (or lower level) *Software Peer Review Plan* should be prepared to define the procedures, data collection

and reporting for Peer Reviews and product evaluations. The implementation of the SPR Plan is the responsibility of the subsystem/element SwIPT software personnel Lead. The SPR Plan typically defines the types of Peer Reviews to be held. Peer reviews should have varying levels of formality, ranging from Formal Inspections to Colleague Reviews and maybe some in between:

- *Formal Inspections*: Formal inspections are performed to verify that software products conform to established technical decisions and applicable standards and procedures. Formal Inspections are the most thorough Peer Review and are conducted initially when the product has reached enough maturity and completeness for a thorough review or when extensive changes are involved. The Software Lead should review all the changes and determine if additional Peer Reviews need to be held.
- *Colleague Reviews*: This is the least formal type of review used primarily to review portions of a System Critical work product during its development to improve the quality of the product during its development. A Colleague Review can be conducted by one person, and it may be used for relatively minor updates to software products that have already undergone a Formal Inspection. Colleague Reviews are not a substitute for required inspections.

6.5.4 Conducting Software Peer Reviews

Software Peer Reviews must be ongoing during the development tasks, prior to work products being released to subsequent activities. The SDP should include a table similar to example Table 6.9 showing the set of development work products that normally require Peer Reviews. The goal of each Peer Review is to identify work product defects as early as possible in the lifecycle. Defects identified and corrected early always results in a lower cost to resolve than finding and fixing them later.

Peer Review Roles. The typical roles include a moderator, reviewers, monitor /coordinator, recorder and the Product Developer or Document Author. Inspection training or skills

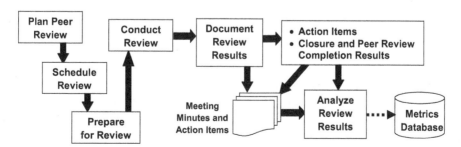

Figure 6.9   Overview of the Software Peer Review Process.

Table 6.9 Software Work Products Requiring Peer Reviews

Development Activity	Work Products Reviewed
System Requirements Analysis	■ Product Specifications ■ Subsystem SEIT test plan
System Architecture Design	■ Architecture and Data Models ■ Interface descriptions ■ Trade study results
Software Requirements Analysis	Software requirements documents(*) and Data Models
Software Design	■ Software Design documents and Data Models ■ Software interface descriptions
Code and Unit Test	■ Source files ■ Unit test cases and procedures
Software Integration Testing	Test cases and procedures
Software Qualification Testing	■ Software Test Plan ■ Software Test Description (cases and procedures)

(*) It is also a good idea to inspect the Requirements Test Verification Matrix (RTVM) during every development activity. The RTVM may be called a Verification Cross Reference Matrix (VCRM).

from past experience should be required for all participants. The following is a brief description of each key role:

- The Peer Review moderator is instrumental in setting up the inspection, and ensuring the key reviewers for the work product are present and prepared for the inspection.
- Reviewers should be trained to perform various inspection roles. These trained inspectors participate to identify defects, improvements and issues in the product(s) being examined and contribute to the determination of the product's ability to proceed to the next development activity.
- The Monitor/Coordinator ensures the Peer Review process is held in accordance with the approved tailored Peer Review process, and they are tasked to compile and analyze Peer Review metrics, maintain a list of qualified inspectors, and oversee the implementation of continuous process improvement within the Peer Review process.
- The Recorder is responsible for creating Peer Review records in hard copy or electronic files.

The Peer Review Package. A Peer Review announcement and review package should be distributed about a week prior to the Peer Review. Participants must review the work product against higher-level requirements, using checklists aimed at finding major system problems and for compliance with templates, standards, and guidelines. Findings must be documented, assigned to responsible individuals as action items for resolution by a due date, and then tracked to closure.

6.5.5 Analyzing Software Peer Review Data

The metrics collected during Peer Reviews can be used to provide visibility into the quality of the produced documents and code, the effectiveness of the SPR, and to indicate the need for corrective or process changes. At the end of each development activity, participating organizations analyze the *root causes* of deficiencies to identify process improvements that can be implemented to avoid future occurrence of those deficiencies.

Defect Data. The number of defects expected from a Peer Review depends on when in the life cycle the review is conducted. Table 6.10 is an example of a defect removal scenario showing the incremental burn down of 300 defects injected during coding to seven defects during field testing.

Analysis of historical SPR data can provide many benefits such as setting expectations for the Peer Reviews conducted on requirements, design, coding, and test products. For software, code count and complexity can be used, as well as experience from prior programs, to predict the number of defects from each life cycle activity. In fact, *fixing defects is one of the largest variables in estimating project cost.*

A Defect Prevention Team can compare predicted versus actual defects to assess the quality of the product at the end of each activity. They can analyze the defect causes to initiate defect prevention process improvements for subsequent builds. Required defect data and associated metrics for Peer Reviews should be described in the *Software Measurement Plan.* See Subsection 3.9.8 for defect estimation.

Peer Review Data. Typical Peer Review data that should be collected at each review and analyzed for potential improvement of the Peer Review process include:

- Identification of the work product(s) under review
- Meeting date, time, location and completion date
- Number of attendees and preparation time spent by each reviewer
- Size of the work product(s) reviewed
- Inspection type and role assignments
- Peer Review Defect List
- Amount of time spent in rework of the software work product and to close action items

Table 6.10   Example Software Defect Removal Scenario

Software Development Activity	Software Defects Injected	Defect Detection Goal (%)	Defects Found and Fixed	Software Defects Remaining
Coding	300	–	0	300
Unit Testing	(*)	75	225	75
Integration Testing	(*)	65	48	27
System Testing	(*)	55	15	12
Field Testing	(*)	40	5	7

(*) Unintended defects might be added during the process of fixing known defects.

6.5.6 Software Product Evaluations

Your contract may require, and some software standards do require, Software Product Evaluations to be performed in addition to the Software Peer Reviews. However, they may not necessarily require two separate processes since Peer Reviews are an acceptable alternative to conducting the evaluations as long as the specific requirements for the evaluation are satisfied. Software work product evaluations are similar to the *Formal Inspection* type of Peer Review.

Product evaluations require the Developers to perform in-process evaluations as well as final evaluations of each software work product before delivery. If your contract requires product evaluations, your *Software Quality Program Plan* or the SDP should list the minimum in-process software products to be evaluated, define the evaluation process to be used, describe the minimum evaluation criteria, and provide clear definitions of the evaluation criteria.

Risk Analysis. If a software work product fails to meet its completion readiness criteria, an analysis can be performed to determine the risk of proceeding to the next process. If the risk is acceptable, and appropriate actions and detailed resolution plans have been put in place, then subsequent process tasks may begin. However, *advancing flawed products should be discouraged* and limited to cases of extenuating circumstances (e.g., the defect does not impact the implementation process until a later part of the schedule allowing ample time for repair). The assessment of product readiness and the decision to proceed (including signatures of agreement from key leadership personnel) must be documented and retained in the evaluation records (Risk Management and Mitigation is discussed in Section 5.2).

Quality Checking Filter. The key gate to in-process quality checking is not allowing products that do not meet the readiness criteria to pass to the next implementation process. This can be achieved by holding a final Engineering Review Board review at the end of each implementation process or activity. Product readiness can be assessed by

determining if output products are complete, have met their completion criteria, and product defects have been resolved.

Records. Software product evaluation records must be maintained for the duration of the contract. The accepted product evaluation comments are usually stored in the Software Development File. A summary of findings should be put in a database for metrics analysis. Closure of product evaluation comments must be verified. The following software product evaluation records should be maintained:

- Product evaluation package, including a copy of the product(s) under evaluation
- Review meeting minutes, including time, date, attendees, technical decisions made, action items captured and evaluation problems found
- Action items resolutions, software work product problem resolutions and closure data

Tracking Action Items. Action items from Software Product Evaluations must be tracked to closure. If the software product under evaluation is under a level of Configuration Management above the individual Author/Developer, then the action items must be documented as an SCR/SDR and handled by the Corrective Action Process (see Section 6.4). If the software product is under control of the individual author or Developer, then the problems should be documented in the SDF, and the subsystem and/or SwIPT Lead is responsible for ensuring the problems and issues are properly closed.

Independence. It is absolutely imperative that *evaluators* of a software work product under review *must not* be the person(s) responsible for developing that software product. However, this does not preclude Software Developers from participating in Software Product Evaluations or reviews to answer questions on products they produced. Document reviews should be coordinated by the Subsystem SwIPT software personnel and may include members outside the Developer's subsystem.

6.6 Capability Maturity Model Integrated

The goal of CMMI is performance improvement through process improvement.

The CMMI is a *framework* for building process improvement systems, related training and process maturity appraisal programs. It was developed by the Software Engineering Institute at Carnegie Mellon University. Many books and reports have been written on utilization of the CMMI for process improvement in a number of industries including (Chrissis et al., 2010; Ahern et al., 2001) as well as the CMMI appraisal methods (SEI, 2006).

Section 6.6 is intended to familiarize Project Managers with the conceptual framework of the CMMI; it is only intended to be a summary and is presented in the following five Subsections:

- CMMI Background and Principles: Roots of the CMMI, why it was needed and what CMMI is and is not (6.6.1)
- CMMI Structure: An overall description of the three CMMI Areas of Interest, the process areas, goals, and practices (6.6.2)
- CMMI Representations: The staged and continuous representations (6.6.3)
- Attributes of the CMMI Staged Representation: A description of the five levels of maturity for the Staged Representation (6.6.4)
- CMMI Appraisals: The method of awarding maturity and capability levels (6.6.5)

6.6.1 CMMI Background and Principles

From a purely business perspective, the purpose of the CMMI, is to *improve operational performance by improving the efficiency of production, delivery, and outsourcing* thereby *lowering the cost.*

Origination. The CMMI originated from the *Capability Maturity Model* (CMM) developed by the SEI specifically for Software Process Improvement in the defense industry. The CMM was used primarily by the software industry from 1987 until the first version of CMMI was released in 2002. The most recent release as of this writing is Version 2.0, released in 2018, replacing Version 1.3, released in 2010. Since 2013 the CMMI product suite has been available from, and managed by, the CMMI Institute at Carnegie Mellon University. Please note that the CMMI product suite is continually being updated and improved, so some of the CMMI descriptions herein eventually will be superseded by newer versions.

Model Expansion. The success of the original CMM Model, focused on software development, spawned the development of several similar models in other related disciplines such as system engineering, software acquisition, and integrated product development to capture concurrent engineering practices. The CMM Integrated (CMMI) project initially was formed to resolve the problem of having multiple versions of the CMM and to integrate them into one comprehensive framework. Furthermore, these divergent maturity models tended to keep the Software Engineering (SwE) and Systems Engineering (SE) communities apart when it became clear that bringing them together was a big part of rectifying problems associated with program failures (the collaboration of SwE and SE is covered in Chapter 12).

Model Consolidation. The task of consolidating the various models generalized the CMMI so it could be applied to many areas beyond software. As a result, the overall CMMI is not as specific to software as its predecessor. But the good news is that the staged representation of the CMMI (discussed in 6.6.4) makes it relatively easy for software organizations to convert from the CMM to the CMMI. The CMMI was developed by a team of professionals from industry, government and the SEI.

Since the CMMI is the product of a "committee," some of the apparent unnecessary complexity of the CMMI may be attributed to the symptomatic outcome of a committee trying to satisfy everyone. The CMMI Product Team reviewed about 2000 Change Requests for the 2010 release.

Performance Improvement. Adapting the CMMI Model approach in order to realize software performance improvement is not a trivial task. If your organization is considering going the CMMI route, you or members of your team, or both, must investigate attendance at CMMI courses. The effort and cost involved, although not trivial, can have a high payback. Results from adapting CMMI are periodically published by the CMMI Institute called the *Published Appraisal Results* (PARS).

The PARS typically show substantial increases in performance including improvements in customer satisfaction and an increase in productivity. However, the CMMI Model deals mostly with *what* process should be implemented and not so much with *how* they can be implemented. The reported improvements cannot be "guaranteed" in every organization. Small companies with limited resources are not likely to realize as much benefit from CMMI as larger organizations. Organizations that perform work for the U.S. government, especially the DoD, may not be able to acquire any software-intensive contract work unless they have achieved a CMMI Maturity Level 3 or above.

CMMI versus Agile Development. A major change in CMMI Version 1.3 was the addition of support to Agile development approaches. A follow-up guide from the CMMI Institute (CMMI, 2016) made it clear that CMMI and Agile development can co-exist. Guidelines are provided that help interpret Agile practices and how to apply them to CMMI practices. An article in the *CrossTalk Journal* (CrossTalk Nov/ Dec 2016) states that: "companies are increasingly turning to CMMI to improve the performance of Agile initiatives. In 2015, over 70% of CMMI appraised organizations reported using one or more agile approaches." In addition, (deSousa, 2015) states that: "Agility is a function of maturity and not the other way around. ... There is a strong correlation between successful Agile projects and CMMI Level 3+ organizations."

CMMI Principles. An original goal of the CMM was to increase the predictability and the confidence that software development projects would be successful. The practices in CMM, followed later by the CMMI, were intended to provide that confidence. The rationale was that *following these practices provided the capability to increase performance*, and the increased performance would, in turn, lead to better results. Some problems began when people viewed the CMMI practices themselves as the intended outcome and not the increase in performance the practices were supposed to provide.

CMMI is a *model for building process improvement systems* from which organizations can create their own process improvement solutions, compatible with their environment, thus helping them to improve their operational performance. *The CMMI describes Process Areas but not the processes themselves and does not contain work instructions: CMMI is not a Development Standard or a Development Life Cycle.*

The CMMI Process Areas were chosen because they are considered the most important operational activities that collectively contribute to a *systematic ability to affect process improvements in, and the management of, those technical process and practices that develop and deliver software products and services*. The fundamental precepts of the CMMI are:

- To improve your processes, you need to manage your requirements, risks and configurations; you need to plan, monitor and control your projects; you need to measure and analyze the output of your efforts; you need to track performance of your project to determine how well the processes are being followed and if they are producing the expected results.
- If you want to improve those objectives, you have to make a focused effort on your processes, and you need standardized process assets, an organization-wide training program and formality to your technical activities.
- If you want to excel, you have to quantify the performance of your projects and processes and be able to improve them by focusing on what the metrics tell you to focus on. Also, if you're going to perform a process, you should have a policy for doing it, a plan for it, resources, assignments, process-level training, stakeholder involvement, and anything else needed to make it effective.

6.6.2 CMMI Structure

CMMI is structured to address three business *Areas of Interest* that are also called *constellations*. The three Areas of Interest are:

- *CMMI-DEV*: CMMI for Development addresses product development and delivery and supersedes previous software and Systems Engineering Areas of Interest since it is a comprehensive integration of them.
- *CMMI-ACQ*: CMMI for Acquisition addresses acquisition of products and services, supply chain management, and outsourcing processes in government and industry.
- *CMMI-SVC*: CMMI for services provides service establishment, management, and guidance for delivering specific services within an organization or to external customers.

All three Areas of Interest share many CMMI elements, but they are all *frameworks for assembling process improvement systems*. Each Area of Interest has Process Areas that targets improvements in specific sub-areas, so you must pick the approach that is consistent with the objectives of your project. Keep in mind the Areas of Interest do not contain processes within themselves. None of them alone can be used to actually develop products, acquire goods or fulfill services. The CMMI likely assumption is that your organization has its own standards, processes and operational procedures by which things get done.

If you are acquiring goods and services from others, then you should choose CMMI-ACQ. Sometimes it may not be easy to determine if you are providing services (CMMI-SVC) or whether you are doing development (CMMI-DEV) as many operations do both. If your software product is specific to a contract specification, or a Statement of Work where the effort is a stand-alone contract, or if your customers pay you for a specific effort over a specified time, then select CMMI-DEV. If your customers pay per request, or if they submit a transaction request into an existing request system where everyone goes through the same process, then CMMI-SVC is likely the best choice.

Overall CMMI Structure. Figure 6.10 is my view of a top-level graphical illustration of the overall CMMI structure. It involves the three *Areas of Interest* (constellations), supported by *Process Areas* (PA) that can be carried out using either the *staged or Continuous Representation* Models. The Process Areas are supported by specific and generic *goals* that are driven by specific and generic *practices* supporting those goals. The basis and inputs for implementing the practices are derived from detailed work products and sub-practices. Each of these elements is briefly described.

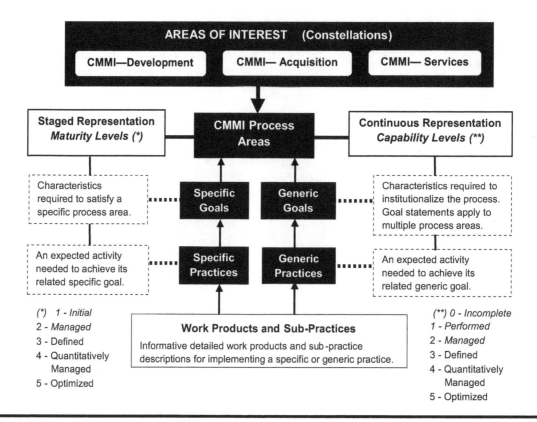

Figure 6.10 CMMI structure.

CMMI Process Areas. *Process Areas* are a cluster of related practices that, when implemented collectively, satisfy a set of goals considered important for making improvements in that area. There are 35 Process Areas in the current version of the CMMI, but this number is very likely to be modified over time. The set of process areas will be different for each Area of Interest and all the PAs are used in both the staged and continuous representations.

The following list shows the distribution of PAs over the three Areas of Interest:

- Core Process Areas shared by all Areas of Interest: 16
- Process Areas unique to CMMI-Acquisition: 6
- Process Areas unique to CMMI-Development: 5
- Process Areas unique to CMMI-Services: 7
- Process Areas shared by CMMI-Dev and CMMI-Svc: 1
  Total number of CMMI Process Areas: 35

In order to stay on topic in this Guidebook, the description of the CMMI Process Areas will be restricted to seven CMMI Process Areas that are *specific to project management* and the Measurement and Analysis PA because I believe it to be critical to successful project management. Process Areas that have been omitted here do not imply that they are not related to good project management practices. Table 6.11 briefly describes these seven Project Management PAs.

CMMI Goals and Practices. CMMI goals are all oriented toward process improvement activities. Each Process Area is made of two kinds of goals, two kinds of practices, and a lot of informative material. The two goal types are *Specific Goals* (SG) and *Generic Goals* (GG) which are in turn supported by Specific Practices (SP) and Generic Practices (GP). Every Process Area has one to three Generic Goals and at least one Specific Goal. The SGs are made up of at least two SPs.

The informative material (work products and detailed subpractices) are very useful and vary from PA to PA. You need to focus on the Goals and Practices because they are the required and expected components of CMMI when it comes time to be appraised (see Subsection 6.6.5). The specific and generic goals and practices are not discussed in this Guidebook.

6.6.3 CMMI Staged and Continuous Representations

The CMMI can be implemented with either of two model representations: *staged* or *continuous* and both models have the same objective of process improvement:

- The *Staged Representation* is a model structure where attaining the goals of a set of process areas establishes a maturity level and each level builds a foundation for climbing to subsequent levels.

Table 6.11 Attributes of the Project Management Process Areas

Process Area	Description
Project Planning (PP)	Includes developing the Software Development Plan, involving the stakeholders, obtaining commitment to the SDP, and maintaining the SDP
Project Monitoring and Control (PMC)	Involves monitoring the software development activities and taking Corrective Actions when progress deviates from the plan. The SDP should specify the monitoring level, the frequency of product reviews, and the measures used to monitor progress
Integrated Project Management (IPM)	Uses a tailored version of your organization's Standard Software Process. The efforts of stakeholders should be coordinated in a timely manner, and critical dependencies identified and tracked and the issues resolved. If you are using the Integrated Process and Product Development (IPPD) approach, as discussed in Section 8.6. integrated teams are established to collectively support and/or perform the work
Risk Management (RSKM)	This process area is broader than risk identification and risk monitoring which are covered in the Project Planning and Project Monitoring and Control process areas identified above. Risk management takes a continuous and forward-looking approach to managing risks including identification of risk parameters, risk assessments, and proactive risk mitigation
Requirements Management (REQM)	Involves managing requirements of the products and product components and identifies inconsistencies between the requirements, work plans and work products
Quantitative Project Management (QPM)	Applies quantitative and statistical techniques to manage process performance and product quality—based on your project's objectives. This may involve the ability to predict process performance but, at a minimum, allows the Manager to understand the variations experienced by the sub-processes critical to achieving the desired quality and process performance objectives
Measurement and Analysis (MA)	Develops and sustains a measurement capability that is used to support management information needs throughout the life cycle

■ The *Continuous Representation* is a model that provides the flexibility to focus only on the Process Areas that are closely aligned with your project or organization's business objectives.

Staged Representation. Originally, the CMM was conceived for the software industry with the concept of fundamental process areas that organizations should excel at to mature their processes and thereby achieve higher capabilities and improved performance. Some process areas were "staged" or grouped together with the belief that such groupings made sense as building blocks where each level of grouped blocks depended on the prior levels of blocks. The result was five levels of increasing maturity where each *staged level* consists of a set of Process Areas.

Continuous Representation. Many CMMI users found that some of the staged capabilities in some Process Areas were of little value, or not even applicable, to them. Thus the idea of a Continuous Representation was developed whereby an organization could choose to get very good at applicable PAs without having to put forth the resources to implement low-value or unused PAs.

The Continuous Representation is important to organizations that want to be able to benchmark themselves (or

need to be formally rated) in only PAs that matter to them. The Continuous Representation allows organizations to pick any number of process areas and decide whatever depth of capability they want to become in those PAs. Table 6.12 is a general summary of the comparative differences and advantages for each approach.

You should decide which representation is best suited to your project and organizational standards. However, it is not unusual for a project to deviate from following either representation exactly as prescribed, but instead, utilizes the best attributes of both representations to address the unique needs of their project. The bottom line is that the Continuous Representation is focused on process area capabilities as measured by capability levels whereas the staged representation is focused on organizational (or project) maturity as measured by maturity levels.

6.6.4 Attributes of the Staged Representation

The staged representation has been a popular approach because it provides an easy migration from the CMM to the CMMI, has been around for a lot longer than the Continuous Representation, and has a long history of being an effective approach especially for the development of large software-intensive systems. Figure 6.11 is one way (of many)

Table 6.12 Comparative Advantages of Continuous and Staged Representations

Continuous Representation	Staged Representation
Utilizes capability levels	Utilizes maturity levels
Offers maximum flexibility when using CMMI for process improvement	A systematic and structured approach to process improvement built up one stage at a time
Can select the specific processes that are considered important for the organization's immediate business objectives and risk reduction	Prescribes a specific order for implementing process areas according to maturity levels defining an improvement path roadmap
Can work to improve performance of a single process-related trouble area or several areas at the same time	A basis for comparing the maturity of different projects and organizations
Good choice if you know the process area, or areas, needing improvement	*Good choice if you do not know where to start or which processes you need to improve*

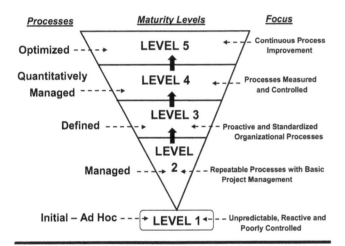

Figure 6.11 Characteristics of the five staged maturity levels.

to graphically depict the *five levels of maturity* for the staged representation and Table 6.13 briefly describes attributes of each of the five staged maturity levels.

6.6.5 CMMI Appraisals

Organizations seeking a maturity rating are *not certified in CMMI*; they are *appraised*. You get appraised by an External Appraisal Team led by a CMMI-Institute-*Certified Lead Appraiser* who determines, with the help of his or her team, to what degree you are performing the practices of the CMMI. Depending on the type of appraisal, your organization can be awarded a maturity level rating, from 1–5, or a capability level achievement profile. Appraisals are typically conducted to:

■ Determine how well your organization's processes compare to CMMI best practices, and to identify areas and improvement opportunities where improvement can and need to be made.

■ Meet the contractual requirements of one or more of your customers.

CMMI appraisals must conform to the requirements defined in the *Appraisal Requirements for CMMI* (ARC) document. Reading the ARC and related documents is appropriate for a Lead Appraiser, but cumbersome for everyone else.

There are three classes of appraisals, A, B and C. The *class A appraisal* is the most formal and is the only one that can result in a formal maturity level rating. The *class B appraisals* are much less formal, vary in scope, performed on implemented practices and are often used to prepare for a class A appraisal. The informal *class C appraisal* is usually performed internally if you have people qualified to do it.

The *Standard CMMI Appraisal Method for Process Improvement* (SCAMPI) is an appraisal method that meets all of the ARC requirements (SEI, 2006; Owens, 2003). Results of a SCAMPI appraisal may be published (if the appraised organization approves) on the SEI's CMMI Web site. SCAMPI also supports the conduct of the *Software Process Improvement and Capability Determination* (SPICE) also published as ISO/IEC 15504 (ISO/IEC, 1993).

Members of the organization being appraised should have some training in the CMMI. Taking the SEI's licensed Introduction to CMMI course(s) may not provide enough of an understanding to determine, without any other direct experience, how closely your company is performing the expected practices of CMMI, or how your particular implementation of the practices will fare in an appraisal.

Sometimes called a "gap analysis," an informal (class C) appraisal can be performed prior to the class A appraisal, and the process areas tagged for improvement should be prioritized. More modern approaches, involving commercially available CMMI-compliant tools, can reduce the time to achieve compliance.

SEI maintains statistics on the "time to move up" to a higher level of maturity. Since the release of the CMMI, the

Table 6.13 Attributes of the Five Staged Maturity Levels

Maturity Level	Description
1	CMMI Level 1 is not really a maturity level because organization at this level exhibits little, if any, maturity. It is an initial level that is often characterized as ad-hoc and chaotic. There is no stable environment to support the processes. In Level 1 organizations, success depends on the extraordinary heroics and competencies of the people who are determined to get the job done despite the chaos. However, these successes most often exceed the budget, do not meet the schedule, and you cannot depend on it being repeatable. Many, if not most, software organizations are at Level 1; they need to read this Guidebook and tailor it to their projects.
2	CMMI Level 2 refers to processes that are managed and repeatable. It is the minimum maturity an organization could achieve by just hiring experienced Software Project Managers and allowing them to do their jobs. Companies struggling to incorporate CMMI Level 2 practices probably have unpredictable and inconsistent on-time delivery records, product quality and profit. Such organizations frequently take on more work than they can handle and then proceed to poorly plan the level of effort and dependencies needed to complete the work.
3	CMMI Level 3 is the next higher level of maturity where the processes are defined, standardized, and management of them is proactive. CMMI Level 3 projects typically include staff members experienced in process management, the operational work of delivering a product, and organizational change who work with the SPM to help make use of the most effective approaches to meet their projects' needs and help understand how well the organization is performing its tasks even if the project is halfway through development. CMMI Level 3 organizations use data and metrics to help understand their internal costs and effectiveness and ask if their processes are good, not just if their processes are followed.
4 and 5	CMMI Maturity Levels 4 and 5 are the high maturity levels and involve staff members who excel in using quantitative techniques to manage the tactical, strategic and operational performance of your organization. They use practices that take data from the lower maturity levels to better help control variation and they use the data to decrease risk and uncertainty and to increase confidence in their forecasting and performance predictions. Organizations operating at CMMI Level 5 are focused on continuously improving performance through the use of process performance models, plus a strategic investment in effective and efficient processes and supporting tools.

median times to move from Level 1 to Level 2 is about 5 months, with median movement from Level 2 to Level 3 another 21 months. An SEI product called the *Accelerated Improvement Method* (AIM) combines the use of CMMI and the *Team Software Process* (TSP).

Keep in mind that the purpose of the CMMI is to improve performance by *improving your processes*. Simply placing a layer of CMMI processes over the top of what you are currently doing is not *automatically* going to result in process improvement. Chances are you now have at least some good processes that have worked in the past. If you find value in some of the CMMI practices, you should implement those practices in a way that continues to provide you with the value of the things you like about your current processes and replace or add to the things in your current practices that need strengthening with the CMMI practices.

A smooth way to gain maximum improvement of your processes is by following CMMI as a *guide to building a systemic process improvement infrastructure tailored to your project*. It will be much easier if you have access to a CMMI consultant or a trained expert on your staff (or available in your company).

SOFTWARE RESOURCES DOMAIN

3

Chapter 7

Managing the Software Project Team

People are your most important resource. Finding, recruiting, nurturing, and retaining a competent staff is the single most important function of a Software Project Manager.

This chapter is focused on *people engineering*. Since software development is a labor-intensive activity, people are a *critical element* of the path to successful program execution. You may have and follow the best processes, you may work in a world class development environment, and you may be an outstanding Manager, but if you don't have a competent staff, both in terms of quantity and quality, the likelihood of problems ahead is very high.

Managing your team is the heart of the Software Resources Domain and is discussed in this chapter in the following eight sections plus several Lessons Learned stories applicable to the related topics:

- Staffing Your Team (7.1)
- Staff Loading and Skill Levels (7.2)
- Social Media and Generational Differences (7.3)
- Motivating and Demotivating Factors (7.4)
- Communications (7.5)
- Managing Programmers (7.6)
- Creating an Environment of Excellence (7.7)
- Software Training Plans (7.8)

A typical Software Development Team was, at one time, composed of a bunch of bearded guys wearing flip flops, who spoke in terms that only they understood, who really were not interested in that thing called *requirements*, had no plans or interest in preparing any type of documentation, and were entirely focused on coding at the computer. Happily, (or maybe, hopefully) those days are gone.

Those days of old have been replaced by years of experiences with successes and failures that have transformed this seemingly unmanageable and uncontrollable *art* into what is rapidly becoming a bona fide *engineering discipline*. Nevertheless, the basic nature of software development may always involve a component of *artistry*—especially for large complex developments.

7.1 Staffing Your Team

In most enterprises, if you have great people, the successes will most likely follow—but the real challenge is *finding* the high performing Software Engineers. Employing the best staff possible is usually *worth the extra cost* since the productivity range for Software Engineers (especially Programmers) is much wider than the range of their salaries. Productivity estimates comparing the best to the worst Programmers can range from 5:1 to 25:1. Barry Boehm et al. (2000) reported that teams of *experienced* Programmers can be expected to be 5.3 times more productive than teams of inexperienced Programmers.

If you apply extra funding and resources to *finding the right people* for your staff, it will provide a high rate of return—exceeded only by your acumen in actually *picking* a compatible team from the candidates you found. Teamwork is an essential ingredient in the successful development of complex software-intensive systems; one weak link in the team's chain can be a recipe for failure.

> **Lessons Learned.** If you are hiring a lower-tier Manager who will report to you, it is important to remember a rule-of-thumb comment presumably made by Steve Jobs who said: *A's hire A's; B's hire C's.* This insightful philosophy addresses the criticality of hiring excellent Managers due to the *long-term combinatorial impact* from the quality of your hires.

7.1.1 Recruiting Paths

There are several paths you can take, most of them simultaneously, to find the right candidates for your team including the following approaches:

- *Personal Contacts and Networks*: Good resumes, and even good interviews, cannot replace *personal knowledge* of the individuals being considered. You should keep a growing active network of known good performers both inside and outside of your organization.
- *Matrix Organizations*: Companies often maintain a centralized organization that provides people on loan (or matrixed) to multiple projects, within the company, who need temporary staffing. The obvious advantage is that your project may need specific expertise for a limited period of time so hiring a full-time employee with that expertise is not justified. Instead, you can "borrow" them temporarily from the matrix organization even though they may be on your team for years.
- *Functional Departments*: Another version of the matrix organization is the use of people from functional departments that are focused on one (or a few) disciplines such as database programming, web development, and many others. If your project is large and complex, your staff may be composed of a potpourri of full-time employees, people from the matrix organization, people from more than one functional department, plus contractors.

    Managing a heavily matrixed team can become difficult because the matrixed people on your team do not report to you directly. The support you receive from functional organizations may be performed on *their* site, and their support is fully controlled by their Manager. As a result, you will have negligible authority and little control over them and your management of them may boil down to negotiations with the Functional Manager for their support.

    Lessons Learned. In my experience, a matrix organization has many more advantages than disadvantages, so it is a recommended approach. It is also an excellent idea to cultivate a good working relationship with the Matrix Department Managers as well as the Functional Department Managers. You need their support.

- *Employment Agencies*: Headhunters are usually a great source for candidates and, even if your company does not allow you to deal with them directly, you can capitalize on their services by working with and assisting your Human Resources (HR) department. There are periods of time, like during a recession, when your company has layoffs which are likely to result in a restriction on outside hiring. However, if you really need an important addition to your staff, and you can justify this need to HR, then be proactive and go for it. It is important to cultivate a good working relationship with your HR Department as they are a big help to you in your recruiting efforts.
- *Trade Shows and Conferences*: Hopefully, your budget will allow you and members of your staff to attend trade shows and technical conferences where you may have a golden opportunity to meet and recruit qualified candidates. Presenting a paper at a conference is a good way to draw attention to your company and to your program or project.

7.1.2 The Job Description

Job descriptions are a valuable marketing tool for attracting qualified candidates so you should give it the attention it deserves. Your HR Department probably has the responsibility for creating job descriptions but most, if not essentially all, of the input will come from you. Make sure you stay involved throughout the entire process and insist on the opportunity to *review the final version* before posting or advertising. Well-meaning non-technical recruiters may make word changes that can alter the context. My suggested elements of good job descriptions are:

- *Job Title and Contact Information*: The job title should be meaningful and descriptive because if you use generic titles you will receive many resumes that are not applicable to your needs. Below the job title should be a Requisition Identification Number for control and communication. Always make sure your full contact information is included on the top of the job description.
- *Job Snapshot*: I recommend placing an overview of the job at the top of the job description. It could contain at least the following bulleted items: Employee type (full-time, part-time, consultant); location of the job (the city or more specific if the job is in a big city); experience (years of experience or range of years); travel (percent of work week or month on travel); security clearance (if applicable); relocation assistance (yes, no or negotiable); and other specific job related topics.
- *Job Responsibilities*: This is the heart of the job description as it contains the principal tasks to be performed. It may be presented as general responsibilities followed by specific responsibilities.
- *Job Qualifications*: The qualifications required might be difficult to determine, because if you are too stringent you may lose good candidates; on the other hand, if you are too lenient you will get many resumes from

under-qualified candidates. One way to handle this is to have two lists: *minimum qualifications* and *preferred qualifications*. In any case, this section must contain the required degrees, skills, related experience and specific knowledge needed.

■ *Company Profile*: As an incentive for the candidate to want to be part of your team, you should include a brief description of your company and/or project, its focus, product lines, and other positive promotional attributes. This "About Us" paragraph could be located at the top, side or bottom of the job description. A marketing profile like this is especially important if your company is new or not well known.

After you prepare the first job description, you will be able to reuse much of it to create a *set of job descriptions* for multiple levels of job capabilities. The time you spend on preparing good job descriptions can be valuable to you later on as a useful tool to help manage your team by comparing their on-the-job performance against their job description.

7.1.3 Staff Selection Filtering Process

The staff selection filtering process that I found to be most effective is performed in stages. A suggested filtering sequence is depicted in Figure 7.1 and each filter is discussed below.

Review Resumes. If you get a truckload of resumes to review, establish a clear first-level selection criteria so HR and some members of your current staff can help you narrow down the candidates. This first-level filtering task is often performed by a computer matching keywords in the resume against the job description. But there are pitfalls here. If the screening criteria is too strict, it can eliminate some very good candidates. The computer may not be able to understand the substantive experience of the candidate and relate it to your specific needs. The following is a personal example of overly restrictive criteria used by a resume reviewer.

Lessons Learned. When I first applied to Lockheed my resume was reviewed by a senior Systems Engineering Manager who rejected it and noted on my resume "He has no previous experience in the aerospace industry." However, the Software Manager, who ultimately did hire me, had the insight to realize that my background and attributes did involve the *type of experience* he was looking for even though it was in other engineering fields. He called me to conduct a telephone interview, it was followed by an interview, the hiring process and my subsequent successful 34-year career in the aerospace industry.

A much more interesting part of this story is that less than 1 year after I was hired into the software matrix organization at Lockheed, I was temporarily assigned to the same Engineering Manager who had rejected my resume, but he did not remember my name. One of my first assignments was to perform an analysis of a complex set of factors in order to make an informed engineering judgment. It was a baffling array of input data to everyone so why not give the task to the new guy who didn't know it was a nearly impossible task!

I identified the salient factors, eliminated the others, synthesized the remaining data, organized it into a simple structure and put it all on *one sheet of paper* making the analysis relatively easy. Needless to say, the Engineering Manager was astonished and told my Manager "I don't know how he did that!" My Home Department Manager showed me his rejection note at my next performance review, told me the story, and gave me a pay raise!

The reason for telling this story is to give an example of being too restrictive or having a too narrow set of guidelines when reviewing resumes. Even well-written resumes do not give you a clear profile of the individual you are interviewing because such things as people skills and personality are not discernable from a resume. Even if the resume is truthful, it may be heavily skewed to the job description so the candidate may not actually possess the level of experience you are looking for. Degrees are important, but if they were obtained more than 10 years before the interview, a candidate's actual experience is usually much more important than the degree.

Likewise, a high college GPA is meaningful for the first 5–10 years after college but beyond that what the person has accomplished is far more important. Also, you need to *read between the lines* because the wording used and the terminology in the resume may be a result of the person's experience in some other field, but if you take the time to understand

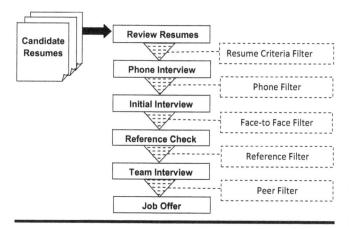

Figure 7.1   Multi-stage staff selection filtering process.

what the words mean, you may find they might match your needs.

When reviewing resumes, you should also make a note of spelling and grammatical errors. This Guidebook emphasizes good software documentation as an important element of the life cycle development process. In most cases, the Software Developers are the authors of that documentation. Here is another personal example:

> **Lessons Learned.** When I was the Software Lead for the NASA Space Station Freedom program at Lockheed Martin, I needed to fill an important role in my Thermal Control System group. During lunch one day I was introduced to a fellow who was leading a Software Team on another program who had a PhD in the applicable engineering field. He even had experience in heat pipes that were an important part of our system.
>
> Wow, what a find! A perfect candidate and he was looking for a change. I asked him to send me his full resume which he did in the company mail system. Attached to his great resume was a three-sentence note he wrote in longhand—two of the three sentences were *barely English* and I am being kind here. At that time we were in the requirements stage of a large program and were focused on writing documentation. Obviously, he was not a good fit for our team at that time. Do not underestimate or undermine the importance of a Software Engineer's *ability to write.* It does not matter whether you love it or hate it—*some level of documentation is part of the job* (Chapter 13 covers software documentation and work products).

If there are a lot of resumes to review, you can speed up the resume review process if you have a streamlined procedure of correlating each resume to the job descriptions. This could be as simple as using a highlighter for the match and mismatch attributes, or as complicated as a spreadsheet with weight factors for each attribute. You could also use this procedure to identify what issues need to be further investigated during the telephone and on-site interviews.

Phone Interview. After you identify a potential candidate from the resume, the next step is to set up and conduct a phone interview. Before conducting the phone interview, make sure the candidate has the pertinent job description and has reviewed it. If not, send it to the candidate by email while you are talking to them. Ask the candidate to explain why he or she believes they are a good match as well as where there is *not* a good fit. Phone conversations should be used to clear up any ambiguities that otherwise could be a show stopper for either side. Offer to answer any questions about the job, the project or the company.

Emphasize to the interviewee that *staffing is a two-way process* for each of you to learn more about the other in order to decide if there is a good match. It is very important to *involve candidates* in decisions that affect them. Unless you, or the candidate, identify a show stopper, and you believe the candidate is indeed a good prospect, set up an interview. The phone call is not a replacement for a face-to-face interview. In some cases, it may be possible to set up a video conference with the candidate in order to combine the objectives of both the phone interview and the initial interview.

Initial Interview. The initial face-to-face interview is like a first date. Both sides are trying to discern the truth about each other while presenting themselves in the best possible light. Start the interview by offering more information about your working environment including topics such as your organization charts, facilities, tools, workload, team members and management style. *Get feedback* as you talk about the candidate's preferences and opinions. While gaining insight about the candidate, you may learn something new about the software industry from them that you were unaware of.

After the ice is broken, and a good dialogue is established, focus on the issue of deciding if there is a good match. Use the marked-up resume or other tools to make connections between the candidate's qualifications and the job description. Keep good notes during the interview since that will prompt your thought process after the interview. Watch out for charm and good talkers; *some people are better at landing a job than doing it.* Also, beware of people who think they are gurus and people who are self-focused rather than task oriented. Your project needs results, not latent capabilities. You also do not need the distraction that can be caused by cynics as discussed in Subsection 7.6.3.

References and Past Work Check. Checking the references provided by the candidate should be done; however, you may not get insightful information since legal reasons prohibit many companies from stating only that the candidate did work there during a specific time span and had a specific job title. You can push for more details and you may get some useful response. Since the reference list was offered by the candidate, you can expect to receive only positive responses but listen for credibility and enthusiasm from the reference and try to evaluate the unexpressed meaning or implications of what is being said.

If you are hiring from within your company, it would be beneficial if you can talk candidly to current or former Managers of the candidate. Beware of a current Manager who may "stretch the truth" to offload a problem employee. In every case, have your questions well outlined before talking to the reference. Be considerate of the person's time, and *listen more than talk.* It is okay to be persistent, but not too pushy, for comments about the candidate's strengths and weaknesses as well as intangibles such as personality, team player and work habits. End by thanking the candidate's Manager and offer to reciprocate as you may be able to enter their name into your network contact database.

In some selective cases, it may be a good idea to ask for work samples, or examples, of the candidates past work especially if there is some concern as to the candidate's level of expertise. The samples could include computer code, documents, presentations, plans or anything else that is relevant to the position. Keep in mind that the work sample may not be a product of the candidate. To help authenticate the pedigree of the sample ask questions about some aspects of the sample face-to-face and judge the explanation and the degree of hesitancy.

Team Interview. You should always give members of your team a voice in the selection process but don't take time from their job to participate in the interviewing process until you have concluded that the interviewee is a very good candidate for the job. Don't waste the team's time on anybody you have already vetoed. When you do get the team involved restrict the size of the jury to those who are most likely to interface with the candidate.

> **Lessons Learned.** I prefer one-on-one interviews; however, group interviews may be needed especially if some of your key members of your team are not very good at conducting an interview. In a group interview, avoid what I call a "sterile" interview where there is a written set of questions and each member of the interview team *reads a question from the script.* This type of interview is a turn off (it is for me). Remember, the interview is a two-way street; a two-way discussion; you must give the candidate ample opportunity to ask questions and make comments.

Follow-up interviews are optional but may be necessary if there is still uncertainty regarding a good match, or if key people on your staff who need to interview the candidate were not available during the initial visit. As the Software Manager, the final decision is yours. However, a consensus is the most effective method. Being closer to the actual work, peers are in a good position to judge the candidate. A member of your team with a negative view of the candidate is likely not going to cooperate with them. When the team rejects a candidate, that decision helps to prove to the members of your team that team membership is a *precious commodity.* Unless you are absolutely certain, you can manage a bad fit, err of the side of caution and don't make a job offer to a candidate rejected by your team.

7.1.4 Round Pegs and Square Holes

When I interview Software Engineers applying for a position on a software-intensive project, one of my key qualifying questions is embedded in the following verbal exchange:

> **Lessons Learned.** I tell the interviewee that although it is somewhat simplistic, let us assume for this discussion that the work performed by Software Engineers fall into *three groups.* The first group can be called the "front-enders" and includes tasks such as requirements analysis, planning and writing plans, Conceptual/Architectural Design, and schedule/cost estimation. The middle group involves the implementers, the Programmers, who develop the Detailed Design, write the code and perform Unit Testing. The third group includes the Testers and Integrators of the system and let's put Configuration Management and Software Quality Engineers in this group even though their involvement is throughout the life cycle. Then I ask the applicant what percentage of his/her background and capability falls into each group.
>
> Once they give me a percentage breakdown of their background into the three groups, I know the type of Software Engineer I am speaking to and if they have a match to the needs of my program. The reason I will know this is because there is a major difference in the *mindset* of the people in group two versus the other two groups and it is a matter of *perspective.*
>
> The people in groups one and three generally have a *big picture* perspective. They are oriented to the interfaces and relationships that pervade big systems. People in group two are much more *introspective* in that they must focus on very precise parts of the total system in order to perform their job. This distinction is very similar to the typical perspective of a mathematician, and that is why mathematicians usually are very good Programmers.

It is important to keep in mind that this dichotomy has absolutely *nothing to do with intelligence*; some of the most brilliant and outstanding Programmers I have met got glassy-eyed once I started talking about external interfaces and the big picture. So, if you are hiring Programmers, select people who fit into the middle group. If you are hiring other members of the team, especially Software System Engineers, System Analysts, Testers and anyone developing customer requirements or plans, then pick only people who fit into groups one and three.

> **Lessons Learned.** The reason this distinction between the two perspectives is so important is because, in my opinion, it is one of the major reasons why so many information systems fail. Namely, Programmers are hired to do the front-end work and it may be considered equivalent to a "fish out of the water." Putting square pegs into round holes is not going to provide successful results.

Now, before you conclude that this is all hogwash, there is *one caveat*. There are some Software Engineers that do good, even very good, work in *all three* groups. These "triple threat" people constitute maybe 5%, or at most 10%, of the Software Engineering community. Even though the valuable people of this type can do "very good" work *in all three groups*, outstanding Programmers are better Programmers, and outstanding Software System Engineers are better at performing work in groups one and three. Obviously, there are many variations to this simplistic dichotomy of perspectives since the background and capabilities of Software Engineers vary significantly. However, it can be a useful gauge.

> **Lessons Learned.** I once had a "triple threat" engineer working for me who needed a lot of managing. He finished his assignments quickly, got bored and ventured off into all kinds of distractions, like inventing things, that had nothing to do with the project. These types of engineers are valuable, but you need to exercise

your management prerogatives to make the most use of their talents. The simplistic bottom line to this discussion is: *make very sure the Software Engineers you hire have the experience, background and mindset for the job they are being hired for*—in other words, put the right pegs in the right holes.

7.2 Staff Loading and Skill Levels

The required time-phasing of software development personnel is normally documented in the program's *Earned Value Management System* (EVMS) database (see Section 8.5). EVMS is the basis for monthly software cost/schedule reporting and tracking for each *Work Breakdown Structure* (WBS) element discussed in Subsection 1.5.2 and in Section 8.2. The WBS, also referred to as the *Work Plan*, is a decomposition of *all* the project activities into smaller, manageable, and controllable tasks that are generally of short duration, and where each WBS task has a measurable completion criteria. Also, it is recommended to avoid assigning a single WBS task element to more than one organization.

7.2.1 Staff Loading

Software staffing varies during the program from an initial buildup to a peak, and then a gradual decline as the majority of the software effort is completed. An estimated staff-loading chart must be included in the SDP, and Figure 7.2 is one example of how it may be depicted. The size of the software task, the schedules in place, and the planned staff loading must form an *executable* software development effort across all the Software Team members.

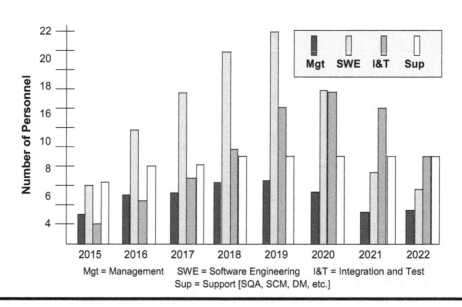

Figure 7.2 Estimated software staff-loading—example.

Table 7.1   Estimated Skill Levels by Location and Function—Example

Able Corporation, Anytown, CA	Skill Level						
Function	Eng-1	Eng-2	Eng-3	Tec-1	Tec-2	Tec-3	Total
Software Engineering	3	10	12	2.5	1.5	0	29
Software Quality Assurance	1	1	2	0	0	.2	4.2
Software Configuration Management	1	2	2	0	.5	0	5.5
Software Testing	1.5	1.4	1.1	.4	.2	.2	4.8
Software Management	2	2	.5	0	0	0	4.5
Total Staff:	8.5	16.4	17.6	2.9	2.2	.4	**48**
Skill Level	**Skill Level Definitions**						
Engineer—1	Senior Level with 12 or more years of experience and …………..						
Engineer—2	Mid-Level with 3–12 years of experience and ……….						
Engineer—3	Entry level with 0–2 years of experience and …………….						
Technician—1	An expert in their discipline with 12 or more years of experience and ………….						
Technician—2	Emerging authority with over 8 years of experience and …………..						
Technician—3	Entry level with an advanced degree and …………….						

7.2.2 Skill Levels

A breakdown of the planned skill levels, geographic locations of the skills needed, and the security clearances of personnel performing the work (if applicable) should all be documented in the SDP. The Software Leads typically estimate and maintain the staffing profiles and work effort loading distributions. This data can be formally reported in what is sometimes called the *Cost Analysis Requirements Description* (CARD). Table 7.1 is a tabular example of how to show the skill level requirements, by company and by software function with a definition of the skill levels.

7.3 Social Media and Generational differences

7.3.1 Social Media

Social media and the internet have had a profound impact on the hiring process. They provide Hiring Managers with a new cost-effective lens through which to evaluate job seekers. The principal tools include Facebook, LinkedIn, Twitter and Google. A 2015 survey (Republic Media, 2015) found that 43% of employers reference social networking sites when researching a candidate, and 51% of the employers admit to *not* hiring a job candidate based on his/her social media posts. Nearly 25% of employers surveyed stated they hired a candidate based on what they found in the internet search. When hiring Software Engineers, an online search will help to locate candidates but much more needs to be done.

That survey also found that about 62% of the candidates check out the social media presence of the companies they are interested in. In order to maximize the attractiveness of your company, it is very important for your company to maintain a high level of social media presence.

7.3.2 Generational Differences

Generations have different personalities that change over time. The experiences and biological impact of aging, plus the changing roles that people play as they grow older, will typically produce changes in their values, attitudes and social behaviors. Definitions vary, but Table 7.2 is a simple interpretation of the differences between four generations. The approach to hiring and managing your staff must take into account their generational background.

Lessons Learned. *A candidate's age is not the only factor you should consider* because there can be a vast difference between a person's "chronological" age and their "real" age due to their health, lifestyle and physiological makeup. There are people in their 30s who are in poor physical and/or mental condition, and there are 70-year-olds who run marathons and win Nobel prizes.

Table 7.2 Generational Differences

Generation	Year Born	Music Media	Use of Social Media and Digital Technology
Silent Generation	1925–1945	Vinyl Records	Social media used primarily with family and friends
Baby Boomers	1946–1964	Cassette Tape	Embraces social media and digital technology
Generation X	1965–1979	CDs	Loves digital technology, social media and independence
Millennials(*)	1980–2000	iPods; DVDs, etc.	Deeply devoted to social media and digital technology. Millennials treat hand-held gadgets as a multitasking *body part*

(*) Also called Generation Y or Generation Next. After the Millennials is Generation Z (2000–Present).

Therefore, there has to be more to the issue of how old you are other than how many years ago you were born! (I could go into a lot more detail on this subject but I believe it is not necessary.)

The Silent Generation tends to be reliable, hard workers and not too hard to manage. Baby Boomers tend to be loyal and often workaholics willing to do more than required to help your project and to advance their careers. Generation X people tend to be a bit cynical, are not as loyal as Baby Boomers, and view time as a commodity they are not willing to give away, in other words, work occupies their time between weekends. These are generic observations and there certainly are wide variations in each group.

7.3.3 Characteristics of Millennials

Software Managers are likely to find that managing millennials is more difficult than managing the other generational groups. However, the valuable attributes of millennials may be exactly what your program needs. Their typical characteristics are listed below along with some recommendations you might consider during the hiring process.

Multitasking. Millennials are multitasking pros and can juggle many responsibilities at once. They are typically easily distracted and find social media and texting hard to resist so keep them busy and don't let them become bored. Keep millennials on track by being upfront about your expectations of them and establish at least weekly goals or, in some cases if needed, daily goals. If your millennial employees have deadlines to meet, you'll be less likely to find them playing on their phones at the office (the key words here are "less likely").

Tech-Savvy. Millennials are more tech-savvy than any other previous generation; they are a different breed. Their inherent capabilities are huge, and it is the task of the Project Manager to tap into this enormous reservoir. Make sure that your company and team stay up-to-date technologically. Also, ensure that your company and career sites are mobile-optimized so that your sites can easily be found online from any device at any time. In addition, make the application process fast and easy by allowing people to apply for positions with content from LinkedIn and other sources. Millennials can be a big help to you in maintaining social media connectivity. Millennials know essentially everything there is to know about social media because they grew up with it. They are constantly perusing Facebook, Twitter, etc. because it is typically how they share and get information.

Instant Gratification and Recognition. Millennials, especially if they are Programmers, need to feel like what they are doing is important. They grew up with constant praise from their parents, so it is what they are accustomed to. During the hiring process, stress the importance of their position and why they will be making a valuable contribution to your project. Once on the job, recognize their accomplishments publicly because recognition encourages them to work hard and it increases their job satisfaction. Also, tell them how the performance review process will be conducted and make sure you follow that process.

Work–Life Balance and Flexibility. Generally, millennials are not as willing as earlier generations to sacrifice their personal life in order to advance their careers. They like to "work hard—play hard" and want to work at a company that appreciates this desire for balance. They also expect a flexible Work Environment so don't force a 9 to 5 work mentality on them. They also prefer to work for a company that supports various causes.

Hopefully, your company values work–life balance so tell them that during the hiring process and mention any sponsored events outside the workplace. Also, mention any fitness or health-related programs that your company provides.

Even though the majority of millennials say they value flexibility over salary, they still want to be paid fairly. A fair approach may be to pay your millennial employees based on an evaluation of the projected impact they will have on your business—and be transparent with them on how you arrived at this figure. It is important to remember that many millennial employees are saddled with a financial burden that previous generations did not have, namely, student loans. By offering a means to help millennials pay off their considerable debt, you will go a long way toward offering them the financial security that allows them to remain with your company and fully focused on their work.

Self-Expression. Most Programmers tend to be individualistic, and they use a variety of ways to express themselves beyond electronics and social media. According to a recent Pew Research Report (Pew, 2010), 40% of the millennials have a tattoo; half of those have 2–5 tattoos and 18% have six or more; however, 70% report that their tattoos are covered. In addition, the report notes, nearly a quarter of the millennials have a piercing in places other than their earlobes.

Such *body art* could leave an unfavorable impression in some business environments. The potential disadvantages of hiring people with visible body art must be evaluated by Project Managers taking into account their technical capabilities, the frequency of customer interface expected for the position they are being hired for, the type of business involved, the makeup of the other team members, and other pertinent factors. It all should be considered on a case-by-case basis.

Collaboration. Millennials are extremely team-oriented and enjoy collaborating and building friendships with colleagues. During the hiring process, let them know that there will be plenty of opportunities for collaboration and team projects. You should also design your office space to allow for teamwork and easy idea sharing (discussed further in Section 9.6).

Transparency. Millennials need an open and honest relationship with their Manager and coworkers, and they do not expect unpleasant surprises when they join a company. They want assurance that their opinion is valued. Make certain that there is valid, up-to-date information about your company available online. During the hiring process let them know about any downsides that the position they are applying for may have. They will appreciate your honesty.

Employment Expectations. It is important to be aware of what recent graduates really want; what really drives their job selection decision. According to the 2018 Student Survey Report by the National Association of Colleges and Employers (NACE, 2018), over 22,000 students with Bachelor Degrees in the 2017–2018 academic year rated the following job attributes of high importance:

- Job security and personal growth (over 82%)
- Opportunity to develop skills, friendly coworkers, and good benefits (76–79%)

7.4 Motivating and Demotivating Factors

Software Managers should understand the key factors that motivate and de-motivate the performance of their staff. It must remain clear that every member of your team has individual traits and personalities, so there are really *no universal, magic wand approaches that works on everyone.*

However, there are some generic factors that influence most workers in every profession. People like Herzberg et al., (1959) researched the issue of job motivation before many of the readers of this Guidebook were born. Mantle and Lichty (Mantle, 2013) revised that research to make it more applicable to current day Programmers.

Table 7.3 is my interpretation of the top 12 motivating and demotivating factors for a typical software staff with the impact of each factor rated as high, medium and low as a motivator or a de-motivator. In the context used here, "demotivating" factors are the *causes for dissatisfaction*—especially by Programmers.

Following Table 7.3 are brief descriptions of the 12 listed key factors to help clarify what is involved in each motivating and demotivating factor and what Software Managers can do to improve motivation and mitigate dissatisfaction.

7.4.1 Achieving Something Important

The motivational impact is very high for jobs that are *perceived as achieving something important,* but if the job is not perceived as important it may not be a major cause for employee dissatisfaction. Most professionals want to believe their work is making a difference and making a contribution. People will have the incentive to work much harder, and to produce a higher quality product, if they believe their efforts really matter—or they will have the opposite incentive if they think what they are doing is unimportant.

If your company, and the system you are developing, can be presented as producing something that will make a positive impact on "improving the world" in some way, then Software Managers can use this pitch to help the recruiting effort and help his/her team to understand the importance of their contribution thereby inspiring them to achieve a performance level they would otherwise not attain. The perception of achieving something important will likely be a greater motivational impact on younger members of your team, but believing that your project is important has a positive impact on almost everyone.

7.4.2 Personal Growth

The software industry is very dynamic since technology changes so rapidly. Like most professionals, Software Engineers must be *lifelong learners* just to keep up-to-date on new tools, techniques and technologies. Software Managers should promote a *proactive learning atmosphere* by holding frequent technical sessions addressing the latest developments in the software industry.

These technical sessions can be conducted by representatives from academia discussing their research, authors of recent books or articles, vendors discussing their products,

Table 7.3 Performance Motivating and Demotivating Factors

Factors Affecting Performance	Impact of Factor as a Motivator	Impact of Factor as a De-Motivator
Achieving Something Important (7.4.1)	High	Medium to Low
Personal Growth (7.4.2)	High	High
New Technology (7.4.3)	High	Low
Recognition and Praise (7.4.4)	High	Low
Job Enjoyment (7.4.5)	Medium	High
Personal Reward (7.4.6)	Medium	Medium
Interpersonal Relationships (7.4.7)	Medium	Low
Promotions (7.4.8)	Medium	Low
Good Working Conditions (7.4.9)	Medium	High
Technical Respect for Manager (7.4.10)	Low	High
Ethical and Realistic Policies (7.4.11)	Low	Medium
Communications (7.5 and 7.7.2)	Low	Medium to High

consultants, customers, Product Managers in your company, members of your staff and Software Managers. In addition, much can be gained from setting up special interest groups and sending members of your staff to technical conferences and trade shows where they, in turn, can convey what they learned to other team members who did not attend.

If your company provides tuition support then, by all means, encourage those on your team to pursue further education that would benefit them. Of course, formal training programs within your company should always be a high priority—especially training for new tools. Finally, your staff should learn from you. If you play tennis, the only way to get better is to play against opponents that are better than you. This analogy applies dramatically to Software Engineers because if you cannot teach them anything *new* your chances of retaining your best employees will be greatly diminished.

7.4.3 New Technology

One of the easiest and most effective ways a Software Manager can motivate their staff is to provide them with the latest and greatest "toys" and technology. An up-to-date *Software Development Environment* will not only motivate your staff to increase their performance and productivity but will also win you some valuable points from your staff. If your company allows multi-year amortization of capital equipment, it may not cost that much to provide equipment such as large dual screen monitors, color printers, stereo equipment, mobile phones, and tablets.

On the other hand, jumping to a new software product release prematurely can be a big distraction as well as a big hit on productivity. If your team is using a *mature version* of a software product, it may *not* be a good idea to upgrade as soon as a new release comes out because it is unlikely the new features will make much of a difference to your project, and it is usually better to wait a while until the major bugs of the new release are found and fixed by other users who decide to go to the latest version. Pioneering and software development are mutually exclusive terms.

Lessons Learned. It is important to keep up with new technology; however, it is sometimes wise to stick with the old "tried and true." A common way to say that is "if it works, don't fix it." As an example, I once was given a tour of a huge Data Center at a major Air Force Base. It was really impressive because they had the latest in technology and it was everywhere. Then we suddenly came upon a 20-year-old little Burroughs computer. I was equally astonished to see it as I was to see the new technology and asked "Why is *that* here." The answer was simple, they got it 20 years ago to perform a specific function, they still need that function performed, and the little old computer still does the job. (I didn't ask about maintenance or spare parts.)

7.4.4 Recognition and Praise

Another high motivator is giving recognition and praise to members of your staff when their work is performed in an exemplary manner. It essentially costs you nothing and takes very little time to do, but it has a profound impact on the work ethic of your staff. Most people love to have their ego enhanced and the more often the better.

> **Lessons Learned.** I have worked for a few (fortunately, very few) Managers who take the approach of always looking for something to complain about; Managers for whom you *cannot please no matter what you do*. There is nothing wrong with pushing your team to excellence; however, when taken to an extreme it is demoralizing to the team and is clearly the wrong approach. This motivational factor is best summed up by the poet Ella Wheeler Wilcox who wrote: *"A pat on the back is only a few vertebrae removed from a kick in the pants, but is miles ahead in results."*

7.4.5 Job Enjoyment

As long as I am quoting Ella Wheeler Wilcox, let me do one more of her famous quotes: "Laugh and the world laughs with you; weep and you weep alone." Successful Software Managers have the ability to loosen up, to enjoy working with their staff, and can find ways to encourage their staff to both *work hard and play hard*. In this context, "playing" can mean having out-of-the-office games such as basketball or soccer lunch breaks, pizza breaks (even better if the Manager buys), company parties, brown bag lunches (also great for learning sessions), holiday parties hosted by the Software Manager and casual gatherings at the local waterhole after work.

It is unreasonable to expect your staff to work hard *all the time*. Outlets are forms of "play" that can pay large dividends. The positive outcomes of these activities include bonding friendships, promoting informal communications, enhancing better health, making your project a fun place to work and making you a terrific Manager.

> **Lessons Learned.** There are also some small, inexpensive fun things you can do. For example, I had a plastic figure of a funny looking guy that you could punch on the head and he made a burping noise. This $3.99 toy was *revered* by my staff since each week it was awarded to the best performer of the week, and it was proudly displayed on the winner's desk during the week. It was once awarded to a senior member on my staff just before he moved to another division of the company and he insisted on taking the toy with him (I gave it to him—and the two of them rode off into the sunset).

Another good approach to enhancing job enjoyment and increasing motivation is to provide food and snacks for your team. Having a well-stocked refrigerator can provide a priceless pay off in on-time product delivery because the Developers can work productively together right through dinner time and often well into the evening. When people leave for dinner, they seldom return that evening. Some larger firms even provide in-house catering to its technical staff all day long.

7.4.6 Personal Rewards

It is generally agreed that Programmers are not highly motivated by their paychecks, however, personal rewards are important. In the high tech software world, this would include salary increases, bonus, stock options, job promotions and increased perks that collectively provide incentives for higher performance. It is important to make sure that the salaries received by your staff are fair and adequate. No one will complain if the best Programmer on your staff is paid the most. Salary and rewards to your staff should be *fair, reasonable* and *understandable*. If your staff members feel fairly compensated they will be focused on doing a good job and salary becomes a non-issue.

Stock options for companies that have not yet gone public can be a significant motivator, especially during recruiting. The value of personal rewards is highly contingent on what is important to each individual, so you have to determine what motivates each person. In some cases, the right perk to the right person may be the greatest impact to increasing their motivation.

7.4.7 Interpersonal Relationships

A good interpersonal relationship usually means that staff members will be much happier if they *like* the people they are working with and for. If they really like the people they daily interface with at work they will be motivated more than they will be de-motivated if they don't like them. Of course, if a staff member simply can't tolerate another staff member, you have a problem that needs immediate attention. Software Managers can and should take positive actions that can prevent, or at least minimize, such problems.

Definitely avoid toxic people who are cynical and abrasive as their negativity can be very disruptive. A serious mistake a Manager can make is to tolerate any unacceptable behavior that threatens team productivity. If you inherit such a person, you will likely need to work with HR to legally eliminate that problem.

Lessons Learned. During the hiring process, here is one trick you can use to help identify potential problems; propose the following scenario to the candidate you are interviewing: "Assume you are working on a project with another Co-Developer, you are both fully qualified to perform the required tasks, and you are both at the same level of seniority.

There comes a point in the design process where it is clear an innovative approach is needed to solve a rather complex problem. You conceive a creative solution that, in your judgment, will fully solve the problem. However, your partner has proposed a completely different approach that you firmly believe is inferior to your solution. How would you (the candidate) handle this predicament?"

There is no one best answer to this question because there are a few very good answers. However, there is *one wrong answer*. If the candidate firmly and resolutely insists on defending his/her approach—no matter what—then you can conclude this candidate is not very likely to be a team player and stubborn enough to cause dissension and disruption to the Development Team. Almost any answer is okay except "my way or the highway."

You may find yourself working for a domineering Senior Manager who believes that there are three ways to perform a task: the right way, the wrong way, and his/her way. If you subscribe to either of the first two approaches, you are forevermore considered a "jerk" by that Manager. Such a Manager believes that the only good ideas are his/her ideas.

Lessons Learned. One way to surreptitiously get your domineering Manager to accept your idea, or your approach to a solution, is to go about it this way. In presentations to your Senior Manager, and even during casual discussions during lunch, lay out the problem, a little at a time, in such a way that he/she will, *on their own*, come up with the solution you have in mind. When that happens, your response should be "that is a great idea!" This works if your Manager is analytical enough to follow a structured thought process leading to the logical conclusion. If your Manager does not think that way… good luck.

7.4.8 Promotions

The management of promotions can be tricky. It helps to have good job descriptions but evaluating an employee's performance can be subjective and debatable when trying to determine if an employee has demonstrated a level of performance equal to or greater than what was expected for his/her job. One common approach is not to promote someone until they have already successfully performed at the level to which they are being promoted.

This approach helps to avoid realization of The Peter Principle—a management concept where the selection of a candidate for a higher position is erroneously based on their performance in their *current role* rather than on their abilities relevant to the higher role. It is named after Laurence J. Peter who co-authored the humorous 1969 book *The Peter Principle: Why Things Always Go Wrong* (Peter and Hull, 2011). The author suggests that people will tend to be promoted until they reach their *level of incompetence*. The generalized Peter Principle is: *Anything that works will be used in progressively more challenging applications until it fails*. In other words, everyone and everything has limitations.

The higher role that the employee is promoted to may not be more difficult than the current role, but it may require different skills the employee does not have. For example, an excellent Programmer may prove to be a poor Manager because of his/her limited interpersonal skills needed by a Manager to lead a large team effectively. The following guidelines can help to mitigate the risk associated with The Peter Principle:

■ Promote based on *proof to succeed in the higher role* rather than the excellent performance demonstrated in the lower-level current role. Progressively add tasks to their current role that they will encounter in the higher role and evaluate how well they performed them.

■ If you must fill a role but you are not sure if the employee you have selected to fill that role is capable of handling it, you can put them in that role in an "acting" status until they prove they can perform the new tasks.

■ Implement training programs in advance for those being considered for promotion.

■ Provide a parallel career path for your technical staff without requiring their promotion to management, similar to a warrant officer in the military.

■ Implement an *Up or Out* approach, similar to policies followed by the U.S. and British armed forces, whereby persons not promoted above certain ranks, within a fixed number of years, are deemed to lack the necessary competence and are then discharged or they resign.

Lessons Learned. The last bullet reminds me of Scott Adams' humorous book, *The Dilbert Principle* (Adams, 1996) where he proposes the least smart people are promoted simply because *they're the ones you don't want doing actual work!* I once worked as a Software Lead on a large software-intensive program where the Program Manager was an old time hardware engineer who called the software group a "cult." After he did sufficient harm to the program, they got rid of him by *promoting* him to another smaller program where he could do less harm. If you work for such a Manager, hang in there until they unravel enough rope to "hang themselves."

It is interesting that some people seem to follow The Peter Principle *in reverse* where their past performance and successes are mediocre (or downright failures) until they achieve a level of great importance and influence where they are somehow inspired to achieve outstanding performance and results. Abraham Lincoln may be an example of this as he failed in most of his endeavors until he became an outstanding President. We could call this the "Retep Principle" (Peter in reverse).

Lessons Learned. There is an old adage that "failure is the mother of success." There are those who would consider the result a failure, for what most people would judge as a reasonable success if their task did not go *exactly* as expected or planned. Such people are striving for unreachable perfection. Sometimes, even if you did everything perfectly, you may still encounter failure for reasons beyond your control. If that happens to you, always remember that losers stay down, but *winners get up, dust off and move on.*

7.4.9 Working Conditions

Most Software Managers have little control over the physical working space for their staff since most companies have standard space allocations and furniture selection choices. Regardless, it is imperative that you provide the best possible working environment for your staff so that they eagerly look forward to going to work. You can allow your staff to personalize their work area, you can provide ample conference areas and whiteboards, and you can procure the best tools and computer equipment to increase the productivity and enjoyment of performing the work. As discussed in Section 9.6, offices with doors, even shared offices, for your technical staff are far superior to cubicles but, realistically, you probably have no choice if cubicles are the company standard.

7.4.10 Technical Respect for Manager

Having technical respect for the Manager is rated a low impact as a motivator probably because it is expected that employees would normally have such respect for their Manager. However, if the employee does *not* have technical respect for the Manager, then the impact as a cause for dissatisfaction is very high. Software Managers must *earn* technical respect from their staff and their peers.

If you are *directly managing Programmers,* you will have a very difficult time managing them if you do not have a very good understanding of the art of computer programming as well as the related tools and processes. It also helps to have a track record as a known and proven outstanding Programmer or Software Engineer. In addition, you will gain technical respect if you have made notable technical contributions, or have advanced degrees, patents, certifications, authored a book (who would want to do *that?*), active membership in professional societies, and up-to-date with the latest technical trends and technology.

In addition to earning *technical respect,* you also need to earn *personal respect* and the best way to do that is to *show respect to your staff.* If you treat them that way, they will treat you that way in return. Showing respect can be demonstrated in many ways including being a good listener, knowing the names of each staff member and greeting them personally, showing genuine interest by learning some things about their non-work life, asking their opinions when appropriate, never reprimanding publicly, and being courteous to them. This may not be easy, especially for members of your staff that are problematic but go out of your way to be respectful because it will have a big payoff.

There are management gurus who claim that a Manager should manage and not perform any technical work. My view is that the percentage of time you should spend on management tasks versus technical tasks depends on the size of your program. If your project is small and you have a small team, it is perfectly reasonable, and probably necessary, for you to participate in the technical work, and your actual responsibilities will be more of a "Programmer Lead" rather than an SPM.

However, if you are managing a *large complex software-intensive system,* you may have little to no time to do any real technical work. If you are an SPM, and performing a substantial amount of technical work on your project, it is almost a certainty that you are shortchanging your management duties at the detriment of the entire project. My notional guideline for the split of your time between management and technical tasks is shown in Table 7.4.

As shown by the guidelines in Table 7.4, on a *small* project you could spend an average of 80% of your time performing technical work; on a *large* project, you should not spend more than 10% of your time on technical tasks. If you find

Table 7.4 Breakdown of a Project Managers Technical and Management Tasks

Project Size	Management Tasks Average/Range	Technical Tasks Average/Range
Small	0.20/0.10–0.30	0.80/0.70–0.90
Medium	0.50/0.40–0.80	0.50/0.20–0.60
Large	0.95/0.90–1.0	0.05/0–0.10

that your time distribution is outside these guidelines, you should re-evaluate what it is you *are doing* versus what you *should be* doing.

7.4.11 Ethical and Realistic Policies

As a Software Project Manager, you may not have much influence on the ethical policies of your organization. However, *you* must always act in an ethical and professional manner. Being an ethical Manager means being honest and sincere with your staff. Sometimes, unrealistic policies or edicts come down from above and, if possible, you need to *intercept them* before they reach your staff to avoid disruption and distraction. Insulating your staff from these *organizational whiplashes* may be necessary so that your Developers can remain focused and productive.

When realistic changes do occur, they must eventually be disseminated to your staff, but you can control the time *when* the announcements are made to your team to avoid interference with the completion of their project milestones.

Lessons Learned. It is usually easy to identify unethical behavior; however, once in a while, the distinction becomes hazy. If you are at a friend's party and he/she tells you something in a private discussion about their company that is proprietary, it is clearly unethical to relate this information to your company because it is a violation of their trust and friendship with you.

What is not so clear (to me) is an episode that happened when I was working on a large government proposal. I was eating lunch alone in a quiet café and my table was against a half wall down the middle of the café, topped with plants that went halfway to the ceiling. Two men sat down at a table on the opposite side of the wall who were talking loud enough for *anyone* to hear. They were working for a different company *on the same proposal,* and they discussed topics that should never be discussed in a public place. When I returned to work, I told the Proposal

Manager what I had heard. Later, I was told this was unethical! I am not sure I learned anything from this episode because, to this day, I do not believe I did anything unethical.

7.5 Communications

Communications is one of the most fundamental skills of life and is a prerequisite to problem-solving. Effective communication is a cornerstone to successful project management.

Abraham Lincoln once said "Not saying anything and being thought of as a fool is better than opening your mouth and removing all doubt." Sure, there are times when remaining silent is the wise choice, but an effective Project Manager must also be an effective communicator. Although good communications is not rated as a high motivator, it is an important de-motivator if your team members feel they are "out of the loop" and not connected to what is going on. If that is the case, it is a problem *you must solve.* The following is a personal example of how a lack of project related communications can be a serious de-motivator.

Lessons Learned. I was the Lockheed Software Group Lead on the NASA Space Station Freedom program on one of the subsystems (called work packages) where Lockheed was a subcontractor to another major aerospace company; they were the prime contractor for our work package. They had frequent meetings and telephone conferences, and good email communications, so *all* of the subcontractors participated, and everyone *knew what was going on*—it was a *full team effort.* Everyone was enthusiastic and productive.

Meanwhile, in another part of our building, there was an additional Software Team working on a different Space Station work package. They were also a subcontractor to a different aerospace company, the prime contractor for their subsystem, who kept them almost *totally in the dark.* Their prime contractor had infrequent meetings and a very serious lack of communication. It is an understatement to say that this Software Team was frustrated and they were not even close to the productivity of my team. They came to me to find out what was going on regarding the overall program.

Both of these prime contractors are large aerospace firms with a long legacy. This experience was an education in the cultural differences in management style by two mature companies in the same industry. The lesson here is you have to expect to encounter, and learn to cope with, extremes in management style.

The importance of communications, and the approaches to enhancing communications in your project, is further illustrated in the following four discussions on the root cause of problems, the importance of honest discussions, the exponential growth in lines of communication as the size of your team expands, and some methods to cultivate communication.

7.5.1 Root Cause of Problems

The lack of good communications is often the *root cause of management problems* in many organizations. Developers need to accept the results of others, and they must communicate their ideas and results verbally and preferably with written documentation. *Constructive criticism* of software development work products is needed and should be encouraged—as long as it is offered in a calm, professional, respectful and non-accusatory fashion.

As the Project Manager, you must communicate regularly and frequently with Software Developers and other stakeholders. Frequent communication is a very important factor in increasing the likelihood of project success and the mitigation of problems. The Development Team should always seek customer and/or end-user involvement and encourage end-user input in the development process. Not having customer and/or end-user involvement can lead to misinterpretation of requirements, insensitivity to changing customer needs, and unrealistic customer expectations.

Digging for the root cause of problems is another important task for Project Managers. The symptoms of a problem are usually easy to see, but the root cause is usually hidden. The problem you see and hear is at the surface; you need to *dig deeper* to find the real root cause. Keep communicating whenever you encounter dysfunctional behavior because if you don't resolve the problem, it can grow in intensity to a point where it may be too late, or too big, to fix.

7.5.2 Honest Discussions

Intellectually honest discussions provide an opportunity to analyze strengths, weaknesses and pitfalls, and to act on that information to minimize potential problems. Even bad news can be helpful if communicated relatively early so that *timely Corrective Action* can be taken. Casual conversations with users, team members, and other stakeholders may surface potential problems sooner than made known at formal meetings, and they help keep the project timely, relevant and within the bounds of what can realistically be completed in a given time period.

7.5.3 Lines of Communication

Progress of your project is highly dependent on the effectiveness and the ability of the team members to *communicate with each other* as well as with end-users and other stakeholders. Software failures can result from a breakdown in understanding, so the ability of people to communicate with one another can easily affect the quality of the product. The reality of this serious problem becomes clear when you realize that the *lines of communication increase exponentially* as your staff grows larger. This exponential increase is demonstrated by the formula:

$L = S(S-1)/2$ *where "L" is Lines of communication and "S" is the "Size" of your staff.*

Table 7.5 shows the results of this compounding communications problem that you must consider and resolve if you have a large staff.

7.5.4 Cultivate Communication

If everyone is thinking alike, then somebody isn't thinking.

—George S. Patton (1885–1945)

There are many ways to foster better communications up, down and across an organization. In larger companies, there should be periodic (often quarterly) all hands meetings, monthly departmental meetings and written communication in the form of bulletins, newsletters, memorandum and email. But for your team, you should (maybe must) have weekly staff meetings, brown bag lunch meetings and off-site meetings.

Also, you should monitor gossip since that can be a major distraction. I read about a Manager that would open his staff meetings with an invitation to share gossip. When part of your team is geographically disbursed, effective communications becomes even more critical, and it is your responsibility to ensure that needed information is flowing to your team regardless of location.

Table 7.5 Growth in Lines of Communication versus Staff Size

Size of Staff (S)	Lines of Communication (L)
2	1
5	10
10	45
25	300
50	1225

Lessons Learned. Walking around the office, unannounced, and saying hello to your staff members can result in useful information, valuable insights and red flags that you can only get by physically being there. For example, one big red flag for me is when a staff member says "Let me show you something neat." The word "neat" is the tip-off. What it usually means is that the person has indeed come up with something really cool. However, it has nothing to do with your project and is not related to their job responsibilities. You need to give that staff member more work, or more difficult tasks, so they will spend more time on your project.

7.6 Managing Programmers

There are many Software Managers who believe that herding a team of wild horses is easier than managing Programmers. Why is that? What is it about Programmers that makes them, or makes them appear to be, so hard to manage? An important element of fitting square pegs in square holes involves placing Programmers into the type of programming disciplines that he/she is most experienced with and most comfortable working in.

The subject of understanding Programmers is addressed in depth in the recent book *"Managing the Unmanageable"* (Mantle and Lichty, 2013) and it is recommended reading. If you have never directly managed Programmers, it will be a great asset to you. The following is my overview of understanding Programmers; it contains some insights and key points to consider when determining how to manage your Programmers. The five selected topics covered, specific to Programmers, are:

- Creativity (7.6.1)
- Individuality (7.6.2)
- Programmer Personalities (7.6.3)
- Lack of Discipline (7.6.4)
- Performance Reviews (7.6.5)

7.6.1 Creativity

Programming takes place in an almost *pure thinking and creative medium* that encourages an undisciplined approach so as not to *hinder the creative process*. For many Programmers, "process" is a negative word because it implies a structured, disciplined, comprehensive environment which is the antithesis to their "free-spirited" mindset. Managing this type of mindset is difficult, but not impossible.

Some really great Programmers act like, and are treated like, prima donnas. In many cases, they deserve that lofty treatment so they often can get away with a lot. For those few really gifted special Programmers, you must decide how much latitude and margin of *freedom you can give to them* that you do not give to the other Programmers. The amount of deviation from the process that you can tolerate largely depends on the level of trust you have in the ability of that special Programmer to proceed with minimal oversight and produce a high-quality product.

Everyone else on your team must realize that *the success and quality of the software product is directly related to the process used to create it.* For large complex projects, a structured and managed software development process, tailored to the needs of the project, must be followed.

7.6.2 Individuality

There are no "standard rules" (or silver bullets) for managing Programmers simply because there is such a wide range of differences between Programmers. Like every professional, each Programmer is a unique individual with different skill sets, breadth and depth of experiences, talents and innate personal characteristics, and they must be managed as individuals. Before you hire the programming staff for your project, you need to determine what *types* of Programmers your project needs. In very general terms, the major types of Programmers include:

- *System Programmers/Architects* tend to be the most individualistic, and the most difficult to manage, but they also tend to be your best Programmers. Many System Engineers were at one time in their career System Programmers or System Architects, so they understand the systems perspective. Programmers of this type are relatively rare as they often are able to transform architectures of very complex systems into elegant and conceptually simple designs. If they can do that it makes the job easier for all the other Programmers leading to a huge leverage gained from one individual. A special Programmer of this type may act like a prima-donna, but they may deserve that distinction so they need to be managed... delicately.
- *Application Programmers* constitute the majority of professional Programmers and produce programs used directly or indirectly by end-users. They are generally more easily managed than System Programmers because they tend to be much less arrogant. Also, their coding progress is typically visible because of the inherent interfaces allowing you to assess and manage their progress. The best Application Programmers can go well beyond just producing code by completely understanding the end-user's needs thus creating a product that is responsive to customer requirements, resulting in fewer problems for you to manage.
- *Client Programmers* perform a task that most Programmers have performed sometime in their

career. Client-type programs typically *reside in the end-user's computer* especially on personal computers and include word processors, games, spreadsheets, other office related products as well as embedded applications (along with microprocessors) that run on mobile phones and numerous consumer devices. Client Programmers are usually not that difficult to manage because their task is generally well-defined with a clear set of deliverables, so their performance is relatively easy to track and manage (please note that I said "usually").

■ *Server Programmers* develop programs that relate both to where the program resides, usually remote from the end-user, but also to programs that *provide information and data to the end-users*. Most server programs are written to handle multiple activities simultaneously from multiple clients creating a level of complexity typically not encountered by client Programmers. Adding to the complexity, server programs often require the ability to add additional resources without changing the fundamental architecture of the program. The arrival and proliferation of the web has resulted in a major change to server programming.

■ *Web Developers* are reliant on the performance of their development tools which are different from the tools used by the other Programmer types. Web Developers mostly use formatting markup tools, such as HTML, XML, CSS and ASP/JSP, plus scripting tools such as Perl, PHP and JavaScript to perform their job along with higher-level tools to facilitate their task.

The proliferation of web applications has caused the client and server types, described above, to become synonymous with web browser interactions and web servers creating complex server programs that can be scaled up to serve many simultaneous users. Combining traditional programming with web development has created the need for a new breed of Programmers who are skilled at both types. If you need this type of multi-talented Programmer, but cannot find enough of them, consider a training program for your top Programmers to create your own Web Developers.

■ *Database Programmers* typically deal almost exclusively with an organization's *data storage and retrieval* for the end-user or a computer application. Database Programmers use different types of tools and schemas than other Programmers. The database tools involve SQL statements and "relational" database tools include Oracle, IBM DB2, MySQL and others.

If there is serious database development work needed on your project, it is highly preferable to hire Database Programmers who are *experts with the database system* used by your project. For simple database tasks, you can usually use any of your Database Programmers. A big challenge in managing Database Programmers is to help them think like Software Engineers rather than custodians of rows and columns of tables accessed by SQL statements.

■ *Computer Assisted Software Engineering (CASE) Tool Users and Scripters* are often not considered *real* Programmers (especially by those who consider themselves real Programmers) because they often use preprogrammed tools and applications rather than developing code. Nevertheless, the CASE tool users and scripters are an important part of your team. They often use Graphical User Interface (GUI) tools to specify program "logic" that drives user-accessible applications, or they may create scripts that produce customized displays.

This type of "Programmer" is growing because CASE tools and scripts are becoming more powerful and useful. CASE tool users may not have the technical skills as the other programming types, but to use a sports analogy, the quarterback, running backs and receivers who score the points cannot win football games without the blocking of the linemen; in a similar way, CASE tool users are important members on your team.

■ *Domain Expertise* is often an important consideration in hiring your best team of Programmers. People with specific engineering domain expertise are typically called *Subject Matter Experts* (SME), but in your organization they may be people with extensive experience in a specific line of business.

Depending on the needs of your program, chances are your team will be a combination of some or maybe all of the above Programmer types. Some highly skilled Programmers can perform the functions of all of the Programmer types listed. However, most Programmers specialize in one type of programming and do their best work in their chosen type of development. A Programmer may be *capable* of doing other types of development, but if they are not *interested* in doing that type of work, and you instruct them to do it, you are most likely creating a disaster waiting to happen.

7.6.3 Programmer Personalities

In addition to understanding the *types* of Programmers, and making sure you are putting square pegs in square holes, you must also understand the related issue of *personality styles* and traits that appear to be common to Programmers. I am *not* going to address "theories" about personalities, or how to categorize and deal with them, because that may not directly help you manage Programmers. However, understanding the typical personalities, you will find in Programmers can be a major asset to you in managing them. Some of the key personality traits include:

■ *Night versus Morning People*: It seems like the majority of the general workforce are morning people; however, a significant percentage of Programmers are night people. Night people tend to arrive past the normal start time but usually work well beyond the normal quit time. If the project is critical, they typically work well into the night. As long as they are producing good results, on time, you should have no issue with their work schedule.

The only requirement you should place on night people is that they have to be available for scheduled staff meetings, important events and milestones, including *core hours* to facilitate communication across the team. Their productivity will be much greater if you do not require them to observe normal working hours. You may, however, have to deal with criticism from other departments.

Lessons Learned. I can relate to this issue as I am a bona fide night person. As a consultant early in my career, I was most creative from 8:00 PM to midnight; in my last 20 years I was most productive and creative from 4:00 PM to 7:00 PM when there were no meetings, no telephone calls and no one around to interrupt my workflow. Nurture us night people; we can be very productive.

■ *Gunslingers versus Planners*: Programmers have a built-in tendency to be gunslingers whose motto seems to be "ready—fire—aim" or, to say it more politely, who jump right in and try to solve problems single-handedly. It is important that you identify the gunslingers on your team because you will have to track them more carefully to prevent them from charging off and developing ad-hoc solutions that could ultimately lead to serious problems.

Generally, you should not tolerate gunslingers; *however*, please note they are typically very good Programmers, and sometimes you may need them for some tasks such as developing quick prototypes. Gunslingers are usually prima donnas and disruptive to your team; they need to be carefully managed, and you need to take Corrective Actions when and if needed.

At the other extreme are methodical planners who basically follow the structured software development process described in this Guidebook, tailored to their need to create functional, reliable and maintainable software. Matching your needs with a Programmer's

intrinsic personality will be a win–win for everyone and will help promote more successful results.

■ *Left Brain versus Right Brain*: The left side of the brain generally specializes in logical, analytical and verbal tasks, whereas the right side of the brain is focused on non-verbal, intuitive, thoughtful and subjective issues associated with creative types such as musicians, writers and artists. Technical people, including Programmers, are considered to be left brain type people; however, *good Programmers need strong skills from both sides of the brain*.

The reason for bringing up this issue is that you will find some of the best Programmers are also musicians or mathematicians. This is not a prerequisite for being a good Programmer, but uncovering this fact during a candidate's interview will give you an insight into their potential. This viewpoint is directly related to the earlier discussion in Section 7.1.4 regarding the proper fitting of square pegs and round pegs based on their mindset.

■ *Heroes versus Introverts*: Heroes are somewhat similar to gunslingers who elect to tackle work tasks that take near superhuman effort to complete. Unlike gunslingers, heroes can work effectively on teams and most often complete their gigantic tasks. Heroes need to be nurtured as they typically rise to become superstars. Your responsibility is to make sure they do not *over commit* and *burn out* from a continuous stream of superhuman efforts. Deciding where to draw that line is an important subjective judgment you must make. It is also not a good idea to overcompensate or over reward heroes as that would encourage others to try to be a hero.

The almost opposite extreme are the introverts who are so quiet and reserved that they are almost invisible. Although they contribute little to team dynamics, they can and do produce very good software. You need to bring them into conversations whenever possible, give them positive reinforcement to build their confidence, acknowledge their contributions and establish a special relationship (on a low level) in order to find ways to connect with them.

■ *Cynics*: Individuals who are deeply cynical should not be hired; if you inherited them you need to work with your HR Department to develop a plan to move them out. A cynic will blow issues out of proportion and convey their radical interpretations to the team in a toxic manner that can cause serious demoralization.

7.6.4 Lack of Discipline

An unprofessional carpenter could *start* to build a house without any requirements or design specifications. As ridiculous as this scenario sounds, it is exactly the type of approach some Programmers (like the gunslingers) follow if they do not have, or are not required to follow, a structured software development process.

> *A major theme of this Guidebook is that there are software development processes that must be followed after tailoring to the needs of your software-intensive system.*

At one extreme are Programmers who start writing code and (maybe) do some design along the way without any requirements and no formal process. At the other extreme are the special Programmers who have a mature level of discipline and generally follow all, or a major part, of the process of producing great software. Always try to hire this special type of Programmer, especially if you are developing a large software-intensive system, because they will help to mitigate your management problems and dramatically improve the probability of a successful outcome.

7.6.5 *Performance Reviews*

As the Project Manager, you have the important responsibility of evaluating the performance of your programming staff and helping them *improve their performance*. Aside from the fact that most organizations require performance reviews, almost everyone wants feedback on their performance. This feedback can occur at any frequency—weekly, monthly, quarterly or annually—or at any time when a Programmer's performance deserves praise or needs some constructive criticism.

Typically, an annual performance review is required for everyone; however, smaller companies usually require performance reviews on the anniversary date of the hire to avoid the financial burden when everyone is reviewed at the same time. An important element in helping staff members improve their performance is to mutually agree on a list of *goals* and a set of *task*s to be completed within a specified time frame. If the goals are general, the time frame should be longer. If the goals are specific, the time frame should be shorter. Make the goals *measurable* as that will make it easier for you to judge the person's progress toward achieving them. Setting goals and objectives is especially valuable for new hires. Also, make sure you conduct the reviews when scheduled.

A key to simplifying the performance review process is to create an easy to use evaluation form (or use your organization's form if it covers your attributes). Such a form facilitates determining a percent complete evaluation. However, rather than waiting for the formal review process, it may be much more valuable for you to do a scheduled periodic walk-around to conduct face-to-face meetings with your staff

members to give them more timely feedback on your evaluation of their progress. Doing this will allow your staff members to make frequent course corrections, will improve their likelihood of meeting their objectives, and will make your evaluation easier.

> **Lessons Learned.** The value of performance reviews are questioned by some, but I am a strong proponent of their value because it encourages good performance, discourages bad performance, and provides a written record in the event a staff member is not performing well and needs an improvement action plan.

7.7 Creating an Environment of Excellence

In 1983, Tom Peters introduced the word "excellence" into the business world with his book *In Search of Excellence* that was republished in 2003 (Peters et al., 2003). He made it clear that the changes leading to excellence *takes time*. It is typically a long process that can result in fundamental infrastructure changes in your company. The overall culture in your corporate organization, and its focus on excellence, is probably (hopefully) already in place. This section will address creating an environment of excellence specific to your project that will *supplement your existing corporate vision and values*.

If your corporate culture for excellence is strong, you can leverage it and flow it down to help create a productive environment for your Software Team. If not, it is imperative for you to *create your own environment of excellence*. Change begins by asking what you want your project to do. Harness the ideas and wisdom of your employees and apply them to your vision of excellence. In other words, *create a continuous feedback culture*.

It is difficult for any organization to achieve a high level of operational performance without a well-defined and well-understood set of cultural expectations. Establishing a culture of excellence, based on clear operational characteristics and cultural expectations, will not only enhance performance and competitiveness, but it will also strengthen the team's unity and support.

7.7.1 *A Healthy Organizational Culture*

Strive to develop what can be called a *healthy organizational culture* in order to increase productivity, growth and efficiency and reduce both counterproductive behavior and turnover of your staff. If there is a strong and healthy organizational culture, people do things because they believe it is *the right thing to do.* The results of a healthy organizational culture include:

- Respect for and fair treatment of each employee's contribution to your project
- Employee pride and enthusiasm for your project and the work performed
- Equal opportunity for each employee to realize their full potential
- Strong communication with all employees regarding policies and company issues
- Company leaders with a strong sense of direction and purpose—including you
- Ability to compete in innovation and exceptional customer service
- Acceptance and appreciation for diversity
- Lower than average turnover rates (perpetuated by a healthy culture)
- Investment in learning, training, and employee knowledge

Figure 7.3 is my perception of 12 key characteristics of an *environment of excellence* for a Software Development Team. It is composed of three major components: the Work *Environment*, the *Work Atmosphere*, and the *Work Infrastructure* as shown in Figure 7.3 and described in subsections 7.7.2 through 7.7.4.

7.7.2 The Work Environment

The Work Environment is the inherent totality of the influences, surroundings, conditions and ambiance affecting the successful development and delivery of your project. The Work Environment exhibits the following five characteristics: communications, quality, customer focus, creativity and innovation, and a cynical free environment as described below:

- *Communications*: Good communications should not be something you add on; it must be a built-in environment in your organization and followed as a standard operating

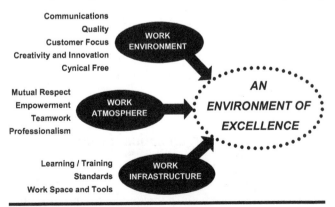

Figure 7.3 Key characteristics of an environment of excellence.

procedure. As discussed in Section 7.5, the likelihood of success, and the mitigation of problems, is highly dependent on the effectiveness and ability of the team members to *honestly, regularly and frequently communicate* with each other as well as with the customer, end-users and stakeholders. Software failures can result from a breakdown in understanding so the ability of people to communicate with one another can affect product quality.

Customer and end-user involvement is important to avoid misinterpretation of requirements and unrealistic customer expectations. A good communications environment would include weekly staff meetings, periodic departmental meetings and documented communications. The lines of communication increase exponentially as your staff grows larger, and if part of your team is off-site effective communications become more difficult and more important.

- *Quality*: Section 6.2 described functions performed by the Software Quality Assurance (SQA) group and the actions they take to help assure that software, and software-related products, satisfy system requirements. The SQA organization has a responsibility to provide project management with *visibility* into the software development process and products by performing *independent* audits and assessments. It is important to note that the SQA group is *not responsible* for software quality; the Software Development Team is responsible for building quality into the product and *the Software Manager is accountable for delivering a quality product.* Quality is the fundamental attribute of a successful software project, and it must become a ubiquitous part of your company's and your project's culture.

- *Customer Focus*: People are your most important resource; however, without satisfied customers, your project and your company's future is at risk. If you don't believe that then read what some luminaries, whom you should listen to, say about the value of customers:
  - "It's customers that made Dell great in the first place, and if we're smart enough and quick enough to listen to customer needs, we'll succeed." (Michael Dell, Dell Computers)
  - "The (most important) result of a business is a satisfied customer." (Peter Drucker, Business Guru)
  - "There is only one boss. The customer." (Sam Walton, Walmart)
  - "If you work just for money, you'll never make it, but if you love what you're doing and you always put the customer first, success will be yours." (Ray Kroc, McDonald's)
  - "For us, our most important stakeholder is not our stockholders, it is our customers." (John Mackey, Whole Foods)

– "Put the customer first and the rest will follow." (Akio Toyoda, Toyota Motor)

For software-intensive systems, the above quotations are actually understatements. A *strong customer focus is absolutely critical for software-intensive systems.* There is no software development project that does not have a customer. You must satisfy the needs of that customer. The undeniable requirement for a powerful customer-focused Work Environment is related to the discussion of the importance of software requirements (see 10.1 and 11.1). Good requirements come from a good interface with the customer—both are critical elements of a successful program. A lack thereof is a primary reason for failure.

■ *Creativity and Innovation*: Creativity is the ability to look at problems or situations from a fresh perspective and to generate or recognize ideas, alternatives, or possibilities that provide innovative or different approaches to a particular issue or task. In order to be creative, you must *think out of the box*, view things from a different perspective, and be able to generate new and insightful solutions to problems.

Innovation is the implementation of a new or significantly improved product, service or process that creates added value for your project or a competitive advantage for your company. Creativity is a crucial part of innovation because there cannot be innovation without creativity. Creativity is the act of turning new and imaginative ideas into reality because, if you have ideas but don't act on them, you are imaginative but not creative.

Software Managers need to realize that creativity is an acquired skill and a process that can be managed and cultivated. Creativity begins with a foundation of knowledge, learning the technical discipline and mastering a way of thinking. Your team can learn to be creative by experimenting, exploring, brainstorming, questioning assumptions, using imagination and synthesizing information. Learning to be creative is like learning a sport. It requires practice to develop the right muscles and a supportive environment in which to flourish.

■ *Cynical Free*: A cynical free environment is important because you have no time to deal with bozos who are abrasive and sarcastic regardless of their technical talents. However, you need to distinguish the difference between good team members who frequently criticize but do so in a positive and constructive manner versus those who criticize in a disingenuous and contentious manner causing dissension among the team members.

Constructive criticism should always be encouraged and the earlier the better.

7.7.3 *The Work Atmosphere*

The work atmosphere is a *pervasive aura* that influences the team member's feelings, tone, spirit and state of mind thus promoting the following four characteristics as shown in Figure 7.3: Mutual Respect, Empowerment, Teamwork, and Professionalism/Integrity.

■ *Mutual Respect*: Team members must *trust and respect* each other and they must learn to communicate their ideas and results verbally and with written documentation. Intellectually honest discussions allow the project team to analyze strengths, weaknesses, opportunities and pitfalls and to act on that information to minimize potential problems. Trust also implies that the team members feel they are being treated fairly. When people do not trust each other, communications is difficult and ineffective. People who work on a team must trust each other to do their share of the work.

■ *Empowerment*: Empowerment in the workplace is an effective catalyst for increased productivity. Conversely, micro-management can be a severe deterrent to productivity. Inexperienced members of your team need advice and direction, but don't tell qualified people *how* to perform a task; tell them *what* needs to be done. Such empowerment will challenge your team members' ingenuity, and the results will likely surprise you.

However, it is important to ensure that team members have the skills to perform their allocated responsibilities and that your company or project has incentives for rewarding them for taking these responsibilities. Empowerment of employees requires a culture of trust in your organization plus an effective information and communication system.

Many years ago, McGregor (1960) proposed a rigid but simple approach to people management called the X–Y Theory. Theory X is an authoritative approach involving a repressive style and tight controls that usually resulted in a depressed culture. Theory Y involves empowerment of the people being managed, along with continuous process improvement, producing improved performance and professional growth.

As a project management rule, you should never adapt the Theory X approach unless you are managing offshore developments in some countries (such as Japan, China and parts of India) where, in their cultures, an authoritarian approach may be the style needed to manage the teams in the companies you have contracted with.

■ *Teamwork*: Teamwork involves cooperation to achieve the common goals of the project. Teamwork requires all team members to contribute their fair share to the workload so that the project, orchestrated by the Project Manager, is accomplished in a timely and satisfactory manner. Your team should realize that they are collaborators in a joint effort to satisfy your customer's system requirements. The willingness to cooperate stems from deep *relationships* that develop between coworkers and between you and your team members. Teams that have worked well together in the past should, if possible, be kept together on future projects. The relationship between you and your peers in other projects is also important in order to establish a long-term cooperative support system.

■ *Professionalism*: Professionalism implies impeccable integrity, strong moral and ethical principles and always pressing for excellence. As the Project Manager, you must personify, and encourage, a culture of professionalism for your entire team. Maintaining a high level of professionalism involves a number of important attributes including:

- Trustworthy/Dependable/Committed/Calm under fire.
- Responsible/Accountable/Technically competent.
- Respectful/Courteous/Considerate.
- Firm/Fair/Non-manipulative/In control or perceived that way.
- Empathetic/Approachable/Supportive.

The intent here is not to recite the Boy Scout oath, but to identify the characteristics you should aspire to and set the example for your team to follow, as well as the characteristics you would hope to see in the Managers you report to. You should also encourage your team to obtain professional certifications especially if that is important in your organization.

7.7.4　The Work Infrastructure

The work infrastructure is the underlying *foundation and basic facilities* on which the continuance and growth of the project depends upon thus promoting the following characteristics as shown in Figure 7.3: timely training and learning, software standards, good working space and effective tools.

■ *Learning/Training*: Professionals, regardless of the profession, must *keep learning*! An adequate training budget for your project is essential for several reasons, and it goes well beyond the importance of just learning how to use new tools. All members of your team should have some training every year. The training objective may be a corporate mandate; for example, one company that I worked for had a training budget and a minimum requirement of 40 hours of training per year for each engineer—and they kept records to assure compliance.

Training can include conferences, workshops, formal courses or brown bag lunch sessions. Also, there are many daily opportunities for learning such as Peer Reviews, Code and Design Reviews and Product Evaluations. *Learning by doing*, also called "On-the-Job Training" (OJT), should be an integral part of your culture. This concept goes way back to Aristotle (384–322 BC) who wrote "What we have to learn to do, we learn by doing."

Training need not be restricted to technical issues. For example, personality clashes between team members are often the result of the arguers' lack of *interpersonal skills*. Most people have never been taught how to sit down and calmly work out their differences with others. Training sessions on how to improve interpersonal skills can have a very high return on investment (ROI). This ROI is hard to prove so training of this type is often overlooked.

When equipment breaks down, there is no hesitation to spend whatever is necessary to fix it. When our precious human resources need fixing, there is typically little or no effort applied to keep them functioning effectively. Try to do a better job at this. For your team members who are recent hires, recent graduates, or anyone unfamiliar with the terminology or the "development process," it is highly advisable for *you* to give them a 2– 3-day training session as that will also have a high payoff.

Another major cause of project failures is tied to *unrealistic expectations* by senior management regarding the results of your project. If your senior Managers do not have experience with the development of software-intensive systems, it is in your best interest to persuade them to attend presentations on the basic principles so they will gain a perspective of what is involved and what to expect. In addition, try to embed such principles in your status reports, include it in your conversations with them, and convince Senior Managers to read Guidebook Chapters 1–3.

■ *Software Standards*: Software standards enable interoperability between different programs created by different Developers. Standards must be specific as well as meaningful to your project. You can adopt existing standards such as those discussed in Section 4.5 and listed in Appendix I. You can create your own standards, or modify existing standards to make them specific to the needs of your project. As the Project Manager, you must ensure that the product delivered is fully responsive to both the documented software requirements as well as the established standards that are imposed on your project.

■ Coding standards are of special importance as they describe the detailed conventions that Developers must follow in creating the systems source code. A common analogy regarding programs that follow coding standards, versus programs that do not follow coding standards, is like a blanket made out of one fabric versus a patchwork quilt made of mismatched scraps. Coding standards should be short (less than 25 pages) and should not be so prescriptive that Developers have a hard time remembering and following them. Hopefully, the standards you are using were developed prior to your project, and the Developers are familiar with them.

■ *Work Space and Tools:* It would seem to be logical, and studies have shown, that productivity rates of Software Developers are affected by their physical environments. The physical environment, including the pros and cons of cubicle "farms" versus private and semi-private offices, telecommuting environment issues, software development equipment, and CASE tools are discussed in Chapter 9.

7.7.5 Conflict Management and Resolution

Some Software Managers believe having a root canal is preferable to resolving conflicts. It may not be fun resolving conflicts between staff members or helping someone with a personal difficulty that is adversely affecting their performance, but it is part of your job. If you follow a little sage advice, it may not be as distasteful as you might think. No world-class software development methodology or ground-shaking process improvement strategy can overcome the serious problems that can result from *mismanagement of interpersonal conflicts.*

One very effective and simple approach is to be a *good listener*—something every great Manager should excel at. Often, the simple process of being able to vent one's feelings, and to express them to a concerned and understanding listener, is enough to relieve frustration and make it possible for the frustrated individual to advance to a problem-solving frame of mind. This approach avoids the Manager from trying to diagnose and interpret emotional problems, which would call for special training. No one has been harmed by being listened to sympathetically.

Lessons Learned. There is an old Chinese proverb that goes something like this: *You have two eyes and two ears but only one mouth—so watch and listen twice as much as you talk.* What is even more astonishing about the value of being a *good listener*, is that often when the aggravated person is describing their problem to you, *they* actually come up with a solution to *their* problem. Even though they did almost all the talking, they will walk away thinking what a wonderful conversationalist you are.

There are other more direct, or more diagnostic, approaches that might be used to manage conflicts in appropriate circumstances such as:

■ *Conflict Avoidance*: Creating a total or partial separation of the staff members having the conflict resulting in minimal or no interaction between them. An example would be moving the office of one of them to a different location.
■ *Harmonization*: Attempting to diplomatically smooth over the issue between disputants to promote teamwork and enhance professionalism.
■ *Dominance Intervention*: The imposition of a mandated solution to the issue by management at a higher level than the level of the conflict.
■ *Compromise*: Seeking a resolution where neither party gets all they want but satisfying at least part of each party's position. This is one of the best methods of conflict resolution.
■ *Confrontation*: Involves an identification and thorough and frank discussion of the root sources for the conflict and achieving a resolution that is in the best interest of the project, even though the resolution may be at the expense of one or all of the conflicting parties.

Lessons Learned. *Compromise* is my strong preference for resolving conflicts; however, reality checking also works. It involves getting the staff members who have conflicts to *address reality*, for example, by making them realize there are *three sides to every story*. As discussed in Chapter 4, the three viewpoints have various names such as the positive, the negative and the neutral; the optimistic, the pessimistic and the realistic; and the best case, the worst case, and the most likely case. The important part of this is to get the staff member with the conflict to agree to the reality that there are other valid viewpoints.

For serious conflicts, a trained conflict counselor may be needed who can try to get the staff members having conflicts to clarify and reaffirm shared goals. If necessary, they can then move through a systematic series of interventions, such as testing the members' ability and willingness to compromise. As a last resort, they may have to go through enforced counseling, reassignment or termination. In any case, there must be a resolution because you do not have time for this distraction that may also likely be distracting to the team members.

Managing Adversity. One of the most serious mistakes any Project Manager can make is to avoid, ignore or minimize the seriousness of conflicts that exist within the team.

You must address interpersonal conflicts, and you must proactively manage their resolution.

Lessons Learned. An analogy of the wrong way to manage adversity is what many doctors do when treating a patient's pain by medicating the symptoms rather than taking the time to identify and treat the cause. *Don't manage adversity like those doctors*; that approach will not mitigate the *source* of interpersonal conflicts. Find and address the root cause of a problem rather than focusing on its easy to fix symptoms. Effective adversity management means doing it the right way, but there is no universal "right way," so you need to work with the aggravated staff members and figure out the right way *for them*.

7.8 Software Training Plans

Training resources, specifically the funds allocated to provide an adequate training program, is often under-funded or possibly overlooked. *Program Training Plans* must be developed to address software training needs. The training plans should be developed, maintained and monitored by training coordinators and/or Software Process Leads in coordination with SwIPT Software Leads.

The plans should address program-specific technical and process training, identify training requirements by job category, provide for a waiver procedure, and require training records to track completion. Training plans should go all the way down to individuals, including the training they have already had, what training they still need to take, and when they are scheduled to take it. The SDP, or a separate training plan if one is produced, should make a clear distinction between:

- Basic training provided by the contracting organization that is funded by that organization
- Program-specific training provided under contract funding

The contracting organizations are responsible for staffing their projects with qualified people and for providing project-specific technical and process training. Each organization must develop coordinated plans to implement training in accordance with its organizational practices.

The SwIPT Leads and SwIPT Software Leads, assisted by their respective functional Staffing Manager, should provide training guidance to their staff. This may include suggestions for either technical enhancement or career development training. Periodic lists of upcoming training classes should be provided by the training coordinators to program personnel. When necessary, the program should request training from the training organization to achieve specific training requirements.

Lessons Learned. There is another important facet to the requirement for a *funded training program* that I encountered a while back with one of our subcontractors. They were a very well respected high tech company that had achieved the Software Engineering Institute (SEI) Level 2 rating and were in the process of going for a Level 3. They had *all the elements in place* including a good set of training plans, and their training program was well planned and well-funded. They had a high level of confidence they would achieve Level 3.

After the formal assessment was concluded for their Level 3 rating, they were shocked to hear that they did not pass because of a lack of training. The reason was, despite all their preparation for training, and a fully funded training program, the assessors learned that very few people actually *attended* the training classes. Their lack of follow through was an important lesson learned.

The best tools are of little value, and a waste of resources, if no one knows how to use them.

Chapter 8

Managing Software Costs and Schedules

The process of controlling software costs and schedules is a critical Software Project Manager (SPM) function.

Chapter 8 covers program plans, schedules, costs and budgets and how the SPM can integrate it all to provide a synergistic set of activities. Chapter 8 is directly related to the discussion of software project planning and oversight in Section 5.1. The management of software costs and schedules are discussed in the following six sections:

- *Program Master Plan and Schedule*: Major programs must have a formal program-level management methodology through a program *master plan* directly mapped to a program *master schedule*. The top-level schedule almost always needs to be augmented with lower-level detailed schedules covering software planning, design, development, integration, and test (8.1).
- *Work Breakdown Structure (WBS)*: The WBS is a valuable software budgetary planning and control approach. The WBS organizes and decomposes, in a hierarchical structure, *all* the project tasks into smaller, more manageable and controllable components (8.2).
- *Managing Software Budgets*: Includes budgeting for labor, contractors, consultants, capital equipment, travel and training. As the SPM, you should become proficient at managing technical budgets even though many Technical Managers consider it to be an unpleasant task (8.3).
- *Managing Software Schedules*: Involves the use of various techniques and charts such as the Gantt, Program Evaluation and Review Technique (PERT) and Critical Path Method (CPM) charts (8.4).
- *Managing Software Cost and Schedule with Earned Value*: The objectives of controlling project costs are

to ensure that the costs incurred stay within the budget while the work is completed on time and with the required functionality. One effective method to accomplish this task is called the *earned value management control system*. Earned value tracks the *actual work* accomplished and is generally referred to as *what you got for what you paid for* (8.5).
- *Integrated Product and Process Development (IPPD)*: The IPPD technique is a program management approach that *simultaneously* integrates *all* essential development activities (not just software) through the use of multidisciplinary teams to optimize the design, manufacturing, and supportability processes (8.6).

8.1 Program Master Plan and Schedule

As introduced in Section 4.4, every major program must have, *and follow*, a formal program management methodology by maintaining an approved program-level *master plan* directly mapped to a program-level *master schedule*. The master plan and master schedule are hereafter called the *Integrated Master Plan* (IMP) and *Integrated Master Schedule* (IMS). All major programs must have an IMP and an IMS, or equivalent documents regardless of their names, to provide a complete schedule and activity network for *all* program activities. The key elements of the IMP and IMS typically are:

- On large programs, the IMP and IMS must be maintained electronically and available through an electronic data management system (see 5.1.10 and 9.5.1).
- The IMP must describe the overall program organization, responsibilities, and management approach.

■ The IMS must provide program-level schedules, time-lines and required resources and is often referred to as the *Program Master Schedule*.

■ The IMP and IMS must be organized by a systematic *WBS*, introduced in Subsection 1.5.2, and described below. The WBS is an enumeration of *all work activities* to be performed by the contract, structured into a hierarchy that organizes the work activities into short, manageable, and trackable tasks.

■ The IMS must include software activities showing the time-phased interrelationships of major milestones and accomplishments for software builds.

■ The Software Integrated Product Teams (SwIPTs) must *manage and control their respective schedules within the IMS structure*. The SwIPT functions are discussed in Subsection 1.6.5.

The program-level IMS is not prepared by the Software Team although software representatives should always be involved in its preparation. Since the IMS is at a relatively high level, it should be augmented with lower-level detailed subsystem schedules covering software planning, design, development, integration, and test. These detailed subsystem schedules must be maintained and monitored at, and by, the subsystems with oversight by the SPM and the Chief Software Engineer (CSwE). Subsystem schedules must be consistent with the IMS. If any conflicts between the IMS and subsystem schedules occur, the program-level IMS always prevails.

An overall master schedule may be included in the initial SDP submitted with the proposal. However, once the contract starts, the schedules, especially the detailed software schedules, are typically updated so frequently that they should not be part of the updated SDP, but their location should be referenced in the SDP.

8.2 Work Breakdown Structure

A Work Breakdown Structure is a decomposition of all the project tasks into smaller, manageable and controllable components or tasks. As depicted earlier in example Figure 1.4, the WBS, sometimes referred to as the *Work Plan*, is portrayed as a *tree structure* with multiple levels, providing a hierarchical and incremental decomposition of the project to achieve the objectives of a program, project or contract.

The WBS is a comprehensive product-oriented hierarchical structure composed of hardware, software, services, data and facilities. It is co-developed with Systems Engineering plus other area experts as appropriate but definitely includes software representatives. The WBS usually includes a *WBS Dictionary*, describing each task or WBS element throughout the life of the contract. The reporting contractor (normally the prime contractor in a large development program)

prepares the WBS Dictionary and must maintain and update it throughout the life of the contract.

The WBS typically is prepared from two perspectives:

■ The *product* hierarchy indicates how the components are organized;

■ The *activity* hierarchy indicates the activities (requirements, design, test, etc.) pertinent to each component.

The hierarchy is only defined down to the *level needed for reporting and control,* and that level varies depending on the size of the program.

The WBS may also be configured as a *Contract Work Breakdown Structure* (CWBS) whereby it becomes the framework for reporting program costs, schedule, and engineering performance on the contract. Contractual work assignments are based upon the *IMP* and *IMS*, and they are the basis for the formal relationship between the customer, the prime contractor, the Subsystem Developer, and the subcontractors. The WBS must subdivide the program into *clearly defined, tracked and manageable discrete tasks*. It provides the necessary framework for detailed cost estimating and cost control, along with technical, schedule, cost, and labor hour reporting, plus guidance for schedule development and control.

Figure 8.1 is an example of a combination of the Project Summary WBS and the Contract WBS. Levels 1–3 are the Project Summary WBS levels and Levels 2–5 contain the Contract WBS levels. The WBS is typically developed by starting with the end objectives of a project and successively subdividing it into manageable components in terms of size, duration, and responsibility (e.g., systems, subsystems, components, tasks, and subtasks) including all steps necessary to achieve each desired objective.

The WBS permits summing of subordinate costs for tasks into costs for their successively higher-level "parent" tasks. For each element of the WBS, a description of the task to be performed should be generated. This technique is used to define and organize the total scope of a project. There should be only *one WBS per contract*. A well-designed WBS makes it relatively easy to assign each project task as well as each team member to a specific element of the WBS. In addition to its function in cost accounting, the WBS also helps map requirements from one level of System Specification to another, for example, a *Requirements Test Verification Matrix* (RTVM) mapping functional requirements to the design documents.

Subcontractor Reporting. For subcontractor reporting, the requirement for reporting must be included in the contract with the subcontractor. Based on the flow down of requirements, the prime contractor and subcontractors must have *identical reporting requirements* (report type, frequency, and method of transmission). The contract and subcontract WBS elements may be different, but they must be complementary.

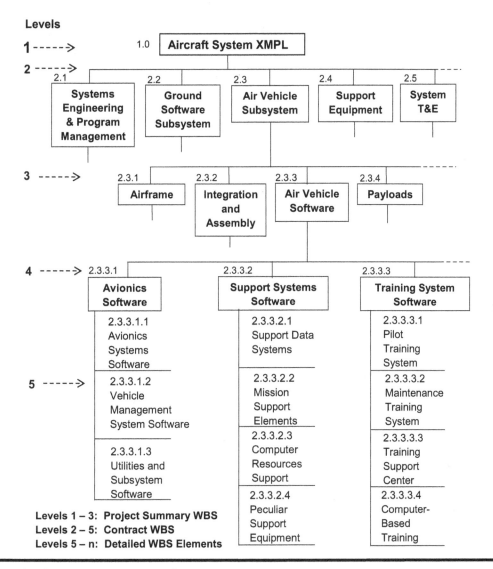

Figure 8.1 Example of a project summary and Contract WBS.

Each subcontractor should be assigned responsibilities for their software portion at the Software Item (SI) level. The subcontractor's reporting requirements should be included in the prime contractor's *Cost and Software Data Reporting Plan* often called the CSDR Plan.

In summary, a WBS is the foundation for:

■ Program and technical planning
■ Technical description of program elements
■ Cost estimation and budget formulation
■ Schedule definition
■ Statements of Work (SOW) and specification of contract line items
■ Progress status reporting and problem analysis
■ Tracking of technical changes such as Engineering Change Proposals (ECPs)
■ Engineering management

8.3 Managing Software Budgets

It seems that many Technical Managers consider budgets to be a necessary evil to be avoided if at all possible! The financial department in your organization usually is responsible for monitoring budgets and tracking costs, however, Managers responsible for projects are allocated a budget for their project, and the Project Manager *must manage it*! Indeed, the ability to manage a technical budget is an important attribute that every Software Manager should master even though it may not be the most pleasant task you have to do.

Budgets are formal statements of financial resources allocated to specific activities over a specified time period. The budget is an important tool for project control. At the start of your project the budget is a plan; during the project, it serves as a control device by which you can track its success in achieving its goals. Financial forecasting, an essential element of planning, is the basis for budgeting and it is the responsibility of your financial department; however, the Software Manager *must* be part of the financial forecasting process and make sure *your* financial department understands that.

This Guidebook will not cover corporate budgets for operations, sales, income, overhead and cash flow or related issues such as balance sheets, value of money, profit margins, etc. It will address budgets for labor, materials, contractors, travel, tools and equipment directly related to the project you are managing that are largely *under your control*. In general, a budget has the following attributes:

- A budget is a financial plan reflecting the goals of an organization flowed down to all the accountable components therein.
- A budget is expressed in time-phased measurable terms so that status can be periodically analyzed in order to take timely Corrective Actions.

Software Managers need flexibility in administering their budget. Flexibility means the corporate financial controllers should *not require approvals for everything*. Software Managers must have the flexibility to use their budget in the most effective manner responsive to the changing needs of the project. Budgets will also be discussed in Section 8.5 where budgets are broken down into cost accounts as part of the earned value management (EVM) process.

8.3.1 Budgeting for Labor

Software development is a very labor-intensive effort, so it is not surprising that salaries and related employee overhead expenses are typically your largest and most important software budget item. Adequate staffing relates to effective estimating (estimating is discussed in Chapter 14). Once the estimating task is completed, a time-phased budget must be prepared because your staff builds up over time, levels off, and eventually tapers off when the project is nearing completion. For multi-year projects, you must also account for projected salary increases, cost of living increases, and bonuses.

8.3.2 Budgeting for Contractors and Consultants

Subcontract management is significant and important enough to warrant its own discussion (see Section 5.5). If all of the software development is performed in-house then, of course, there is no budget needed for contractors. However, large software development projects almost always involve a team of subcontractors and managing them, as well as monitoring their budgets, can be one of the Software Managers most significant and challenging tasks.

The use of software consultants may be required regardless of the size of your project. Acquiring specific expertise, not available from your full-time staff, may be accessible only through the consulting route. Using consultants is especially cost beneficial if their expertise is needed only for a relatively short duration or intermittently. Also, if the right people needed for your staff cannot be hired on a timely basis, the funding available for their salaries can be used to bring in consultants to avoid falling behind schedule.

8.3.3 Budgeting for Capital Equipment

The productivity of your Programmers can be greatly enhanced with the right quantity and quality of computers, related peripherals, and *Computer Assisted Software Engineering* (CASE) tools. Like every professional, Programmers need the best tools available to maximize their performance. When procuring equipment and tools, you must be cognizant of the lead time needed for receipt of orders to make sure the products are available when needed. Equally important is budgeting for the training needed to instruct your Developers how to use new tools.

Software Managers must work closely with their financial controls group to determine, if, when and how to capitalize equipment by depreciating the cost of the equipment over several years thus minimizing its impact during the year of the purchase. It helps if the Software Manager understands the rules for capitalizing equipment. In some cases, even development costs can be capitalized.

8.3.4 Budgeting for Travel

If your entire software Development Team is co-located, you will have a relatively small travel budget to periodically visit the customer, operational sites, software vendors or for training. At the other extreme, large distributed development programs can involve extensive travel to many locations scattered across the USA and maybe in foreign countries as well. Minimize travel by promoting telephone conferences, videoconferencing and through email.

Face-to-face interaction is periodically necessary to attend status updates and milestone meetings. As a Software Manager responsible for tracking performance at all of your scattered locations, you will find yourself spending a lot of time on an airplane to assess and coordinate the multitude of development and Testing Teams. As long as you have your laptop with you, the time on the airplanes and at airports can be put to good use.

> **Lessons Learned.** It amazes me how many times I have conceived great ideas or problem solutions on an airplane trip by taking advantage of the (usually) uninterrupted focus time. Wear earphones even if you are not listening to music because no one will talk to you.

8.3.5 Budgeting for Training

The training budget is often not given the attention, or funding, it needs and deserves. There is no point to procure whizz-bang CASE tools if there is no budget to train the Development Team on how to use those tools. In addition to training the Development Team, there may be a need for budget items to cover software training costs at the installation site. This would include the training of users, operations personnel and maintenance staff.

> **Lessons Learned.** The value of training cannot be denied. However, I have encountered what appears to be a paradox. It seems that the more I learn, the more I realize *how little I know* because I keep encountering massive amounts of new information. Does that mean, as I become more knowledgeable I am becoming progressively dumber? Regardless, you must keep learning in order to "grow" and a funded training program has a high ROI.

If you have been a Programmer for several (or many) years advancing up the ladder of responsibility, and one day find yourself in a management role, having outstanding technical knowledge, but totally unprepared for a management role—then this Guidebook can serve as a crash course. It would, however, be much more effective if, in anticipation of your promotion to management, your parent organization prepares you in advance. This would mean that senior management would (should) have a systematic policy of requiring a training budget set aside to prepare future Software Managers. With a training program like that in place, future Software Managers could take a series of courses, as permitted by their schedule, including topics such as:

- Project planning, scheduling, and controlling
- Requirements specification and management
- Customer focus and managing customers in foreign countries
- Team management and leadership

In addition, certification from the *Project Management Institute* (PMI) can be obtained if that achievement is an important attribute to your career and in your organization.

8.4 Managing Software Schedules

Summary and detailed schedules for the software project activities can be updated weekly or monthly to be consistent with overall program schedules. The software development schedules must show the details of the proposed builds and how they relate to overall program milestones. Eventually, the software schedules may get all the way down to the detailed "inch-stones" with tasks identified at the level of individual engineers. Typically, the schedules are updated by "*rolling waves*" which can be for a 3 or 6 month period or build-by-build.

To properly account for software-related costs, WBS elements must be created that allow software costs to be properly assigned to the correct categories. SwIPT leads should status the schedule, perform analysis and trending, identify problem areas, develop action plans, and periodically update management.

Tracking Charts. Techniques to assist in developing, tracking and presenting schedules include the ageless *Gantt Chart*, and one of the two activity network charts: the *Program Evaluation and Review Technique*, and *the Critical Path Method*. In addition to the Gantt, PERT and CPM tracking charts described below, I have added a simple management control chart; I call it the *Tri-Chart*, useful in tracking smaller projects.

PERT and CPM. The PERT and CPM charts are very similar in their approach but there are definite differences. **CPM** is used for projects that assume *deterministic* activity times meaning the length of time each activity will take are either known or an accurate estimate can be made relatively easily. **PERT**, on the other hand, allows for *stochastic* activity times where the time it takes for each activity is *uncertain or not easily estimated*. Because of this core difference, CPM and PERT are used in different contexts as described in Subsection 8.4.2.

> **Lessons Learned.** Many years ago I had a consulting firm that provided computer-based services including CPM scheduling services. I obtained a contract to provide a CPM schedule for new construction of a very large apartment complex. I conducted detailed meetings with the two construction supervisors to obtain the construction sequence of tasks and the estimated duration of each task. I entered all of this data into a CPM package and reported to them that their project would be completed 9 months later than the announced completion date. They looked at me like I was crazy, concluded that I knew nothing about construction, and ignored the news. Sure, I knew very little about construction, but they failed to realize it was not my estimate—it was *their estimated data* that I keyed in.
>
> That project was actually completed 12 months later than their date (it slipped 3 months more). The next time I prepared a CPM schedule for that company they took me seriously. The lesson I learned was to make sure the customer

understands the results come from their estimates, not yours. PERT and CPM are powerful tools and they work.

8.4.1 The Gantt and Milestone Diagrams

The Gantt Chart is basically a bar chart. It was developed during World War I by pioneer Henry L Gantt, and it exists in many variations. Figure 8.2 is a simple Gantt Chart example.

The Gantt Chart lists the full set of tasks in the vertical axis and includes horizontal hollow rectangular bars to depict the time frame for each of these tasks. As the tasks are completed, the hollow bars are filled in to show progress as of the date of the chart. An option is to portray two bars, one showing the planned duration and the second showing the actual duration. This option provides the additional feature of identifying task overruns and underruns. The example in Figure 8.2 indicates that the design and fabricate tasks are behind schedule at the end of time interval 5.

The chart in Figure 8.2 is actually a combination of the Gantt Chart and the Milestone (or event) chart as it also shows the planned and completed milestones by the empty and filled in triangles. The milestones in Figure 8.2 are shown as triangles, but they can be represented in a variety of ways including dates, letters or numbers.

Lessons Learned. Gantt Charts are simple to understand so they are likely the most widely used of all schedule types. However, the big problem with Gantt Charts is that they do not reflect the interrelationships or dependencies between the tasks. To overcome this deficiency, you must use one of the network schedules, discussed below, or try a trick I have used by drawing connecting arrows between the end of one bar and the beginning of another in the Gantt Chart.

8.4.2 Comparing the PERT and CPM Network Charts

The *PERT* chart is a *milestone or event* network where the nodes (usually boxes) on the chart represent project activities and their durations. The links (arrows) between the activities represent precedence. In other words, if there is an arrow between node X to node Y, then activity X must be completed before activity Y can start. PERT charts, once called PERT/Time, are also known as "precedence diagrams."

On many developments, there is often an overlap of some activities, so to represent the real process, the PERT chart can show concurrent activities. For example, some coding could begin before detail design is fully completed and some software integration could begin before all coding is completed. The PERT chart is a useful tool to approximate the software development process for the purpose of planning and control because it provides additional attributes such as:

- The *Critical Path* is the longest path (or duration) through the activities in the network. If any activity on the Critical Path takes longer than planned, the completion date will slip by an equal amount of time. If any activity on the Critical Path can be accomplished in less time than planned, the project end date will be shortened by that same amount of time.
- Activities *not* on the Critical Path will have *slack time,* so those activities can be delayed, lengthened in duration, or both, without affecting the project end date.

A *Critical Path Method* chart is similar to the PERT chart except CPM is an *activity*-oriented network plotted on a timeline. An example of each chart for the same 13 tasks, and their graphical differences, are displayed in Figure 8.3. From a management perspective, both PERT and CPM will help to identify which activities need the most attention. Any type of project with interdependent activities can apply the CPM method of mathematical analysis. Software packages

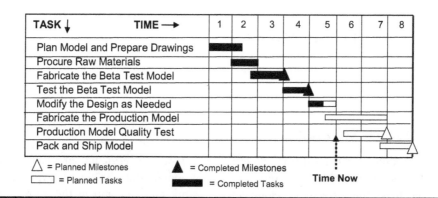

Figure 8.2 Combined Gantt and milestone chart example.

Summary of Tasks to Build a House

1. Lot excavation and plumbing rough-in:
   2 weeks
2. Foundation and soil treatment: 1 week
3. Slabs and framing: 3 weeks
4. Exterior siding, trim and roofing: 3 weeks
5. Rough-ins: 2 weeks
6. Insulation and flooring: 1 week
7. Drywall: 2 weeks
8. Driveway and walkways: 1 week
9. Interior trim and cabinets: 2 weeks
10. Painting: 2 weeks
11. Final plumbing, mechanical and electrical:
    2 weeks
12. Clean-up and carpet: 1 week
13. Landscaping and final inspection: 2 weeks

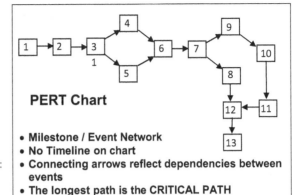

PERT Chart

- **Milestone / Event Network**
- **No Timeline on chart**
- **Connecting arrows reflect dependencies between events**
- **The longest path is the CRITICAL PATH**

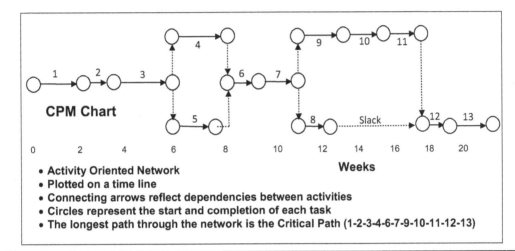

CPM Chart

Weeks

- **Activity Oriented Network**
- **Plotted on a time line**
- **Connecting arrows reflect dependencies between activities**
- **Circles represent the start and completion of each task**
- **The longest path through the network is the Critical Path (1-2-3-4-6-7-9-10-11-12-13)**

Figure 8.3   Comparison of the PERT and CPM charts.

are available to facilitate the use of both PERT and CPM, and there are many such packages on the market.

> **Lessons Learned.** Take note that network management (with both PERT and CPM) can have an unintended pitfall. It seems to have the effect of encouraging activities with slack (not on the Critical Path) to be purposely delayed and start at the latest start date with the justification that it will not affect the end date. This is playing "Russian roulette" and must be avoided. As the Software Project Manager, you should do everything in your power to *start activities at the earliest possible start date.* That approach will give the schedule the slack time, or buffer, it will need to contend with the unexpected delays that almost always seem to show up.

8.4.3 Other Charts and Tools

The Tri-Chart. Figure 8.4 is an example of a very *simplistic management control chart* useful for small projects or small

tasks that could be part of a large project. I call it a *Tri-Chart* because it combines tasks, time and cost all on one page.

The Tri-Chart example in Figure 8.4 displays a *task* schedule in the Gantt type format, along with planned versus actual *time* charges allocated to and performed by the identified team members, plus the cumulative *cost* showing planned versus actual charges. It is a convenient *one-page overview* of the project and is easy to prepare and maintain.

Scheduling Techniques and Tools. A useful variation of PERT and CPM charts is to superimpose the performing organization(s) on the chart. This is easily done by vertical lines creating sectional columns on the PERT or CPM chart indicating the group responsible for performing the activities in that portion of the chart. This is sometimes called an "activity-responsibility" network.

Project Management Tools. Microsoft Project is a popular PC-based project management software product. It assists Project Managers in developing a Project Plan, assigning resources to the tasks, tracks progress, manages budgets, and analyzes workloads. In addition, Microsoft Excel is a very powerful spreadsheet developed for Windows, MacOS, Android and iOS platforms. Excel features calculation,

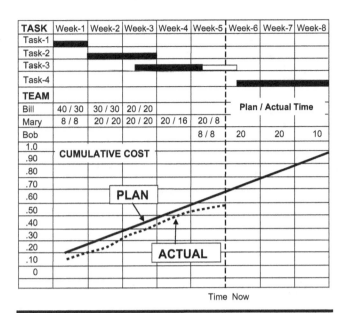

TASK	Week-1	Week-2	Week-3	Week-4	Week-5	Week-6	Week-7	Week-8
Task-1								
Task-2								
Task-3								
Task-4								
TEAM								
Bill	40 / 30	30 / 30	20 / 20			**Plan / Actual Time**		
Mary	8 / 8	20 / 20	20 / 20	20 / 16	20 / 8			
Bob					8 / 8	20	20	10

Figure 8.4 Tri-chart example for management control of small projects.

graphing tools, and the programming language "Visual Basic for Applications."

8.5 Managing Software Cost and Schedule with Earned Value

What might be called the "conventional cost management approach" involves plotting and comparing two cumulative cost curves—planned versus actual. The *planned cost* curve reflects the work that management hopes to accomplish, time-phased over the length of the contract. The *actual cost* curve represents the cumulative cost of the actual work accomplished up to the date of the report.

It is often assumed that the *actual cost represents actual work accomplished*. This assumption is *false* because the actual cost curve shows only the cost incurred and usually does *not reflect the amount of real work performed*. In order to determine the real status of the project, from a cost and performance point of view, a Project Manager needs additional insight and visibility.

An approach that works to effectively give you the detail and insight needed to obtain a true status of your project is the *earned value management control system* (EVMS). The *actual* work accomplished is generally known as *earned* value or "what you got for what you paid for."

A formal *Cost/Schedule Control System* (C/SCS) was developed by the DoD in the late 1960s to justify partial payments to their contractors. Even though there is a government connotation to this discussion, earned value has also been proven to be a very effective management tool for *non-government* developments of large systems as described

by (Fleming, 2010). The C/SCS is based on the approach of planning and controlling programs through the use of *earned value techniques*—sometimes called *variance analysis*.

Something similar to a C/SCS is (or should be) currently operational at all major government contractors. Not all government contractors will conduct C/SCS in *exactly* the same way, but the basics are very similar, and everyone uses computer applications to assist and facilitate the process. It can be assumed that most organizations have some method for fiscal control of cost, schedule and project performance. However, organizations that do not have something similar to the C/SCS can use the following description as an initial model from which to construct a tailored system to provide them with an earned value management control capability that can be used for many projects.

The following brief description of the C/SCS is an overview and introduction to the C/SCS as it is *not* intended to cover all the nomenclature and tasks involved. It is presented in this Guidebook to explain the basic elements of the C/SCS to Software Managers who are not familiar with the EVMS technique and to help them plan, monitor, manage and control their software project.

8.5.1 Cost Account Manager

The *Cost Account Manager* (CAM) plays a primary role in the implementation, operation, and reviews for the C/SCS. Elements of the C/SCS are described below. The CAM is the most significant contributor to the successful operation of the C/SCS and to the successful completion of internal audits and customer project reviews. Day-to-day management of the project takes place at the CAM level. If each cost account is not managed competently, program performance will be degraded regardless of the sophistication of higher-level management systems in the organization.

The CAM should be skilled in the areas of cost account performance. It is not enough to be technically proficient; the CAM must also be familiar with all aspects of the C/SCS as well as internal and customer reporting. Internal audits should be made periodically to ensure the CAMs are properly trained. Appropriate and timely training classes must also be made available for all new CAMs. If the project is small to medium-sized, it is common for the Software Project Manager to be the CAM, but on large systems the SPM may assign CAM responsibilities to senior members of his/her staff and then work closely with the CAM to gain the benefits provided by the C/SCS.

8.5.2 Elements of the Cost/ Schedule Control System

Work Scope and Authorization. When a contract is received, or when the contract changes, Senior Management, Financial Operations and the Contracts Department will release an *authorization to proceed* (ATP) to the Program (or Project) Manager via a *Scope of Work Order*. The Program Manager, working with the Program/Financial Controls Group, prepares an "authorization" document, sometimes called the *Resource Allocation Notice* (RAN), to accomplish specific WBS effort and (this is the most important part) to *allocate budget resources* to accomplish each authorized WBS task.

Cost Account. In the Cost/Schedule Control System, a *Cost Account* (CA) is the focus for planning, managing, and controlling effort because it represents work within a single WBS element, and it is the responsibility of a single responsible organization—essentially, the CAM in that organization. The scope of effort planned for a CA must be carefully considered to ensure that work will be properly refined or divided into "manageable units" with responsibilities clearly defined. Major changes to the CAs are authorized by revisions to the RANs.

Work Package. A *Work Package* (WP) is a detailed set of tasks that are defined by the CAM for performing work within a Cost Account. Each WP is the responsibility of a single performing organization. A WP has scheduled start and completion dates, with interim milestones if applicable, that are representative of physical accomplishments. The WP has an assigned budget and its duration is typically relatively short. The WP schedule must be integrated and in harmony with all other schedules. In some cases, the WP might be a subsystem.

Planned Package. If a CA cannot be completely, or realistically, subdivided into detailed WPs, the *Planning Package* (PP) is used for budgeting and scheduling. The PP work is usually effort to be performed *sometime in the future*. A budget reserve is created for the PP along with scheduling considerations.

8.5.3 Software Performance Measurement

Software performance measurement in the C/SCS consists of providing status, evaluating performance, and forecasting future activity at the CA level. The earned value terms are:

■ *Planned Values*: Budgeted Cost of Work Scheduled (BCWS)

■ *Earned Value*: Budgeted Cost for Work Performed (BCWP)
■ *Actual Cost*: Actual Cost of Work Performed (ACWP)
■ *Cost Performance Index (CPI)*: BCWP ÷ ACWP
■ *Cost Variance (CV)*: BCWP – ACWS
■ *Schedule Performance Index (SPI)*: BCWP ÷ BCWS
■ *Schedule Variance (SV)*: BCWP – BCWS

Table 8.1 is an overview of the basic elements of earned value performance analysis. The complete C/SCS involves several other formulas, but those covered in the table are the major ones. Table 8.1 also shows the meaning of the positive or negative numbers generated by the formulas.

To help clarify the relationship between the three major elements of the C/SCS (BCWS, BCWP and ACWP), Table 8.2 provides an example comparison of results obtained from these three factors and the resulting status of the project for the example results shown in the table.

8.5.4 Graphical Performance Analysis

To help in the analysis of earned value cost and schedule performance, it can be displayed graphically as shown in the example Figure 8.5. The graph can be used for an individual CA or the entire program. Use of a graphic of this nature is highly recommended and is almost indispensable when making presentations. The formulas in Table 8.1 should be used in conjunction with the graphic. The example shown in Figure 8.5 indicates that there will be both a cost overrun and a late delivery projected based on extrapolated past performance.

8.5.5 Example of Earned Value in Practice

An example of the conventional cost management approach is shown in Figure 8.6. It involves a hypothetical contract to build ten luxury boats in 10 months at a cost of $100,000. per boat. In this example, management decided not to build one boat at a time in favor of building them all at once *from bottom-up*, starting with the *keel*, under the assumption that it is less expensive to have each subcontractor to come in one

Table 8.1 Basic Elements of Earned Value Performance Analysis

Name	Formula	If it is a Positive Number	If a Negative Number
CPI	BCWP ÷ ACWP	Project is Below Planned Cost	Cost Overrun Forecast
CV	BCWP – ACWS	Project is Below Planned Cost	Cost Overrun Forecast
SPI	BCWP ÷ BCWS	Project is Ahead of Planned Schedule	Late Delivery Forecast
SV	BCWP – BCWS	Project is Ahead of Planned Schedule	Late Delivery Forecast

Note: If the result is 1.0: Project performance is at the planned cost and schedule.
ACWP = Actual Cost of Work Performed; CV = Cost Variance; BCWP = Budgeted Cost for Work Performed; SPI = Schedule Performance Index; BCWS = Budgeted Cost for Work Scheduled; SV = Schedule Variance; CPI = Cost Performance Index.

Table 8.2 Comparison of Example Earned Value Results and Project Status

BCWS	BCWP	ACWP	Status of Project
$ 1000	$ 1000	$ 1000	On Schedule–On Cost
$ 2000	$ 2000	$ 1000	On Schedule–Cost Underrun
$ 1000	$ 1000	$ 2000	On Schedule–Cost Overrun
$ 1000	$ 2000	$ 2000	Ahead of Schedule–On Cost
$ 1000	$ 2000	$ 1000	Ahead of Schedule–Cost Underrun
$ 1000	$ 2000	$ 3000	Ahead of Schedule–Cost Overrun
$ 2000	$ 1000	$ 1000	Behind Schedule–On Cost
$ 3000	$ 2000	$ 1000	Behind Schedule–Cost Underrun
$ 2000	$ 1000	$ 3000	Behind Schedule–Cost Overrun

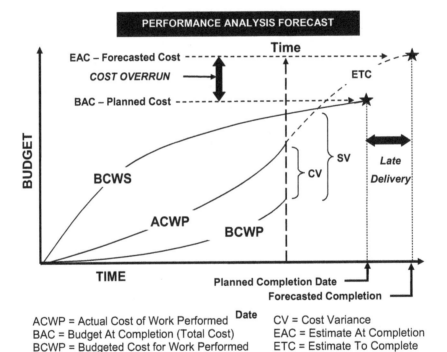

ACWP = Actual Cost of Work Performed
BAC = Budget At Completion (Total Cost)
BCWP = Budgeted Cost for Work Performed
BCWS = Budgeted Cost for Work Scheduled

CV = Cost Variance
EAC = Estimate At Completion
ETC = Estimate To Complete
SV = Schedule Variance

Figure 8.5 Cost and schedule graphical performance analysis example.

time to accomplish their specific expertise on all ten boats rather than have them come back ten times.

Using the conventional approach, Figure 8.6 shows the actual cost curve under-running the planned cost by about $100,000 at month 6. If this is all the information available to management, it can be concluded that *the contract is doing very well*. However, if the Manager has access to an earned value cost reporting system a very different and serious *conclusion is evident* as displayed in Figure 8.7.

Figure 8.7 shows the same planned cost curve and actual cost curve but adds a third curve—earned value (EV) that

is calculated. Figure 8.7 indicates a severe cost and schedule overrun because $500,000. was expended to date, but the *EV was $300,000. or the equivalent of three boats. That is $200,000 more to build three boats than what was planned.*

8.6 Integrated Process and Product Development

The IPPD technique is a program management approach that *simultaneously* integrates *all* essential development

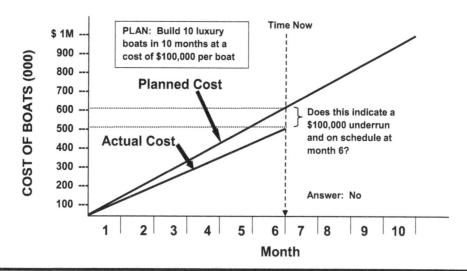

Figure 8.6 Conventional cost status report.

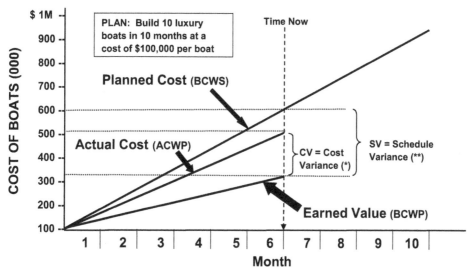

(*) Cost Variance indicates $500K was expended to build 3 boats or $166,667 per boat compared to the $100,000 planned cost per boat

(**) Schedule Variance indicates 3 months behind schedule because only the equivalent of 3 boats were completed instead of 6 boats

Figure 8.7 Earned value cost status report.

activities (not just software) through the use of multidisciplinary teams to optimize the design, manufacturing, and supportability processes. The IPPD technique was developed for and used by the DoD, however, the concept has wide applicability.

The IPPD technique establishes a clear structure and unambiguous responsibility for program participants. This approach helps to meet cost and performance objectives from product concept through development and into field support. The overall IPPD concepts can be summarized by the following five principles which are all strongly supported by this Guidebook:

■ Customer focus (if you can only remember one principle, this is the one)
■ Concurrent development of products and processes
■ Early and continuous life cycle planning
■ Proactive identification and mitigation of risk
■ Maximum flexibility for optimization and use of in-house company approaches/resources

Chapter 9

Managing Software Facilities, Reuse and Tools

The adequacy of your facilities has a significant impact on the productivity of your team.

Software Facilities. A *Software Development Environment* (SDE), and a *Software Integration and Test Environment* (STE), must be established to meet your project's software development and testing requirements. In addition, the appropriate libraries and related files, plus the important physical facilities for your staff must be planned for and provided when needed. Managing Software *Facilities* is described in the following Sections:

- *Software Development Environment:* The hardware, software and documentation needed to support the software development task (9.1).
- *Software Integration and Test Environment:* The hardware, software and documentation needed to support the integration and test of the Software Items (9.2).
- *Software Development Libraries:* Controlled collections of software, documentation, tools and procedures used in the development of software-intensive systems (9.3).
- *Software Development Files:* Organized project data repositories (9.4).
- *Software Data Management:* Controlled access to project data repositories (9.5).
- *Software Development Physical Working Environments:* Guidelines on the physical working environments (9.6).

Software Reuse and Tools. Incorporating reusable software has become much more important and more widely used than in the past. Software reuse is not a trivial issue. Projects that plan to use a significant amount of *Commercial*

Off-the-Shelf (COTS) software, or any reusable software product, must address COTS/Reuse in considerable detail in their SDP. Two options are suggested for addressing the COTS/Reuse issues: cover all of the topics in the SDP, or include an informative overview in the SDP and refer to a *Software COTS/Reuse Plan* for the details. If you are using a significant amount of COTS/Reuse software, you should take the second option.

Managing reused software and *Computer Assisted Software Engineering* (CASE) tools are covered in the following Sections:

- Managing Reusable Software Products (9.7)
- Managing CASE Tools (9.8)

9.1 Software Development Environment

The *SDE* may also be called the Software Engineering Environment (SEE). In either case, the SDE consists of *the hardware, software, procedures, documentation and facilities* necessary to support the software development effort. A key management and communications element of the SDE is an *Electronic Data Interchange Network* (EDIN) for remote access which includes the ability for storage, retrieval and distribution of program related software documentation and work products (see 9.5.1).

An overview of the data network must be included as a figure in the SDP or referenced to its location. In addition to a data network diagram, the major operational software development sites should be listed in an overview table similar to the example in Table 9.1. This table could be expanded to incorporate additional data columns important to your project.

Table 9.1 Software Development Sites—Example

Developer	Site Location	Network ID	Function	Function Name
Alpha Co.	City, State	A1	FSS	Flight Software Subsystem
Beta Co.	City, State	B1	TT&C	Telemetry, Tracking & Command
Delta Co.	City, State	D1	MPS FSE	Mission Processing and Services Field Station Element
Gamma Co.	City, State	G1	MMC	Mission Management Center

9.2 Software Integration and Test Environment

Each software development subsystem must have an *STE* that supports integration and testing of its SIs as part of its integrated SDE. These test environments should be defined by subsystem test personnel and described in their *Software Test Plan* (STP) and their SDP Annexes.

Care must be taken at all levels, including System Integration, to procure the needed integration and test tools far enough in advance to assure they are available when needed and that there is enough time for user training. Some SIs may be developed at multiple sites. All software developed at geographically dispersed sites must be fully tested at each development site, preferably on the target hardware, prior to final installation and qualification testing at the integration location(s).

All planned Integration and Test Environments should support testing using *Test-Like-You-Fly* (TLYF) principles (described in Section 10.4.9). This includes high-fidelity simulators and target test beds and test facilities that are representative of the operational environments.

9.3 Software Development Libraries

As described in Subsection 6.3.1, three generic levels of Software Development Libraries are normally used to implement Software Configuration Management as follows:

■ *Master Software Development Library (MSDL)*: The MSDL is a *single master program-level* repository of software information covering your entire project (or, possibly, the entire program).
■ *Subsystem Software Development Library (Subsystem SDL)*: Each software development subsystem should maintain a subordinate library for control of its subsystem software products before transmitting the work products to the MSDL.
■ *Software Development Library (Site SDL)*: Each software development site should maintain a subordinate library at its site for *local control* of its software products before transmitting their work products to their Subsystem SDL.

The MSDL, and the SDLs, must provide repositories for products resulting from software requirements definition, design, implementation, and test in accordance with the requirements of the SDP. The MSDL and SDLs are controlled collections of software, documentation, and associated tools and procedures used in the development of software-intensive systems. The SDL for each development site must be defined in its respective SDP Annex. The MSDL and the SDLs must be maintained throughout the contract duration. Also, electronic items must be maintained in a restricted environment and access controlled by login procedures and/or other measures. Each SDL contains code, test cases, and the electronic version of the software documentation specific to that site.

Logical Partitioning. Figure 9.1 is an example graphical depiction of typical logical partitioning for the electronic SDL format. In Figure 9.1 the SDL is shown as three primary library logical partitions: the software *development area*; the *controlled library area*; and the *Verification Area*. Specific organization of Subsystem SDLs should be defined and described in the subsystem annexes to the SDP. Each of these logical work areas are described below.

9.3.1 SDL Work Area

The Software Development Library Work Area is maintained and controlled by the Programmers as a *working area to develop the software*. This working area is used to create new code and/or documents, modify previously released code and/or documents, maintain databases, perform Unit Testing by the Software Developers, and for other users to review the products electronically. At the end of each build, the finished products, and the Software Development Files (SDFs), are transferred from the Work Area to the Controlled Library Area.

9.3.2 SDL Controlled Library Area

After Code and Unit Test is complete, the software source code goes under subsystem *Configuration Management (CM) control* (meaning the Software Unit is baselined). The CM group copies finished products received from the Software

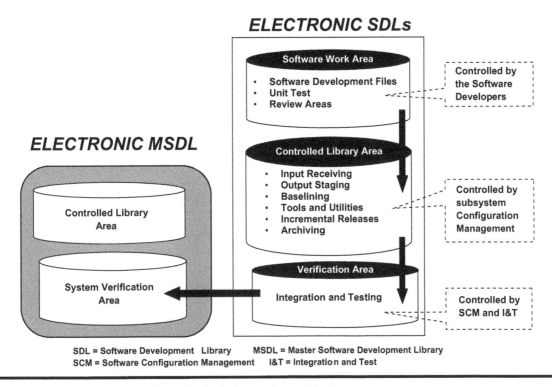

Figure 9.1 Software Development Library logical electronic partitioning.

Developers into this library file area. CM will always rebuild the executables from the source files before transferring them. CM has ownership, full accountability, and full access privileges; all other users have *read-only* privileges in the controlled library area. Software products are held here for CM to verify that the necessary files have been received by doing a preliminary validation of the product. When all files are received, the executable software products will be built from the source code and transferred to the Verification Area.

9.3.3 SDL Verification Area

The SDL Verification Area is owned, maintained and controlled by the integration and testing group. They have Write privileges and no other users may modify the data in this area. Before products are transferred to the Verification Area from the Controlled Area, CM verifies that the product is complete, and is ready for integration with other parts of the system; when that is done, the code and executables are copied into this area. All software products promoted into the SDL Verification Area are under the strict control of the chosen Configuration Management tool. Following Software CCB approval, files are transferred to the System Verification Area in the MSDL.

9.3.4 Software Process Assets Repository

In addition to the MSDL and SDLs, software documentation should be provided, via electronic access to a

program-level library that may be called the *Software Process Assets Repository* (SPAR). The SEPG should be responsible for defining and maintaining the SPAR; this repository (library) usually consists of both electronic and hard copy materials.

9.4 Software Development Files

The development of software-intensive systems generates a great deal of information. Poor record keeping is an easy pitfall that Software Project Managers can fall into primarily due to neglect. *SDFs*, also called *Software Project Workbooks*, are essential organized project data repositories that captures the knowledge generated on your project. SDFs should be required for all software categories at the subsystem, SI, and SU levels. SDFs must be prepared and *kept current throughout the program duration*. If an SI or SU is deleted, its data should be retained in an inactive file. SDFs are intimately involved throughout the software development process. Formats and audits of the SDF are described below.

9.4.1 Format of the Software Development Files

SDFs are normally maintained in electronic format *and* non-electronic format (hard copies) if that format is of value to your project. Electronic information should always be the preferred format. Information can either be placed directly into the SDF or provided by pointers to an external location. SDFs should be initiated during software requirements definition and remain under control of the Subsystem Development Teams from the time they are created until completion of the contract. Table 9.2 is an example tabular version of an overall SDF organization.

9.4.2 Audits of Software Development Files

SDFs should be inspected and audited throughout the program, to determine compliance with the SDP, with at least one inspection performed during each build and prior to each major review. Deficiencies identified during these inspections normally result in Corrective Action through the Corrective Action system. The frequency of SDF audits should be defined in the *Software Quality Program Plan* (SQPP). After an SDF inspection by SQA, the CSwE, or the customer, the SDF must be updated to note that it has been audited.

9.4.3 Non-Deliverable Software

Non-deliverable software consists of non-operational software developed, purchased, or used during software development but *not required by the contract to be delivered* to the customer or other designated recipient. It is identified as Category SS-3 (described in Subsection 1.8.2). Software can be classified as non-deliverable software only if the full operation and sustainment of the contracted deliverable software system does not depend on its use, or if the customer either has the software or can readily obtain it themselves. Examples of non-deliverable software that is excluded from formal control includes:

- Test data generators not deliverable to the customer
- Internal demonstration software not planned to be reused
- Prototype versions of operational software that has been superseded
- Engineering analysis software

Table 9.2 Electronic SDF Organization—Example

SDF Folder Item	Description
Assessment Reports	Assessment support materials, reports and management information
Design Materials	In-work design specifications; design definition; agreements/decisions; databases from CASE tools; interface data
Lessons Learned	Lessons learned support materials and reports/meeting minutes/action items
Meeting Minutes	Minutes not already stored in another location
Metrics	Data, analysis and reports not already stored in another location
Notes	Unique and miscellaneous information about the SUs
Plans/Tracking/Supporting Materials	Administrative, risks, low-level schedules; trade studies and evaluations; BOE; prototype plans; earned value schedule status supporting artifacts
Problems/Issues	Generated discrepancy reports and Change Requests with current status
Process Improvement	Plans, minutes, and reports
Referenced Locations	File(s) with pointers or links to SDF related materials in other locations
Requirements	Software requirements and requirements traceability data
Reviews/Presentations	Formal briefing and presentation materials and minutes
SI Level Peer Reviews	Peer Review materials (checklists, forms, notification; materials; log)
Source Code	Organized by SUs within subsystems—or whatever is meaningful to the project
SQA Reports	SQA Audit support materials and reports
SU Level Peer Reviews	Peer Review materials (checklists, forms, notification; materials; log) for each SU
Test Artifacts	In-work plans, procedures; unit test and integration plans, procedures, results, reports
Tools	Common scripts and tools used that are not stored in another location
Training Materials	Training and orientation materials and records not recorded in another location

9.5 Software Data Management

Data Management (DM) provides the *interchange and access of controlled project data* to program personnel and the customer, supports timely delivery of contract deliverables and addresses key issues such as disaster recovery and data rights. Software DM and its related concerns covering disaster recovery, proprietary rights, and international issues, are often not addressed in the software standards. DM is not an integral part of the software organization since its functions typically span the entire program. If your project is part of a program, the DM function may be carried out at the program level.

The DM group, and the organization it is part of, should be responsible for the repository and central access point for all program and software documentation, the data accession list, storage media control, and informal documents. A *Data Management Plan* should detail the guidelines for preparation, identification, filing, retrieval, training, and standards for all program documentation. The DM Plan should be updated in accordance with the evolving requirements of the contractual phases. The DM Plan is not a formal part of the SDP, but there is no restriction preventing it from being an SDP addendum. Even if the DM is not under your purview, as the Software Project Manager you must ensure that software documentation follows all the applicable DM guidelines.

The *Data Center* is usually the hub of the DM task and the source for all configuration controlled documents including publication and distribution. Software documentation should be made available to the Program Team and the customer on the program's electronic website. Software development documentation must be retained in an SDL typically located at each development site.

9.5.1 *Electronic Data Management and Cloud Storage*

Reasonaby sized programs must have the ability to provide continuously available, secure, encrypted remote access such that any *authorized* individual can view the data (documents, code, databases, or other information) using a standard web browser. The data must be safeguarded at multiple levels (such as Unclassified, Contractor Proprietary and government classified levels if appropriate) in accordance with proprietary and government requirements, and negotiated *restrictions to rights* in technical data and software. Access must include data generated by Developers and all subcontractors on the program.

This Guidebook assumes that the Developer's parent organization has in place an effective and comprehensive electronic data management system, that may be called the *EDIN*, for the storage, retrieval and distribution of program related software documentation and work products. The EDIN must be described in the SDP or in a separate document.

Cloud Storage. With the emergence of *cloud computing*, the entire approach to the storage and retrieval of data is changing. *Cloud storage* is an industry term for managed data storage through hosted networks in which data is stored on remote servers, typically accessed from the Internet, referred to as the "cloud." Data storage on the cloud can be maintained, operated and managed by cloud storage service providers (referred to as a public cloud) or in your organization's internal cloud (called a private cloud).

Most users today prefer a hybrid cloud approach with a unified management tool that can manage both public and private cloud resources from a single interface. The concept of a cloud management platform emerged initially as a response by IT departments to the requirement of provisioning resources more quickly and efficiently. A true enterprise-ready Cloud Management Platform (CMP) can include both provisioning with ongoing management plus optimization of computer, storage, network, and application services as well as improving the tracking and management of deployed resources.

In addition to efficient data storage and retrieval advantages, if the cloud concept persists, it may also have a major impact on future software development efforts. A full discussion on those impacts is reserved for a future edition of this Guidebook. Some current cloud-related challenges include:

■ Delivering a complete application environment to Development Teams upon request so teams can be immediately productive by providing "as needed" governance and control over resources without negatively impacting Developer productivity

■ Adopting concepts such as continuous delivery and continuous integration that may be able to automate more of the development process

■ Streamlining the development process, including the ability to quickly get to the development assets

■ Enhancing the ability to collaborate on development efforts with subcontractors

■ Providing tight security controls for handling proprietary and classified information

■ Defining exactly how, and if, the cloud can assist Software Project Managers in managing contracts during the full System Development Life Cycle such as using the cloud as an automated workflow and progress tracking process in real-time

9.5.2 *Disaster Recovery*

Plans for disaster recovery should be included in the SDP or in an external plan referenced by the SDP. Disaster recovery provides an alternate repository and backup system of software, databases, documentation, and equipment (if necessary). Disaster recovery plans also provide an alternate development/operational capability in case of a catastrophic situation after initial delivery.

The disaster recovery plans also ensure protection against loss of, or damage to, organizational assets and data. They should describe a smooth transition from normal to backup operations and ensure an expeditious restoration of the site capabilities.

> **Lessons Learned.** Like life insurance, you hope you don't need it, but if there is a catastrophic event and you do not have an effective recovery plan, it can be a career-ending disaster for you. Pay attention to this issue.

9.5.3 Proprietary and Users Rights

Rights restrictions apply as identified in the contractual Technical Data Restrictions. Vendor trademarked or copyrighted items must be used in accordance with applicable licenses. Users of the vendor products must have the right to use these items in accordance with the applicable licenses. Restrictions on these products or tools (if any), other than those dictated by commercial practices, must be clearly described in the SDP and/or in the IMP. *Proprietary concerns can be major issues in contractor source selection.* Data rights apply to all software products—not just code or COTS software. The SDP needs to specify what standard level of data rights applies to each category of software on the program.

9.5.4 International Traffic in Arms Regulations (ITAR)

If you plan to use or provide technology and products *from or to* foreign countries, ITAR is likely to apply, and this issue must be addressed. The local legal or ITAR Compliance office must be consulted for specifics because this is a program-wide issue and the topic needs to be addressed sooner rather than later.

9.6 Software Development Physical Working Environments

The physical working environment, and other related aspects of the environment for Software Developers, can have a major impact on their productivity, morale, retention and even recruiting. Although physical environments vary, there are three basic formats: cubicles; private (or semi-private) offices; and telecommuting. Providing cubicles for some

Developers on your staff and private offices for others is a recipe for potential problems and should be avoided. It is better to do one or the other but not a combination unless there is a clearly understood justification for it.

9.6.1 Cubicle Environment

Cubicle farms are the most common working environment. The advantages of cubicles are:

- Face-to-face communication is facilitated by elimination of doors
- There is increased visibility of project plans, schedules, etc. on the walls
- Cubicles are a compromise between private offices and open work bays
- Cubicles are cost beneficial
- Cubicles provide more privacy than open work bays but can be just as noisy

Software Developers working in cubicle environments often use stereo headphones or other ways to minimize disruptive noises and they can block entrance to their cubes when they do not want to be interrupted. In many cases the Developers don't listen to anything through the headphones—they are just a way to block out the distractions. In addition, cubicle environments need to have team meeting rooms and conference rooms available to support collaborative work without distracting others working in the cubes. It is also preferable to have some private office areas available when needed for private conversations and "deep" thinking. I recall once we had a small conference room available in the basement that had this name above the door: "Whine Room."

9.6.2 Private and Semi-Private Office Environment

Having worked in all of the office environments, I believe that the open bay is by far *the worst* environment for productivity because of the noise. Some people do not like private offices because they feel isolated, i.e., the privacy can become too much of a good thing. I believe semi-private offices (two persons) may be the best choice and it is more cost-effective than a private office for everyone.

> **Lessons Learned.** At one time I had to work in an open bay that was very noisy. I had to learn to focus so deeply on what I was doing that I developed the ability to *tune out the noise.* This approach worked well at work. The problem was when I was home, and concentrating on

something, I did not hear my wife talking to me (some might consider this an advantage). She thought I was being rude or had a hearing problem. Fortunately, I did not work in an open bay for very long, so my marriage was saved.

Cubicles are cost-effective and used widely, but private or semi-private offices are the best choice if that is affordable and allowable in your company. Private or semi-private offices have several advantages including:

- A door provides a good opportunity for uninterrupted work time and concentration resulting in a higher productivity rate.
- Private offices can double as team meeting and mini-conference rooms, resulting in a reduced need for additional meeting and conference rooms.
- Private offices, especially if they have wood furniture, have an aura of professionalism that can be ego boosting and a valuable recruiting tool.

Preferences about these options depend on your company culture and personality attributes of the people your company tends to attract. Some technical staff prefer a more social approach to programming, whereas others prefer a more solitary approach. The key to creating an excellent working environment is to be sure it is consistent with other elements of your company's culture. If your company emphasizes collaboration, cubicles would be a good match; if your company is primarily composed of individual contributors, then private or semi-private offices would be best suited to them.

9.6.3 Telecommuting Environments

The advantages of telecommuting may be in the "eyes of the beholder" as many organizations do not allow it at all, others are unsure how much of it to allow, and some are not sure it has justifiable advantages. The advantages of telecommuting are somewhat suspect but include:

- Reduced commute time thereby improving the quality of life for the telecommuter
- Uninterrupted work time at home compared to an office environment at the office where achieving an uninterrupted workflow may be difficult
- Retaining the services of a key staff members who live near your geographic area but beyond a reasonable commuting time
- A huge potential for vastly increasing your off-site staff with key contributors who otherwise will not, or cannot afford to, move to your geographic area or who live in foreign countries

On the other hand, there are numerous disadvantages of telecommuting including:

- Inhibits the team's identity, comradery and face-to-face communication
- Creates obstacles to technical assistance and mentoring
- May provide a less productive work environment (if team members are reluctant to call other team members at home or if children are at home during the telecommuting hours)
- Creates many opportunities for abuse requiring increased management oversight
- Must schedule periodic (often weekly) face-to-face team meetings
- Potential increased cost for telecommunications equipment support

9.6.4 Tools to Increase Productivity

Dual flat-panel monitors are now near universal. Productivity can increase dramatically when going from 1 to 2 monitors. Going from 2 to 3 monitors seems to increase productivity a bit more, but not as dramatically as the move from 1 to 2. There are potential benefits gained from providing Developers with very large monitors. Developer laptops or a laptop and a desktop are now common. Most staffs seem to view having a laptop as a clear benefit.

Ping pong tables, X-Boxes, and other similar recreational equipment are still common and commonly used for decompression times in organizations that provide them. It is essentially a necessity to provide plenty of meeting space (including informal conversation areas) including sound suppression of some kind—such as doors, white noise generators, or company provided headphones.

9.7 Managing Reusable Software Products

The normal definition of a "reusable software product" can include any existing software product, such as specifications, designs, test documentation, executable code, and source code that can be effectively used to develop and test the software system. Reusable software products include *Commercial Off-the-Shelf* and *Government Off-the-Shelf* (GOTS) software products, as well as reuse libraries from the program you are working on, or from other reuse library databases internal or external to your organization. Reusable software products may include software that does not need to be modified or migrated software that may require some changes. From a cost perspective, software Development

Teams should consider the use of reusable software products wherever possible as long as the amount of modifications required is justifiable.

The opportunity to incorporate reused products can be a major cost savings. The management of reused software products is described in the following six subsections:

■ Incorporating Reusable Software Products (9.7.1)
■ Justifying Use of Software COTS Products (9.7.2)
■ Functionality of COTS Products (9.7.3)
■ Integration of COTS Products (9.7.4)
■ Management of COTS Implementation (9.7.5)
■ Developing Reusable Software Products (9.7.6)

9.7.1 Incorporating Reusable Software Products

The approach to be followed for identifying, evaluating and incorporating reusable software products must be described in the SDP. It should include the scope of the search for such products, the criteria to be used for their evaluation, and address the related contractual clauses. If reusable software products have been selected, or identified at the time the SDP is prepared or updated, they must be identified and described including their known benefits, risks, constraints, and restrictions. As an alternative, the SDP can contain an overview and point to a *COTS/Reuse Plan* for the details. This plan can be an addendum to the SDP and should include a discussion of at least the following reuse topics:

■ Responsibilities for developing and maintaining the Software COTS/Reuse Plan
■ Heritage reuse base programs
■ Controlling, testing, and upgrading COTS/Reuse baselines
■ Developing and integrating reusable software products
■ Approach to managing the COTS/Reuse software implementation
■ COTS/Reuse software selection criteria and responsibilities

Reusable Software Criteria. Reusable software products must meet the specified functional and contractual requirements and be cost-effective over the life of the system. The following factors should be considered in an evaluation and selection of candidate reusable software products:

■ Technical capabilities of the product or functionality applicable to your project
■ The need for required changes and the feasibility of making those changes
■ Demonstrated product reliability and maturity
■ Testability and availability of test cases

■ Short- and long-term cost impacts of using the software product
■ Technical, cost, and schedule risks and trade-offs in using the software product
■ Product data rights are transferable from the product vendor
■ Interoperability with the target software environment
■ Ability of the product to meet safety, security, and privacy requirements
■ Availability and quality of documentation and source files
■ Supplier maintainability, warranty and restrictions on copying and distribution

The SDP or the *COTS/Reuse Plan* should cover the entire COTS/Reuse life cycle process, including identification, investigation, evaluation and selection, implementation and sustainment as shown in the example in Figure 9.2. Note that a bad evaluation report on *any single factor* can be a sufficient condition to reject a reuse candidate. Appendix D contains an example of the criteria for evaluating COTS and reusable software products.

Responsibilities for Software Reuse. The SwIPT should be responsible for the identification and evaluation of reusable software products for the SIs and SUs of the system. Beginning with the software requirements definition activity, and continuing through the testing activity, the SwIPT should identify appropriate candidate reuse products for each software activity.

Depending on the specific functionality being considered for reuse, the SwIPT may need to perform trade studies or perform some modeling or analysis with the candidate products to obtain sufficient information to make an evaluation. If any technical or non-technical issue is not fully resolved prior to the time when the product is selected for use, the SwIPT must define the issue as a risk and resolve it before a final selection is made.

9.7.2 Justifying Use of Software COTS Products

Using *Commercial Off-the-Shelf* software allows you to be selective in what functions and capabilities can be *acquired* without having to pay the price for custom development. The use of COTS software can also have a major impact on the reduction of schedule risk and cost risk. However, the process of *including COTS components is often difficult,* and care must be taken to avoid a number of potential risks.

Lessons Learned. If a COTS product requires significant modification of the code, the product becomes your responsibility unless the vendor is hired to make the modifications and that can be

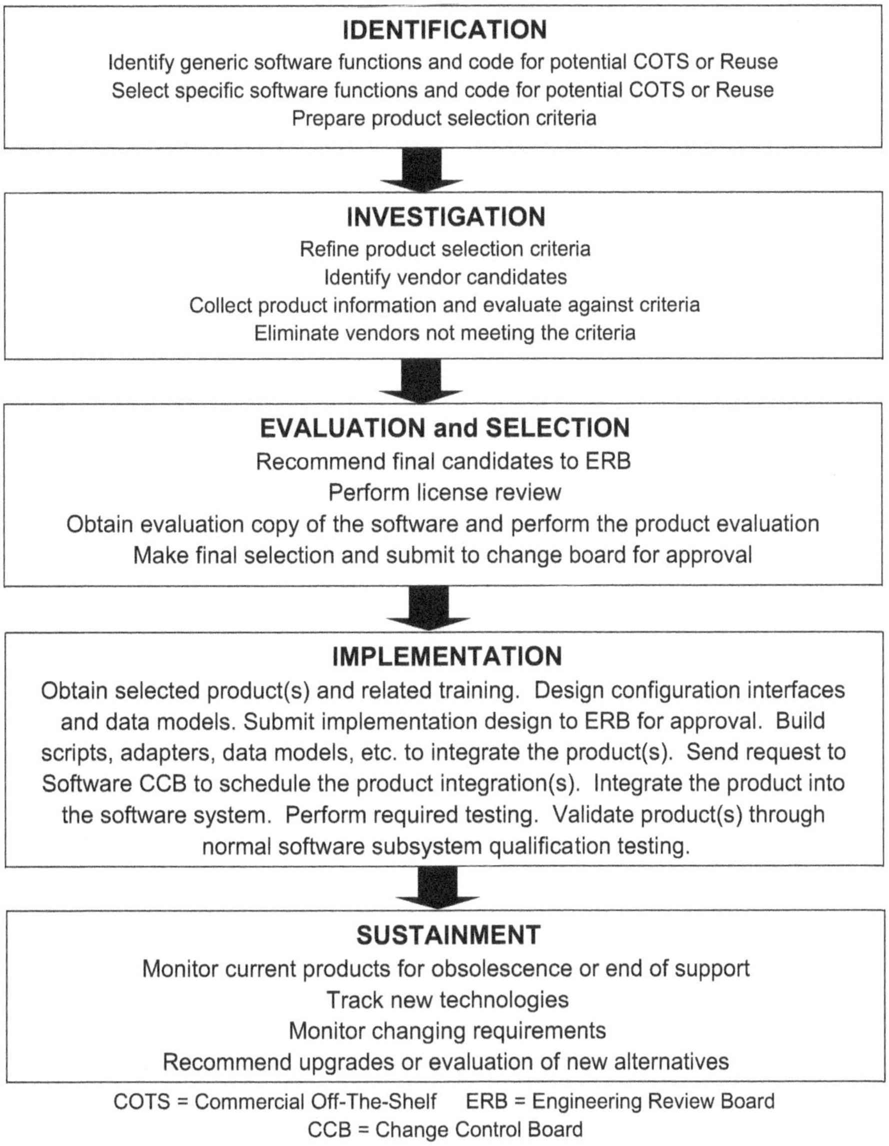

IDENTIFICATION
Identify generic software functions and code for potential COTS or Reuse
Select specific software functions and code for potential COTS or Reuse
Prepare product selection criteria

INVESTIGATION
Refine product selection criteria
Identify vendor candidates
Collect product information and evaluate against criteria
Eliminate vendors not meeting the criteria

EVALUATION and SELECTION
Recommend final candidates to ERB
Perform license review
Obtain evaluation copy of the software and perform the product evaluation
Make final selection and submit to change board for approval

IMPLEMENTATION
Obtain selected product(s) and related training. Design configuration interfaces
and data models. Submit implementation design to ERB for approval. Build
scripts, adapters, data models, etc. to integrate the product(s). Send request to
Software CCB to schedule the product integration(s). Integrate the product into
the software system. Perform required testing. Validate product(s) through
normal software subsystem qualification testing.

SUSTAINMENT
Monitor current products for obsolescence or end of support
Track new technologies
Monitor changing requirements
Recommend upgrades or evaluation of new alternatives

COTS = Commercial Off-The-Shelf   ERB = Engineering Review Board
CCB = Change Control Board

Figure 9.2   COTS/reuse management process—example.

an expensive and risky approach. Generally, if any COTS or reused product requires *more than 30% recoding*, it is usually more cost-effective to build it from scratch. Industry estimates for this threshold range from 15% to 35%. When other related factors are carefully considered, it may be *more cost-effective to build than to buy*.

When COTS products are acquired the principal risk is the potential loss of control over the formalized development process. Software product vendors have their own agendas that are different from those who adopt their tools. Therefore, an analytical trade-off must be made in order to gain the benefits from using COTS software. To be successful in using COTS software, the major factors described in the following three subsections (9.7.3–9.7.5) must be considered.

9.7.3 Functionality of COTS Products

A selected COTS product may not have the *exact* functionality required to be responsive to specific allocated requirements. The COTS product may have more capability than is needed, or may not provide all the required functionality, thus necessitating integration with other components or making potentially sophisticated modifications. Key COTS-related questions include:

■ How mature is the COTS product and how easy is the COTS product to use?

- Are the COTS product capabilities and operation fully understood?
- How much, if any, modification is needed to incorporate allocated requirements *not* satisfied by the COTS product?
- How will unneeded capabilities of the COTS product be handled?
- How have known problems with the COTS product been rectified by the vendor?

9.7.4 Integration of COTS Products

Trade-offs may be necessary because the constraints and requirements imposed by the selected COTS products often result in less flexibility available to the software architect. The method of integrating selected COTS components may impose additional constraints on the software architecture, and planners must account for the additional effort required to understand the behavior of the COTS products. Key COTS-related questions may include:

- Was your software architecture designed first and the COTS products selected to fit it or do you need to modify the architecture to accommodate the COTS product?
- Is the Development Team trained and qualified to integrate the COTS product?
- Does the COTS product have an Application Programming Interface (API), and does the Development Team understand the API's capabilities, limitations and complexities?
- Have the impacts of the COTS product on system resources been analyzed?
- Has the size of the integration effort for the COTS product been estimated, and what is the level of confidence for that estimate?

Loosely Coupling. How you integrate, or couple, the COTS product into your architecture has a strong influence as to how closely you are committed to that product. If you integrate or deeply imbed the COTS products into your architecture flow, you are essentially committed to it as it will be too costly to replace them. If the vendor goes out of business, you will likely have an expensive rework to perform. However, if you loosely couple the COTS product with simple Application Program Interfaces (API), you can relatively easily replace the COTS products without significantly affecting your architecture.

9.7.5 Management of COTS Implementation

The incorporation of COTS products introduces new issues that do not exist when an entire system is developed in-house. For example, licensing will have to be considered as well as other vendor relationships. Also, the cost of adopting and adapting the components must be considered as well. Key COTS-related management implementation questions include:

- Was the COTS product selected using a defined selection and evaluation process?
- Have all the integration and related costs been properly estimated?
- How long has the vendor been in business and what is their financial stability?
- What experience (if any) does your organization have with the vendor?
- Have the vendor's technical support capabilities been fully evaluated?
- Is the vendor willing to modify the product to meet the requirements?

Requesting the vendor to modify their COTS product to meet the needs of your program is generally considered high risk and is definitely not recommended.

- Have mutual non-disclosure agreements and data rights been negotiated?
- Have cost-effective licensing agreements been worked out with the vendor?
- Has Configuration Management of the COTS product been properly planned for?
- Has integration testing of the COTS product been thoroughly planned for?
- Have the risks related to using the COTS product been identified and can they be managed?

9.7.6 Developing Reusable Software Products

In addition to reusing existing software products, there may be opportunities for your developed software products to be used elsewhere. The SwIPT should carefully review the SIs under development for opportunities where software products can be used elsewhere in your project, in the program or in your organization to improve efficiency of the organization's software development effort.

The use of Object-Oriented Design naturally produces cohesive objects that encapsulate functionality and data have well-defined interfaces and are therefore suitable for reuse in many instances. In addition, class hierarchies and design patterns capture commonality and provide for abstractions that can lead to reuse. Specific activities in software analysis and design processes identify opportunities for not only design and code reuse but also use case and scenario reuse for requirements traceability and testing. These opportunities for reuse should be recorded in the design documentation.

The task of identifying, evaluating and reporting opportunities for developing reusable software products is often tailored out of the contract, however, with proper management it can be a path to considerable long-term savings for your organization, and maybe a big help to you on your next project.

9.8 Managing CASE Tools

Computer Assisted Software Engineering tools have been promoted by their vendors for decades as a method to enhance the efficiency of software development. Indeed, there are many examples where CASE tools have had a positive impact. When you select CASE technology for your program, it is important to remember that new tools will not make an *ineffective* process more effective; *new tools are not a panacea for fixing problems—but they can make an effective process more efficient.* CASE technology used *across the program* must be identified in the SDP by a table like the example in Table 9.3.

The mechanism for making changes to the program-wide *Software Development Environment* toolset should be by approval of the Software Change Control Board (SwCCB). Additional SDE tools used only by a subsystem must be defined in a similar table in their SDP Annex.

The CASE tool industry has always been dynamic and there is no reason to believe that will change. The examples in Table 9.3 can be expanded to include all tools and the subsystems using each tool. This table can become lengthy (e.g., 100 tools or more) in large programs so it may be put

in an SDP appendix. If it is lengthy, it is recommended that the tools be grouped in usage categories such as: operating systems, compilers, configuration/change management, requirements traceability, documentation, metrics collection and analysis, performance analysis, test tools, etc.

9.8.1 CASE Technology Types

In general, the use of CASE software can be classified into three types: tools, workbenches and life cycle environments although in practice the distinctions are somewhat flexible—each are briefly described below:

CASE Tools. In the past, CASE tool vendors did not produce tools that addressed the entire Software Development Life Cycle. Instead, they focused their tools entirely on single activities, such as Software Design or software testing, leaving it up to the user to integrate the tools set into a full SDE. That is changing but even today there are CASE tools that support specific tasks in the Software Development Life Cycle such as:

- Business and analysis graphical modeling tools
- Tools for the development and design phases of the life cycle (e.g., debuggers)
- Verification and Validation tools (e.g., code analyzers)
- Configuration management tools (e.g., ClearCase)
- Measurement collection and analysis tools (e.g., DataDrill)
- Project management tools (requirements, planning, scheduling, etc.)

Table 9.3 Program-wide SDE CASE Tool Set—Example

Purpose of Case Tool	Name of Tool	Vendor
Object-Oriented Analysis and Design	Rational Rose	IBM/Rational
Code Development and Testing	SparcWorks	Sun Microsystems
Large Relational Database	Oracle	Oracle
Small Relational Database	Access	Microsoft
Problem Tracking Reports	ClearQuest(*)	IBM/Rational
Planning and Scheduling	Project	Microsoft
Configuration Management	ClearCase(*)	IBM/Rational
Requirements Management	DOORS(*)	Telelogic
Software Estimation	SEER–SIM	Galorath
Software Metrics	DataDrill(*)	Distributive Software
Test Automation	Rational Robot	IBM/Rational

Note: Trade names used as examples in this document are not intended in any way to infringe on the rights of the trademark holder.

(*) Core software management tools used across the program.

Workbenches. Workbenches integrate two or more CASE tools and support specific software process activities in order to achieve a more seamless integration of tools (and toolchains) as well as consistent interfaces. One example workbench is Microsoft's "Visual Basic" programming environment that incorporates several development tools: a GUI builder, smart code editor, debugger, etc. Most commercial CASE products tend to provide workbenches that seamlessly integrate two or more tools.

Life Cycle Tool Environments. An "environment" is a collection of CASE tools and/or workbenches that *attempt* to support the complete software process. They can be grouped into the following categories:

- *Toolkits*: These are loosely coupled collections of tools that typically build on operating system workbenches such as the Unix Programmer's Workbench or the VMS VAX set. They typically perform integration via some basic mechanism to share data and pass control.
- *Fourth Generation Language Environments (4GL)*: 4GLs were the first environments to provide deep integration of multiple tools. Typically these environments were focused on specific types of applications such as user interface driven applications that performed standard transactions to a relational database (e.g., Informix 4GL, and Focus).
- *Language-Centered Environments*: These are based on a single (usually Object-Oriented) language (e.g., Lisp or Smalltalk). In these environments, all the operating system resources were "objects" in the Object-Oriented language. This provides powerful debugging and graphical opportunities but the code developed is mostly limited to the specific language. For this reason, these environments are used mostly for prototyping and R&D projects.
- *Integrated Environments*: These toolsets attempt to cover the complete life cycle from analysis to maintenance and provide an integrated database repository for storing all artifacts of the software process (e.g., AD/Cycle, Foundation, Cades, and Cohesion). The integrated software repository was the defining feature for these kinds of tools. They provided multiple different design models as well as support for code in multiple languages, plus the ability to make changes at the design level and have the changes automatically reflected in the code and vice versa. These environments were also typically associated with a particular methodology for software development.
- *Process-Centered Environments*: These environments are the most ambitious type of integration. They attempt to not just formally specify the analysis and design objects of the software process, but specify the actual process itself, and to use that formal process to control and guide software projects. These environments were by definition tied to some methodology since the software process itself is part of the environment and can control many aspects of tool invocation (e.g., East, Enterprise II, Process Wise, Process Weaver, and Arcadia).

The CSwE must coordinate implementation of the common tool suite among all subsystems to ensure effective information transmittal and maximum commonality. The CSwE, or designee, should be responsible for monitoring the implementation of the SDE to ensure that all requirements are implemented, for periodically assessing the continuing adequacy of the environment, and for identifying additional tools needed. Details of the SDE configuration for each subsystem should be maintained in a current inventory list and available from the System Administrator at each site or subsystem.

9.8.2 CASE Technology Risks

Despite the significant potential advantages for your project, or your organization, to adopting CASE technology, there are some risk factors that you, as the Software Project Manager, must be aware of including:

- *The Silver Bullet*: CASE tool marketers often hype expectations that their approach will be a silver bullet that solves all of your problems. No such technology can do that, so if you approach CASE with unrealistic expectations, you will inevitably be disappointed. CASE tools can also fail to perform if the initial project attempting to use this new technology is itself a risky undertaking.
- *Inadequate Training*: Utilizing CASE technology requires time to train your staff in how to use the tools as well as time for the learning curve. If you cannot afford the required training, don't get the tool.
- *Inadequate Standardization*: Typically, you have to tailor and adopt methodologies and tools to meet your project's specific requirements. Doing so may require considerable effort to integrate both divergent technologies as well as divergent methods. For example, before the adoption of the UML standard, the diagram conventions and methods for designing Object-Oriented Models were profoundly different among followers of Jacobsen, Booch, Rumbaugh and others.
- *Inadequate Process Control*: CASE technology provides new capabilities and utilizes new types of tools in innovative ways. Without adequate process guidance, and implementation controls, these new capabilities could create significant new issues or complications of their own.

SYSTEMS AND SOFTWARE ENGINEERING DOMAINS

4

Chapter 10

Systems Engineering Domain

The success or failure of large software-intensive systems is based upon a clear understanding by the customer and the Developers as early as possible, as to what is being built even though the precision of that understanding will change during development.

The System and Software Development *Life Cycle processes* were defined in Chapter 3 and described in Figures 3.3 and 3.4. Figure 10.1 depicts the system and Software Development Life Cycle processes from a "big picture" perspective. Chapter 10 is focused on the Systems Engineering left half of Figure 10.1. Chapter 11 will focus on the Software Engineering right half with a much-expanded version of the big picture overview shown in Figure 10.1.

In this Guidebook, the *Systems Engineering Domain* includes four principal activities directly involving Software Engineering. The first two activities (10.1 and 10.2) are performed sequentially early in the life cycle, and the last two activities (10.3 and 10.4) are performed sequentially late in the life cycle, with appropriate iterations. *Systems Engineering* (SE) and *Software Engineering* (SwE) activities are intimately interrelated and that will be discussed in depth in Chapter 12. Tasks related to the "system" are normally applicable to the subsystems within the system. The

key Systems Engineering activities related to software-intensive systems are described in the following four sections:

- *System Concept and Requirements Development*: Covers development, specification and documentation of system-level and subsystem-level *requirements* (10.1)
- *System Design*: Involves System and Subsystem *Design and configuration* of required software, firmware and hardware interfaces (10.2)
- *Software and Hardware Items Integration and Testing*: Describes the integration of Software Items (SI) with other interfacing SIs (SI/SI I&T), and with Hardware Items (SI/HI I&T), until all interfacing SIs and HIs in the system are integrated and successfully tested (10.3)
- *System Qualification Testing*: Involves the integration of the entire system and verification that the full *system requirements* have been met including the system interface requirements (10.4)

Performing Requirements Development. Documenting an *understanding* by the customer and the Developers as to *what* is being built is called *requirements development*. The degree of understanding is based on what is known at the time the requirements analysis takes place. As the depth of knowledge and understanding grows during development, the requirements will change and become more precise. Without documented requirements, there will be serious implications including:

- Requirements created on-the-fly by Developers and Testers with frequent redirection
- No way to perform top-down design
- Undetected errors with haphazard and meaningless testing
- No way to make reliable estimates
- Confusing and irrelevant code will be prepared
- Lower productivity and higher stress levels

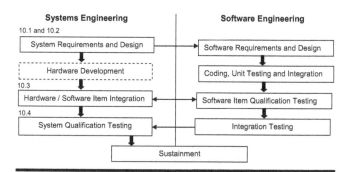

Figure 10.1 Overview of the System and Software Life Cycle process.

- Forced workarounds and retrofitting in the field
- No customer confidence or meaningful system buyoff
- No realistic management control and severe overruns

Poor definition of requirements at the start of a project normally results in a poor implementation and a high probability of failure. Documenting the requirements does not produce a design description, but it often does become a basis for a *contract* between the customer and the development organization. If properly performed, requirements definition answers the following:

- What specific work needs to be performed to implement this system?
- Why is the system needed and what are the system's objectives?
- What are the alternative methods of performing this work?
- What generic functions should be allocated to hardware, software and human operators?
- How can development of the system functionality be organized into subsystems?
- Are all of the proposed functions technically feasible?
- What is the extent of testing needed to be fully responsive to the system requirements?

Types of Requirements. There are various types of requirements that a Software Project Manager (SPM) may be involved with including:

- *Functional (or Design) Requirements*: What will be built
- *Performance or Operational Requirements*: How well it will be built; Operational Concept Description (OCD), Concept of Operations (CONOPS), etc.
- *Interface Requirements*: Boundary relationships or conditions
- *Organizational Requirements*: Staffing, training, subcontracting, etc.

- *Acceptance Requirements*: Integration and testing criteria and approvals

Writing Requirements Statements. The way a requirement is phrased is just as important as the requirement itself. If a requirement statement is vague it will be interpreted in different ways by different people; vague requirements are essentially useless. There are guidelines for writing requirement statements, sometimes referred to as "spec writing." As the SPM you must make sure the requirements analysts (and you may be one of them) follow those guidelines. A good requirement statement *must* be:

- Crystal clear and unambiguous
- Concise, simple and understandable
- Quantitative (meaning it has numbers) and must be testable
- Composed of *one* requirement per statement
- Feasible from a technical, cost and schedule perspective
- Free from an implementation bias (meaning *what* not *how*)
- Coordinated, reviewed and approved

The word "will" is a non-binding informational term, whereas "shall" and "mandatory" means it must be done. Don't write "shall be capable of…" but write "shall provide…" instead. Also, using the term "design goal" is no assurance it will be done. Other vague words to avoid are: achievable, adequate, flexible, expandable, efficient, instantaneous, etc. Table 10.1 is a short example of poorly versus clearly written requirements statements derived from (Endsley, 2015):

Requirements Development for Smaller Systems. The discussion of requirements development in this Guidebook is focused on medium to large software-intensive systems. Large systems involve a lot of pre-planning and concept analysis resulting in the preparation of an extensive amount of preliminary requirements. Small systems present an entirely

Table 10.1   Good versus Bad Requirements Statements

Poorly Written Requirements	Good Testable Functional Requirements
Be user-friendly.	Users shall be able to perform tasks at 99% accuracy and reliability.
The System Design should effectively direct the user's attention by means of alerting and coding techniques.	The system shall alert users to cautionary and critical conflicts using redundant shape, color and auditory cues allowing operators to detect all alerts within 2 seconds.
Weather information shall be provided to the user.	Weather information shall include temperature in Celsius, dew point, winds at altitude, and a graphical display of convective weather at a focal point on the user's display that is within 30 degrees of the normal viewing position.
Reduce personnel requirements from the current levels.	Total personnel required to operate the system during full mission execution shall not exceed 80 (maximum) with a goal of 40.

different situation because the customer often (too often) does *not know exactly what they want.*

Small system customers normally have at least a vague idea of what is needed, but the perplexing problem facing requirements analysts is: Customers know *what they don't want* when you show them your best guess as to what you think they do want. To resolve this conundrum, you need to prepare several iterations of prototypes until you provide something the customer agrees that that is what they want.

Much has been written about the elicitation of good requirements, and some of that documentation is listed in Appendix I. A complete discussion of the requirements development process is beyond the scope of this Guidebook. However, the following steps of the requirements development process, specific to *interactive* software applications, were derived from (McConnell, 1998, Chapter 8) which is a good general description of a requirements development process:

- Identify a set of key system users, customer representatives and other stakeholders
- Interview all the identified stakeholders
- Create a list of preliminary system requirements
- Build a simple interactive user interface prototype
- Demonstrate the prototype to stakeholders and solicit feedback
- Make prototype adjustments and obtain stakeholder concurrence
- Create a Software Requirements Specification (SRS) document and put it under change control
- Expand the prototype to provide a full system demonstration (not implementation)
- Ensure the developed software matches the final prototype and treat the fully extended prototype as the baseline specification
- Update the Software Requirements Specification and prepare an Interface Requirements Specification (IRS)

Early in the life cycle you are developing, defining, and gathering system requirements from numerous sources. *Specifying* the requirements means they are documented. The *analysis* of requirements could be considered part of the design process performed during development, but analyzing the *feasibility* of implementing identified requirements can take place at almost any time during development.

10.1 System Concept and Requirements Development

As the name implies, the major objective of system concept and requirements development is the specification and documentation of *system-level* requirements. The tasks in this activity are equally applicable to subsystems or any other level of

requirements above software. The principal tasks performed in this activity should be led by the *Systems Engineering Integration and Test* (SEIT) organization with support from the subsystem *Software Integrated Product Teams* (SwIPT).

This system-level activity is based on inputs generated by the customer and the gathering of requirements. Typical inputs from government customers could include the: *Statement of Objectives* (SOO); *Initial Capabilities Document* (ICD); *Capabilities Development Document (CDD)*; *Technical Requirements Document* (TRD); and the *Request for Proposal* (RFP). Typical inputs from non-government customers could include customer interviews, the contract, *Statement of Work* (SOW); work orders; questionnaire to stakeholders, or review of vision, scope and related documents.

If you are given these user-provided requirements, a lot of your requirements-related work is done for you; however, they may or may not be complete and certainly may change during project development. Working with the SEIT, the SwIPT members will have to do a lot of data collection by asking questions, doing research, preparing a questionnaire, and most important, conducting interviews with the end-users.

> **Lessons Learned.** There are times when you are collecting requirements, or collecting information of any kind, that you realize you have an *overwhelming* amount of data to deal with. If you don't *plan in advance how to organize that data*, it is amazing how quickly it can become out of control. I failed to do that the first time but never again. As the data is collected, file the data items in an organized fashion to help you understand what you have collected and enable you to conduct a meaningful analysis in a reasonable amount of time. A simple example of how to organize data items, so they are more meaningful and useful is Appendix I. Instead of having a very long list of bibliographic references, Appendix I is organized by subject as that is the likely way users will access that database.

The major output documents resulting from the requirements analysis activity are typically preliminary versions of the *System/Subsystem Specifications* (SSS); the *Operational Concept Description* (OCD); and top-level *Interface Specification* (IS). The system verification and test plans, usually produced by the SEIT, may also be revisited and updated if necessary.

In addition, *interface definitions* must be provided to enable the further definition and management of the computer software and computer equipment resources. This should be documented in the *Interface Control Document* (IFCD). The acronym "IFCD" is used in this Guidebook in order to avoid confusion with the ICD defined above. Depending on contract provisions, interface definitions may also be included in the SSS or in the *Interface Requirements Specification*. The IRS is essentially a software document and may be contained within the *Software Requirements Specification* or as an annex to the *Software Development Plan* (SDP).

Inputs can also be derived from Systems Engineering studies. An early draft version of the SSS might also be provided to the Developers by the customer. The example in Table 10.2 summarizes the *readiness criteria* for this activity with the entry and exit criteria, verification criteria to ensure completion of the required tasks, and the measurements usually collected.

10.1.1 System Operational Concept Requirements Process

The Subsystem SwIPTs should support the SEIT in developing and defining the system's *Operational Concept Description*, or similar documentation, by:

- Identifying and evaluating alternative concepts for technical feasibility, user input, cost-effectiveness, schedule impact, risk reduction and critical technology limitations

- Analyzing the operational concepts and other inputs to derive any software requirements that are not specifically stated
- Supporting the refinement of the operational system concept based on current analyses and updating the operational concepts with user interface analysis material

The system and subsystem requirements definition process is shown in example Figure 10.2 in terms of the input documents used, tasks performed, and output products.

10.1.2 System/Subsystem Requirements Analysis

System-level requirements must be documented in the *System/Subsystem Specifications* or similar document. The SSS specifies system capabilities and allocates functional requirements to the subsystems. Requirements for subsystem-to-subsystem interfaces, subsystem-to-hardware interfaces, and system-to-external system interfaces are initially defined and documented during this activity. All system interfaces should be maintained in a *Requirements Database* (see Subsection 11.1.2). Subsystem *Software Integrated Product Team* personnel should participate in working group discussions and joint

Table 10.2 Readiness Criteria: System/Subsystem Requirements Development

Entry Criteria	Exit Criteria
■ External system interfaces have been identified and the related documentation has been reviewed. ■ Preliminary concept of operations and system capability definition have been completed. ■ Systems Engineering notifies Software Team of the need for their support.	■ System-level requirements development and subsystem requirements analysis are complete. ■ Performance allocation, interface requirements, and user interface analysis are documented. ■ Joint technical and Management Reviews related to system requirements are successfully completed. ■ Software representatives have reviewed system requirements and the concept of operations. ■ System/subsystem requirements are allocated to software. ■ Bi-directional traceability is completed from customer requirements to/from system specifications and from system specifications to/from subsystem specifications.

Verification Criteria
■ SwIPT personnel participate in the review and approval of the system and subsystem requirements and interface requirements documentation. ■ Program and Senior Management are provided the status of ongoing product engineering tasks (including subsystem requirements analysis and management) on a periodic and event-driven basis. ■ The System Requirements Review (SRR) is successfully completed.

Measurements
■ Status of the requirements analysis task schedule. ■ Number of system/subsystem requirements allocated to software. ■ Planned versus actual level of effort. ■ Requirements traced versus untraced orphan requirements (see Chapter 15).

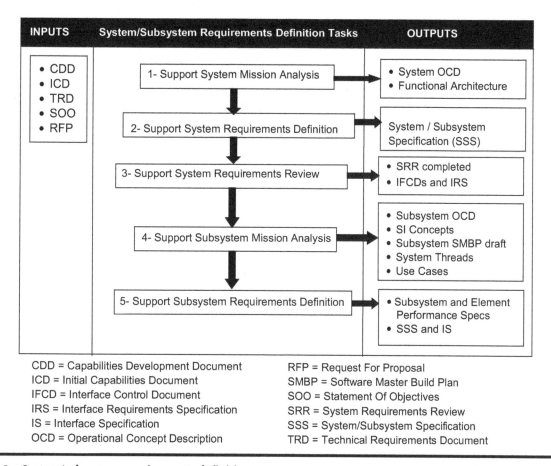

INPUTS	System/Subsystem Requirements Definition Tasks	OUTPUTS
• CDD • ICD • TRD • SOO • RFP	1- Support System Mission Analysis	• System OCD • Functional Architecture
	2- Support System Requirements Definition	System / Subsystem Specification (SSS)
	3- Support System Requirements Review	• SRR completed • IFCDs and IRS
	4- Support Subsystem Mission Analysis	• Subsystem OCD • SI Concepts • Subsystem SMBP draft • System Threads • Use Cases
	5- Support Subsystem Requirements Definition	• Subsystem and Element Performance Specs • SSS and IS

CDD = Capabilities Development Document
ICD = Initial Capabilities Document
IFCD = Interface Control Document
IRS = Interface Requirements Specification
IS = Interface Specification
OCD = Operational Concept Description

RFP = Request For Proposal
SMBP = Software Master Build Plan
SOO = Statement Of Objectives
SRR = System Requirements Review
SSS = System/Subsystem Specification
TRD = Technical Requirements Document

Figure 10.2 System/subsystem requirements definition process.

SwIPTs to review and comment on the parent specification requirements related to software. This seems logical and obvious but is not always done.

Subsystem SwIPTs support subsystem requirements analysis through the identification and derivation of the software-related aspects of functional performance, interfaces, constraints and quality requirements. These requirements must be analyzed for completeness, consistency, verifiability and feasibility. Subsystem SwIPT participants also must identify and recommend requirements that could be feasibly allocated to, and implemented in, software and identify possible software verification methods and traceability for those subsystem requirements.

System-Level Requirements. The Systems Engineering Integration and Test organization has primary responsibility for the system-level tasks performed during this activity. The CSwE and/or the Chief Software Architect are often (and should be) part of the SEIT Team. They directly support these tasks so that: system decisions involving software can be made with input from the appropriate expertise, and interface requirements are consistent across the system.

Software Developers must also participate in analyzing user input to help ensure that all involved users maintain ongoing communications regarding their needs throughout

development of the system. In addition to the Developers, involved parties may include the customer, Test Engineers, the ultimate end-users and the maintenance organization(s). Work product outputs of this task may include "need" statements, surveys, *Software Change Requests* (SCRs), *Software Discrepancy Reports* (SDRs), the results of prototypes, and documented interviews.

Subsystem-Level Requirements. Subsystem SwIPT personnel support system requirements analysis to ensure that requirements involving software are adequately addressed. Subsystem SwIPT members assist and support the SEIT in the identification and capture of the software needs by actively participating in system-level working groups.

Lessons Learned. Conducting interviews with the users sometimes brings hazards of its own because some people are still "afraid" of what automation will do to their job. Today, this issue is nowhere near what it was when automation was in its infancy. For example, years ago we had a contract to automate the departments of a medium-sized municipality. In order to do that we had to interview the Managers of each department.

One of the key Department Managers refused to talk to us because he was terrified as to what was going to happen to his department. We treated him with respect. He finally agreed to sit in on our interview with another Department Manager. After he realized we were not monsters out to destroy his fiefdom, he ultimately became one of our strongest supporters.

Subsystem requirements analysis can be accomplished by analyzing allocated subsystem requirements from the System Specification and interface requirements. The Systems Engineering groups in the subsystems have primary responsibility for the subsystem tasks performed during this activity. The Subsystem SwIPTs should assist and support Systems Engineering in the derivation of their subsystem's specific requirements from the system-level requirements. This is normally done by participating in the Systems Engineering working groups. If the subsystem does not have its own Systems Engineering group, then the Subsystem SwIPT would have the primary responsibility for subsystem-level requirements.

10.2 System/Subsystem Design

Software Integrated Product Team personnel representing the subsystems must support the SEIT in developing the system and Subsystem Design and the specific configuration of hardware, software, and firmware to meet performance and reliability requirements. Table 10.3 summarizes the readiness criteria for the System Design activity in terms of the entry and exit criteria, verification criteria to ensure completion, and the measurements usually collected.

During the System Design activity, major system characteristics should be refined through trade studies, analyses, simulation, and prototyping. This activity must perform *functional decomposition* to define subsystem *Software Items* and *Hardware Items* as well as SI interfaces to other SIs and HIs. System requirements and interfaces should be refined, allocated (i.e., decomposed) and flowed down to the SI level. In addition, make, buy and reuse trade studies can be performed during this activity.

The results of these tasks should be used to determine system characteristics (performance, cost and schedule) and to provide confidence that risks are being resolved or reduced in impact and severity. During the System Design activity, the maturity of technology can also be evaluated to make decisions about the use of new technology. This activity is led by the SEIT group. The test group must review the System/Subsystem Design to determine if the requirements allocated are feasible, testable and verifiable.

10.2.1 System and Software Architecture Overview

The Software Development Plan (SDP) is a process document, not a design document; however, it is recommended that your SDP contain some high-level architectural design overviews of the system and the software. The SDP should identify the subsystems comprising the system, contain at least one paragraph describing top-level software functions for each subsystem, and it should include a graphical overview of the system similar to the example in Figure 10.3.

Table 10.3　Readiness Criteria: System/Subsystem Design

Entry Criteria	Exit Criteria
Preliminary versions of the: ■ System and subsystem requirements. ■ System OCD and Requirements Database. ■ System Test approach and the Requirements Test Verification Matrix (RTVM).	■ System architecture, SI and HI definitions, software system architecture decisions, and non-developmental software analysis are documented. ■ System architecture baseline established.
Verification Criteria	
■ Software participates in review and approval of the system architecture, SI definitions and interfaces. ■ Program and senior management are provided the status of ongoing product engineering tasks (including System Design) on a periodic and event-driven basis. ■ SQA performs process and product audits for ongoing product engineering tasks. ■ System Functional Review (SFR) successfully completed.	
Measurements	
■ Product Engineering schedule. ■ SI SLOC and SI requirements estimates (see Chapter 15).	

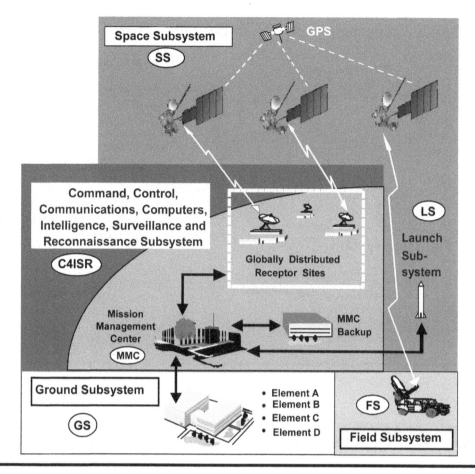

Figure 10.3   System overview diagram example.

In order to help convey the big picture to your staff so that they will better understand what they are building, the SDP should state the purpose of the system to be developed, identify the customer and/or project sponsor, user(s), Developers, the planned sustainment (maintenance) organization, and the proposed operating sites. The hypothetical system example shown in Figure 10.3 contains five subsystems; they are all software-intensive.

The SDP should also contain an overview of the software system (or functional) architecture, definition of the software categories, identification of the Software Items and responsibilities. The overall software architecture can be depicted in a diagram similar to the overview example Figure 10.4. An optional approach would be to include a *functional matrix table* showing software functionality for each subsystem, Software Item, or build. A physical overview of the entire system may also be needed.

10.2.2 System/Subsystem Design Process

Six principal tasks are recommended for the System Design activity as depicted by the example process flowchart in Figure 10.5. Details of the six tasks in the design activity are described

in its directly related Task Table 10.4 that shows the inputs, subtasks, and outputs to each task identified in Figure 10.5.

10.2.3 System-Wide Design Decisions

System-wide Software Design decisions and their rationale should be documented by the SEIT in *Engineering Memoranda* (EMs) and the *System/Subsystem Design Description* (SSDD). EMs are stored in an Electronic Data Interchange Network (EDIN: see 5.1.10 and 9.5.1). System requirements are often generated from EMs and SSDDs and are flowed down to the subsystem product specifications.

10.2.4 Subsystem Design Decisions

Subsystem-wide Software Design decisions and their rationale should be documented in EMs residing in the EDIN. Subsystem EMs must be evaluated by the Subsystem SwIPT to determine if they impact the software requirements. However, the principal product is the SSDD. The Subsystem SwIPT should record, in the subsystem's Software Development File (SDF), or equivalent location, the rationale for selecting the COTS/Reuse and other Non-Developmental

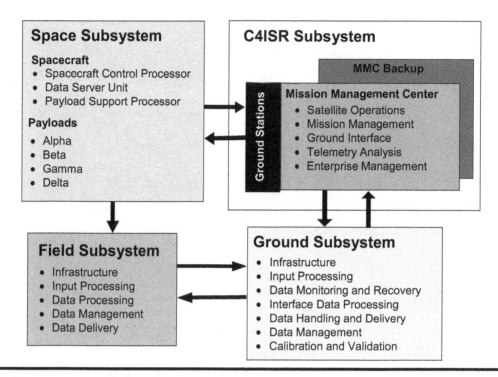

Figure 10.4   Software system architecture overview example.

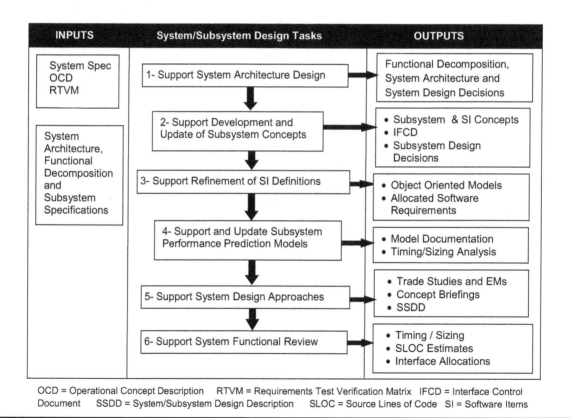

Figure 10.5   System/Subsystem Design process flow.

Table 10.4 System/Subsystem Design Tasks

Tasks	Inputs	Subtasks	Outputs
1. Support System Architecture Design	■ System OCD ■ System Specification ■ RTVM	■ Assist SEIT to identify system-level architecture ■ Assist SEIT in functional decomposition ■ Coordinate subsystem definitions with SEIT	■ System Architecture ■ System Functional Decomposition
2. Support Development and Update of Subsystem Concepts	■ System Architecture ■ System Functional Decomposition ■ Subsystem Specifications	■ Describe subsystem capabilities in the context of system specs, HIs and SIs ■ Describe subsystem interfaces to other SIs and elements ■ Describe individual SI capabilities, including plans for reuse of non-developmental software	■ Subsystem concepts ■ SI concepts ■ External interfaces captured in the SRS or IRS ■ IFCD
3. Support Refinement of SI Definitions	Subsystem and SI concepts	■ Develop appropriate OO diagrams reflecting all software objects needed to achieve scope with interfaces to other SIs ■ Allocate subsystem system requirements to SI classes; verify traceability of system requirements to SIs	■ OO-based Models ■ Allocated software requirements
4. Support and Update Subsystem Performance Prediction Models	Subsystem and SI concepts	Develop and update subsystem models to support: ■ Updated timing and sizing analysis ■ Algorithm development ■ Interface analysis	■ EMs documenting timing and sizing analysis ■ Performance prediction models and documentation
5. Support System Design Approaches	■ Subsystem Specs and SI concepts ■ Performance prediction models ■ Subsystem OO Models	■ Create subsystem-level behavior diagrams for key design approaches ■ Verify approach satisfies associated system requirements ■ Support documentation of technical approaches and the System/Subsystem Design Description	■ Trade Study EMs ■ Concept briefings ■ SSDD
6. Support System Functional Review	■ Concept briefings ■ Subsystem OO Models	Conduct analysis to: ■ Allocate timing and sizing budgets to SIs ■ Establish and update SLOC estimates ■ Allocate system and external interfaces ■ Flow up changes to system requirements	■ Timing and sizing budgets ■ Updated SLOC estimates ■ Interface allocations

Items (NDIs)—including rejected approaches and the studies and analyses that led to the selected approach.

Software system/subsystem architecture decisions made during System/Subsystem Design must also be recorded for later use in developing software requirements and design. Decisions are usually recorded using EMs. The Subsystem SwIPT Lead should participate in establishing the rationale for software architecture, definitions, interfaces, COTS/reuse/other NDIs approach, and should be responsible for ensuring that the following data are recorded:

■ The overall software architecture that was selected, including the studies and analyses that lead to the selected architecture

■ The Software Item definitions and interfaces, including the studies and analyses that lead to the selected SI definitions and interfaces

■ The software COTS/Reuse and other NDIs selected, including studies and analyses that led to the selected approach

10.2.5 *System/Subsystem Design*

This task involves organizing a system into subsystems, then decomposing subsystems into elements (or modules), and further decomposing them into Hardware Items, Software Items, human interactions and other operations needed to make the fully integrated system work successfully.

System Design. During the System Design task, required subsystems must be identified along with subsystem-to-subsystem, and subsystem-to-external systems interfaces, plus a concept of system operational execution often described in a *Concept of Operations* (CONOPS) document. The interfaces must be documented by the SEIT in the *Interface Control Document*. The Subsystem SwIPT personnel should support the SEIT by:

- Participating in the system architecture and design as well as the specific configuration of hardware, software, and firmware to meet performance and reliability requirements
- Assessing the software impact of implementing the operational concept and system requirements in terms of technical suitability, cost, and risk
- Participating in trade studies to select processing, communications, and storage resources
- Reviewing the System Test approach and test philosophy to ensure testing compatibility
- Identifying how each requirement will be tested, what test support software will be needed for System Test, identifying the System Test environment, and developing a *System Test Plan*
- Reviewing and analyzing the System Design to determine testability of the requirements allocated to software
- Recommending requirements changes as necessary

Subsystem Design. Subsystem Design should be documented as part of the subsystem architecture baseline process. The basic responsibilities, typically assigned to Subsystem SwIPT personnel during Subsystem Design, are the same seven bullets listed for System Design, but with a software focus, *plus* the following five activities:

- Support the identification of test support software and test environment needed for subsystem-level test and development of a subsystem test plan.
- Recommend requirements changes to subsystem system engineering.
- Support subsystem system engineering in creating definitions of Software Items, in allocating subsystem requirements to the SIs and in review and refinement of the interfaces.
- Identify potential candidates for reuse and COTS software products at the SI level.
- Review and refine the definition of Software Items.

When using the Object-Oriented (OO) methodology, the subsystem software high-level Architecture Design must be captured in SI OO Models and placed in the electronic Software Development File. These architecture models should be documented with the applicable OO methodology products. These products should then be utilized to refine and update the development of timing and sizing budgets, SLOC estimates, and to prototype algorithmic approaches.

10.3 Software and Hardware Items Integration and Testing

The software and hardware *Integration and Testing* (I&T) process involves integrating Software Items with other interfacing SIs (*SI/SI I&T*), and with Hardware Items (*SI/HI I&T*), testing the resulting groupings to make sure they work together as intended, and continuing this process until all interfacing SIs and HIs in the system are integrated and tested successfully.

Software Item/Hardware Item I&T is usually the first integration of the software system with the *target hardware*. SI/HI I&T may involve the entire system, or a portion thereof, building up to the entire system. For activities involving hardware integration with software, the Software Team is usually in a *support role* to the Systems Engineering Integration and Test group.

Objectives of SI and HI Integration and Testing. The principal objectives of SI and HI Integration and Testing are to:

- Perform the individual SI-to-SI integrations and SI-to-HI integrations to produce the complete software build for each successive level of test and verify its successful integration.
- Integrate software into the target hardware system and verify integration success.
- Verify SI-to-SI and SI-to-HI interface compliance with requirements.
- Support and complete integration and qualification testing at each level of integration.

Approach to SI and HI Integration and Testing. The *Software Master Build Plan* (SMBP) should be kept current to define the SI functionality planned to be operational for each build. The integration sequencing should be documented in the SMBP and the *overall approach to I&T* documented in the *Master Test Plan* (MTP). Although integration of software with the hardware is a critical objective of this activity, some aspects of the hardware integration may not be able to be performed until the full System Integration. Table 10.5 is an example of the readiness criteria for this activity in terms of entry and exit criteria, verification criteria to ensure completion of the required tasks, and the measurements typically collected during this activity.

Test Beds. A "testbed" is a system testing environment containing the target hardware, operational software, instrumentation, simulations, software tools and any other

Table 10.5   Readiness Criteria: Software and Hardware Item Integration and Testing

Entry Criteria	Exit Criteria
■ Hw/Sw integration approach is defined and approved in the Master Test Plan. ■ SwIPT software personnel are requested by SEIT to support the Hw/Sw integration activities. ■ The executable software product has completed the SIQT process and is capable of supporting Hw/Sw integration. ■ The software release and the integration database are under control of the Software Development Library and integration tools are available.	■ Hw/Sw integration and testing is successfully completed including an action plan to close remaining SDRs. ■ Sw test and Sw Management Reviews and approves the SDR/SCR closure plan. ■ Regression Testing is completed and accepted by the Sw test Lead.

Verification Criteria
■ SI-SI and SI-HI integration is verified and accepted by the Sw Test Lead. ■ SQA performs process/product audits for ongoing product engineering activities. ■ Sw Test Lead reviews regression test logs and accepts completion of Regression Testing.

Measurements
■ Test Coverage: Number of requirements tested and passed. ■ Number of test cases—planned versus actual and percent of interfaces tested. ■ SDRs and SCRs opened, closed, aging data, origin and root cause analysis (see Chapter 15).

supporting element needed to conduct a test of the full or partial system *on the ground*. The following steps are an example scenario of the SI and HI I&T process for spacecraft hardware and software integration using laboratory testbeds:

1. Software Item Qualification Testing (SIQT) for the spacecraft bus software is performed in the *flight test bed,* and SIQT for the payload software is performed in the *payload test bed*.
2. The two test beds are connected, and the software-to-software interfaces between the spacecraft bus and the payload are tested.
3. The spacecraft bus software and the payload software are integrated into the actual vehicle.
4. The flight and payload software and hardware are integrated and tested.

There may be early progressive integration points using the two example testbeds so that all the spacecraft software and payload software interfaces do not have to wait until the software is completely finished to be integrated and tested.

Roles and Responsibilities for SI and HI Integration and Testing. Generally, the SEIT Team is responsible for performing SI/HI I&T and the software role in this activity consists primarily of support tasks. The SI-to-HI integration tasks should involve both Hardware and Software System Engineers and Testers. A comprehensive description of the tasks involved in the full SI/HI integration and testing activity is not addressed here—focus is on the support provided by the SwIPT personnel. All testing must be run using

documented test descriptions developed collaboratively by the software and SEIT engineers.

Work Products for SI and HI Integration and Testing. The documentation produced during this activity is involved with testing the integration of software and hardware at *various levels of the progressive integration*. During this activity the SI/HI I&T test cases, test procedures, test drivers, test scenarios, test stubs, databases, and other needed test data are produced.

The SI/HI I&T activity concludes with the SI and HI integration and test results and preparation of *Software Discrepancy Reports* for all problems encountered while testing the integration of software and hardware. *Software Change Requests* may be produced as described in Subsection 6.4.1. The following products are developed during the SI/HI I&T activity:

■ A baselined *Software Product Specification* (SPS) (See 11.9.1)
■ A baselined *Software Version Descriptions* (SVD) supporting the current release (See 11.9.2)
■ An updated *Software Test Description* (STD) as baselined during software qualification testing (See 11.7.7)

10.3.1 Software and Hardware Items Integration and Testing Process

The process flowchart in Figure 10.6 shows the inputs, outputs and relationships between the four SI/HI I&T tasks: *prepare, perform, revision/retest,* and *analyze/record results*.

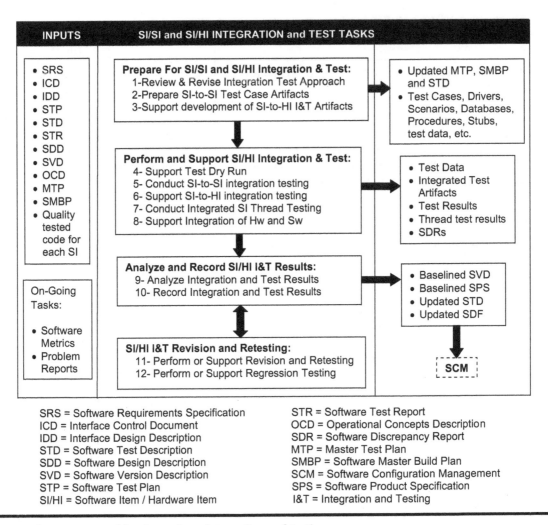

Figure 10.6 Software Item and hardware item integration and testing process.

These four tasks are described in more detail in the following four subsections and by their directly related task tables (Tables 10.6–10.9).

The Software Developers, and Software System Engineers, assigned to perform integration of the SIs, must develop an *integration strategy* that defines a systematic *approach for integrating the SIs into the complete software releases*. Issues such as SI-SI interfaces, inter-SI timing and sequencing, and simulations or emulations (for external interfaces) are examples of the issues that go into determining the proper order of integrating the SIs.

10.3.2 Preparing For SI and HI Integration and Testing

The objectives of preparing for SI/SI integration and SI-HI integration testing are to finalize the *Master Test Plan*, *Software Master Build Plan*, and *Software Test Description*, and to develop integration test artifacts (test cases, test procedures, test drivers, test scenarios, test stubs, databases and test data) necessary to verify the success of the integration

effort. Table 10.6 contains examples of the tasks applicable to preparing for the SI/SI and SI/HI I&T activity.

Hardware and Software System Engineers must collaborate in the preparation of appropriate test description information for the hardware and software integration accomplished during this activity. They review in-progress SI/HI I&T *Software Test Descriptions*, provide recommendations to Test Engineers (including recommended software test equipment needed), and ensure that test cases and corresponding test procedures are sufficiently defined in the updated STD to verify the success of each partial integration. An updated STD for SI/HI I&T should contain at a minimum:

- The overall test descriptions and test environment needed for SI-to-SI and SI-to-HI I&T
- Specific test cases and corresponding test procedures to verify correct execution of SI-to-SI and SI-to-HI interfaces including:
  - End-to-end functional capabilities
  - Sequencing and timing of events and data flows
  - All requirements allocated to software

Table 10.6 SI and HI Integration and Testing Preparation Tasks

Tasks	Inputs	Subtasks	Outputs
1. Review and Revise Integration Test Approach	See inputs to Figure 10.6	■ Review SI/SI and SI/HI I&T approach in the MTP, the test sequence in the SMBP, test requirements in the STD, and update if necessary. ■ Define functional capability threads. ■ Develop an integrated schedule and activity/dependency network. ■ Identify assumptions/constraints. ■ Document planning results; obtain approvals.	■ Updated MTP, SMBP and STD ■ Updated SI/SI & SI/HI integration plans ■ Updated integration schedule ■ Action items
2. Prepare SI-to-SI Test Case Artifacts	Outputs and Action Items from Task 1	■ Prepare SI-to-SI test cases, test procedures, test drivers, test scenarios, test stubs, databases, and other test data as needed. ■ Update MTP, SMBP and STD. ■ Define integration test, threads, and test cases.	■ Test Cases, test procedures, test drivers, test scenarios, test stubs, databases, etc. ■ Functional Capability Thread Descriptions ■ Updated MTP, SMBP and STD
3. Support Development of SI-to-HI I&T Artifacts	Outputs from Tasks 1 and 2	■ Support development of SI-to-Hi test cases, test procedures, test drivers, test scenarios, test stubs, databases, and needed test data. ■ Update MTP and SMBP if necessary.	SI-to-HI test cases, test procedures, test drivers, test scenarios, test stubs, databases, test data, MTP and SMBP as applicable

- Stress testing including worst case scenarios
- Start-up, termination and restart procedures
- Fault detection, isolation, and recovery handling
- Performance testing including input/output data rates, timing and accuracy requirements
- Operation of multiple SIs on a single computer platform, where applicable
- Integrated error and exception handling capabilities
- Testing of limit and boundary conditions
- Resource utilization measurements (e.g., CPU, memory, storage and bandwidth)

■ Input data definitions (e.g., data files, databases, etc.)

■ Required simulations and emulations needed for external or hardware interfaces

■ Specific output data to be collected and recorded in the appropriate SDF

■ The expected results and success criteria

10.3.3 Performing SI and HI Integration and Testing

The SI/HI I&T must be performed using the target hardware in a configuration that is as close as possible to the operational configuration. All reuse software, including legacy reuse and COTS software, must also undergo the SI/HI I&T process. The SwIPTs provide Software System Engineers and Test Engineers along with applicable software test support items, expertise and training in using the software.

The objective of performing SI/HI I&T is to integrate and test the software in accordance with integration and test strategies in the approved *Master Test Plan* and the *Software Master Build Plan*. The five principal tasks for performing integration of hardware and Software Items are:

■ Support the Test Dry Run
■ Conduct SI-to-SI Integration Testing
■ Support SI-to-HI Integration Testing
■ Conduct Integrated SI Thread Testing
■ Support Integration of hardware and software

Integration testing may be called *Factory Acceptance Test* (FAT) or *Element Qualification Test* (EQT). Table 10.7 contains examples of the tasks applicable to performing the SI/HI I&T activity. Software integrators normally begin the integration of SIs by ensuring that all SIs to be integrated and all necessary data and tools are available and ready. Software Test Engineers support needed Corrective Actions on any soft-related errors, request re-execution of build procedures as required, and accept SI builds upon satisfactory verification.

When all required elements are assembled, the integration proceeds. The Software System and Subsystem Test Engineers run test cases using the test procedures, specified in the STD.

Table 10.7 Performing SI and HI Integration and Testing Tasks

Tasks	Inputs	Subtasks	Outputs
4. Support Test Dry Run	■ Approved test Integration procedures, plans and schedules ■ SRS for the build ■ Integrated Sw–Hw build ■ Integrated SI thread test cases ■ Integration test data and tools ■ Hardware test equipment	■ Ensure all test Sw and Hw are available, the correct version and under CM control. ■ Perform test cases & procedures. ■ Document Sw–Hw dry run integration test results in the test log. ■ Generate SDRs/SCRs as applicable. ■ Redline procedures and obtain SQA approval.	■ Approved integration procedures ■ Integrated test stubs, drivers, and scenarios ■ Test results in Log ■ Integrated build
5. Conduct SI-to-SI Integration Testing	■ Approved STP ■ SW plans and schedules ■ SRS for the build ■ SIs from SCM ■ Test data and tools	■ Integrate SIs in accordance with integration test procedures in the MTP and SMBP. ■ Develop SI thread test cases in accordance with the SRS and MTP.	■ Integrated builds ■ Integrated SI thread test cases ■ SCRs/SDRs
6. Support SI-to-HI Integration Testing	Same as Activity 5 plus: Integrated Hw–Sw test drivers, scenarios, and stubs	Same tasks as Activity 5 plus: ■ Record test logs. ■ Document status for interface requirements. ■ SQA audit and review test status.	■ Integrated Hw/Sw builds ■ FAT completion ■ SCRs/SDRs ■ Code and test procedure revisions ■ SQA audit report
7. Conduct Integrated SI Thread Testing	■ Approved Test Integration Procedures ■ Sw plans and schedules ■ SRS for Build ■ Integrated SIs ■ Integrated SI thread test cases ■ Integration test data and test tools	■ Develop SI thread test procedures, stubs, drivers and scenarios in accordance with integration plan and SI test cases. ■ Develop Hw/Sw integration test cases and procedures. ■ Perform SI thread test procedures. ■ Document SI thread test results. ■ Generate SCRs/SDRs if applicable.	■ Integrated build thread test procedures, drivers, stubs and scenarios. ■ Thread test results ■ Integrated and tested builds ■ Test Cases
8. Support Integration of Hw and Sw	■ Approved test Integration procedures, plans and schedules ■ Integrated and tested SIs ■ Target hardware ■ Hw/Sw Integration Test Cases ■ SRS for Builds and SDRs/SCRs ■ Integrated test data & tools	■ Get Integrated and tested builds from SCM. ■ Integrate Software build with the target hardware. ■ Rework source code if required in response to approved SDRs and SCRs.	■ Hw/Sw Integrated build ■ Hw/Sw integration test cases ■ SDRs/SCRs

They collect or record the outputs, logs, test notes, and results. All problems, errors, and discrepancies are noted.

10.3.4 Analyzing and Recording SI and HI Integration and Test Results

The objectives of analyzing and recording SI/HI integration and test results are to: analyze integration tests results to ensure the tests have been successfully completed; and document the respective test data and results as required. Table 10.8 contains examples of the tasks applicable to analyzing and recording of the SI and HI I&T activity.

After all SIs and HIs have been successfully integrated and tested, the Integration and Test Team *must review the test results for consistency and completeness* and verify that the integration test data and results have been documented. If discrepancies

Table 10.8   Analyzing and Recording SI and HI Integration and Test Tasks

Tasks	Inputs	Subtasks	Outputs
9. Analyze Integration and Test Results	■ DRs ■ Approved Test Integration Procedures ■ Integration Test Results	■ Collect test results. ■ Analyze test data to ensure proper processing of input data by each procedure and correct output data.	■ Analysis results
10. Record Integration and Test Results	■ Integration and test results ■ Analysis results	■ Collect test and analysis results. ■ Ensure that results are correctly and completely recorded.	■ Build Integration Release Notice ■ Released build for site and system testing ■ Document test and analysis results

Table 10.9   Revision and Retesting SI and HI Integration and Test Tasks

Tasks	Inputs	Subtasks	Outputs
11. Perform or Support Revisions and Retesting	Same as Activity 5 or 6	Same as Activity 5 or 6 plus: Perform DR fixes	Same as Activity 5 or 6
12. Perform or Support Regression Testing	Same as Activity 5 or 6	Same as Activity 5 or 6 plus: Perform DR fixes	Same as Activity 5 or 6

or problems are found, then the portion of the integration in question must be *retested* (see 10.3.4). It is also a good idea to perform an independent review by an independent reviewer not involved with the subsystem hardware/software integration testing, but not a team member of the SEIT. Note that independent reviews are not required by most standards.

Once the independent reviewer signifies that the integration and testing is complete, and the integration testing was successfully completed, the release is baselined. At the last stage of integration and testing, the test results are normally documented by SCM in a *Build Integration Release Notice*, and the build is then ready for System Qualification Testing (SQT) as discussed in Section 10.4.

10.3.5 Revision and Retesting of the SI and HI Integration and Testing Results

The objectives of revision and retesting are to verify that changes and applicable SDR/SCR modifications have been implemented correctly and that the functionality is performing in accordance with requirements *after the fixes* have been completed. Changes can also involve test procedures, test data, etc. as well as code changes. The documented problems must be evaluated by Software Developers to determine the necessary changes to SIs, SUs or to the test descriptions.

Retesting is performed to show that a problem is fixed and the test case executes properly. Regression testing (see Subsection 3.9.3) is performed to show that the fix did not break anything that was previously tested and working

properly before the fix. Changes to software require SDRs or SCRs to be generated and the changes are handled by the Corrective Action Process (CAP). In cases where test descriptions require modification, the changes identified in the SDRs or SCRs, must be made by Software Test Engineers and a *version history* included in the test description to record the changes made. Modified software requires retesting of the integration tests that previously failed and for any tests that are dependent on the failed tests. Table 10.9 contains examples of the tasks applicable to revision and retesting of the SI/HI I&T activity.

This process must be repeated until all SIs and HIs have been successfully integrated, and all tests have been completed. If the element SwIPT Lead determines it is impractical to complete some changes until a later build, then SCRs/SDRs must be used to document and control the modifications, but integration testing still must be performed. The Change Control Board (CCB) at the element level must approve all such delays.

10.4 System Qualification Testing

System Qualification Testing, often called System Acceptance Testing, involves verifying that *all* system requirements have been met—including the system interface requirements. The Systems Engineering Integration and Test Team is responsible for SQT and the SwIPT is in a support role but directly involved.

Subsystem Integration and Testing. The full SQT cannot occur until all of the subsystems comprising the system have been integrated. SQT, as described here, is also applicable to the verification of requirements at all levels above verification of the software requirements. Those levels include the *integration of all Subsystems and the elements within each subsystem*. The major qualification tasks at the subsystem and system levels are similar, but details of the required tests, procedures and documentation may be different. The subsystem I&T task includes a *Subsystem Acceptance Test* (SAT). If a system is developed in multiple builds, qualification testing of the full system will not occur until the final build.

The system qualification and acceptance testing activities are described in the following nine paragraphs. Subsections 10.4.4–10.4.8 are the sequential processing steps of the SQT activity:

■ Independence in System Qualification Testing (10.4.1)
■ System Testing on the Target Computer (10.4.2)
■ System Qualification Test Planning (10.4.3)
■ Preparing for System Qualification Testing (10.4.4)
■ Dry run of System Qualification Testing (10.4.5)
■ Performing System Qualification Testing (10.4.6)
■ Analyzing and Recording System Qualification Test Results (10.4.7)
■ Revision and Retesting System Qualification Test Results (10.4.8)
■ Test-Like-You-Fly/Test-Like-You-Operate (10.4.9)

System Qualification Testing Objectives. System Qualification Testing is the comprehensive formal test demonstrating that the system software functional and interface requirements have been met for each release of the system. At the system level, SQT is focused on testing the integrated hardware/software system against the system requirements.

The *Technical Requirements Document* and *Interface Specifications* define the system requirements, and the *Software Master Build Plan* defines what SI functionality is to be operational for each release and what subsystem releases are used for each system release. The SQT activity must fully test the integrated software with the system hardware it interfaces with. This activity also retests those portions of the hardware/software integration that have been previously completed.

Approach to System Qualification Testing. The SQT activity consists of the following tasks:

■ Prepare the SQT software test data.
■ Conduct an SQT Test Readiness Review (TRR) including a test readiness on the target computer hardware to ensure that the tests described in the STDs are complete and accurate.
■ Perform formal Subsystem Acceptance Testing and the SQT.

■ Execute tests using SQT test procedures and record the test results, problems, and anomalies.
■ Analyze test results and document the test data and results.
■ Record test results in the *Software Development File* or *Software Work Book*.

Table 10.10 is an example summary of the readiness criteria in terms of entry and exit criteria, verification criteria to ensure completion of the required tasks, and the required measurements to be collected during the SQT activity.

At the system level of testing, *Discrepancy Reports and Change Requests* (DR/CR) usually replace their software counterparts (SDR/SCRs) and discovered software problems are allocated to the SwIPT. If DR/CRs fixes are incorporated into the software under test, constituting a new sub-release, then portions of the SQT test procedures are re-run to verify that applicable fixes are implemented and working correctly. In addition, it must be determined that selected pre-existing functionality is *still performing* per software and interface requirements after the fixes have been implemented.

Roles and Responsibilities for System Qualification Testing. Depending on where in the Specification Tree hierarchy the testing is performed, SQT is the responsibility of the system, subsystem or element integration and test teams. Software Developers and Software Test Engineers have no *formal* role in System Qualification Testing but provide vital technical support as needed. Test description preparation, test execution, and test results documentation are performed by the System, Subsystem or Element Test Engineers.

Software Developers implement software changes resulting from DR/CRs generated during this activity and support the System Test Engineers in these activities. Also, Software Engineers may support the System Functional Configuration Audit (System FCA) and the System Physical Configuration Audit (System PCA), if required, as outlined in the *Master Test Plan*. The MTP is sometimes called the *System Test and Evaluation Plan* (STEP) or *Integrated Test and Evaluation Plan* (ITEP).

System Qualification Testing Work Products. SQT must be performed using documented test descriptions developed by the Test Engineers. The *Software Version Description* and *Software Product Specification* documents are updated concurrently if required. Additional products may be required at other development sites as specified in their site-specific SDPs.

10.4.1 Independence in System Qualification Testing

System Qualification Testing demonstrates that the system, subsystem and related elements meet the performance and interface requirements allocated to it for each release. System

Table 10.10   Readiness Criteria: System/Subsystem Qualification Testing

Entry Criteria	Exit Criteria
■ System/subsystem test plan and approach is defined and approved. ■ The SwIPT is requested by SEIT to support SQT activities. ■ The executable software product is capable of supporting SQT. ■ The software release to be integrated and the integration database are under control of the Master Software Development Library (MSDL). ■ The System Test database requirements have been defined. ■ All subsystems have been integrated.	■ The System Test Readiness Review is successfully completed. ■ Verification of test cases and procedures (e.g., Peer Reviews) have been completed. ■ The release being tested is ready and accepted. ■ Required test databases are created, populated and accepted by the test conductor. ■ System/subsystem testing is successfully completed with an action plan generated to close remaining DRs/CRs. ■ Regression Testing is completed and accepted. ■ Software and System Management Reviews and approves the DR closure plan.

Verification Criteria
■ Releases provided by the SDL to the MSDL are verified and accepted by the test conductor. ■ SQA performs process / product audits for ongoing product engineering tasks. ■ Test conductor reviews the test database for completeness. ■ Test Conductor and Software Test Lead reviews Regression Test logs and accepts completion of the Regression Testing.

Measurements
■ Test Coverage: Number of requirements tested and passed. ■ Percent of paths tested. ■ SDRs, DRs, SCRs and CRs opened and closed (see Section Chapter 15).

Qualification Testing is normally the responsibility of the program-level SEIT; however, at the lower levels, SQT can be performed by the Subsystem or Element Test Engineers. To ensure objectivity, the tests *must be performed by "independent" Test Engineers—not the Developers*. System Test Engineers would normally have no role in the software development process, so they may be considered as inherently independent testers of the software. In any case, Software Engineers support the SQT process.

10.4.2 System Testing on the Target Computer

System Qualification Testing should be performed on the target hardware system, in the *operational configuration* to the maximum extent possible, to demonstrate that there are no hardware/software incompatibilities. Testing on the target hardware, often in a test bed, verifies a successful hardware/software integration and interoperability. Operation on the target computer, or target comparable systems, also enables the collection and analysis of measurements of the computer resource utilization. If qualification and acceptance testing is *not* performed on the target hardware, or a reasonable facsimile, as the Project Manager you must clearly inform your management and the customer of the risks involved, and you must *document* your objection to that approach.

10.4.3 System Qualification Test Planning

The project schedule must address test planning and show the period of time during which the qualification and acceptance testing will be conducted. It must include the time needed to analyze the test data. If applicable, the schedule should allow a reasonable amount of time for the customer to review test results and make a determination as to whether the system meets the overall acceptance criteria. Table 10.11 is a list of the types of typical software acceptance tests that can be performed. All types of software testing planned for your project should be identified and addressed in the Software Test Plan.

Depending on the type of test involved, the SEIT role may range from observer, to participant, to approver. The SEIT role for each type of test should have been agreed to in the contract or before SQT begins. Whatever the role, plans must call for the SEIT to be composed of experienced and trained staff in order to witness all system and subsystem software testing.

10.4.4 Preparing for System Qualification Testing

The objective of preparing for the dry run, before the actual System Qualification Test, is to prepare and finalize through

Table 10.11 Types of Software Acceptance Testing that Can Be Performed

Test Type	Testing Description
Integration Testing	Performs integration testing for multiple interrelated SIs.
Interface Testing	Internal and external testing of system interfaces including communications and COTS interfaces.
Usability Testing	Operational testing to ensure that the system is usable by the expected users.
Verification Testing	Confirms that the system meets the specified system requirements.(*)
Validation Testing	Confirms that the system effectively meets operational needs.(*)
Robustness Testing (Stress Testing)	Multiple testing approaches that demonstrate the system's ability to perform under exceptional circumstances (e.g., peak loading, error detection/correction, etc.).(**)
Interoperability Testing	Determines the ability of the software to provide data to and from other systems for mutually effective operations.
Modeling and Simulation Testing	Testing to determine the bounds of usefulness.
Security Testing	A type of robustness testing that ensures the system can restrict access to only authorized personnel, and that the authorized personnel can only access the functions available at their security level.
Fall-back Testing	Fall-back testing ensures functionality of failure modes, maintenance modes, etc.
Roll-back Testing	Removing the new version and restoring old version to avoid a stop in operations.
Regression Testing	Verifies correctness of existing functionality in each new release of the software.
System Testing	Provides end-to-end testing of the full system functionality.

(*) Verification and Validation testing are tightly coupled to the system requirements in order to understand what the system was intended and designed to do.

(**) Robustness testing focuses on testing the system and software under realistic operational conditions with users who represent those expected to operate and maintain the system when it is deployed.

reviews or inspections, the subsystem test description and test data. Once the review changes are incorporated into the documents, they should be baselined and submitted to documentation control. In addition, all supporting test software (simulations/emulations) and data must be prepared. System Test plans, procedures, and test data should be prepared at the appropriate level of testing but is often prepared by the SEIT.

Approach to SQT Preparation. Separate test descriptions should be generated for each release. They should contain test cases and procedures for the requirements of the current release, plus those Safety Critical requirements from previous releases. For the final release, the documents provide test cases and descriptions for the final release; software system requirements; regression test cases; and descriptions of the software requirements from previous releases.

SwIPTs should review in-progress test plans and test descriptions; provide recommendations to System, Subsystem or Element Test Engineers; assist in determining needed software test data, equipment and test support items; support applicable readiness reviews; and provide expertise

and training in using the software. The test data and the software must be placed under CM control prior to testing.

10.4.5 Dry Run of System Qualification Testing

Objective of the SQT Dry Run. The objective of the System Qualification Testing dry run is to exercise the test cases and test procedures to ensure that they are complete and accurate and that the subsystem is *ready for formal witnessed testing*. The Test Engineers normally begin by ensuring that all necessary data, equipment, materials and tools are available. Test Engineers must support SwCCB Corrective Actions on any release errors, request re-execution of release procedures as required and accept software releases upon satisfactory verification. (The availability of materials includes the seemingly inevitable $4.99 missing cable required to make the hardware work!)

Approach to the SQT Dry Run. When all the required elements are assembled, testers execute the procedures and collect or record the outputs, logs, test notes and results. They execute all tests specified for each release, and perform Regression Testing on all Safety Critical requirements from previous releases. All problems, errors and discrepancies must be noted by the tester. No modification to the SIs, hardware, configuration, test data, or environment should be made until after the dry run is complete and the results are documented and reviewed.

There is no *formal* Software Developer role during the SQT dry run, except the important role of assisting Test Engineers in analyzing test discrepancies and generating DRs/CRs. Also, if software code requires changes, SDRs/SCRs must be generated. In cases where test procedures require modification, the procedures must be redlined and approved by Software Quality Assurance (SQA). *Retesting* should be required for all modified software, test cases, test descriptions, and test cases that directly interact with the modified software and test cases. (In other words, make sure everything works before the SQT!)

10.4.6 Performing System Qualification Testing

Objective of Performing SQT. The objective of performing SQT is to *execute the test procedures* in a formal and witnessed test environment using products under CM control. This task normally begins with the *Test Readiness Review*. This review ensures that all necessary test documentation, equipment, materials, personnel are available and ready, and the coordinated test schedules are in place.

Approach to Performing SQT. The actual execution of the SQT is the same as the dry run, except that the performance of the testing must be witnessed by SQA and optionally by the customer's Program Office and/or its representatives. Reasonable notice of the tests must be provided to permit the Program Office an opportunity to attend.

Like the SQT dry run, there are no *formal* Software Developer participation requirements for this System Test, except for analyzing software discrepancies, generating SDRs/SCRs as needed and implementing needed software code changes resulting from the SDRs/SCRs. However, the SwIPT normally supports the performance of the Functional Configuration Audit (discussed in Subsection 12.6.5), System Verification Review (SVR) (discussed in Subsection 12.6.6), and the Physical Configuration Audit (discussed in Section 12.7.1).

10.4.7 Analyzing and Recording System Qualification Test Results

The objectives of analyzing and recording SQT test results are to review SQT tests results to ensure the tests have been successfully completed and to document test results in the SDF and in the *Software Version Description* if required.

Approach to Analyzing and Recording SQT Results. Again, there are no *formal* Software Developer roles in System Qualification Testing other than supporting the SIQT at the level being tested. However, results of the qualification tests must be analyzed for completeness and then recorded in the project's Software Development File by the Test Engineers who performed the tests. For software developed in multiple releases, test results must be prepared, reviewed and recorded after each release unless a program decision has been made to defer the higher-level (system) test until all the software releases are complete. Subsystem SVDs should be updated after each release.

10.4.8 Revision and Retesting System Qualification Testing Results

Objective of SQT Revisions and Retesting. The objective of the revision and retesting activity is to rework the source code or test descriptions to eliminate problems identified during the qualification testing, and then to retest the appropriate portions of the system to verify that the changes have been successful and have not produced side effects. The test results and documented problems should be evaluated by Software Developers to determine the necessary changes to the software and test descriptions.

Approach to SQT Revisions and Retesting. Unit-level retesting is required for all modified procedures and functions. Modified software releases require retesting of all Safety Critical requirements and previously failed test cases. Although there are no *formal* Software Developer roles for this task, software code changes resulting from the change control process must be implemented. All modifications to the source code must be handled as SDRs/SCRs.

Revision and retesting must be repeated as needed until all test cases have met the test case success criteria decided in advance. In some cases, resolving incomplete or failed tests can be postponed until a later release if: no subsystem external interface is involved; specific functionality is not required by another SI for the release; and the delay is approved by the Change Control Board.

10.4.9 Test-Like-You-Fly/ Test-Like-You-Operate

Test-Like-You-Fly (TLYF), also called *Test-Like-You-Operate*, is an approach to testing that does not necessarily have anything to do with actual flying unless your project concerns airplanes or aerospace. As described in Subsection 3.9.6, TLYF really means that even if your system is fully compliant with the customer's requirements, and the quality of the software is world class, it still might fail if you failed to test it in its intended real operational (flight-like) condition and expected environment.

Table 10.12   Root Cause of Failures from Not Following a TLYF Approach

Vehicle	Complete Loss of Mission—Root Cause
Arianne V	Hardware changes were made to Arianne V, but no software changes were made so the software that worked perfectly for Arianne IV was not retested for Arianne V. The Inertial Reference System was disabled resulting from "dead code" used in the preceding Arianne IV flight so Arianne V had to be destroyed.
Mars Polar Lander	Modifications made to both the hardware and software were not fully retested. Faulty logic in the touchdown sensor caused the vehicle to crash.
Space Payload	Vehicle exploded in space after a leaking battery electrolyte caused a short circuit. Battery had only passed the non-flight qualification tests.
Titan Launch	An error in wiring prevented satellite separation. Ground testing, using only non-flight software, did not identify the problem.
Mars Climate Observer	Software modifications contained an English-Metric conversion error. The modifications were considered non-critical, so the software was not retested (bad decision) and the mission was lost.
Terriers	Hardware and software were never tested *together*; a torque coil that was installed upside down was never identified during testing.
	Significant Loss of Functional Capability—Root Cause.
Genesis Return Capsule	Four deceleration switches were installed backwards causing the parachute failure; the vehicle was never tested in a real flight configuration.
Hubble Space Telescope	A complete flight-like configuration was not tested on the ground prior to launch, so a mirror grinding anomaly was not identified. The lost functionality was restored after an expensive fix while in earth orbit.

TLYF is *not* a replacement for *any* testing function described in this Guidebook: performance, environmental and calibration testing do not change. TLYF is an additional and optional testing approach when dealing with very complex systems where *all* the elements, or components of the system that interface with each other, must all be tested—together—*before* the system goes operational. Let me say that again, tested *before* the system goes operational in its intended real operational and expected environment—or as close as you can get to it.

Post-mortem analysis of failed missions, or missions with significant loss of its intended functional capability, indicates that a consistent absence of the TLYF approach was a major contributor to the failure. Table 10.12 is a brief review of a few past examples of failures that could have been prevented if testing had followed a TLYF approach. Most of these are good examples of management trying to save a little time and cost, and making a bad decision by "cutting corners" resulting in a disaster. From my perspective, most of these examples are hard to believe and demonstrate a total *lack of common sense*.

The TLYF approach has a philosophical underpinning that makes it distinct from other types of testing. This approach includes a unique assessment process derived primarily from mission failure lessons learned.

It is important to remember that one principal purpose of all testing is to *find faults* that have escaped detection by quality and other inspection activities—the purpose of testing is *not* to prove that no flaws exist. It is not possible to do that. You can only show that you have not found any, or any more faults beyond those identified, with the tests you have executed. Finding flaws requires multiple levels of robust testing.

Ignorance, anywhere, is also a flaw! The bottom line is, if you *change anything* in a complex system that has previously been tested—*retest*!

Chapter 11

Software Engineering Domain

The Software Engineering Domain is the core of your software-intensive system. If you are going to manage the software development process, you must first understand it.

Chapter 10 described the *Systems Engineering* (SE) activities involving software participation; Chapter 11 describes the activities comprising the *Software Engineering* (SwE) Domain and the software development process. The flowchart in Figure 11.1 is a greatly expanded version of the SE and SwE Development Life Cycles overview depicted in Figure 10.1. Figure 11.1 is a comprehensive flowchart that identifies the typical sequential activities performed by SE and SwE, the activities that are shared, technical reviews conducted, and locations where each is discussed in this Guidebook.

Format of the Process Descriptions: In order to clearly *describe and display each software development activity* being addressed in the SwE Domain, five interrelated items (*four tables and a flowchart*) are used in Chapter 11. These five items collectively provide a consistent and comprehensive definition and format to describe the software tasks involved in each software development activity of the SwE Domain. The five items are:

- *Readiness Criteria*: This table contains the Entry Criteria, Exit Criteria, Verification Criteria, and Measurements collected (for example, see Tables 11.1 and 11.5).
- *Software Work Products*: A table containing a list of software work products required, or typically produced during each activity (for example, see Tables 11.2 and 11.6).
- *Roles and Responsibilities*: A table that identifies the groups responsible for each software development activity (for example, see Tables 11.3 and 11.7).
- *Input/Process/Output (IPO) Process Tasks*: This flowchart shows for each activity the principal input documents,

the software development tasks being addressed, and key outputs (for example, see Figures 11.3 and 11.4).
- *Task Table*: The items in this table are *directly mapped* to the same numbered process activities in the IPO flowchart, but the Task Table contains more details including important subtasks (for example, see Tables 11.4 and 11.8).

The four tables and the flowchart are important companions of the text in order to fully explain the tasks that must be performed during each activity in the SwE Domain. During the Software Development Life Cycle, members of the Development Team can use the sections in Chapter 11 as a *reference guide during each of the principal activities* described in this chapter.

The Software Engineering Domain is described (in this Guidebook) as consisting of the following ten *principal activities* which include the Software Implementation Process discussed in Chapter 3:

- *Software Requirements Analysis*: Defines *what* the software subsystems must do but not *how* to do it (11.1).
- *Software Item Design*: An introduction to the generic design activities covering defining and recording design decisions; work products typically produced, plus roles and responsibilities (11.2).
- *Software Item Architectural Design*: Describes the high-level organization of the Software Items (SIs) in terms of the Software Units (SUs) within the SIs and their relationships (11.3).
- *Software Item Detailed Design*: Defines the implementation details for each SU. It involves decomposing the SUs from the SI Architectural Design into the lowest level Software Units in sufficient detail to map the design to the features of the selected programming language, the target hardware, operating system and network architecture (11.4).

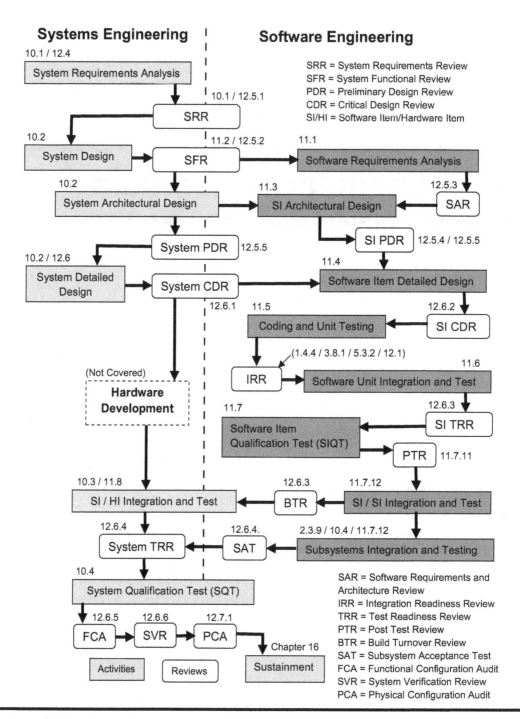

Figure 11.1   SLC and SDLC process flowchart.

■ *Coding and Unit Testing*: Converts the SU Detailed Design into computer code and databases that must be inspected, unit tested and confirmed as responsive to the design (11.5).

■ *Software Unit Integration and Testing*: A systematic and iterative series of integration builds of Software Units that have successfully completed Coding and Unit Testing, and have been built up to a higher level SU, or SI, for the current build (11.6).

■ *Software Item Qualification Testing*: Demonstrates that the developed SIs meet the system and interface requirements allocated to the SIs being tested (11.7).

■ *Software/Hardware/Subsystem Integration and Testing*: Integrates and tests Software Items with other interfacing SIs, and related Hardware Items (HI), plus the integration and testing of the subsystems comprising the full system. It involves testing the resulting groupings to determine if they work together as intended, and continuing this process until all interfacing SIs, HIs, and subsystems are integrated and tested (11.8).

■ *Software Transition to Operations*: Software testing on the target system at the user's site (11.9).
■ *Software Transition to Sustainment*: The software and documentation preparation work that must be completed to transition the system and support software and related products required for the performance of long-term System Sustainment (11.10).

11.1 Software Requirements Analysis

Software requirements are the foundation of successful software-intensive systems.

Lessons Learned. Based on my experience, software requirements analysis is *the most important activity* in the entire Software Development Life Cycle. *The critical importance of establishing sound software requirements, agreed to by the customer, cannot be overemphasized.* An analogy can be made to building a tall building and making sure you have a solid foundation upon which to build the structure. However, the big difference is that the building's foundation is constructed with concrete—your customer's requirements cannot be in concrete as they will almost certainly change during the development process. A full set of requirements, compiled and refined over time, must cover the needs of your customer. If those requirements do not accurately reflect your customer's needs, the resulting system you deliver, no matter how elegant it is, will be headed for the grave.

11.1.1 Software Requirements Analysis— Objectives and Approach

Software requirements analysis must be accomplished by analyzing the system and subsystem requirements in depth to identify *allocated and derived* software requirements. *Allocated* software requirements are flowed down, often called *decomposed*, from the system engineering requirements allocation process (i.e., decisions are made regarding specific functions performed by software and specific functions performed by hardware).

Derived software requirements are not specifically flowed down from the system requirements but, as the name implies, are arrived at through logic and reasoning. In other words, *your analysis concludes that the software system must perform additional secondary tasks in order to comply with specific allocated software requirements.*

Lessons Learned. The resulting requirements should define *what* the software system must be able to do while avoiding implementation bias. Requirements analysis should *not* address *how* to implement the requirements. Instructions regarding how to implement generic tasks may be included in your corporate *Software Procedures Manual*. I have seen corporate procedure manuals ranging from a few three-ring binders to three linear feet of documentation.

Origin of Requirements. As depicted in Figure 11.2, software requirements can originate from several sources. The types of requirements include, but are not limited to, capabilities, behavior, processing, control, interfaces, performance, sizing, timing, packaging, human factors and software qualification. Requirements should be analyzed for completeness, traceability, consistency, testability, criticality, feasibility, correctness and accuracy.

In addition, requirements must be specifically evaluated for safety, security, privacy protection, dependability, reliability, maintainability and availability. Critical requirements need additional tracking and monitoring as discussed in Section 5.4. In addition, requirements identified that have significant risk associated with them must be evaluated by the SwIPT for risk assessment and mitigation as discussed in Section 5.2 and elsewhere.

Software Requirement Attributes. Requirements must possess four important attributes: *completeness, correctness, non-redundancy* and *unambiguity*. The general approach for preparing software requirements specifications is to make each requirement:

■ Clear and concise in a single statement with a single "shall" where the word "shall" implies a mandatory requirement
■ Testable or verifiable and traceable via a unique product identifier

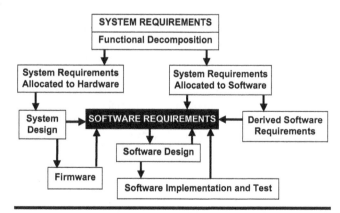

Figure 11.2 Origin of Software Requirements.

- Consistent with all system requirements
- Understandable statement and independent of other software requirements
- An unambiguous statement of *what* the software will do but *not how* it should do it

All software requirements may not (but should) meet all of these guidelines, however, these guidelines should be considered in defining each requirement. In addition to being clear and concise in documenting and testing the requirements, System Analysts should:

- Use data dictionary terms and approved acronyms.
- Use consistent terminology and avoid the use of lists.
- Limit the use of the words "and" and "or."
- Use positive requirements; and do not use the term "and/or."
- Make sure there is *no ambiguity*.

Requirements Database. A single project Requirements Database should be used to capture all requirements. Systems Engineering and Software Engineering should use the same Requirements Database for documenting and maintaining requirements to assure full compatibility between these tasks at the system, subsystem and software levels. Portions of this database can be partitioned out and updated by the various SwIPTs; however, a single master copy should be maintained

to ensure consistent communication of requirements among the teammates—including subcontractors.

Table 11.1 summarizes the readiness criteria for this activity in terms of the entry and exit criteria, verification criteria to ensure completion, and the measurements usually collected.

11.1.2 Work Products for Software Requirements Analysis

Software Requirements Specification (SRS). *Only one SRS should be developed for each Software Item.* The Subsystem SwIPT must ensure that there is consistency between SRSs for each SI. SS-2 Software Items do not require an SRS document, but all software, interface requirements and traceability data for SS-2 software must be captured in the *Software Development File* (SDF).

Software Requirements Database. The software subsystems should document *allocated and derived* software requirements in a Requirements Database, as discussed in Subsection 2.3.1, for all software categories. Each specific requirement must be assigned a *Program Unique Identifier* (PUI) number for *individual requirements traceability*. Requirements for SCS-1 and SS-1 software should be traced bi-directionally (up) to system requirements and (down) to software builds.

Traceability Matrixes. The traceability data must be documented and the recommended format is a *Software*

Table 11.1 Readiness Criteria: Software Requirements Analysis

Entry Criteria	Exit Criteria
■ System requirements allocated to software are available. ■ Software system architecture is available. ■ System Operational Concept Description (OCD) is available. ■ Appropriate Software Development Environment (SDE) elements are available for use (see Section 9.1).	■ Software system architecture and software system interfaces are documented in the SDF, SRS and IRS. ■ Software requirements, Requirements Test Verification Matrix (RTVM), and the Software Requirements Traceability Matrix are documented in, or their location referenced by the SRS, or in a traceability or requirements management tool. ■ Lessons learned are recorded.

Verification Criteria
■ Software Management Reviews and approves: software system architecture, software system interfaces, software requirements, RTVM and SRTM as documented in the SRS, or their location referenced by the SRS, or in a traceability or requirements management tool.   ■ Program and senior management are provided with the status of ongoing product engineering tasks (including software requirements) on a periodic and event-driven basis.   ■ All software products are Peer-Reviewed and the Requirements Database is inspected.   ■ A Software Specification Review or Technical Interchange Meeting has been completed.

Measurements
■ Software requirements added, modified and deleted during reporting period.   ■ Product Engineering schedule.   ■ Requirements traceable versus untraceable (see Chapter 15).

Requirements Traceability Matrix (SRTM). In addition, a *Requirements Test Verification Matrix* (RTVM) should also be prepared. The RTVM may also be called the *Verification Cross Reference Matrix* (VCRM).

The SRTM and the RTVM may be part of the SRS or reside in a traceability and requirements management tool. It is extremely important to provide time for stakeholders and users to review these documents to assure that simulators are developed with required functionality and that they incorporate the needed fidelity (precision).

Software Development File. Subsection 9.4.1 describes examples of items typically contained in the SDF. Also, diagrams for algorithm models and simulations should be captured in SDFs as well as the appropriate software tools. For Commercial-Off-The-Shelf (COTS)/Reuse (C/R) software, the only documentation available may be from the software supplier. However, SCS and SS-1 categories of COTS/Reuse software (defined in Section 1.8) once integrated must be fully documented in order to pass the documentation reviews.

Electronic Data Interchange Network (EDIN). All of these software products must be made available via an EDIN as described in Subsections 5.1.10 and 9.5.1. The EDIN capability is a critical capability for decentralized developments. For Developers in a single location, working on small systems, the *Software Development File* may be sufficient.

Software Master Build Plan (SMBP). A draft version of the SMBP may be prepared during the requirements analysis activity (see 5.1.4).

Interface Control Document (IFCD). The IFCD must be provided to enable the definition and management of software and computer equipment resources.

The typical software work products during the software requirements analysis activity, for each software category, are summarized in example Table 11.2.

11.1.3 Requirements Creep

Walking on water and developing software from a specification are easy if both are frozen.

—Edward Bernard

Uncontrolled changes or continuous growth in the planned scope of a project is called requirements creep (it may also be called *scope creep*, *function creep* or *feature creep*). It usually occurs when the scope of a project is poorly defined or not well documented or controlled. There will be necessary changes that as the Project Manager you have no control of, but you must manage them because *uncontrolled changes* can have a significant impact on cost, schedule and quality.

Lessons Learned. Requirements creep is a common risk. It is difficult to avoid and remains a challenge for even the most experienced Project Managers. You should have a mitigation plan ready to go because changes and growth are a certainty for large and complex projects. Requirements creep usually results in cost

Table 11.2 Software Requirements Analysis Work Products

Software Requirements Analysis Products	Software Category				
	SCS-1	SS-1	SS-2	SS-3	C/R
Requirements Database	Required	Required	Required	Required	Required
Software Requirements Specification (May include IRS, SRTM & RTVM)	Required	Required	Information Required (*)	Information Required (*)	Information Required (*)
Interface Requirements Specification	Required	Required	Information Required (*)	Information Required (*)	Information Required (*)
Software Requirements Traceability Matrix	Required	Required	Information Required (*)	Information Required (*)	Information Required (*)
Requirements Test Verification Matrix (RTVM)	Required	Required	Information Required (*)	Information Required (*)	Information Required (*)
Software Master Build Plan	Required	Required	Information Required (*)	Information Required (*)	Information Required (*)
Software Work Products (see 2.3.7)	Required	Required	Information Required (*)	Information Required (*)	Optional

SCS = System Critical software; SS = Support Software C/R = COTS/Reuse Software.
(*) Document not required, but applicable information is developed and retained in the SDF.

overruns because many Project Managers do not plan for it even though they know it will happen. If you plan in advance for it, you can have budget reserve plus the resources and schedule that can be increased along with the requirements increases. These changes could be considered an *acceptable addition* and not an item of risk because you "planned" for it (but you may be accused of rationalizing the risk).

11.1.4 Responsibilities for Software Requirements Analysis

The Subsystem SwIPT personnel should be responsible for the SI requirements analyses, the generation of the work products, and the documentation of the requirements in SRSs or SDFs, as appropriate to their level of responsibility on the project. Table 11.3 is an example of the roles and responsibilities of the SwIPT personnel and other groups during the requirements analysis activity.

11.1.5 Process Tasks for the Software Requirements Analysis Process

The software requirements analysis activity typically involves ten tasks as depicted in the example input/process/output Figure 11.3. Details of these ten tasks are described in

Table 11.4 in terms of the inputs and outputs for each task in the flowchart and a description of the related subtasks.

Verifying completion of the ten tasks described in Figure 11.3 is accomplished by a combination of approvals by Software Leads, Peer Reviews, Software Quality Assurance (SQA) audits, periodic audits by the Chief Software Engineer (CSwE) and Joint Technical Reviews (JTR) or Technical Interchange Meetings (TIM) as determined to be necessary for each task.

11.1.6 Exit Criteria for Software Requirements Analysis

There is no requirement for the software requirements analysis activity to be entirely completed prior to the start of the Software Design activity. When following iterative life cycle models, the software requirements analysis activity may be repeated for each build so the software requirements would be developed iteratively. In this case, the *Software Specification Review* (SSR), or a *TIM* for support software, would be held on a build-by-build basis.

In recent years, the name of the Software Specification Review was changed to *the Software Requirements and Architecture Review* (SAR). When following the Object-Oriented Analysis (OOA) methodology, the software requirements analysis and software architecture definition can be concurrent resulting in combining the SSR, or TIM, with the *Preliminary Design Review* (PDR). The software

Table 11.3 Roles and Responsibilities during Software Requirements Analysis

SwIPT	Performs software requirements analysis, definition and documentation.
	Generates initial traceability products for SRS requirements to parent specification requirements in the Requirements Database.
	Identifies Software Item risk areas.
	Initiates a Software Development File for each Software Item.
	Collects and reports requirements metrics.
	Submits problem reports after the requirements documentation is baselined.
Software Test	Identifies the verification methods in the Requirements Database; prepares RTVM.
Chief Software Engineer	Supports the SwIPT in the identification and specification of critical requirements, reviews the SDFs and the activity products, and attends all formal activity reviews.
Software Quality Assurance	Evaluates the subsystems for adherence to the documented policies and procedures, product quality criteria and the Requirements Database. Findings are reported to appropriate levels of management.
Software Configuration Management (SCM)	Manages software requirements baseline and processes all Software Discrepancy Reports/Software Change Notices (SDRs/SCRs) for documented software requirements changes to the SRS or SDF as they are generated by the software subsystems or elements.
SwIPT CCB	Addresses all subsystem internal change notices and SDRs/SCRs.

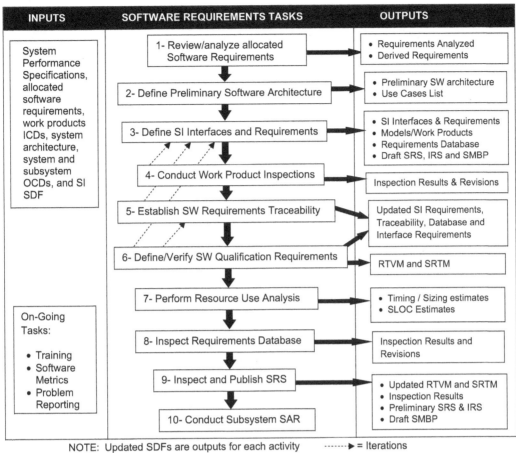

INPUTS	SOFTWARE REQUIREMENTS TASKS	OUTPUTS

System Performance Specifications, allocated software requirements, work products ICDs, system architecture, system and subsystem OCDs, and SI SDF

1- Review/analyze allocated Software Requirements
- Requirements Analyzed
- Derived Requirements

2- Define Preliminary Software Architecture
- Preliminary SW architecture
- Use Cases List

3- Define SI Interfaces and Requirements
- SI Interfaces & Requirements
- Models/Work Products
- Requirements Database
- Draft SRS, IRS and SMBP

4- Conduct Work Product Inspections
Inspection Results & Revisions

5- Establish SW Requirements Traceability
Updated SI Requirements, Traceability, Database and Interface Requirements

6- Define/Verify SW Qualification Requirements
RTVM and SRTM

7- Perform Resource Use Analysis
- Timing / Sizing estimates
- SLOC Estimates

On-Going Tasks:
- Training
- Software Metrics
- Problem Reporting

8- Inspect Requirements Database
Inspection Results and Revisions

9- Inspect and Publish SRS
- Updated RTVM and SRTM
- Inspection Results
- Preliminary SRS & IRS
- Draft SMBP

10- Conduct Subsystem SAR

NOTE: Updated SDFs are outputs for each activity   ------▶ = Iterations

OCD = Operational Concepts Description
SLOC = Software Lines of Code
IRS = Interface Requirements Specification
RTVM = Requirements Test Verification Matrix
SRS = Software Requirements Specification

SDF = Software Development Folders
ICD = Interface Control Document
SMBP = Software Master Build Plan
SRTM = Software Requirements Traceability Matrix
SAR = Software Requirements and Architecture Review

Figure 11.3   Software Requirements Analysis process flow.

requirements analysis activity formally ends upon completion of the SSR (or SAR) and baselining of all work products. Lessons learned should be captured in both an Engineering Memoranda (EM) and in the Software Development File.

11.2 Software Item Design

The description of the *Software Item Design* activity begins with an introductory overview of defining and recording overall design decisions; the work products typically produced during SI Design; and the roles and responsibilities during this activity. Specifics of the Software Item Design activity are described in the two sections covering architectural and Detailed Design:

- *Software Item Architectural Design*: Software Designers develop an Architectural Design that partitions each SI into SUs; these SUs may be subdivided into

smaller SUs. SI Architectural Design is also called *SI Preliminary Design* (covered in 11.3).
- *Software Item Detailed Design*: Software Designers perform a Detailed Design on individual Software Units and produce a description of the Software Items down to the level of detailed algorithms and logical procedures (covered in 11.4).

The processes specified in these tasks pertain to SCS-1 and SS-1 software only. For the SS-2 software category, these tasks are simplified and may be combined into a single activity.

11.2.1 Software Item-Wide Design Decisions

The objective of this task is to define and record all Software Item Design decisions that affect the selection and design of the Software Units comprising each SI. These decisions constrain how the Designers partition the SIs into SUs as well

Table 11.4 Software Requirements Analysis Tasks

Task	Inputs	Software Requirements Subtasks	Outputs
1. Review and Analyze Allocated Software Requirements	■ See inputs in Figure 11.3	Review and analyze allocated software requirements	■ Allocated requirements analyzed ■ Derived requirements developed
2. Define Preliminary Software Architecture	■ Allocated software requirements	■ Define software architecture components ■ Develop / update SI-to-SI interfaces ■ Identify subsystem Use Cases	■ Preliminary Software architecture at SI level ■ Subsystem Use Cases List
3. Define SI Interfaces and Requirements	■ Interface Control Documents ■ Software Requirements ■ Software Architecture At SI Level	■ Refine SI level Sw architecture model ■ Model software architecture in OO ■ Develop Sw requirements and interface requirements, including data items ■ Develop software work products (see 2.3.7, Work Products Table in 11.1 to 11.6 and Chapter 13) ■ Identify software risks ■ Enter software and interface requirements into database ■ Review database for completeness	■ High-level analysis and class models ■ SI and interface requirements ■ Work products ■ Draft SRS, IRS and SMBP ■ Populated Requirements Database
4. Conduct Work Product Inspections	■ Work products ■ Requirements Database	■ Schedule inspection; distribute review package ■ Conduct Peer Reviews; verify feasibility, completeness of Sw requirements and consistency between Sw requirements and update work products ■ Document results and post to SDF ■ Fix inspection deficiencies	■ Peer Review results ■ Work products updated ■ Deficiencies recorded
5. Establish Software Requirements Traceability	■ Requirements Database	■ Update software requirements tables to add traceability between system and software requirements ■ Verify both downwards and upwards traceability between system and software requirements ■ Add software requirements or flow up recommended changes to system requirements as necessary to complete traceability ■ Conduct Peer Review of Requirements Database	■ Updated SI requirements ■ Updated traceability ■ Updated Requirements Database ■ Updated interface requirements
6. Define/Verify Software Qualification Requirements	■ Draft Software Requirements	■ Update software requirements tables to add a qualification method (inspection; analysis; test; demonstration; other) for each software requirement ■ Verify qualification method satisfies verification plan ■ Create and Peer Review the RTVM and SRTM	■ RTVM and SRTM ■ Updated SI requirements. traceability, interfaces and database

(Continued)

Table 11.4 (Continued) Software Requirements Analysis Tasks

Task	Inputs	Software Requirements Subtasks	Outputs
7. Perform Resource Use Analysis	■ Software Work Products ■ Requirements and Preliminary Architecture	■ Conduct timing and sizing analysis ■ Develop/update SLOC estimates ■ Post information to the SDF ■ Verify that timing and sizing meets requirements	■ Timing and sizing estimates ■ SLOC estimates
8. Inspect Requirements Database	■ Requirements Database	■ Announce inspections, disseminate schedule & review products in advance ■ Conduct inspection to verify correctness, completeness, and consistency of data ■ Document results and post to SDF ■ Fix inspection deficiencies	■ Requirements Inspection results ■ Baselined Requirements Database
9. Inspect and Publish SRSs	■ Requirements Database	■ Update the preliminary SRS and IRS ■ Conduct Peer Reviews utilizing SRS inspection criteria ■ Update the draft SMBP ■ Obtain board approval	■ Preliminary SRSs and IRSs ■ Updated SI SDF ■ Updated RTVM and SRTM ■ Draft SMBP
10. Conduct Subsystem SSR	■ SRSs/IRSs ■ Agenda ■ Presentation slides	■ Conduct SAR ■ Publish Minutes and action items list ■ Resolve action items ■ Deliver documentation per contract	■ SAR minutes ■ Action item results ■ SRSs and IRSs ■ Requirements Database

as the overall design of the SUs. These are global decisions about the structure of the design that impact the SIs.

Approach to Software Item Design Decisions. The Software Design activity normally begins by performing an examination of the requirements relative to the SI plans, environment and interfaces to determine if there are any design issues affecting all SIs. Where such issues are identified, Subsystem SwIPT personnel should analyze the issues and determine an appropriate design constraint or decision for each. These design decisions must then be documented and communicated to the Software Designers as a set of design constraints in conjunction with the requirements of *what they are to design*.

Design decisions are program specific; however, key factors that may be considered in determining SI Design issues include:

■ Computer hardware platform and resource utilization requirements.
■ External SI constraints and interfaces and use of applicable standards.
■ Safety, Security and Privacy Critical requirements
■ Algorithms and Application Program Interfaces (API) to be used
■ Uniform exception handling, recovery methods, data storage and access methods
■ Major architectural trade-offs

■ Performance characteristics including response times, software maintainability, reliability and availability not allocated to individual architecture components
■ Human factors, training requirements and SI operational constraints

Key design decisions, and the rationale for making those decisions, must be recorded in *applicable* documents for the SCS-1 and SS-1 software class and in the SDF for the SS-2 software class. The applicable documents could include:

■ *Software Test Plan* (STP)
■ *Software Architectural Description* (SAD)
■ *Software Design Description* (SDD)
■ *Interface Design Description* (IDD)
■ *Software Master Build Plan* (SMBP)
■ *Data Base Design Description* (DBDD)

Key design decisions are those that could impact or constrain the SI Architectural Design, SI to external interfaces, software requirements, cost or schedule. Design decisions for SS-2 software should be reviewed during design inspections. For multiple build Software Items, design decisions should be addressed prior to completion of the Detailed Design for the first build.

11.2.2 Readiness Criteria for SI Design

The Software Design activity tasks are intended to be performed as consecutive steps of increasing levels of design specificity. However, there is no requirement for each activity to be completed for an entire SI before the next design activity is started. The tasks usually overlap each other. Table 11.5 summarizes the readiness criteria in terms of entry and exit criteria, verification criteria to ensure completion of the required tasks, and the measurements usually collected during the architectural and Detailed Design activities.

11.2.3 Work Products for SI Design

The documentation produced during the Software Design activity for each SI includes the software architecture and design and interface design descriptions; test plan, models and diagrams; traceability products in the Requirements Database; and the *Software Master Build Plan* that maps each build to the capabilities provided by the build and specific requirements allocated to the build.

Software interface design descriptions must be documented in the *Software Design Description*, *Software Architecture Description* and the *Interface Design Document* (IDD). The SDD documents the SI Design decisions and the SAD documents the SI Architectural Design and design of each SU. All software and interface design work products must be recorded in the SDF as defined in Subsection 9.4.1.

Design documentation for the COTS/Reuse (C/R) category of software is usually limited to data provided by the COTS vendor.

Diagrams for algorithm models and simulations, initiated during the software requirements definition activity, should be expanded and refined during the Software Design activity. Revised work models and diagrams must be maintained in the appropriate software tools and recorded in the SDF.

An example of required work products for the Software Design activity is summarized in Table 11.6. These software products must be made accessible through an Electronic Data Interchange Network.

11.2.4 Roles and Responsibilities for SI Design

The Subsystem SwIPT personnel must be responsible for both the Software Item Architectural and Detailed Designs, the revision of design work products and the documentation of the design in SADs, SDDs, IDDs, DBDDs and SDFs, as appropriate. Table 11.7 is an example of the roles and responsibilities for SwIPT personnel and other groups during the Software Design activity.

SCS-1 and SS-1 software classes require the traceability of the software architecture and design elements from the SAD, SDD and DBDD to the software requirement unique project identifiers in the SRSs. Also, IDD elements are required to be traced to SRS or Interface Requirements Specification

Table 11.5 Readiness Criteria: Software Item Design

Software Design Entry Criteria	Software Design Exit Criteria
■ Software requirements are allocated to an SI and approved. ■ System architecture and the OCD are available. ■ System verification matrix is available in the Requirements Database. ■ Software system architecture has been approved. ■ Software use cases and scenarios, SI definitions, interface design, updated Requirements Database, and preliminary database architecture are all documented in the SDF.	■ Software Architecture and Design are captured in design models. ■ Performance and sizing analyses are documented in Engineering Memos. ■ SI SLOC estimates are updated. ■ For SCS-1 and SS-1 software, a baselined STP, SAD, SDD, SMBP, DBDD and IDD are ready. ■ Design is baselined and placed under SCM control.

Software Design Verification Criteria
■ Management is provided with the status of ongoing tasks on both a periodic and event-driven basis. ■ SQA performs process / product audits for ongoing product engineering tasks. ■ All software architecture and design work products are Peer-Reviewed and measurements documented. ■ A preliminary technical review of the architecture and design has been completed, and the software PDR and CDR (or TIMS) are completed.

Software Design Measurements
■ Product Engineering schedule (including software architecture and design tasks). ■ Results from Peer Reviews. ■ SLOC estimates (see Chapter 15).

Table 11.6 Required Software Item Design Activity Work Products

Software Design Products	SCS-1	SS-1	SS-2
SAD and SDD (per Software Item per build)	Required	Required	Information Required (*)
IDD and SMBP (per build)	Required	Required	Information Required (*)
STP (per Software Item)	Required	Required	Information Required (*)
DBDD (if required)	Required	Required	Information Required (*)
SDF capturing revised models and diagrams	Required	Required	Required
Software design elements traced to SRS requirements in the Requirements Database	Required	Required	Optional

SCS = System Critical Software; SS = Support Software; SAD = Software Architecture Description; SDD = Software Design Description; IDD = Interface Design Document; STP = Software Test Plan; SMBP = Software Master Build Plan; DBDD = Data Base Design Description.
(*) Document not required, but applicable data is developed and archived in the Software Development File.

Table 11.7 Roles and Responsibilities during Software Design

SwIPT	Conducts reviews of the software architecture and design process and develops outputs
	Updates and maintains the SDF
	Addresses critical software requirements in the software architecture and design
	Generates traceability products for design elements to SRS requirements unique project identifiers in the Requirements Database
	Identifies and reports software architecture and design risk areas
	Collects and reports software architecture and design activity measurements
	Generates computer hardware resource utilization estimates, compared to the required threshold values, and addresses estimates that exceed the requirements
	Submits SCRs and SDRs, as necessary, after design documentation is baselined
Software Test	Initiates the STP
CSwE	Supports the Subsystem SwIPT software personnel in the handling of security and critical requirements in the Software Design, audits SDFs, reviews activity products, attends all formal activity reviews, and monitors and analyzes software metrics
SQA	Evaluates the Subsystem SwIPT software activities for adherence to documented policies and procedures, evaluates Subsystem SwIPT software architecture and design work products for product quality, and documents and report findings to upper-level management
SCM	Processes all SCRs and SDRs for software architecture and design changes to the baselined SAD, SDD, IDD, DBDD, SMBP and STP documented design as generated

(IRS) requirements. This traceability information must be documented in a *Software Requirements Management and Traceability Database*.

11.3 Software Item Architectural Design

Software Item Architectural Design is the first focused design activity at the SI level. Architectural decisions applicable to all SIs should have been made during the System Design activity (see 10.2). Architecture at the SI level must determine the design for interfacing to other SIs and to hardware units (if appropriate). Human interfaces may be designed (e.g., using prototypes to validate the designs with end-users). The tasking or operating system process structure for the SI should be determined in this activity. Additional site-specific products and design reviews may be specified in the subsystem's site-specific SDP Annexes.

11.3.1 Objectives and Approach to SI Architectural Design

The objective of *SI Architectural Design* is to describe the high-level organization of the SIs in terms of Software Units and their relationships. The SwIPTs developing the SI must prepare an architecture that is compatible with the system requirements. The main objectives are to:

- Decompose the SIs Design into SUs.
- Allocate requirements from SRSs to SUs.
- Complete allocation of requirements from the SRS to Use Cases (for OOD).
- Describe the Architectural Design and requirements allocation in a preliminary SAD and SDD.
- Update as applicable the SDF, SRS, IFCD.
- Prepare, as applicable, preliminary versions of the STP, SAD, SDD, IDD, SMBP and DBDD.

Approach to Software Item Architectural Design. Software Item Architectural Design must be performed by Subsystem SwIPT personnel. Using the documented software requirements, and the initial work products (models

and diagrams) from the requirements definition activity, the software architecture models are refined and the architectural components, including Software Units, are identified. SUs are logical constructs for classes and associations in OOA/OOD or specific capabilities in a structured development. Use of graphical architecture modeling techniques, for example, the Unified Modeling Language (UML), is often required.

11.3.2 Process Tasks for SI Architectural Design

The principal tasks, suggested for the SI Architectural Design Process are depicted in the Figure 11.4 flowchart. Task Table 11.8 is directly mapped to flowchart Figure 11.4 in terms of the inputs and the outputs of each task shown in the flowchart. Draft document versions precede Preliminary versions.

The Software Test Plan is usually produced concurrently with the Software Item Architectural Design Activity. Production of the STP is actually a product of the Software Item Test Planning activity (see Subsection 5.1.7) and it is prepared by the Software Test Engineers.

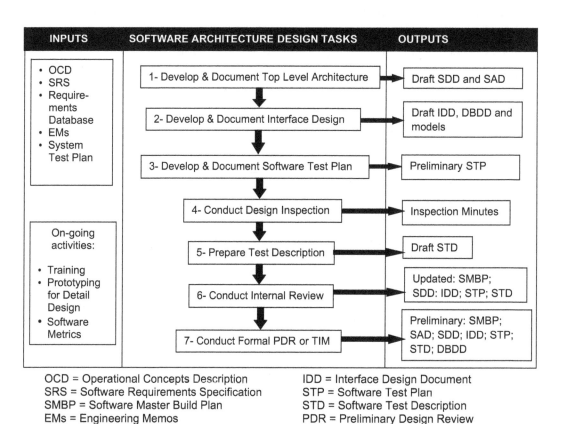

Figure 11.4 Software Item Architectural Design process flow.

Table 11.8 Software Item Architectural Design Tasks

Task	Inputs	SI Architectural Design Subtasks	Outputs
1. Develop and Document Top-Level Software Architecture	■ SRS ■ Requirements Database ■ OCD ■ EMs ■ System Design ■ Draft SMBP ■ Software system architecture	■ Determine modes of operation and architectural approach for Software Items ■ Perform analysis of reusable software: allocate to SIs ■ Define software functions, behavior, error conditions, services, and controls ■ Identify architectural components including SUs ■ Prepare applicable OO or SA/SD models ■ Perform resource use analysis of timing and sizing budgets ■ Allocate requirements from SRS to SUs and Use Cases ■ Allocate SUs to processors and determine protocols ■ Update RTVM with links to design components ■ Prepare draft SDD and SAD ■ Conduct internal review of software architecture	■ Draft SDD and SAD ■ OO Models ■ SA/SD Models
2. Develop and Document Interface Design	Preliminary SDD	■ Allocate requirements to SUs and Use Cases ■ Define software internal interfaces ■ Update software external interfaces and RTVM ■ Prepare draft IDD ■ Define database logical design and the draft DBDD ■ Conduct internal review of software interface design	■ Draft IDD ■ OO Models ■ SA/SD Models ■ Draft DBDD
3. Develop & Document STP	Draft SDD, SAD and IDD	Prepare preliminary Software Test Plan based upon System/Subsystem Test Plan	Preliminary STP
4. Conduct Design Inspection	■ SAD, SDD, STP and IDD ■ Requirements Database ■ Design Models	■ Inspect links to the design in the Requirements Database ■ Inspect design work products ■ Perform document reviews of the SDD and the IDD	Inspection minutes
5. Prepare Test Description	Preliminary STP SAD, SDD, IDD	Identify threads and prepare the preliminary schedule for integrating threads on the target hardware	Preliminary STD &Test Schedule
6. Conduct Internal Review and Update Documents	■ Preliminary SAD, SDD, IDD and STP ■ Draft DBDD	■ For SCS-1 software, Subsystem SwIPT conducts an internal subsystem software SI PDR with management ■ For SS-1 & SS-2 software, Subsystem SwIPT conducts an internal subsystem software SI TIM. Update the SAD, SDD, IDD, DBDD and STP as required	■ Updated SDD, IDD, STP and SMBP ■ Incorporate SS-2 data into SDF
7. Conduct Formal PDR (SCS-1) or TIM (SS-1)	Updated SDD, IDD, STP and SMBP	■ Schedule the PDR/TIM, identify attendees, and finalize agenda for the PDR or TIM ■ Conduct PDR/TIM; generate minutes and action items ■ Ensure closure of action items; generate final outputs	Preliminary SAD, SDD, IDD, STP, SMBP and DBDD

11.4 Software Item Detailed Design

11.4.1 Objectives and Approach to SI Detailed Design

The objective of SI Detailed Design is to determine and define the *implementation details* for each Software Unit. Designers define the specifics of the algorithms or processes an SU is to perform and determine details of the data structures used by the SU internally—and for interactions with other SUs. The resulting SU Detail Design descriptions are normally sufficient for Code Developers to implement the design into code. The main objectives are to:

- Complete identification of design components including Software Units.
- Complete a description of the design for each SU and record all results in the SDF.
- Complete baseline documentation with applicable SAD, SDD, STP, IDD, SMBP and DBDD.

The Detailed Design activity involves decomposing the SUs from the SI Architectural Design into the lowest level SUs. The design must be developed in sufficient detail to map the design to the features of the selected programming language, the target hardware, operating system, and network architecture. As the Software Project Manager, especially on large systems, you may not be heavily involved with the Architectural Design, and not at all likely to be deeply involved with Detailed Design. However, you must always preach to your team to keep the design as simple as possible. A lesson learned example:

> **Lessons Learned.** Back in the 1950s, there was a program called Project Vanguard; it was a rocket launch vehicle that blew up at the launch pad at almost every attempt. The design of the Vanguard was a remarkable engineering masterpiece. The problem was (in my opinion) the "plumbing" of the rocket was so complex and intricate that it was very difficult to get everything to work properly *at the same time*. When all the components worked together, it was a marvelous feat. Over the years I have always told System Designers "don't build a Vanguard" meaning don't make it so complicated that it is likely to fail.
>
> *Simplicity beats complexity every time.*

Critical Design Review. For System Critical Software (SCS-1), the Detailed Design activity ends with a formal *Critical Design Review* (CDR) in which the baselined design documents are evaluated. For software that is developed in multiple builds, only a subset of the SUs may undergo

Detailed Design. The SUs that undergo Detailed Design should be only those units necessary to meet the SI requirements for each build, as specified in the Requirements Database.

For SS-1 software, the CDRs are normally replaced with Technical Interface Meetings. For SS-2 software, the CDRs are typically Peer Reviews held for each build. At the conclusion of the Detailed Design activity, all work products must be placed under configuration control (see Section 6.3).

11.4.2 Human System Integration Design

Software Design must include understanding and consideration of the human aspect of the "man-machine" interface, also referred to as the Human System Integration (HSI) as discussed in Subsection 5.4.3. In order to maximize system usefulness, users must easily and efficiently monitor and control software functionality. The software system may be designed to control itself—most of the time, but great care and consideration must be given to the interaction between the computer and the user. This includes such things as intuitiveness, color, light, noise, redundancy, input methods, output display of the interface, plus fatigue, stress, environment and distraction of the user.

Software Designers should have a basic understanding of information theory and human information processing. Failure to incorporate human factors into the design process may result in just poor sales for a computer game, but in operational systems it could lead to people not using your system, or it could result in injury or death in an avionics or military system. Human factors are very important!

> **Lessons Learned.** You cannot always expect intelligent people with college degrees to have common sense. For example, an engineer once demonstrated a device he built that involved a unique type of screen with a dial control. It looked great, but I did not see a dial. When asked, he pointed to a dial in the rear side of the device. A seven-foot basketball player could reach it and see the screen at the same time, but it is not possible for the rest of us. HSI never entered this Engineer's mind because, in his defense, he was never schooled in the tenants of the human system interface. If there is a human element in your system make absolutely sure your team understands and follows the HSI guidelines and use the expertise of human factor engineers.

Through many years of research on human factors, specific HSI Design characteristics have been identified, and they are documented in design guidelines and standards. However, many of the existing standards do not provide needed support for rapidly developing technologies such as mobile computing devices and voice recognition systems, or sufficient guidance for the design of computer information displays. Some of the human–computer interaction design standards are listed in Appendix I.

11.4.3 Work Products for SI Detailed Design

During SI Detailed Design, the Designers complete the refinement of the work products (for example, models and diagrams of the SI). General operations identified in earlier versions of the products must be defined to the SU level of functions and procedures, and then defined as to how specific algorithms and support services are implemented in software. This process should occur repeatedly with each build.

Details of the data structures must be defined, including temporary data items. The physical database design, if any, must also be defined, including data entities, attributes, relationships and constraints. Interfaces determined in Architectural Design, including user interfaces, are refined and elaborated. The software Detailed Design tasks must refine the software system architecture until the lowest level classes and interfaces have been identified and described.

Detailed Design must be performed for each software increment in the current build. There may be multiple builds and design components concurrently in various overlapping stages of completion. For an Iterative Software Life Cycle Process, components may have been partially designed during prior software development builds, and only the additional design details for the current build must be added. At the conclusion of this activity, the Detailed Design products must be baselined and placed under software configuration control as described in Section 6.3.

11.4.4 Process Tasks for Software Item Detailed Design

The principal tasks recommended for the Detailed Design process are described by the example flowchart in Figure 11.5 and in its related Task Table 11.9 in terms of the inputs and outputs of each task shown in the flowchart. Major tasks performed during SI Detailed Design should include:

■ *Refining the Design Model*: Adding additional details to the Preliminary Design Model to accommodate Detailed Design decisions and constructs necessary for implementation.

■ *Defining Implementation Details*: Refining internal design to add data structures, attribute types, visibility, interfaces and usage mechanisms. Factors to consider include execution time, memory usage, development time, complexity, maintainability and reusable

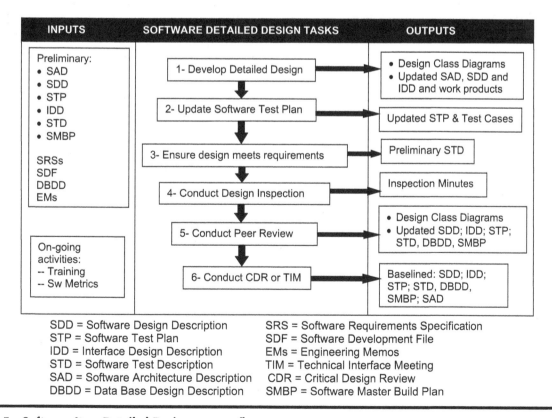

Figure 11.5   Software Item Detailed Design process flow.

Table 11.9 Software Item Detailed Design Tasks

Task	Inputs	SI Detailed Design Subtasks	Outputs
1. Develop Detailed Design	■ Preliminary SAD, SDD, IDD, STP, SMBP and DBDD ■ Baselined SRS ■ Requirements Database ■ EMs	Define Detailed Design including: ■ Analyze models to identify additional requirements. ■ Define, describe and decompose SU Detailed Design. ■ Design and develop algorithms, prototypes, control mechanisms and support services. ■ Determine applicability of COTS/Reuse software. ■ Prepare Design Class Diagrams. ■ Prepare dynamic behavior diagrams showing the sequencing of component iterations, states and modes, and transitions. ■ Prepare SDD containing Detailed Design data. ■ Update IDD with Detailed Design data. ■ Ensure conformance with architecture. ■ Refine database physical design and the DBDD. ■ Perform resource use analysis of timing and sizing budgets. ■ Review requirements and update the Requirements Database. Define Interface Design including: ■ Allocate and decompose architecture and user interface requirements to a Detailed Design level. ■ Define interface design external to the SI and between SUs within each SI. ■ Define information flow between SUs. ■ Develop the design of user screens. ■ Apply human factor standards to user interfaces. ■ Coordinate and review interface design updates.	■ Design Class Diagrams ■ Updated SAD and SDD ■ Updated IDD ■ SA/SD work products such as data flow diagrams and structure charts (if applicable) ■ Data Dictionary ■ DBDD
2. Update STP and Integration Testing Approach	Outputs of Task 1	■ Generate test software requirements. ■ Document traceability between software test cases and software test requirements in the STP. ■ Update schedules for conducting each test case. ■ Identify needed integration information (input data, scenarios, data analysis, etc.).	Updated STP, Test cases and schedules
3. Ensure Design Meets Requirements	Outputs of Task 1	■ Design software performance and reliability models and develop simulations. ■ Conduct analysis to determine if design meets requirements.	Models and simulations Verify performance, timelines and reliability
4. Conduct Design Inspection	■ Design documents. ■ Requirements Database. ■ Design work products.	■ Inspect for links of requirements to Detailed Design components. ■ Inspect Object-Oriented products. ■ Perform document reviews.	Inspection Minutes

(Continued)

Table 11.9 (Continued)   Software Item Detailed Design Tasks

Task	Inputs	SI Detailed Design Subtasks	Outputs
5. Conduct Peer Review	■ Preliminary SAD, SDD, IDD, STP, DB DD and SMBP. ■ OO Products. ■ SA/SD Products.	■ For SCS-1 software, subsystem SI SwIPT conducts an internal design review for the subsystem software. ■ For SS-1 and SS-2 software, subsystem SI SwIPT conducts an internal subsystem software TIM. ■ Update design documentation as required.	■ Updated SAD, SDD, IDD, STP, DBDD and SMBP ■ Detailed Design Diagrams ■ Peer Review Minutes
6. Conduct Software CDR (SCS-1) or TIM (SS-1 & SS-2)	■ Updated design documents. ■ Detailed design diagrams.	■ Schedule the CDR/TIM, determine attendees, and update the evaluation criteria. ■ Conduct the CDR/TIM and generate minutes and action items. ■ Ensure closure of action items and generate final outputs.	Baselined SAD, SDD, IDD, STP, SMBP and DBDD

software and hardware. Analysis and modeling may be necessary to determine the best design approach.

■ *Generating Class Stubs*: Generate code header files and class stubs based on the object model definitions, design class algorithms or logic.

■ *Prototyping and Simulations*: Performing prototyping and simulation to validate critical processing areas, mitigate implementation risks, or to identify optimizations.

■ *Generating and Reviewing Products*: Holding Peer Reviews on Detailed Design products and adding the Detailed Design information to the SDD, IDD and DBDD.

Other tasks that may be performed (if applicable) in the Detailed Design activity include:

■ Define detailed software user interfaces to the Architectural Design level and validate it with software prototypes, working models, simulations and/or display layouts.

■ Identify concurrency in threads or capabilities, global resources and determine mechanisms for access control.

■ Choose the implementation method of control in software (e.g., procedure-driven, event-driven).

■ Determine methods for handling boundary conditions (i.e., initialization, termination and failure) and establish trade-off priorities.

■ Prepare computer system hardware diagrams including the purpose of each component, including its interfaces and physical processing characteristics.

■ Describe how and where (and if) the architecture supports Modular Open Software Architecture.

■ Analyze and document the availability of Non-Developmental Items (NDI), incorporate NDI into the design, and allocate requirements to it (NDI is discussed in 1.8.3).

■ Consider reusable architecture designs for all or portions of an SI; trade-off studies and analyses may be necessary to determine the best design approach.

■ Create draft versions of *Software User's Manual* (SUM) and *Software Transition Plan* (STrP).

11.5 Coding and Unit Testing

The software *Coding and Unit Testing* (CUT) activity of the Development Life Cycle may also be called *Software Implementation and Unit Testing*. That may be a better name, but I prefer to call it CUT. The objective of the software CUT activity is to convert the SU Detailed Design into computer code and databases that have been inspected, unit tested and confirmed. The term *coding* is used throughout this process to mean the generation of computer-readable instructions and data definitions in a form that can be acted upon by a computer.

Section 11.5 addresses the objectives, approach, readiness criteria, software work products, roles and responsibilities and tasks specific to the software CUT activity. Software Coding and Unit Testing is described in the following Subsections:

■ Software Coding and Unit Testing—Objectives and Approach (11.5.1)
■ Work Products for Coding and Unit Testing (11.5.2)
■ Roles and Responsibilities for Coding and Unit Testing (11.5.3)
■ Process Tasks for Coding and Unit Testing (11.5.4)
■ Computer Code Development (11.5.5)
■ Preparing for Unit Testing (11.5.6)
■ Conducting Unit Testing (11.5.7)
■ Verifying and Recording Software Unit Test Results (11.5.8)
■ Revision and Retesting of Software Units (11.5.9)

The requirements specified in these sections are for SCS-1, SS-1 and SS-2 software only (SS-3 compliance should be optional).

11.5.1 Software Coding and Unit Testing—Objectives and Approach

Major tasks of the Software Coding and Unit Testing process includes:

- The *Detailed SU Design* must be converted into computer code in accordance with the coding standards for the selected programming language. This may include partial units or modifications to those created in prior builds.
- Specific *test descriptions* must be generated in order to unit test the SU that define the test cases, test procedures, test input, support data, and expected test results.
- The completed *source code*, test description data on all developed units and documentation, should be reviewed through a Peer Review inspection, which may include SQA participation.
- The *test cases* should be executed against the executable code to determine the success of the coding effort. White Box (internal structure) and Black Box (external functions) tests should be performed on the individual units. Successful completion of unit level testing is a prerequisite for promotion of units to software integration.
- The *results of the test cases* must be reviewed, and the code reworked and retested until all unit tests have been successfully completed.

- The test results must be *independently reviewed* by someone other than the Developer to confirm successful completion of the test, and that test data and results have been recorded in the SDF.

These steps are highly iterative, in that the code and test tasks are performed for each SU (class). Groups of SUs may be coded, reviewed and tested as a set, according to the development plan and schedule for the increment. SUs may also be incomplete, in that only the functionality required to support the current increment (build) is implemented. Table 11.10 summarizes the readiness criteria in terms of entry and exit criteria, verification criteria to ensure completion of the required tasks, and the measurements usually collected during the software CUT activity.

The CUT activity for a single SU formally ends upon completing confirmation of the test results and recording of the test data and results in the SDF. After these actions have been completed, the SU must be brought under configuration control. Changes to the SUs are handled using Software Change Requests (SCRs) or Software Discrepancy Reports (SDR).

11.5.2 Work Products for Coding and Unit Testing

The principal product produced during this activity is the SU source code as shown in example Table 11.11. The format for the source code is established by the coding standards for the particular language used. During the software CUT activity, SU Test Cases, test procedure data, sizing and timing are prepared and updated. The SRS, STP, IDD and DBDD may also be updated as required.

Table 11.10 Readiness Criteria: Software Coding and Unit Testing

CUT Entry Criteria	CUT Exit Criteria
<ul><li>SU Detailed Design has been completed.</li><li>Software coding standards are established.</li><li>The Software Development Environment (SDE) has been established.</li><li>The Requirements Test and Verification Matrix (RTVM) and Software Requirements Traceability Matrix (SRTM) is available.</li><li>The SDL has been established.</li></ul>	<ul><li>Software Unit Test Cases is completed and accepted by the Software Lead.</li><li>SU test procedure data is recorded in the SDF.</li><li>SU Source Code is developed, compiled, debugged, and accepted by the Software Lead.</li><li>SU test results are recorded in the SDF.</li><li>Software is put under software integration CM control.</li></ul>

CUT Verification Criteria
<ul><li>Peer Reviews of SU Test Cases, Source Code and SU test data and results are completed and recorded.</li><li>Software Development Librarian accepts the source code.</li><li>SQA performs process and product audits for ongoing product engineering tasks.</li></ul>

CUT Measurements
<ul><li>Actual Thousands of Lines of Code (KSLOC) coded versus KSLOC planned.</li><li>Unit testing planned versus actual test progress.</li><li>Number of defects found in Peer Reviews.</li><li>SCRs and SDRs opened versus closed (see Chapter 15).</li></ul>

11.5.3 Roles and Responsibilities for Coding and Unit Testing

Table 11.12 is an example of the roles and responsibilities of the SwIPT Developers and other groups during the software CUT activity.

11.5.4 Process Tasks for Coding and Unit Testing

An example of the Software CUT development process is shown in Figure 11.6.

Since this is an iterative process, there are no vertical arrows in the process flowchart. The seven process tasks described in Figure 11.6 are expanded with more detail in its related Task Table 11.13 containing the inputs and outputs to each task identified in the flow chart.

In addition, during the CUT activity, draft versions may be prepared (if applicable) of the *Computer Programming Manual*

(CPM) and the *Firmware Support Manual* (FSM) if your system has firmware; both are described in Subsection 11.10.8.

11.5.5 Computer Code Development

The objective of software implementation is to implement requirements by converting the Software Unit Detailed Design into executable computer code. The major tasks are:

- Develop the code for each SU based upon the design requirements and Detailed Design.
- Code the software using the coding standards required by your project.
- Create executable source code and debug using applicable tools.
- Update source code estimates with actual measurements of the SU.
- Document rationale to reuse code and identify reuse code modules.

Table 11.11 Required Software Coding and Unit Testing Work Products

Software CUT Products	SCS-1	SS-1	SS-2	SS-3
Source code or reference to source code in the Software Development Files	Required	Required	Required	Optional
SU Test Cases, procedures, data and test results	Required	Required	Required	Optional

Table 11.12 Roles and Responsibilities during Software Coding and Unit Testing

SwIPT and Developers	Codes SUs to the appropriate coding standards
	Develops the unit test description and executes the unit test
	Conducts the required inspection of the source code and test documentation
	Reworks and retests the SU when problems are identified
	Confirmation of the test results (The SU author participates in the inspection, but someone else performs the inspection and confirmation of test results)
	Updates and maintains the SDF with source code, unit test descriptions, test code, and unit test results
	Ensures critical software requirements are traced to SUs
	Collects and reports software CUT task metrics
Software Test Personnel	Continues development of the STP
CSwE	Reviews the SDFs, reviews task products, and attends activity reviews
SQA	Evaluates the Subsystem SwIPT software products for adherence to documented policies and procedures, evaluates Subsystem SwIPT products for product quality, and reports findings to upper-level management
CCB	Addresses subsystem SCR/SDRs as they are generated
SwCCB	Addresses SU SCR/SDRs involving external interface changes or SU changes from prior build releases as they are generated
SCM	Processes all SCR/SDRs for SU changes to the code and documentation

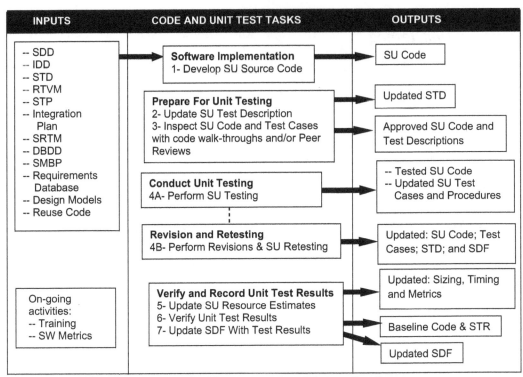

NOTE: There are no vertical flow arrows since this is an iterative process

SDD = Software Design Description   RTVM = Requirements Test Verification Matrix
IDD = Interface Design Description   STP = Software Test Plan   SU = Software Unit
SRTM = Software Requirements Traceability Matrix   SDF = Software Development Files/Folders
STD = Software Test Description   SRS = Software Requirements Specification
DBDD = Data Base Design Description   SMBP = Software Master Build Plan

Figure 11.6   Coding and Unit Testing process flow.

Approach to Computer Code Development. The Software Developers must generate the SU source code using the appropriate programming language, based on the Detailed Design, interface requirements, and supporting design information. Changes made to the executable code must be accomplished through modification of the source code and subsequent recompilation or reassembly. Once successfully compiled and executed, a Peer Review inspection of the source code should be conducted.

Refactoring. Adding, removing or changing source code in order to make the code easier to read, understand and maintain is called *refactoring*. Refactoring changes the design and improves it but does not alter the behavior or functionality of the software. Many Programmers find opportunities to improve the design of the system, but these improvements usually become apparent only during development of the code. There are many different improvement techniques that are called refactoring.

Lessons Learned. Trying to make your system perform more efficiently, and making the code easier to understand and maintain is a good thing. However, allowing your team far too much time trying to make it progressively better and better can be detrimental to the project's cost and schedule targets. As the Software Project Manager, after the functional and contractual requirements are met, you must decide when to draw the line and decide *when good enough is good enough*. Perfection is too expensive unless people's lives are involved (see Subsection 5.1.6). I have encountered situations where I almost had to physically take the software product away from Developers who wanted to continue making it better. This can be risky—especially if the Developer is a foot taller than you.

Table 11.13   Coding and Unit Testing Tasks

CUT Task	CUT Inputs	Coding and Unit Testing Subtasks	CUT Outputs
1. Develop SU Source Code	See Fig 11.6 inputs	▪ Check SUs against input documents, SCRs and reused code to confirm definitions and requirements of SUs ▪ Code SUs	SU Code
2. Update SU Test Description Data	SU Code	▪ Address SU requirements in Test Descriptions ▪ Develop test inputs and outputs ▪ Develop SU Test Cases and test exceptions ▪ Ensure adequate Test coverage and number of iterations	SU Test Descriptions and Test Cases
3. Inspect SU Code and Test Cases	▪ SU Source Code ▪ SU Test Cases	▪ Schedule SU for code walk-through and/or Peer Review after successful compilation ▪ Hold inspection of SU source code and Test Cases ▪ Close Peer Review findings	Approved SU Code and test description data
4A. Perform SU Testing 4B. Perform Revisions and Retesting	▪ SU Code ▪ SU Test Cases	▪ Follow procedures in SU Test Cases ▪ Record test results in the SUs SDF ▪ Fix source code problems ▪ Modify and approve unit test procedures and results ▪ Retest; repeat until Unit Testing is successful	Updates to: ▪ Tested Code ▪ SU Test Cases and Procedures ▪ SDF
5. Update SU Resource Estimates	Tested Source Code	▪ Measure the sizing, timing and complexity of SU as required ▪ Record SU SLOC count and productivity metrics in the SDF ▪ Update metrics with measurements from SU testing	▪ Sizing and Timing in SDF ▪ Updated Metrics
6. Verify Unit Test Results	SU Test Results	▪ Verify correctness of test results ▪ Capture test procedures, inspection results and unit test results ▪ Place tested code under configuration control ▪ Update SRS as required ▪ Release SU for Unit Integration	▪ Baseline Source Code ▪ Updated SRS as required
7. Update SDF with Test Results	Update information from 1 to 6 above	▪ Update SDF and document Lessons Learned ▪ Record test results in the SDF ▪ Prepare Release Notice to inform availability of SUs ▪ Conduct Integration Readiness Review (IRR)	▪ Updated SDF ▪ Release Notice ▪ IRR completed

11.5.6 Preparing for Unit Testing

The principal objectives of preparing for Unit Testing are to:

▪ Develop overall test objectives and assumptions including constraints.
▪ Define, develop, and document the Unit Test Cases and Unit Test Procedures.
▪ Develop input test data including data files, databases, algorithm and simulation data.
▪ Identify support resources, including required drivers and stubs.
▪ Test preparation (including hardware and software).

▪ Describe the inputs, expected results, success criteria and evaluation criteria for each test case.
▪ Allocate software requirements to each test case and ensure that all SU requirements are tested.
▪ Define data files, databases, simulation programs and additional resources required.
▪ Layout a preliminary schedule of when the unit test cases are to be performed.
▪ Execute statements and branches of the Software Unit at least once.
▪ Identify and define interfaces and dependencies between the test cases.
▪ Identify start-up, termination, restart, error and exception handling procedures.

- Verify that the Software Unit performs its intended operations using nominal plus boundary upper and lower limit input values.
- Record the above information in the SDF or Software Project Workbook.

11.5.7 Conducting Unit Testing

The objectives of conducting Unit Testing are to:

- Perform Unit Testing of the developed source code in accordance with test cases and procedures.
- Verify the unit level functional, interface, and SU performance requirements.
- Verify the SUs exception handling capability.
- Maintain unit test logs to verify and track SU test execution and completion.
- Update the unit source code to correct errors detected during the Unit Testing.
- Record the unit test results and performance measures in the SDF.

Approach to Conducting Unit Testing. Software Testers normally begin by ensuring that all necessary data, tools, test environment and unit test configuration are available. When all required pieces for the test are assembled, the test can proceed per the test procedures. The Testers must verify the unit level functional, interface and performance requirements. The Testers must collect and record the test outputs, logs, notes, results and discrepancies found.

Although Software Developers must make sure each SU satisfies its requirements, Unit Testing may be considered principally "white box" testing, i.e., where the testing results are compared to the design. The Software Developers must identify test cases and procedures to be performed on a Software Unit. For cases where tests cannot be developed to adequately verify that functionality has been demonstrated, verification by analysis may be permitted. This situation can occur when an event to be tested is infeasible to create or involves prohibitively extensive testing or extreme cost to conduct the test.

Reused SUs that have been modified require complete retesting of the Software Unit. If the reused SU is deemed *critical,* it must be unit tested even if it has not been modified. The completed test description, including both test case definitions and test procedures, must be retained in the SDF. The inspection package for modified reused code should also include a *code difference listing*.

11.5.8 Verifying and Recording Software Unit Test Results

The objective of analyzing and recording unit test results is to finalize the Unit Testing for an SU by ensuring that:

- The unit satisfies the expected results of the test cases.
- The test data, test results, unit test dependencies and supporting analysis material have been recorded in the SDF.
- Root cause analysis of problems has been performed.
- The SU is ready to be released for Unit Integration and Testing (see Section 11.5).

Approach to Analyzing and Recording Unit Test Results. After completing Unit Testing, the Software Lead must perform an independent confirmation of the test results and ensure the results have been recorded in the SDF. If discrepancies or problems are found, then appropriate Corrective Actions must be performed.

Once the independent review signifies that the SU has successfully passed the verification process, the SU can be baselined and brought under configuration control. The SU source code is then submitted for incorporation into software integration builds. In addition, Developers should incorporate supporting analysis material and unit test dependencies information in the appropriate SDF. The related measurements obtained during SU testing should also be updated and recorded.

11.5.9 Revision and Retesting of Software Units

The objectives of revision and retesting of Software Units are: to modify or rework the source code and test description to eliminate any problems identified during Unit Testing; and to retest the SU to verify that the changes have been successful and have not produced side effects. If a unit test fails, the problem must be fixed and the test(s) repeated. The standard design inspection process must be invoked again and the SDFs updated. Regression testing of affected SU test cases must be performed after any modification to previously tested software. Changes must be made in accordance with the Corrective Action Process (see Section 6.4).

Approach to SU Revision and Retesting. Test results, and the documented problems, must be evaluated by Software Developers to identify needed changes to the SU and test description. This task should be repeated until all the SU test cases have been successfully completed.

11.6 Software Unit Integration and Testing

This Section addresses the objectives, approach, readiness criteria, software work products, roles and responsibilities and tasks specific to the Software *Unit Integration and Testing* (UI&T) activity portion of the Software Implementation Process. It is described in the following nine Subsections:

- Software Unit Integration and Testing—Objectives and Approach (11.6.1)
- Work Products for Software Unit Integration and Testing (11.6.2)
- Responsibilities for Software Unit Integration and Testing (11.6.3)
- Process Tasks for Software Unit Integration and Testing (11.6.4)
- Prepare for Software Unit Integration and Testing—includes updating the STP and test procedures (11.6.5)
- Perform Software Unit Integration and Testing—includes performing the integration and test of a build in accordance with integration test procedures (11.6.6)
- Revision and Retesting Software Unit Integration and Testing—includes reworking the source code; perform Regression Testing for changes and documenting the discrepancies (11.6.7)
- Verify and Record Software Unit Integration and Testing Results—includes analyzing test results, documenting the Software UI&T results in the SDF, and identifying who decides an Integration Build is ready for release to SI Qualification Testing (11.6.8)
- Fault Detection, Isolation and Recovery—includes managing the error resolution effort (11.6.9)

Subsystem Software Integration Teams must develop the integration plans, integration test cases, and integration test procedures and test data in preparation for the actual integration and test. The SUs should be checked out of the Controlled Area of the *Software Development Library* (SDL) by the integrators. As the builds are successfully integrated, the SUs are typically returned to the SDL to be elevated to a higher level of control. Discrepancies must be recorded using SDRs.

11.6.1 Software Unit Integration and Testing—Objectives and Approach

The objective of the Software *Unit Integration and Testing* activity is to perform a systematic and iterative series of integration builds on Software Units that have successfully completed Code and Unit Test, and build them up to a higher level SU (formerly called a Software Component), or Software Item, for the current build.

The *Software Test Plan* and *Software Test Procedures* should be reviewed for consistency with the *Software Master Build Plan*, and revised if necessary. In addition, preparation of a draft *Software Version Description* (SVD) and a draft

Software Test Description (STD) for qualification testing should begin during this activity. The UI&T activity consists of the following major tasks:

- The software integration plans, test cases and test procedures are Developed and Peer-Reviewed.
- Test data, tools, drivers, simulators, etc. must be in place before the start of testing.
- The integration test procedures must be executed against the executable code.
- Needed corrections to the software, and the integration test procedures, must be made and the affected integration iteration retested; this activity should be repeated until all SUs have been successfully integrated and have met the test acceptance criteria.
- Test results for integrated SIs must be independently analyzed, or with a Peer Review, to verify successful integration and recording of results in the *SDF*.
- The SI *Software Test Plan* should have been baselined prior to start of actual testing.
- Regression testing must be performed as needed to incorporate SUs from prior builds.

The UI&T activity formally ends with the verification of the test results and the recording of the test data and test results in the SDF. The build must then be baselined and moved to the Verification Area of the *Software Development Library*. All changes to an SI thereafter must be handled through the process described in the *Software Configuration Management* (SCM) Plan.

UI&T Readiness Criteria. Table 11.14 summarizes the readiness criteria in terms of entry and exit criteria, verification criteria to ensure the completion of the required activities, and the required measurements normally collected during the Software UI&T activity.

11.6.2 Work Products for Software Unit Integration and Testing

Examples of software work products for the UI&T activity are summarized in Table 11.15.

11.6.3 Responsibilities for Software Unit Integration and Testing

Software Developers must be responsible for development of the integration test plans, procedures, data, actual integration of the SUs and the execution of the tests. When problems are identified, the Software Developers must be responsible for reworking SUs and retesting the integration of those units. Table 11.16 summarizes typical responsibilities and roles for the UI&T activity.

Table 11.14 Readiness Criteria: Software Unit Integration and Testing

UI&T Entry Criteria	UI&T Exit Criteria
■ Software Test Plan is available. ■ Coding and testing of the SUs are completed. ■ Software test procedures data is available. ■ Integration builds are available from the SDL. ■ The RTVM, SRTM and SMBP are available.	■ SI build is successfully integrated, accepted by the Software Team Lead, and turned over to the SDL. ■ The draft Software Version Description is approved by the Software Team Lead. ■ The STP is updated and ready to support SIQT.
UI&T Verification Criteria	
■ Software Peer Reviews have been successfully completed. ■ Unit integration plans, test cases and UI&T procedures developed and successfully Peer-Reviewed. ■ Software Units successfully integrated in accordance with the integration plans. ■ The Software Team Lead reviews and approves the integration test reports and integration release notice. ■ All SUs in the build per the SMBP are successfully integrated and tested and the results stored in SDFs. ■ SQA performs process / product audits for ongoing product engineering activities.	
UI&T Measurements	
Defects found from Peer Reviews. SDRs opened versus closed. Units integrated—planned versus actual. SLOC count—planned versus actual (see Chapter 15).	

Table 11.15 Software UI&T Work Products

Sw UI&T Work Products Per Build	SCS-1	SS-1	SS-2	C/R
Updated STP	Required	Required	Data in SDF (*)	Data in SDF (*)
Draft Software Version Description	Required	Required	Data in SDF (*)	Data in SDF (*)
Draft Software Test Description	Required	Required	Data in SDF (*)	Data in SDF (*)
SI test cases traced to SRS requirements in the Requirements Database	Required	Required	Data in SDF (*)	Optional
UI&T Products: ■ Unit integration plans ■ UI&T test cases and procedures ■ UI&T scripts, drivers and test data ■ UI&T test results	Required	Required	Data in SDF (*)	Data in SDF (*)

SCS= System Critical Software; SS = Support Software; C/R= COTS/Reuse Software.
(*) Document not required, but applicable information is developed and retained in the SDF.

11.6.4 Process Tasks for Software Unit Integration and Testing

Figure 11.7 shows a typical Software UI&T process. The four UI&T process tasks are expanded in its related Task Table 11.17 containing more details to the Software UI&T subtasks.

11.6.5 Preparing for Software Unit Integration and Testing

The principal objective of preparing for the Software UI&T task is to establish test cases, test procedures, and test data for conducting Unit Integration and Testing in order to define a systematic and iterative approach for integrating a subset of

SUs until the entire set of SUs are integrated into the complete SI (for that build). As a minimum, the test cases must cover a description of:

■ Execution of all interfaces between Software Units—including limit and boundary conditions
■ Integrated error and exception handling across the SUs under test
■ End-to-end functional capabilities through the SUs under test
■ All software requirements allocated to the SUs under test
■ Performance testing—including operational input and output data rates plus timing and accuracy requirements

Table 11.16   Software UI&T Responsibilities

Group Roles	UI&T Responsibilities
Software Development Personnel	Conducts the required Peer Reviews of the UI&T documentation
	Updates and maintains the SDF with test procedures and test results
	Address safety, security, privacy and critical software requirements in the test cases
	Collects and reports SU integration and testing activity metrics
	Submits SDRs as necessary
Other Groups	Software Test: Update the STP and prepare drafts of the STD and SVD
	Chief Software Engineer: Reviews the SDFs, reviews the activity products, attends activity reviews, and monitors and analyzes software metrics
	SQA: Evaluates the SwIPT for adherence to the documented policies and procedures, evaluates SwIPT products for product quality, witnesses testing and documents and reports findings to upper-level management
	Subsystem CCB: Addresses internal SCR/SDRs as they are generated
	Software/CCB: Addresses SU SDRs involving external interface changes and prior build release SU changes as they are generated by the SwIPTs
	SCM: Processes SDRs for SU changes to the source code and test documentation

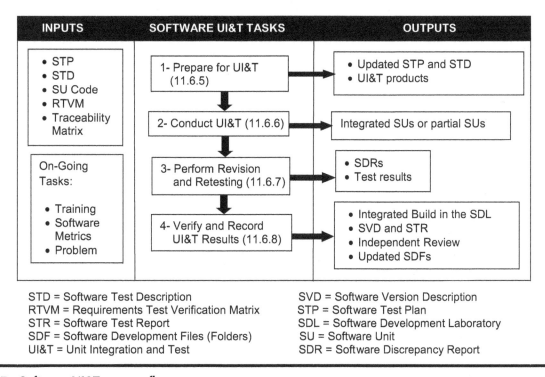

Figure 11.7   Software UI&T process flow.

- Stress testing—including worst case scenarios
- Start-up, termination and restart
- Fault detection, isolation and recovery handling
- Resource utilization measurement

Whenever possible, SU integration should be performed on the target hardware in a configuration as close as possible to the operational configuration. All COTS/Reuse software, whether modified or unmodified, must undergo Software UI&T. Software Developers must define integration test cases that are mapped to use cases and then mapped to requirements and corresponding test procedures, in an integration test description to *verify the success of each partial integration before proceeding to the next iteration*. The specified

Table 11.17 Software UI&T Tasks

Tasks	Inputs	Software UI&T Subtasks	Outputs
1. Prepare for UI&T	See input documents in Figure 11.7	■ Update STP and STD with the current plan for build integration. ■ Link requirements to integration and test cases. ■ Create and populate the traceability database. ■ Prepare Unit integration plans, UI&T test cases, procedures, scripts, drivers and test data.	■ Updated STP and STD ■ UI&T test products
2. Conduct SU Integration and Testing	■ SU code ■ Test Cases ■ Test Data ■ Test scripts and drivers ■ Test ■ Procedures	■ Integrate the SUs per the integration plans. ■ Conduct integration testing of the current build based on the test cases. ■ Record and document anomalies and errors detected during testing in the Software Discrepancy Report.	■ Integrated SUs or partial SU build ■ SDRs ■ Test Results
3. Perform Revision and Retesting	■ Integrated SUs ■ Updated SUs to fix SDRs ■ Test Results	■ Perform Regression Testing to accommodate new functions or changes to the previously integrated code. ■ Perform retesting after fixes to test procedures and/or code. ■ Record and document anomalies and errors detected during testing of the SDR fixes.	■ SDRs ■ Test Results
4. Verify and Record UI&T Results	■ SDRs ■ Test Results	■ Verify results and document findings of the integration tests in the SDF. ■ Inform software development and the SDL that the current build has successfully completed integration testing. ■ Prepare Draft SVD and STD. ■ Conduct Independent Review of test results.	■ Integrated build in SDL ■ SVD & STD Draft ■ Updated SDF ■ Review completed

integration sequences are: to verify that the SUs operate together using nominal and exception conditions; and to exercise all interfaces for the SUs that have been integrated.

11.6.6 Conducting Software Unit Integration and Testing

The principal objectives of conducting the Software UI&T are to:

■ Integrate and combine Software Units.
■ Execute the integration plan and corresponding test procedures as documented in the integration test description to produce the integrated Software Item.
■ Execute the integration runs and verify the complete integration.
■ Verify that SUs within the SI *provide the functionality required* for that build.
■ Record test results for this level of testing.

During this task, coded and tested SUs should be integrated into SIs by a Software Integration Team in a series of integration builds. The SUs should be obtained from the

development Controlled Area of the Software Development Library. The software personnel performing the integration usually begin by ensuring that all the SUs to be integrated and all necessary data and tools are available. The integration build to be tested should be generated by the Developers using baselined SUs obtained from the SDL.

When all required pieces for the integration are assembled, the integration should proceed per the procedures specified in the test description. During integration and testing, the Software Developers collect or record the outputs, logs, test notes, and test results. All problems, errors, and discrepancies must be recorded as SDRs.

11.6.7 Revise and Retest Software Unit Integration and Testing

The primary objectives of Software UI&T revision and retesting are to:

■ Revise the source code and regression test in response to problems identified in SDRs, first at the SU level and then at a combined SU level.

- Perform regression tests to accommodate new functions or changes in the current build to ensure that existing functionality has not been impaired.
- Document and track integration problems and test errors.

The Integration Team must perform the integration tests, and any necessary Regression Testing and record discrepancies on SDRs which are placed into the Corrective Action Process for disposition and rework. The documented problems must be evaluated by the Software Developers to determine the necessary changes to SUs or test descriptions. In cases where SUs require changes, SDRs and SCRs must be generated and the changes handled by the Corrective Action Process.

In cases where test descriptions require modification, the appropriate changes must be made and a version history included in the test description. Retesting must also be performed when test procedures are changed. *Retesting must be progressively repeated as needed until all of the SUs are successfully integrated and tested.* Software personnel determine the necessary modifications to the source code, SDFs, and/or documentation. Source code and documentation are modified based on approved changes by the Software Change Control Board (SwCCB typically focused at the SU level) and at the subsystem's Configuration Control Board (CCB) or equivalent (typically at the SI level). Changes must be handled with SDRs (see Subsection 6.4.1).

11.6.8 Verify and Record Software Unit Integration and Testing Results

The primary objectives of analyzing and recording Software UI&T results are to:

- Verify that the tests have been successfully completed and that the test data and results have been recorded in the Software Development File.
- Handle changes to SIs after being brought under SCM control using SDRs.
- Complete SI level integration by the successful execution of all of the defined integration test procedure runs.
- Meet integration completion criteria and perform root cause analysis for deficiencies.
- Document SI level integration, and modify SDFs, source code and documentation.

Upon successful completion of a round of integration, the Software Lead should authorize the release of the build by providing the Software Development Librarian with a release notice. The Integration Team must document the results of each round of integration and file it in the SDF.

After all SUs have been successfully integrated and tested, the SwIPT Lead should perform the independent review of the test results and ascertain that the integration test data and results have been recorded in the SDF. Peer Reviews may also be used.

The SI can then be submitted to the SCM Librarian and baselined in the SDL Verification Area in preparation for SI Qualification Testing. Procedures for recording, analyzing, verifying and storing Software UI&T results should be included in the SDP.

11.6.9 Fault Detection, Isolation and Recovery

The *Mean Time Between Failure* (MTBF) can be calculated for hardware because hardware can fracture or wear out; for software, all we have in the failure category is scope, design and coding *errors*—software does not wear out. Software or system failure can result from an error, causing a *fault*, and that fault can cause a *failure*. It may help you manage the error resolution effort if you view the process from this perspective:

- *Software Error*: A software error is a discrepancy between a performed software behavior versus its specified behavior. In other words, an error occurs when the software does not do what it *should do* or does something it *should not do*. An error can cause a component fault. Keeping track of the total number of errors detected may show progress but does not help you track where you are because you don't know how many total errors there are.
- *Component Fault*: A fault occurs when a component does not perform its intended function. The component fault can occur from a software error. If the component that failed is essential to the system functionality, the fault can result in a system failure.
- *System Failure*: A system-level failure results when a system operation deviates from its specified system requirement. It may be produced when a fault is encountered. A good System Design allows for backup systems and a mechanism for graceful failure recoveries.

11.7 Software Item Qualification Testing

Section 11.7 addresses the objectives, approach, documentation, staff responsibilities and tasks for the *Software Item Qualification Testing* (SIQT) activity. Passing SIQT is a major milestone. The SIQT activities are important enough to warrant the 12 paragraphs below needed to cover this activity. Five of these activities (11.7.7 through 11.7.11) are the sequential processing steps of the SIQT activity:

- SIQT Objectives and Approach (11.7.1)
- Work Products for SIQT (11.7.2)
- Responsibilities for SI Qualification Testing (11.7.3)
- Process Tasks for SIQT (11.7.4)
- Independence in SIQT (11.7.5)
- SIQT on the Target Computer System (11.7.6)
- Preparing for SIQT (11.7.7)
- Readiness Review for SIQT (11.7.8)
- Conducting SIQT (11.7.9)
- Revision and Retesting During SIQT (11.7.10)
- Verify and Record SIQT Results (11.7.11)
- Integration of Multiple SIs and HIs (11.7.12)

11.7.1 SIQT Objectives and Approach

The objective of SIQT is to demonstrate that the SI meets the system, performance and interface requirements allocated to the SI being tested. SIQT must be a *controlled and documented activity* assigned to Software Test Engineers who are *independent* of the Software Development Team. SIQT must demonstrate that: the software performs correctly; contains the features prescribed by its requirements at the SI level, and properly interacts and performs its specified functions within the total system as documented in the Software Requirements Specification for each build.

Approach to SIQT. For software developed in multiple builds, the SIQT for each build must address the software and interface requirements allocated to the current build being tested. The SIQT for the SI being tested will not be completed until the final build for that SI. Regression tests must be performed as needed throughout the iterative process. The software test results must be documented after each test.

A *Software Test Report* (STR) must be published to document the final test results. Any discrepancies noted must be recorded in *Software Discrepancy Reports*, analyzed and dispositioned in accordance with the Corrective Action Process. If the corrections are deferred for a future release, then the STR, and all related release documentation (e.g., the Software Version Description), must reflect SI constraints or workarounds needed.

The SIQT activity ends when documentation of the software test results is completed and the open SCRs/SDRs that can be resolved are resolved for the current release. All test materials and results must be *impounded to establish the as-conducted archive*. A post-test debrief should be conducted to evaluate preliminary results, to analyze anomalies that occurred and to collect lessons learned.

SIQT Readiness Criteria. Table 11.18 summarizes the readiness criteria in terms of entry and exit criteria, verification criteria to ensure completion of the required activities, and the measurements normally collected during the SIQT activity. If SCR or SDR fixes are incorporated into the software, then portions of the SIQT test procedures must be *rerun to verify that applicable SCR/SDR fixes are implemented and working correctly*. In addition, it must be determined that selected pre-existing functionality and interface requirements are still performing correctly after the fixes have been implemented. Subsection 11.7.10 covers the details on performing revision and retesting activities.

11.7.2 Work Products for SI Qualification Testing

Documentation products normally produced during the SIQT activity for each SI include *Software Test Description* (STD), *Software Test Report(s)* (STR), an updated *Software Test Plan* (STP), and traceability products from the

Table 11.18 Readiness Criteria: Software Item Qualification Testing

SIQT Entry Criteria	SIQT Exit Criteria
SRS, IFCD, SMBP, SDD, SAD, IDD, DBDD and the STP have been baselined. The Requirements Traceability Verification Matrix (RTVM) is available.	Formal Software Item qualification tests are successfully completed including an action plan generated to close remaining SDRs. STDs and STRs are completed. The STP is updated as required.

SIQT Verification Criteria
Software Peer Reviews have been successfully completed for all required documentation. SQA and customer witness test execution. SQA performs process / product audits for ongoing product engineering activities.

SIQT Measurements
Number of Peer Review defects. Number of SCRs and SDRs opened and closed, aging data, origin, and root cause of problems. Number of test cases completed and number of requirements verified (see Chapter 15).

Requirements Database. For SCS-1 and SS-1 software, these software products are required. The traceability products contained in the *Requirements Database* are required for each SI. For SS-2 software, the SI test description and test results may be documented in the Software Development File.

The general recommended contents of the STD and STR are:

- ▪ The STD describes the SI-specific test cases and corresponding software and interface requirements, test environment, test procedures, input data, simulations or emulations, expected results, and success criteria.
- ▪ The STR specifies or references the test outputs, logs, notes, and test results.

Work products for this activity are summarized in Table 11.19. The documentation must be made available via an electronic data repository system. Test Logs, describing the results of the tests, are not listed in Table 11.19 but are required.

In addition, preliminary versions of the following software documents may be prepared concurrently with the SIQT activity. Each document has a specific purpose; only those documents having value to your program should be produced. All of these documents are described in sections 11.9 and 11.10.

- ▪ *Software Product Specification* (SPS)
- ▪ *Software Version Description* (SVD)
- ▪ *Software User's Manual* (SUM)
- ▪ *Computer Programming Manual* (CPM)
- ▪ *Firmware Support Manual* (FSM)
- ▪ *Software Transition Plan* (STrP)

11.7.3 Responsibilities for SI Qualification Testing

The Software Test Lead, supported by software test personnel, should be responsible for the development of the SI test plan and test description. Execution of the SI test should be performed by the Software Test Engineers. Table 11.20 is a summary of responsibilities for Software Developers and roles of other groups in the SIQT activity.

For support software, the Software Test Lead may have Software Developers prepare and run the tests, provided they are not the same individuals who performed SI integration. Where problems are identified during SI testing, Software Developers should work with test engineers to analyze problems to determine if it is a software issue or a test procedure issue.

11.7.4 Process Tasks for SIQT

Figure 11.8 is an example of a flowchart of the SIQT process. As shown in the figure, the SIQT process involves 17 recommended tasks that are organized into five groups covering the preparation, dry run, conducting the SIQT, revision and retesting, and the verification and recording of SIQT results.

Task Tables 11.21– 11.25 contain more details and are mapped directly to the 17 tasks within the five groups in Figure 11.8. The tables are included in Subsections 11.7.7–11.7.11.

11.7.5 Independence in SI Qualification Testing

Software Item Qualification Testing must be performed to demonstrate that the SI meets the software and interface requirements allocated to that build. To ensure objectivity, the tests must be performed by *independent* Software Test Engineers. They may be either personnel not involved in any of the development activities up to this point or other Software Developers on your team who have not been involved with the coding and integration activities for the SI being tested.

Table 11.19 Software Item Qualification Testing Work Products per Build

SIQT Documentation	SCS-1	SS-1	SS-2	663575-698500SS-3	C/R
STP and STD (separate document for each SI)	Required	Required	Required (*)	Optional	Optional
SI test description and test results entered in the SDF	Required	Required	Required	Required	Required
STR	Required	Required	Required (*)	Optional	Optional
SI test cases traced to requirements in the requirements databases	Required	Required	Required (*)	Optional	Optional

STP = Software Test Plan; STD = Software Test Description; STR = Software Test Report.
(*) Document not required but applicable information is developed and retained in the SDF.

Table 11.20 SIQT Responsibilities

Software Development	Implements software changes as a result of SCRs/SDRs.
	Addresses critical software requirements in the software test cases.
	Generates traceability products for SI test cases to the Program Unique Identifier (PUI) for each SRS requirement.
	Collects and reports software metrics; Updates and maintains the SDF.
CSwE	Monitors SI tests, reviews SDFs, reviews software products, attends reviews (or designee), and monitors and analyzes software metrics. The CSwE must concur that the SI is ready for qualification testing.
Software Test	Executes the SI test and submits SCRs/SDRs as necessary for detected problems.
System Safety	Supports verification, testing and tracking of Safety Critical requirements.
SQA	Witnesses program and/or Element SQA and audits the tests, attends reviews and inspections, evaluates adherence to documented policies and procedures, monitors the quality of output products, and documents and reports findings to management.
Element CCB	Addresses all current build subsystem internal SCR/SDRs, as they are generated by the SwIPT software personnel or Software Test Engineers.
Element SwCCB	Addresses all SCR/SDRs involving external interface changes and prior build release SU changes as they are generated by the SwIPT software personnel and approves all requests to postpone qualification test cases to later builds.
Element SCM	Manages test database, the source code, SCR/SDRs for SI baseline, changes to source code, and the test documentation generated.

11.7.6 SIQT of the Target Computer System

To the maximum extent possible, SIQT should be performed on the target computer system, or as close as possible to the operational target hardware configuration, to demonstrate that there are no hardware or operating system incompatibilities. Operation on the target computer can also provide useful measurements of the computer resource utilization. If the SIQT is to be performed on a compatible system, an analysis must be conducted to determine that the computer resource utilization requirements can be met. All target hardware/computer system(s) used for testing must follow your project's Configuration Management Policies and Procedures.

11.7.7 Preparing for SI Qualification Testing

The objective of preparing for SIQT is to finalize, through reviews or inspections, the STD and the STP. The Software Developers must define and record the test preparations, test cases, and test procedures to be used for SIQT as well as the traceability between the test cases, test procedure steps, and the SI and software interface requirements.

In addition to writing the STD, the test engineers must run all or selected portions of the STD, update test scenario procedures and databases, and perform other necessary test activities to prepare for the SIQT dry run. The STP and STD must be baselined and placed under Configuration Management control prior to the run for record.

Reference to verification *testing* during SIQT should not be confused with the verification *methods* of testing. Software qualification testing may require the use of all *verification methods* including *inspection, analysis, demonstration and test*. The STD, supported by the STP, must provide test case descriptions with test procedures for each test case, special test environments, and test sequencing requirements for all SRS and IRS requirements allocated to the SI build.

Approach to Preparing for SIQT. All requirements in preparation for SIQT must be addressed including verification of the following (as applicable) software requirement issues:

- Requirements under conditions as close as possible to those that the software will encounter in the operational environment
- Interface requirements using the actual interfaces wherever possible or high-fidelity simulations
- Specialty engineering requirements such as supportability, testability, reliability, maintainability, availability, safety, security, and human systems integration, as applicable

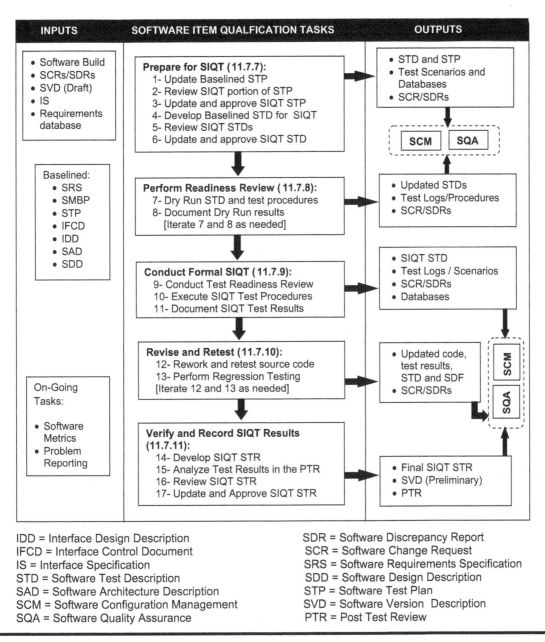

INPUTS	SOFTWARE ITEM QUALIFICATION TASKS	OUTPUTS

INPUTS
- Software Build
- SCRs/SDRs
- SVD (Draft)
- IS
- Requirements database

Baselined:
- SRS
- SMBP
- STP
- IFCD
- IDD
- SAD
- SDD

On-Going Tasks:
- Software Metrics
- Problem Reporting

SOFTWARE ITEM QUALIFICATION TASKS

Prepare for SIQT (11.7.7):
1- Update Baselined STP
2- Review SIQT portion of STP
3- Update and approve SIQT STP
4- Develop Baselined STD for SIQT
5- Review SIQT STDs
6- Update and approve SIQT STD

Perform Readiness Review (11.7.8):
7- Dry Run STD and test procedures
8- Document Dry Run results
[Iterate 7 and 8 as needed]

Conduct Formal SIQT (11.7.9):
9- Conduct Test Readiness Review
10- Execute SIQT Test Procedures
11- Document SIQT Test Results

Revise and Retest (11.7.10):
12- Rework and retest source code
13- Perform Regression Testing
[Iterate 12 and 13 as needed]

Verify and Record SIQT Results (11.7.11):
14- Develop SIQT STR
15- Analyze Test Results in the PTR
16- Review SIQT STR
17- Update and Approve SIQT STR

OUTPUTS
- STD and STP
- Test Scenarios and Databases
- SCR/SDRs

SCM   SQA

- Updated STDs
- Test Logs/Procedures
- SCR/SDRs

- SIQT STD
- Test Logs / Scenarios
- SCR/SDRs
- Databases

- Updated code, test results, STD and SDF
- SCR/SDRs

SCM
SQA

- Final SIQT STR
- SVD (Preliminary)
- PTR

IDD = Interface Design Description
IFCD = Interface Control Document
IS = Interface Specification
STD = Software Test Description
SAD = Software Architecture Description
SCM = Software Configuration Management
SQA = Software Quality Assurance

SDR = Software Discrepancy Report
SCR = Software Change Request
SRS = Software Requirements Specification
SDD = Software Design Description
STP = Software Test Plan
SVD = Software Version Description
PTR = Post Test Review

Figure 11.8   SIQT process flow.

- Reliability requirements including fault detection, isolation, and recovery
- Requirements for stress testing to be performed including worst case scenarios

Commercial-Off-The-Shelf software, reuse code, or newly developed software can be used to satisfy and verify software requirements during qualification testing. During SIQT, applicable resource measurements must be collected typically including CPU usage, memory, storage, and bandwidth data.

For testing some SRS or IRS requirements, it may not be possible, or practical, to fully test the requirement at the SI level for the current SI build. As a result, it may be necessary to satisfy the requirement using unit integration or SU tests rather than an SI test. This situation can result when data associated with an SI requirement is not accessible at the SI level, or the test requires an inordinate amount of time or costs to perform. If system hardware or special test environments are not ready for the current build test, the tests could be deferred to subsystem or system testing.

The STP and STD should be evaluated at a Document Peer Review. Once the review changes are incorporated into the documents, they must be baselined and further modifications handled via change control. Supporting test software (simulations/emulations) and data should also be prepared. The updated STP and STD must also contain test cases and descriptions for

Safety Critical requirements from previous builds. For the final build, the documents must provide test cases and descriptions for the final build plus regression test cases and descriptions of software requirements from all previous builds.

Table 11.21 is an example description of the six tasks applicable to the SIQT preparation:

11.7.8 Readiness Review for SIQT

The objectives of the SIQT readiness review are to exercise the test cases and test procedures in a "dry run" to ensure that they are complete and accurate and that the SI is ready for witnessed testing. SIQT readiness testing also verifies that all necessary test data and the test environment are under proper SCM control and are adequate for verifying the software requirements. Table 11.22 contains an example of the tasks applicable to the SIQT dry run.

Approach to Dry Run of SIQT. Testers must obtain the appropriate software SI build from the SDL. When all required elements are assembled, the tests, including required Regression Tests, should be accomplished following the procedures specified in the STD. The Testers must collect, analyze

Table 11.21 SIQT Preparation Tasks

SIQT Tasks	SIQT Inputs	SIQT Subtasks	SIQT Outputs
1. Update Baselined STP for SIQT	■ See Figure 11.8 Inputs ■ System and Element/Subsystem Use Cases	■ Define test environment and test schedule ■ Develop test categories ■ Identify processes for conducting the SIQT ■ Identify assumptions and constraints ■ Document items in the STP	STP
2. Review STP	STP	■ Announce the Peer Review and disseminate schedules and STP in advance ■ Conduct the Peer Review and document Peer Review results ■ Schedule next revision and review dates	■ Comments against the STP ■ Action Items
3. Update and Approve STP	■ Comments against the STP ■ STP ■ Action Items from the previous review ■ Updates to the STP	■ Update STP based on review comments ■ Provide STP for re-review ■ If re-review is required, conduct, document and schedule next revision and review ■ Get the proper approvals ■ Provide to SCM and to Document Control for distribution	Approved STP
4. Develop Baselined STD for SIQT	■ See Figure 11.7 Inputs ■ Integration test cases ■ SCRs/SDRs	■ Map SRS ad IRS requirements to test cases ■ Populate RTVM with requirement to test case mapping data ■ Identify requirements to be verified at the SU or SI level ■ Develop automated test scenarios and databases ■ Develop test procedures including post-test analysis steps ■ Update scenarios and test databases ■ Document anomalies in an SCR	■ STD ■ Test scenarios and databases ■ Updated Requirements Database ■ SCRs/SDRs
5. Review STDs	■ STDs ■ STP ■ SCRs/SDRs	■ Announce review and disseminate schedules and SIQT STDs ■ Conduct review and document review results ■ Schedule next revision and review dates	■ Comments against the STD ■ Action Items
6. Update and Approve STD	■ Comments against the STD ■ SCRs/SDRs ■ Completed Action Items from the previous review	■ Update STD based on review comments ■ Provide STD for re-review ■ If re-review required, conduct and document it ■ Schedule next revision and review if needed ■ If no more review time is required, get approvals ■ Provide STD to SCM and Document Control for distribution	Approved STD

Table 11.22 SIQT Readiness Review Tasks

SIQT Tasks	SIQT Inputs	SIQT Subtasks	SIQT Outputs
7. Dry Run STD and Test Procedures	■ Approved STD ■ SCM controlled scenarios and databases ■ SCM controlled software build ■ CM controlled HW test bed and test environment	■ Ensure all test software and hardware are available and are the correct version ■ Ensure that all approved test software and the software to be tested are under SCM control ■ Execute test and post-test analysis as documented in the STD ■ Redline procedures and obtain SQA approval ■ Update scenarios and databases	Approved STD with SQA approved redlines
8. Document Dry Run Test Results	■ STP ■ Approved STD with approved redlines	■ Document Dry Run results in the test log ■ Document all anomalies in an SDRs	■ SCRs/SDRs ■ Test logs

and record the outputs, logs, test notes and results (problems, errors, and discrepancies noted by the Tester). No modification to the SI, test data, or environment should be made until after the dry run is complete and the results documented in the SDF. SQA may audit the dry run and all test results.

After the test procedure is executed, and SQA captures the redlines for anomalies, data from the test execution must be analyzed to determine if the software under test produced the correct results and whether the SRS and IRS requirements allocated to the test procedure were actually verified. Results of the post-test analysis may be software changes (documented in SDRs), procedure changes, test data/scenario changes, or test environment changes.

In cases where requirements are not satisfied, SCRs/SDRs must be generated and the changes handled by the appropriate Corrective Action Process. In cases where test descriptions require modification, the procedures must be redlined and approved by SQA. Retesting must be required for all modified SUs, test cases, and test descriptions, and any additional SUs and test cases that directly interact with the modified SIs and test cases.

11.7.9 Conducting SIQT

The objective of conducting the SIQT is to *formally execute the test procedures* as documented in the STD, using products under SCM control, and in a witnessed test environment. This task should begin with the *Test Readiness Review* (TRR) that should be described in an SDP appendix covering software reviews. The material presented at the TRR should include: SU testing, SU test results and SI dry run results; formal test environment description (hardware, test tools, and associated software); formal test approach; SI requirements verification at a lower level; test schedules; and SIQT tasks as described in the SI STP and the SI STD.

Approach to Conducting SIQT. The TRR ensures that all necessary test documentation, materials and personnel are ready, and that coordinated test schedules are in place. The actual execution of the SIQT should be essentially the same as the dry run. The Testers usually begin by ensuring that all necessary software test data and tools are available. Testers obtain the appropriate SI test build from the SDL. When all required elements are assembled, the tests, including required regression tests, must proceed in accordance with the procedures specified in the STD.

The Test Team must collect and record the outputs, logs, test notes and results. They must execute all tests specified for the current build, and they must perform Regression Testing on all Safety Critical requirements from previous builds. The primary difference from the dry run testing is that the performance of the SIQT is normally witnessed by SQA and the CSwE (or designee) and optionally by the customer and the Independent Verification and Validation (IV&V) agent.

Reasonable notice of the tests must be provided to the customer and the IV&V agent to permit them the opportunity to attend. Table 11.23 is an example of the three tasks applicable to conducting a formal SIQT.

11.7.10 SIQT Revision and Retesting

The objective of the revision activity during SIQT is to rework source code or test descriptions to eliminate any problems identified during qualification testing. Appropriate portions of the SI must be retested to verify that the changes have been successful and that other problems have not been produced as side effects. The objective of retesting is to verify that applicable SCR and SDR fixes are properly implemented and that selected existing functionality is still performing per software and interface requirements after the SCR and SDR fixes have been implemented.

Approach to SIQT Revision and Retesting. The test results, and documented problems, must be evaluated by Software Developers to determine if changes need to be made to the SUs or the STD. Regression testing of affected SIQT test cases must be performed after any modification to previously tested software. In addition to modifications made to SUs to fix defects, Regression Testing can also include regression tests of SIQT test procedures from the last build to show that the current build has not broken any software requirements that were previously verified.

All modifications to the source code must be handled as SCRs/SDRs by the appropriate Change Control Board. Unit-level retesting must be required for all modified procedures and functions. Modified SIs requires retesting of Safety Critical requirements and previously failed test cases. Products from previous activities must be reviewed for possible changes resulting from the implemented software changes, and then updated as appropriate.

This activity must be repeated as needed until all test cases have met the test case success criteria. In some cases, uncompleted or failed tests can be postponed until a later build if approved by the SwCCB. Retesting objectives must be reviewed by the appropriate Systems Engineering Integration and Test (SEIT) team, and SwIPT prior to the testing and preparation for testing. An updated set of the STP or STD should not be mandatory for each iteration of retesting. STRs must be provided at the end of SIQT testing. Table 11.24 is an example of the two tasks applicable to revision and retesting of SIQT results.

Table 11.23   Perform Formal SIQT Tasks

SIQT Tasks	SIQT Inputs	SIQT Subtasks	SIQT Outputs
9. Conduct SIQT Test Readiness Review	■ TRR entrance and exit criteria ■ Test environment ■ Approved STD ■ SDFs ■ SRS and IRS ■ SCM controlled software build, test software & hardware ■ Open SCRs/SDRs ■ Test logs	■ Review SU test and integration test status. ■ Review SIQT dry run status and open SCR/SDR status. ■ Review test environment status and STD status. ■ Review test limitations and test schedule. ■ Prepare TRR test log. ■ Ensure all test software and hardware are available and are the correct version. ■ Ensure that all test hardware and software are under SCM control. ■ Assess SIQT test readiness based on the above.	■ Pass/fail status of TRR exit criteria ■ Test Logs
10. Execute SIQT Test Procedures	■ Approved SIQT STDs ■ Test environment ■ TRR results (test logs) ■ SCM controlled build ■ Open SCRs/SDRs	■ Perform test steps as documented in the STD. ■ Perform analysis steps as documented in the STD. ■ Perform retesting as required.	■ Completed SIQT testing ■ Test Logs
11. Document SIQT Test Results	Completed SIQT testing	■ Prepare the test log. ■ Document anomalies in SDRs and rework source code or STDs to eliminate problems. ■ Document verification status for SRS requirements and obtain SEIT approval. ■ SCM updates to baseline documents.	■ SCRs and SDRs ■ Test logs ■ Revisions to code or test procedures ■ SQA audits and reports

Table 11.24   SIQT Revise and Retest Tasks

SIQT Tasks	SIQT Inputs	SIQT Subtasks	SIQT Outputs
12. Rework and Retest Source Code	Outputs of Task 11	■ Revise source code and STD as needed. ■ Retest revised source code. ■ Update the SDF.	Updated code, test results, STD and SDF
13. Perform Regression Testing	■ SI builds ■ Test Results ■ SCRs and SDRs	■ Perform Regression Testing to accommodate changes in the current build. ■ Document anomalies.	■ SCRs / SDRs ■ Test Results ■ SQA Audits

11.7.11 Verify and Record SIQT Results

The objective of verifying and recording SIQT results is to finalize the SIQT activity by:

- Documenting test results in the *Software Test Report* (for SCS-1 and SS-1 software) or in the SDF (for SS-2 software)
- Performing a review of the STR, or verifying the capture of the test results in the SDF
- Conducting an optional *Build Turnover Review* (BTR) also called a Test Exit Review (TER)

Approach to Verifying and Recording SIQT Results. The results of the SIQT must be analyzed for completeness and documented in an STR or captured in the SDF. This documentation should be prepared by Software Test Engineers who performed the SI tests. The completed documentation is subject to a document review.

Software Test Reports. STRs must be baselined once all review modifications have been incorporated. Any SI documentation, notes or data that are not incorporated into the STR should be captured in the SDF. For SCS-1 and SS-1 software developed in multiple builds, an STR must be prepared and reviewed after each build. The STR, when approved, must be maintained under CM control. The intent of recording SIQT test results, in the SIQT STRs, is to document and finalize the test activity and effectively capture test results. Table 11.25 is an example of the four tasks applicable to analyzing and recording SIQT results.

11.7.12 Integration of Multiple SIs

Most systems, and all large systems, have more than one Software Item. Therefore, after completing the process for conducting SIQT on the first SI, the SIQT process is essentially repeated for merging in each additional SI until the full system is assembled. In Figures 2.5 and 3.6, this step is referred to as *Software Items Integration and Testing* (SI/SI I&T). The only open question is *when* should the hardware be integrated and tested with the software?

There are also multiple subsystems that must be integrated prior to any system testing as discussed in Section 10.4. The subsystem integration and testing activity is normally followed by a Subsystem Acceptance Test (SAT) as discussed in Subsection 12.6.4 and Section 10.4.

Lessons Learned. I recommend that *all of the SIs are integrated before* the software integrations with the hardware and subsystems take place. With this approach, you can identify and resolve conflicts when more than one SI interfaces with the same hardware at the same time. However, as the Software Project Manager, you must decide, or come to a team consensus, as to what is the best approach for your project. If each SI has a high level of interaction with hardware, system testing may be more efficient if the full SI/HI integration takes place on each SI before merging in the other SIs.

Table 11.25 Verify and Record SIQT Results

SIQT Task	SIQT Inputs	SIQT Subtasks	SIQT Outputs
14. Develop STR	■ SIQT test logs ■ SCRs/SDRs ■ "As run" STD	■ Review SIQT test logs, as run STD and SCR/SDRs. ■ Review verification status of SRS requirements. ■ Document all of the above in STR.	■ STR ■ SCRs/SDRs
15. Analyze Test Results	■ STR ■ SCRs/SDRs	■ Perform root cause analysis of test anomalies as documented in the SCRs/SDRs. ■ Obtain CCB approval of SCR/SDR resolution plan.	■ Root cause analysis ■ Resolution Plan
16. Review STR	■ SIQT STR ■ SCRs/SDRs ■ SIQT test logs ■ "As run" STD	■ Announce review and disseminate schedules and STR. ■ Conduct a BTR (or TER) and document the review results.	■ Comments against the SIQT STRs ■ Action Items
17. Update And Approve STR	■ Comments against the preliminary SIQT STRs and STR ■ Action Items from previous reviews ■ STR updates	■ Update STR based on review comments. ■ If re-review required, conduct and document review. ■ Schedule next revision and review if needed. ■ If no more review time is required, get approvals. ■ Provide STDs to SCM and Document Control.	■ Approved SIQT STR ■ SQA audits and reports ■ Preliminary SVD

11.8 Software/Hardware/Subsystems Integration and Testing

During hardware integration with software, the Software Team is usually in a *support role* to the SEIT group. As a result, the Software Item/Hardware Item Integration and Testing process is considered the responsibility of the SEIT group and is covered in the Systems Engineering Domain, Section 10.3. The following SI/HI I&T discussion is presented as part of the Software Engineering Doman for continuity and is focused on the *technical support provided by the SwIPT personnel to the SEIT group*.

11.8.1 Objectives of Software/Hardware/ Subsystem Integration and Testing

As described in Section 10.3, SI/HI I&T involves integrating Software Items with other interfacing SIs and interfacing Hardware Items, testing the resulting groupings to determine if they work together as intended, and continuing this process until all interfacing SIs, HIs and subsystems are integrated and tested. The principal objectives are to:

- Perform the individual software/hardware/subsystem integrations to produce the complete software build for each successive level of test and verify its integration success.
- Integrate software into the target hardware system and verify integration success.
- Verify SI-to-SI, SI-to-HI and subsystem-to-subsystem interface requirements compliance.
- Support and successfully complete integration and qualification testing at each level.

The objective of performing SI/HI I&T is to integrate and test software in accordance with integration and test strategies in the approved Master Test Plan (MTP) and SMBP. Integration testing may also be called *Factory Acceptance Test* or *Element Qualification Test*. The five principal tasks are:

- Support the Dry Run
- Conduct SI-to-SI Integration Testing
- Support SI-to-HI Integration Testing
- Conduct Integrated SI Thread Testing
- Support Integration of Subsystems

11.8.2 Approach to Software/Hardware/ Subsystem Integration and Testing

The Software Master Build Plan, discussed in Subsection 5.1.4 should be updated to define the specific SI functionality that is planned to be operational for each build. The integration sequencing should be documented in the SMBP and the overall approach to I&T documented in an *MTP*—sometimes called the *System Test and Evaluation Plan* (STEP).

Integration of software with the hardware is a critical objective of this activity; however, some aspects of the hardware integration may not be able to be performed until the full System Integration. In addition, there may be early integration points using two testbeds so that all the software and interfaces do not have to wait until implementation is completely finished to be integrated and tested.

Although the software role in this activity consists primarily of support tasks, *the Software Test Team has a vital role, especially for the SI-to-SI integration tasks*. The SI-to-HI integration tasks should involve both hardware and Software System Engineers/Testers. All testing must be run using documented test descriptions developed collaboratively by the software and SEIT Test Engineers.

The Software Developers, and Software System Engineers, assigned to perform integration of the SIs, must develop an *integration strategy* that defines a systematic *approach for integrating the SIs into the complete software release*. Issues such as SI-SI interfaces, inter-SI timing and sequencing, and simulations or emulations (for external interfaces) are examples of the issues that go into determining the order of integration of the SIs.

Hardware and Software System Engineers must collaborate in the preparation of appropriate test description information for the hardware/software integration that needs to be accomplished during this activity. They review in-progress SI/HI I&T *Software Test Descriptions*, provide recommendations to test engineers (including software test equipment needed), and ensure that test cases and corresponding test procedures are sufficiently defined in the updated STD to verify the success of each partial integration.

The SI/HI I&T must be performed using the target hardware in a configuration that is as close as possible to the operational configuration. All reuse software, including legacy reuse and COTS software, must also undergo the SI/HI I&T process. The SwIPTs provide Software System Engineers and Test Engineers the expertise and training, as required, in using the software along with applicable software test support items.

11.8.3 Software/Hardware/Subsystem Integration and Testing Process

Software integrators normally begin the integration of SIs by ensuring that all SIs to be integrated and all necessary data and tools are available and ready. Software Test Engineers support Configuration Control Board (CCB and SwCBB) Corrective Actions on any soft-related errors, request re-execution of build procedures as required, and accept SI builds upon satisfactory verification. When all required elements are assembled, integration proceeds.

The Software Test Engineers must run test cases using the test procedures, as specified in the STD. They collect or record the outputs, logs, test notes, and results. All problems, errors, and discrepancies must be noted. Similarly, subsystem test engineers run the hardware/software integration tests as defined in the test descriptions, collect and record test results and problems.

The purpose of analyzing and recording SI/HI integration and test results are to: (a) analyze integration tests results to ensure the tests have been successfully completed; and (b) document the respective test data and results as required. The purpose of retesting is to verify that changes and applicable SDR/SCR modifications have been implemented correctly and that the functionality is performing in accordance with requirements after the fixes have been completed.

Changes can involve most anything including test procedures, test data, and code changes. Re-integration and retesting must then be performed to verify that the changes have been successful and have not caused side effects. The documented problems must be evaluated by Software Developers to determine the necessary changes to SIs, SUs or to the test descriptions.

Retesting is performed to show that a problem is fixed and the test case executes properly. *Regression Testing* is performed to *show that the fix did not break anything* that was previously tested and working properly before the fix. In cases where software requires changes, SDRs or SCRs are generated and the changes are handled by the Corrective Action Process (CAP). In cases where test descriptions require modification, the changes, identified in the SDRs or SCRs, must be made by Software Test Engineers and a version history included in the test description to record the changes made.

Modified software requires retesting for the integration tests that previously failed and for any tests that are dependent on the failed tests. Similarly, test description changes require retesting of the changed tests plus tests that are dependent on the results of the changes. This process must be repeated until all SIs and HIs have been successfully integrated and all tests have been completed.

If the SwIPT Lead determines it is impractical to complete certain changes until a later build, then SCRs/SDRs must be used to document and control the modifications and integration testing that still needs to be performed. The CCB must approve all such delays. After all SIs and HIs have been successfully integrated and tested, the Integration and Test Team *must review the test results for consistency and completeness* and verify that the integration test data and results have been documented. If discrepancies or problems are found, then the portion of the integration in question must be retested. Once the independent reviewer signifies that the integration and testing is complete, and the integration testing was successfully completed, the release is baselined.

At the last stage of integration and testing, the test results are normally documented by SCM in a *Build Integration Release Notice* and the build is then ready for system testing. Systems Engineering is responsible for System Qualification Testing (SQT) and it is covered in Section 10.4.

11.9 Software Transition to Operations

This activity is concerned with the preparation, installation and testing of the executable software, on the *target system*, at a customer or user site. Upon successful completion of the System Qualification Testing for the final build, all software-related SDRs/SCRs (Software Discrepancy Reports/ Software Change Requests as discussed in Subsection 6.4.1) allocated to software that can be resolved have been dispositioned and *all* SDRs/SCRs *that can be closed are closed*. The initial software development cycle is then completed, and the software is ready for transfer to the customer for system testing. It may also be necessary to provide interim releases to development sites if needed to facilitate their development and testing process.

Prior to actually releasing the software for use, there remains software and documentation preparation work that must be completed. This Section addresses the tasks necessary to prepare the software and software-related *products necessary for a user to run the software*. Software Transition to the Operations activity is described in the following four Subsections:

■ Preparing the Executable Software (11.9.1)
■ Preparing Version Description Documents for User Sites (11.9.2)
■ Preparing Software User Manuals (11.9.3)
■ Installation at User Sites (11.9.4)

Approach to Transition to Operations. Although ensuring a smooth transition to operations takes place at the end of the Development Life Cycle, consideration of these tasks should occur much earlier and preferably concurrently with design, development, and testing throughout the life cycle. During each design period, new or updated user and operations manuals can be prepared for review by the customer and users. Draft versions of the *Software Transition Plan* should be started during the Software Design activity.

For final deliveries, the tasks and products of this activity must be in compliance with the *Master Test Plan*. This planning must be coordinated with the hardware installation schedules. Schedules are established and resources and personnel required for installation and support are identified. This activity also involves the planning, preparation, and presentation of required training.

Software installation and checkout tasks are performed by software test personnel at the user site. When SQT has been completed, SCM prepares the software product(s) for use in accordance with the CM Plan. For example, non-flight software products arc stored on media formatted as required for installation at the operational site. For onboard flight software, the preparation of the executable software usually includes downloading it into the actual flight hardware. Similarly, software for user equipment is usually downloaded into the target processors.

11.9.1 Preparing the Executable Software

Preparing software for use ensures that there is a smooth transition of software into the actual operational system. This activity includes preparation of the specific executable code and source files for each SI, batch files, COTS, command, data, or other software files needed to install and operate the software on the target computer(s). This data is documented in a *Software Product Specification* as described in Subsection 11.10.4. The SPS should clearly specify the data to be prepared.

11.9.2 Preparing Version Description Documents for User Sites

Each software release requires a *Software Version Description* document. Format and contents of the SVD should be described in the *Data Item Description* (DID) or a similar source. The completed SVD requires a document review prior to release. The SVD is primarily composed of lists that should include, as applicable:

- Complete identification of all material released including numbers, dates, abbreviations, version and release numbers, physical media and documentation
- Inventory of all computer files that make up the software being released
- History of all changes incorporated since the previous version
- Site-unique data contained in the new version
- Related documentation not included in the current release
- Installation instructions
- Possible problems and known errors

11.9.3 Preparing Software User Manuals

Software customer user manuals are required to be prepared for software in user equipment with a *human interface*. However, not all of the user manuals need to be produced by all programs because the full set of user manuals normally has some duplication. Onboard flight software does not require user manuals nor does equipment with embedded software (Firmware). The customer and the Developer must determine which user manuals are appropriate for each system. User manuals or user guides should be produced for SS-1 and SS-2 software. Existing vendor documentation can be used for the COTS/Reuse software.

There are various types of user manuals as described below. For each of the required user manual types, a separate document should be written for each subsystem; however, subsystems can optionally write multiple user manuals covering one or more SIs, rather than a single user manual for the entire subsystem. Multiple user manuals are recommended when different users run selected SIs within the subsystem. All of the documents below require a document review prior to release.

Software User Manual. SUMs must be written to provide information needed at the customer site if required by the customer. The SUM describes, in depth, *how to use* the software and includes, as applicable:

- An inventory of software required to be installed for the software to operate
- Resources needed for a user to install and run the software
- Software overview including logical components, performance characteristics, etc.
- Procedures to access the software for first-time or occasional users
- Detailed procedures for using the software including organization, capabilities, conventions used, backup, recovery and messages

Computer Operation Manuals (COM). COMs are written to provide information needed by the customer site to operate the target computers. A COM is typically needed only if the hardware is unique or new. The COM describes the computer system operations data including, as applicable:

- Computer system preparation, power on/off, initiation and shutdown.
- Operating procedures including input/output, monitoring, and off-line procedures.
- Diagnostic features, procedures and tools.

Other User/Operator Software Product Descriptions. There are two optional user/operator manuals: the *Software Input/Output Manual* (SIOM); and the *Software Center Operations Manual* (SCOM). The SIOM and SCOM are used for software systems installed in a computer center or other centralized or networked software installation. There is some overlap in these documents, as with other user manuals, so the appropriate set must be determined for each system. The SIOM describes how to prepare input to, and interpret output from, the software including, as applicable:

- An inventory of software files and databases needed to access the software
- Resources needed to access the software
- Organization and operation of the software from a users' point of view
- Contingencies, security and problem reporting procedures
- Input conditions, formats, rules, vocabulary and examples of each type of input
- Output descriptions, formats, vocabulary, use, examples and error diagnostics
- Query procedures including file formats, capabilities and instructions
- Terminal processing procedures covering capabilities, displays, updates, retrieval, error correction, and termination

The SCOM describes required installation procedures including:

- An inventory of software required to be installed for the software to operate
- Resources needed for a user to install and operate the software
- Software overview including logical components, performance characteristics, etc.
- Detailed description of runs to be performed including run inventory, phasing, diagnostic procedures, error messages, control inputs, input and output files and reports, and procedures for restart and recovery

11.9.4 Installation at User Sites

Preparation for installation of the system at customer sites should be handled by the development organization. The Developers should be responsible for the system setup and checkout, development of user training, provision of user training and initial user assistance.

Software Installation Plan (SIP). The SIP is the plan for installing software at user sites. It includes preparations, user training and conversion from existing systems. It is prepared only when the Developer is involved in the installation of software at user sites, and when the installation process is sufficiently complex to warrant the need for a SIP. If the software is embedded in a hardware-software system, the installation plan for the system usually includes a *System Installation Plan* covering both hardware and software.

The installation of the system at user sites for government requirements verification testing and the final delivery after requirements verification is handled as specified in the contract. Software Developers and SCM support this process by providing technical support and implementing changes that result from the testing prior to final delivery.

Software Product Specification. The SPS is a compilation of the "as built" version of the software, available at the user site, and includes or references as applicable:

- Executable software and source files for the SI
- The "as built" versions of the SDD, IDD and DBDD
- Compilation, build and modification procedures and Packaging Requirements
- Measured utilization of computer hardware resources
- Requirements traceability data

11.10 Software Transition to Sustainment

Chapter 16 is devoted to management of the sustainment activity (also called maintenance) over a long period of time. This section addresses the software and documentation preparation work that must be completed to *transition* the application and support software required for the performance of System Sustainment and is described in the following nine Subsections:

- Updating the Executable Software (11.10.1)
- Preparing Source Files (11.10.2)
- Preparing the Version Description Documents for Sustainment Sites (11.10.3)
- Preparing the "As Built" Software Item Design and Related Information (11.10.4)
- Updating the System/Subsystem Design Description (11.10.5)
- Updating the software requirements (11.10.6)
- Updating the System Requirements (11.10.7)
- Preparing Software Maintenance Manuals (11.10.8)
- Transition to Sustainment Sites (11.10.9)

If the software is developed in multiple builds, planning should identify what software builds are to be transitioned to the maintenance organization and when. Preparing for software transition is interpreted to include those activities necessary to carry out the transition plans for each build.

Software transition involves considerable advance planning and preparation that must start early in the life cycle to ensure a smooth transition to the sustainment organization. The tasks and products of this activity must be in compliance with the *Software Transition Plan* that defines the plans for transitioning the software, test beds and tools to the maintenance center's facilities (see Subsection 5.1.8).

Approach to Transition to Sustainment. The Preparation for Software Transition to Sustainment includes preparation of the documentation and software products required by maintenance personnel at the sustainment center to perform their maintenance tasks. This includes preparation of:

- Source code and executable code for each Software Item
- Release build files and documentation
- *Software Version Description* for the executable code
- Applicable test beds and tools
- Contents of the Master Software Development Library (MSDL)

The actual Software Sustainment activity may be performed by the software development organization or by another organization such as a different organization within the company that developed the software, another development contractor or by a government maintenance organization.

The preparation of maintenance manuals, such as the *Computer Programming Manual* and *Firmware Support Manual*, should begin early in the Software Design activity and continue into subsequent activities as pertinent information becomes available. The final updated versions of the system and software requirements and design descriptions are also involved. The preparation may also include the planning, preparation, and presentation of maintenance training as required by the contract.

11.10.1 Updating the Executable Software

The executable software was prepared for its transition to operations; however, if some anomalies are detected during operations, it may need to be updated before the transition to sustainment. The executable software is prepared using the SVD for each SI or related collection of SIs and includes the executable code, batch, command, data, test, support or other software files needed to install and operate the application and support software on the target computers. The results must include all applicable items in the *Software Product Specification*.

11.10.2 Preparing Deliverable Source Files

The final versions of all source code files should be assembled from the MSDL and transferred to the desired transfer media per the agreement with the selected maintenance center in accordance with the *Software Configuration Management Plan* (see Section 6.3). The results must include all applicable items in the source file Section of the SPS.

11.10.3 Preparing Version Description Documents for Sustainment Sites

A *Software Version Description* must be prepared for each software delivery to the maintenance site. The SVD provides an inventory of the software contents for the final build. It also provides a history of version changes, unique-to-site data, related documentation, installation instructions, and possible problems for each SI within the subsystem. The maintenance center's version requires the additional version information related to the maintenance center's tools, SDFs, test software and documentation.

11.10.4 Preparing the "As Built" SI Design and Related Information

Software Item Designs must be documented in SDDs and IDDs for SCS-1 and SS-1 software. In addition, software development work products for Systems Critical (SCS-1) and Support Software (SS-1) are documented in SDFs. These documents and products must be updated with each build to maintain them as the "as built" software throughout the development process. The *Software Product Specification*, outlined in Subsection 11.10.4, provides additional information needed by the maintenance center to maintain the application and support software. In addition, the SDP must define and record:

- The methods to be used to verify copies of the software
- The measured computer hardware resource utilization for the SIs
- Other information as needed to maintain the software
- Traceability between the Software Item's source files and Software Units
- Traceability between the computer hardware resource utilization measurements and the SI requirements concerning them

The SDP must indicate that the results of this task are placed in the qualification, software maintenance, and traceability sections of the SPS. A document review prior to release is required.

11.10.5 Updating the System/Subsystem Design Description

The System Design must be documented in the *System/Subsystem Design Description* (SSDD). Once baselined, the SSDD must be maintained under configuration control throughout the development process. Software development work products evolve from the system requirements definition activity from the software requirements and design activities to the "as built" system. If throughout the development process, these products are maintained as the "as built" system, there is no special updating required at the end of the development process—except for modifications that result from finalizing SCRs/SDRs after the last system build qualification test is complete.

11.10.6 Updating the Software Requirements Documents

The baselined SRSs and IRSs must be maintained under configuration control throughout the Software Life Cycle process and should require no special updating in preparation for software transitioning. If this is not the case, the SRS and IRS must be updated to contain the current set of approved requirements that the software transition to maintenance is to meet.

11.10.7 Updating System Requirements Documentation

The system specifications should be controlled throughout the System Life Cycle Process and should require no specific updating in preparation for software transitioning. If this is not the case, the *System/Subsystem Specifications* (SSS), and system-level IRSs, must be updated. Systems Engineering is responsible for updating system requirements with the support of the SwIPTs.

11.10.8 Preparing Software Maintenance Manuals

Preparation of a *Software Maintenance Plan* (SMP), early in the development process, is highly recommended. A comprehensive SMP is a major asset in assuring the timely availability of adequate facilities, support software, personnel, and documentation so that the software can be maintained in an operational and sustainable condition.

Also, if contractually required and applicable, the subsystems should prepare *Computer Programming Manuals* and *Firmware Support Manuals* for the maintenance center. For each of the required maintenance manuals, a separate document should be written for each subsystem. In addition, a *Software Center Operators Manual*, produced as described in Subsection 11.9.3, may be useful to the maintenance center. Subsystems can optionally write multiple documents for each maintenance manual type covering individual computers or firmware devices, rather than single maintenance manuals for the entire subsystem. This approach is recommended when computers and firmware devices are maintained in separate locations.

Computer Programming Manual. CPMs provide information needed by the maintenance center to program and reprogram the computers, and peripheral devices, on which the application and support software run. The CPM requires a document review prior to release to the MSDL. *The CPM is primarily intended for unique or newly developed*

computers. The CPM basically describes the programming environment and includes, as applicable:

- The components and configuration of the computer system
- Equipment needed including operating characteristics, capabilities and limitations
- Programming features, input and output control programming instructions, subroutines, interrupt procedures, timing, memory protection, etc.
- Description of instructions including use, syntax, execution time, etc.
- Special programming techniques including error detection and diagnostic features

Firmware Support Manual. FSMs provide the information needed by the customer's sustainment (maintenance) center to program and reprogram firmware devices with application or support software. The FSM requires a document review prior to release to the MSDL. The FSM describes the details of a programmed firmware device and includes, as applicable:

- Relevant vendor information plus model number and a complete physical description
- Operational and environmental limitations
- Software to be programmed into the device and equipment and software needed
- Procedures for programming, reprogramming, installation and repair

11.10.9 Transition to Sustainment Sites

The sustainment center staff should be responsible for installation and check out of the system application and support software in the sustainment center(s) and demonstrating that the software can be regenerated in each sustainment center. The SwIPT supports these tasks including the transition of licenses, providing training, and other software support to the sustainment center(s).

Software Transition Plan. The completed software, support environment, and the above-mentioned documentation must be delivered to the sustainment center facility in accordance with the STrP. The STrP identifies and describes all resources needed to maintain the deliverable software. It includes, as applicable, the following resources needed to maintain the deliverable software:

- The facilities, hardware, software, documentation, personnel and other resources
- Interrelationships of components and recommended procedures
- Training plans, anticipated changes and transition plans

Chapter 12

Collaboration of Software and Systems Engineering

The Software Engineering and Systems Engineering domains are tightly coupled because that collaboration is essential to the project's success.

Software considerations need to be involved in every system-related decision made by Systems Engineering (SE), and Software Engineering (SwE) expertise should be utilized by System Engineers *throughout the Development Life Cycle*. At the same time, Software Engineers need to be continually looking at the impact *their decisions* make on the system and they must convey this information to the System Engineers.

Approach and Purpose: This chapter approaches the essential collaboration of SE and SwE from two perspectives.

1. There are system development activities that are the responsibility of Systems Engineering but these activities are *supported by Software Engineering*.
2. There are software development activities that are the responsibility of Software Engineering but these activities are *supported by Systems Engineering*.

The purpose of Chapter 12 is to identify and describe those collaborative activities so that you, the Software Project Manager, can take whatever actions are needed to make sure the collaborations described actually takes place on your project. Don't underestimate the value of this collaboration.

Lessons Learned. In the past, when a system was in the testing stage and problems arose, the System Engineers would declare it was a software problem and the Software Engineers would say it was a hardware problem so the System Engineers

should fix it. This is a "them" versus "us" mentality prevalent because there was little if any collaboration between SE and SwE during development. Software was metaphorically "thrown over the wall" to the Systems Engineering Integration and Testing (SEIT) group. Those days are gone (I hope). Collaboration as a team is essential for success and Project Managers must make sure it takes place from the start. Collaboration will not only result in fewer problems, but when problems do occur, the SE-SwE Team will be better prepared to handle them efficiently and effectively.

Organization of Chapter 12. The first three sections of Chapter 12 are *overviews* as they address the collaborative efforts of SwE and SE in terms of the:

- System and Software Development and Verification Process (12.1)
- Software Life Cycle Phases (12.2)
- Software Development Documentation (12.3)

The remainder of Chapter 12, sections 12.4–12.7, addresses the collaborative efforts in terms of how System Engineers and Software Engineers *support each other* during the five software acquisition periods that includes logical *management approval points* during the System Life Cycle:

- *Pre-Milestone A:* Concept Studies (12.4)
- *Milestone A*: Concept Development (12.5)
- *Milestone B:* Engineering and Manufacturing Development (12.6)
- *Milestones C and D:* Production, Deployment and Sustainment (12.7)

The terminology used to describe the major software Acquisition Milestones in Sections 12.4–12.7 has a government orientation for the acquisition of large software-intensive systems. However, the process described is the foundation of a successful implementation for *any large software-intensive system* regardless of the customer or industry. Change the names to match the terminology of your industry.

12.1 System and Software Development and Verification Process

Figure 12.1 is an overview of the collaborative system and software development and verification activities including the reviews related to each activity. Suggested exit criteria for key software reviews (e.g., Software Requirements and Architecture Review (SAR), SI Preliminary Design Review (PDR) and SI Critical Design Review (CDR)) are listed in Appendix G. Also, the software process needs to be coordinated with SE and incorporated, as applicable, into their *Systems Engineering Plan* (SEP).

12.2 System Life Cycle Phases and Reviews

Chapter 3 described the overall system and software development process. Chapter 11 provided a detailed description of the Software Implementation Process (SIP). Figure 12.2 is an example summarization of the *System Life Cycle (SLC) Phases*, sometimes referred to as the "System Acquisition Life Cycle." Figure 12.2 is an important companion chart to the

overview Figure 3.4 discussed in Chapter 3. The five phases shown are a logical sequence of phases followed by many large system developments; change the names to suit your terminology.

Milestones. The example in Figure 12.2 shows four milestones, A–D, that are logical *management approval points in time* during the full System Life Cycle—from cradle to grave. Milestone A takes place before the start of Phase A; Milestone B takes place before the start of Phase B, etc. The life cycle phases, milestone names and formal reviews are consistent with the "Integrated Defense Acquisition, Technology and Logistics Life Cycle Management System," Chart Version 5.3.4, dated June 2009 (DAU, 2009). Over time the terminology will change.

Figure 12.2 also shows the major reviews and (approximately) when they typically take place over the timeline. The following is a brief description of the four major milestones:

■ *Milestone A*: After the concepts of the system are defined and refined, management approval is needed to either proceed to the Concept Development Phase A, revise the concept or reject the concept.

■ *Milestone B*: After the concept is developed to a point where overall system requirements and plans are prepared, management approval is needed to advance to the execution Phase B. That key approval kicks off *Engineering and Manufacturing Development* (EMD) involving the development of the Preliminary Design and the Preliminary Design Review.

■ *Milestone C*: At the end of Phase B, management approval is needed to embark on full-scale development and production of the hardware and software system. The *Software Implementation Process*, described

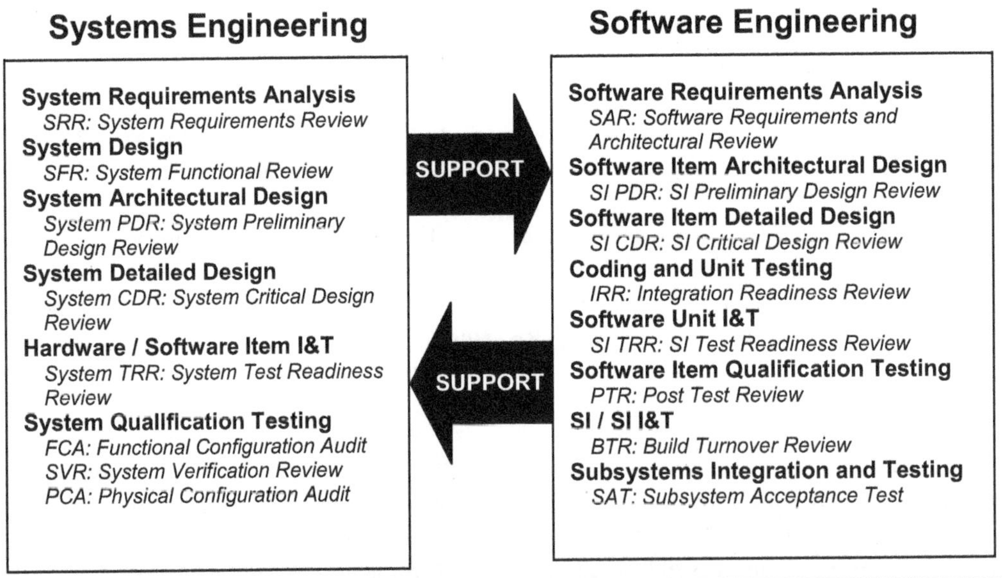

Systems Engineering   Software Engineering

System Requirements Analysis
  SRR: System Requirements Review
System Design
  SFR: System Functional Review
System Architectural Design
  System PDR: System Preliminary Design Review
System Detailed Design
  System CDR: System Critical Design Review
Hardware / Software Item I&T
  System TRR: System Test Readiness Review
System Qualification Testing
  FCA: Functional Configuration Audit
  SVR: System Verification Review
  PCA: Physical Configuration Audit

SUPPORT

SUPPORT

Software Requirements Analysis
  SAR: Software Requirements and Architectural Review
Software Item Architectural Design
  SI PDR: SI Preliminary Design Review
Software Item Detailed Design
  SI CDR: SI Critical Design Review
Coding and Unit Testing
  IRR: Integration Readiness Review
Software Unit I&T
  SI TRR: SI Test Readiness Review
Software Item Qualification Testing
  PTR: Post Test Review
SI / SI I&T
  BTR: Build Turnover Review
Subsystems Integration and Testing
  SAT: Subsystem Acceptance Test

Figure 12.1 System and software development and verification overview.

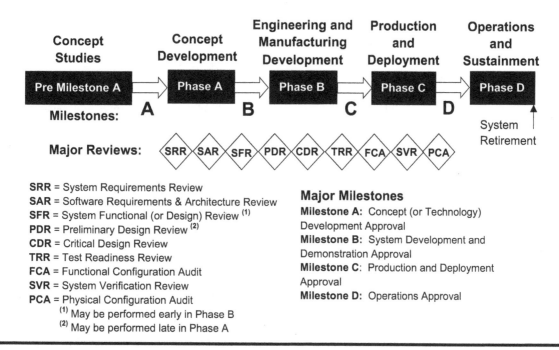

SRR = System Requirements Review
SAR = Software Requirements & Architecture Review
SFR = System Functional (or Design) Review [(1)]
PDR = Preliminary Design Review [(2)]
CDR = Critical Design Review
TRR = Test Readiness Review
FCA = Functional Configuration Audit
SVR = System Verification Review
PCA = Physical Configuration Audit
[(1)] May be performed early in Phase B
[(2)] May be performed late in Phase A

Major Milestones
Milestone A: Concept (or Technology) Development Approval
Milestone B: System Development and Demonstration Approval
Milestone C: Production and Deployment Approval
Milestone D: Operations Approval

Figure 12.2   Major software phases, milestones and reviews during the System Life Cycle.

in Section 3.5 and Chapter 11, takes place during the production Phase C. Normally, serious software coding should not begin until the Milestone C approval, however, some coding can begin beforehand as long as you, the Project Manager, can identify basic modules that will be needed regardless of the Milestone B design decisions.

■ *Milestone D*: When it is confirmed that the system developed meets the requirements, approval is needed to deploy the system and move it from development to operations and then into long-term sustainment (maintenance) until the system is eventually replaced or retired.

12.3 Software Development Documentation

During development of the system and software process, there is an essential series of software documents that are produced. Software documents are produced during each software activity. The software documents referenced in Chapter 12 are listed below and are further discussed in Chapter 13 which is devoted entirely to software documentation and work products:

■ **CPM** = Computer Programming Manual
■ **DBDD** = Data Base Design Document
■ **FSM** = Firmware Support Manual
■ **IDD** = Interface Design Description
■ **IFCD** = Interface Control Document

■ **SAD** = Software Architecture Description
■ **SDD** = Software Design Description
■ **SMBP** = Software Master Build Plan
■ **SPS** = Software Product Specification
■ **SRS** = Software Requirements Specification
■ **STD** = Software Test Description
■ **STP** = Software Test Plan
■ **STR** = Software Test Report
■ **STrP** = Software Transition Plan
■ **SUM** = Software User's Manual
■ **SVD** = Software Version Description

Table 13.1, to be discussed in Chapter 13, is a Software Documentation Production matrix identifying the software documentation that *may be produced* during each activity as well as the maturity of the document (Draft, Preliminary, Baselined or Updated) at each activity. Many of these software documents require the support of Systems Engineering and some are closely tied to the *Systems Engineering Plan*. Some of the aforementioned documents are required, some may be combined, some may be reduced in scope, and some should be eliminated if they do not add value to your project.

12.4 Collaboration during Pre-Milestone A: System Concept Analysis

For most system acquisitions, the process begins during Pre-Milestone A. For government acquisitions, that is typically when an Acquisition Decision Memorandum (ADM) from the Materiel Development Decision (MDD) is released with an order for the development of a "materiel solution." Pre-Milestone A concludes when the Milestone Decision Authority (MDA) authorizes the start of Milestone A.

The focus of the Pre-Milestone A phase is on activities performed by the System Developers. However, Software Engineering and Systems Engineering must collaborate during this phase because complex systems, especially military systems and space-related systems, are *high assurance, real-time information systems with sophisticated reliability and security requirements; software always plays a critical role in their operation*. The key collaborative tasks that may need to be addressed during Pre-Milestone A, and related reviews that may take place, are shown in Table 12.1 and are discussed in the following subsections:

- Software Engineering's Role during Pre-Milestone A (12.4.1)
- Pre-Milestone A Data Products (12.4.2)
- Development of Pre-Milestone A Capabilities and Requirements (12.4.3)
- Development of Pre-Milestone A Plans (12.4.4)
- Development of Pre-Milestone A Strategies (12.4.5)

12.4.1 Software Engineering's Role during Pre-Milestone A

During Pre-Milestone A, Software Engineering (SwE) is involved almost exclusively with system issues and activities. Structuring your SwE Team similar to the customer's Systems Engineering Team, so they can work closely together, is *highly* recommended.

Pre-Milestone A is focused on all aspects of the system solution being developed including technology development strategies, operational concepts and modes; alternative architectures; operational capabilities; candidate solutions; command and control schemes; risk reduction activities; initial software development planning; and generating estimated costs and benefits of various support structures.

Software Subject Matter Experts (SSMEs). System Engineers need the support of SSMEs for estimating software development and maintenance costs. SSMEs determine the size of the software effort typically expressed as Source Lines of Code (SLOCs) or Equivalent Source Lines of Code (ESLOCs), described in Subsection 14.1.3. They decompose the size of each type of software, including ground, flight, databases, script files, etc., because the cost of software development differs for each type.

Early Cost Estimate. Cost estimates include an analysis of the amount of software for new code to be developed and reused software including Commercial Off-the-Shelf (COTS) and Government Off-the-Shelf (GOTS). Once the software size is estimated for each type of software, an independent analysis translates software size estimates into cost estimates, using software cost estimation parametric models such as the Constructive Cost Model (COCOMO) or the System Estimation and Evaluation of Resources–Software Estimation Model (SEER–SEM) covered in Chapter 14.

For government procurements, the Program Office Estimate (POE) is an estimate of the cost of a system. It is used by the planners in the *early* portion of the Pre-Milestone A for establishing the cost estimate of the acquisition and preliminary expectations. When the POE is mature, it is submitted to the Independent Cost Analysis Team (ICAT) leader.

Table 12.1 Collaborative Activities and Reviews during Pre-Milestone A

Pre-Milestone A Concept Analysis: Collaborative Activities	Related Reviews
Analysis of Alternatives (AoA)	
Configuration Management Plan (Initial)	Alternative Systems Review (ASR)
Cost Analysis Requirements Description (CARD)	
DoD Architecture Framework (DoDAF)	
Human System Integration Plan (HIS)	Initial Technical Review (ITR)
Initial Capabilities Document (ICD)	
Request For Proposal (RFP)	
Acquisition Plans and Strategies (see Appendix E)	Technology Readiness Assessment (TRA)
Software Acquisition Management Plan (SWAMP)	
Software Acquisition Process Improvement Plan (SWAPI)	
Technical Requirements Document (TRD)	

The ICAT is normally part of the Defense Acquisition Board (DAB). The DAB uses this cost estimate to decide if the program should proceed to the next milestone. *For software-intensive systems, the DAB process is effective and meaningful only if software is adequately and realistically addressed in the POE.*

For non-government systems, a similar early system cost estimate can (and should) be prepared.

12.4.2 Pre-Milestone A Data Products

Table 12.2 is a summary of key data products that *could* be produced during Pre-Milestone A including the formal plans, strategies and related documentation. Table 12.2 provides a *notional* Level Of Effort (LOE) normally required by SwE during Pre-Milestone A. Table 12.2 also shows who is responsible for producing the listed data products.

12.4.3 Development of Pre-Milestone A Capabilities and Requirements

As the system evolves, the *Initial Capabilities Document* (ICD) can be replaced by a draft *Capability Development Document* (CDD) that will ultimately support the Milestone A "go-no go" decision. The ICD and CDD documents typically have limited software content, but SwE must identify the major software development implications of the required user capabilities.

The challenge is the early recognition of system capabilities that *will require software contributions* in their development or implementation, as well as assessing the feasibility of proposed software solutions. System and Software Designers evaluate actual alternatives in later milestones; however, infeasible or unviable capabilities or implementation options should be identified, negotiated or mitigated during Pre-Milestone A.

Table 12.2   Potential Pre-Milestone A Data Products

Pre-Milestone A Data Products Produced by Software Engineering with Support from Systems Engineering	
Clinger-Cohen Compliance Table	
Human System Integration (HSI) Plan	
Data Management (DM) Strategy (DMS)	
Information Assurance (IA) Strategy	
Modular Open Systems Approach (MOSA) Implementation Plan	
Net-Centric Data Strategy	
Software Acquisition Process Improvement (SWAPI) Implementation Plan	
Software Acquisition Management Plan (SWAMP)	
Pre-Milestone A Data Products Produced by Systems Engineering with Support from Software Engineering	*SwE LOE*
Acquisition Strategy Panel (ASP) Briefing and Concept Study Results	M
Analysis of Alternatives (AoA) Study Plan	H
DoD Architecture Framework (DoDAF)	H
Initial Capabilities Document (ICD)	M
Life Cycle Management Plan (LCMP) and System Training Plan	M
Request For Proposal (RFP)	H
Risk Management Plan (RMP)	H
Systems Engineering Plan (SEP)	M
Technical Requirements Document (TRD)	H
Technology Development Strategy (TDS)	H
Test & Evaluation (T&E) Strategy (TES)	M
Pre-Milestone A Data Products Produced by Other Organizations with Support from Software Engineering	*SwE LOE*
Cost Analysis Requirements Document (CARD)	H
Configuration Management Plan (CMP)	M

(Continued)

Table 12.2 (Continued)   Potential Pre-Milestone A Data Products

Integrated Master Plan (IMP) and Integrated Master Schedule (IMS)	M
Program Management Plan (PMP)	M
Acquisition Decision Memorandum (ADM)	L
Budget Estimate Submission (BES) and Milestone A Exit Criteria	L
Defense Acquisition Board (DAB) Request Letter	L
Enabling Concept (EnCon) and Integrated Program Summary (IPS)	L
Independent Cost Assessment (ICA)	L
Materiel Solution Recommendation and Program Office Estimate (POE)	L
System Threat Assessment Report (STAR)	L
Technology Maturation Plan (TMP)	L

LOE = Level Of Effort; H = Heavy involvement by Software Engineering (SwE); M = Moderate Level Of Effort by SwE; L = Low Level Of Effort by SwE.

During Pre-Milestone A, the customer's Program Office is usually responsible for the preparation of the *Technical Requirements Document* (TRD). Technical requirements will evolve during the Development Life Cycle. However, during Pre-Milestone A, SwE should support SE in verifying the relationships between the ICD and the TRD via bi-directional traceability and verify that all required capabilities are translated into executable requirements in the TRD.

12.4.4 Development of Pre-Milestone A Plans

Pre-Milestone A *plans* may be developed in coordination with the development of the *Systems Engineering Plan* and, in most cases, are tightly coupled with the SEP. *Appendix E*, derived from Harrell (2009), contains a brief description of each of the following *Pre-Milestone A plans*:

- *Cost Analysis Requirements Description* (CARD)
- *Configuration Management Plan* (CMP)
- *Human Systems Integration Plan* (HSI)
- *Modular Open Systems Approach* (MOSA) Plan
- *Risk Management Plan* (RMP)
- *Software Acquisition Management Plan* (SWAMP)
- *Software Acquisition Process Improvement* (SWAPI) Implementation Plan

12.4.5 Development of Pre-Milestone A Strategies

Pre-Milestone A *strategies* are developed in coordination with the development of the SEP and, in most cases, are tightly coupled with the SEP. *Appendix E*, also derived from Harrell (2009), contains a brief description of each of the following Pre-Milestone A strategies:

- Data Management Strategy (DMS)
- Information Assurance (IA) Strategy (IAS)
- Technology Development Strategy (TDS)
- Net-Centric (NC) Data Strategy
- Test and Evaluation (T&E) Strategy (TES)

12.5 Collaboration during Acquisition Phase A: System Concept Development

The collaborative *Systems Engineering* and *Software Engineering* (SwE) activities, performed during acquisition Milestone A, are shown in Table 12.3 along with the reviews related to each activity. Systems may also be called Segments.

12.5.1 Collaboration during System Requirements Analysis

The system requirements analysis activity is led by Systems Engineering—typically the *Systems Engineering Integration and Test* organization. However, support from the Software Integrated Product Team (SwIPT) is essential, and the SEIT staff in the Program Office must ensure this support is provided. If software is not involved in the upfront decision-making process, bad choices may be made (and often are) resulting in a system architecture that is difficult to implement or not cost-effective.

During the system requirements analysis and review, the Chief Software Engineer, or the Chief Software Architect, or both, should be members of the SEIT Team. They directly support this activity so that: (a) decisions involving software can be made with the appropriate expertise; (b) interface requirements are consistent across the system; and (c)

Table 12.3   Collaboration Activities during Milestone A: System Concept Development

Acquisition Milestone	Key Software and System Collaborative Activities	Related Reviews
MILESTONE A CONCEPT DEVELOPMENT (aka Technology Development)	*System* Requirements Analysis	System Requirements Analysis and Review—*see 12.5.1*
	System Functional Design	System Functional Design Review—*see 12.5.2*
	Software Requirements Analysis (*)	Software Requirements Analysis and Architecture Review (SAR)—*see 12.5.3*
	Software Item Architectural Design (*)	Software Item Preliminary Design Review (SI PDR)—see 12.5.4 (**)
	System Preliminary Design Review	System PDR—*see 12.5.5* (**)

(*) These software activities are iterative.
(**) Your customer's Program Office may place these reviews in Milestone B.

Table 12.4   Major Input and Output Documents for System Requirements Analysis

Major Input Documents	Major Output Documents
Initial Capabilities Document	System Specification (SS)
Capability Development Document	Operational Concept Document
Statement of Objectives (SOO)	System verification plans
Technical Requirements Document	System test plans
Statement of Work (SOW)	
Request for Proposal (RFP)	

testability of system and software requirements can be determined. Major input and output documents for the system requirements analysis activity are shown in Table 12.4.

Support from the SwIPT must be provided for interface definitions between computer system functions, communication functions, personnel functions, etc., to enable the further definition and management of the computer software and computer equipment resources (the SwIPT is described in Subsection 1.6.5). Interface definitions must be documented in the ICD and in the SS.

The SwIPT should support the SEIT in defining the Operational Concept Document (OCD), or similar document regardless of its name, by identifying and evaluating alternative concepts for technical feasibility, user input, cost-effectiveness, schedule impact, risk reduction and critical technology limitations. The SwIPT should also: analyze the operational concepts and other inputs to derive any software requirements that are not specifically stated; and support the refinement of

the operational concept based on current analyses, and update it with the user interface analysis material as appropriate.

A critical task for the SwIPT is supporting interface requirements analysis through the identification and derivation of software-related aspects for functional performance, interfaces, constraints, fault management, alternative modes, and quality requirements. These requirements must be analyzed for completeness, consistency, verifiability and feasibility. In addition, the SwIPT should identify and recommend system requirements that could be feasibly allocated to, and implemented in, software and identify possible software verification methods for the system requirements.

The system requirements analysis activity concludes with a formal System Requirements Review (SRR). From a software perspective, Table 12.5 lists the entry and exit criteria for the SRR. System Engineers responsible for the SRR, the *customer* and the system's Project Manager, must make sure these criteria are met.

12.5.2 Collaboration during System Functional Design

The System Functional Design activity is led by Systems Engineering—typically the SEIT group (the SEIT is discussed in Section 1.6). The SwIPT must support the SEIT in developing the System and Subsystem Design as well as the configuration of hardware, software and firmware to meet performance and reliability requirements. This support is *essential* so that critical performance and reliability issues related to software, computer systems and interfaces are addressed in as much detail as possible.

In addition, SwIPT personnel must review the System and Subsystem Designs to determine if the requirements allocated are *verifiable* and provide an estimate of the level of system testing support that will be needed. They must also support

Table 12.5 System Requirements Review—Entry and Exit Criteria

SRR Entry Criteria	SRR Exit Criteria
System specifications are available including the CDD, ICD, TRD, SOW and SOO.	The draft System Specification includes allocation of system requirements to software subsystems or elements.
Software is covered in the draft Acquisition Management Plan and in the IMP.	Draft interface specifications and interface control documents include software interface requirements.
Candidate software technologies identified.	Key software technologies to be applied are assessed.
Initial software trade studies and effectiveness studies are completed.	Initial software risks are identified with mitigation approaches.
A draft Software Development Plan including definition of the software development process, is available.	Programming languages, architectures, initial hardware, operating systems, security requirements, and operational concepts are identified.
	System/Software Development Environment (SDE) defined.
	Preliminary software development process and software architectures are defined.
	Software trade-offs addressing COTS, reuse, risks, and architectures are identified.
	Software training and support needs are defined.
	EXIT EVENT: The formal SRR is conducted and the exit criteria is satisfied or waivers have been approved.

the System Test group in the development of system verification plans, identification of System Test tools, and definition of software test suites. The rationale and the decisions should be captured in the system Architecture Design documentation.

System Decisions. System-wide Software Design decisions and their rationale should be documented by the SEIT, often in Engineering Memoranda (EMs), which are maintained in the Development Contractor's electronic data management system and fully accessible by the customer. System requirements are generated from the EMs and are flowed down to the subsystem product specifications.

Software *architecture* decisions made during the System Design activity must be recorded for later use in developing software requirements and design. Decisions are usually recorded using EMs or software documents. The SwIPT Lead should participate in establishing the rationale for software architecture, definitions, interfaces and reuse approach. The SwIPT Lead should also be responsible for ensuring that this rationale is recorded including the studies and analyses that lead to the selected architecture, the selected SI definitions, and the software reuse approach.

System Architectural Design. Required subsystems must be identified along with system-to-system, and system-to-external system interfaces, plus a concept of system execution. The interfaces must be documented by the SEIT in the ICDs. The SwIPT personnel should support the SEIT in performing the following tasks:

■ Assessing the software impact of implementing the operational concept and system requirements in terms of technical suitability, feasibility, cost and risk.

■ Participating in trade studies to select processing, communications, and storage resources. The list of trade studies can include operating systems, middleware and development languages.

■ Reviewing the System Test approach and test philosophy to ensure testing compatibility.

■ Identifying how each requirement will be tested. This includes the test and Support Software that will be needed for system testing and a description of the System Test environment.

■ Developing a System Test Plan and reviewing and analyzing the System Design to determine testability of the requirements allocated to software.

■ Recommending requirements changes as necessary.

■ Supporting subsystem system engineering in creating or refining definitions of Software Items, in allocating subsystem requirements to the SIs, and in review and refinement of the interfaces among the defined software products.

■ Identifying and evaluating the benefits and risks of potential candidates for reused software and COTS software products at the SI level.

The System Functional Design activity concludes with a formal System Functional Review (SFR) that was formerly called the System Design Review (SDR). From a software perspective, Table 12.6 lists the entry and exit criteria for the SFR. System Engineers responsible for the SFR, from the customer and the development contractor, must make sure these criteria are met.

12.5.3 Collaboration during Software Requirements Analysis and Architecture Review

This activity establishes the allocated developmental *baselines* for each Software Item (SI) and it concludes with the *Software Requirements and Architecture Review* (SAR)—formerly called the Software Specification Review (SSR). Demonstrating the adequacy of the internal development baselines includes adequacy of key software documents such as the *Software Requirements Specification* (SRS) and the *Interface Requirements Specification* (IRS).

Another objective of the SAR is to identify and establish relevant SAR exit criteria. The SEIT Team must assist SwE in ensuring that:

■ The SAR exit criteria is consistent with the Integrated Master Plan (IMP) and the Integrated Master Schedule (IMS).

■ All allocated and derived requirements have been identified and clearly documented.

■ The requirements as stated are testable, traceable and measurable.

■ SRS performance requirements are feasible, complete, and consistent with the higher-level specification requirements.

■ There are complete, verifiable requirements for all performance requirements.

SARs must be conducted for each SI or collections of related SIs. Each SAR is held after the SFR and should be completed prior to the initiation of top-level design for the individual SIs. This review is part of the overall system engineering process of allocating and formally defining requirements and must occur after the system-level hardware/software functional allocation decisions have been made.

Systems Engineering support to SwE is essential in evaluating the areas listed below to avoid disconnects that can easily arise in requirements, margins and program direction:

■ Agreements on system interfaces and boundaries
■ Requirements allocation decisions
■ Software risks and proposed mitigation methods

Table 12.6 System Functional Review—Entry and Exit Criteria

SFR Entry Criteria	SFR Exit Criteria
System specification is completed with software functional and performance requirements, including software verification requirements.	Draft subsystem and allocated functional specifications, including software requirements, are defined and baselined in the system specification.
Planned reviews are completed with all software issues resolved or a documented resolution plan.	Draft preliminary Software Requirements Specifications, including verification requirements, and Interface Requirements Specifications, are defined.
Software architectures are defined as integral to the system architecture.	Development Specification Tree is defined through subsystem and configuration items, including interface specifications.
Planned software trade studies completed.	Software requirements traceability is defined.
Subsystem are identified, including a draft list of all Software Items.	The planned System/Software Development Environment is defined.
	Software Design and development approach is defined and confirmed by analysis, demonstrations, or prototyping.
	Software process IMP/IMS events, schedule, task definitions and metrics are defined.
	Software development estimates are updated.
	Preliminary software risk management strategy, approach and process is defined and integrated with the system risk management process.
	EXIT EVENT: The formal SFR is concluded and the exit criteria is satisfied or waivers have been approved.

- Trade studies and feasibility analyses
- Design constraints and human factors considerations
- Reserve capacity requirements and scenarios
- Procedures for measurement collection, analysis and reporting
- Results of the functional analyses

The SEIT Team plays an important supporting role by assisting SwE in complying with the SAR entry and exit criteria as shown in Table 12.7.

12.5.4 Collaboration during SI Architectural Design

The objective of the Software Item Architectural Design is to describe the high-level organization of the SIs in terms of Software Units (SUs) and their relationships. Systems Engineering must support the SwIPT in preparing an architecture that meets the system requirements. The main objectives of SI Architectural Design are to:

- Decompose the SI Architectural Design into SUs.
- Allocate requirements from the Software Requirements Specifications to SUs.
- Complete allocation of requirements from the SRS to Use Cases (for Object-Oriented Design).
- Describe the Architectural Design and requirements allocation in a preliminary Software Architecture Description (SAD) and Software Design Description (SDD).
- Update the SI Software Development Files (SDF) for the SI and update the baselined SRS if necessary.

Table 12.7 Software Requirements and Architecture Review—Entry and Exit Criteria

SAR Entry Criteria	SAR Exit Criteria
Higher-level (system and subsystem) specifications are baselined.	Software requirements are traceable to higher-level requirements.
Draft versions of SRSs and IRSs for all SIs to be reviewed are completed.	Internal baselines are established (e.g., SRS and IRS).
Relevant Software architecture studies and analysis are available.	Requirements incorporate the functionality that must be implemented by software.
Software risk management approach is defined and analysis results are available.	The software risk management process is defined, cost, schedule and performance risks are identified, integrated with the system risk management process.
Functional analysis results are available.	Life cycle support requirements are compatible with, and incorporated into, system lifecycle resource requirements.
Software work packages defined and draft allocation of requirements to SUs completed.	Software and interface requirements are allocated to SIs, Software Units and planned builds or spirals.
Formats and tools for establishment and maintenance of the SDFs are defined.	The S/SEE required and the needed software development and integration and test tools and facilities are available.
	Planned software measurements are defined and metrics collection, analysis and reporting mechanisms are in place.
	Software development schedules reflect and accommodate Developer-selected processes and defined IMP events.
	Identified software risks are acceptable and risk mitigation plans are developed.
	Software development size, schedule, cost, and staffing estimates are updated.
	Compliance with relevant documentation requirements in the documentation production matrix (Table 13.1) and the Software Review Exit Criteria in Appendix G.
	EXIT EVENT: The formal SAR is concluded, SI requirements are approved, and Preliminary Design is ready to start.

- Prepare the preliminary versions of applicable documents that may include the: *Software Architecture Description*, *Software Design Description*, *Software Test Plan* (STP), *Interface Design Description* (IDD), *Software Master Build Plan* (SMBP), the *Data Base Design Document* (DBDD) and the *System/Subsystem Design Description* (SSDD).
- Conduct a Software Item Preliminary Design Review (SI PDR) to determine whether a preliminary architecture and design approach for each SI is acceptable to start the detailed Software Design activity.

The SI PDR must ensure that the process used to arrive at functional and performance requirements for each SI:

- Is complete including trades and allocations
- Demonstrates a balanced and integrated approach
- Establishes an audit trail from functional baseline to customer requirements (substantiating changes as necessary)
- Ensures that design approaches are confirmed through evaluation of risk mitigation and resolution

During the Software Item PDR, the physical architecture and design must be assessed to determine adequacy, completeness and realism of the proposed development functional requirements. Higher-level configuration items (i.e., integration and interface) must be assessed to ensure they satisfy their portion of the functional baseline. The software PDR also confirms that the integrated system meets functional baseline requirements and that the System Preliminary Design meets customer requirements. Entry and exit criteria for the Software Item PDR are shown in Table 12.8.

12.5.5 Collaboration during the Preliminary Design Reviews

The *System Preliminary Design Review* (System PDR) determines if the hardware, software and human-related preliminary designs are complete and if Detailed Design is ready to begin. The System and Software PDRs should be accomplished in accordance with plans established and committed to in advance by the Developer in the IMP/IMS and the Software Development Plan (SDP). The System PDR, or its equivalent review, is normally held after completion of all software and other lower level PDRs and before Detailed Design begins. Each project must determine the best sequence of system-level and software-level PDRs.

The System PDR Contractors (or Developers) are responsible for conducting design review activities appropriate for the current stage of the project; however, they must also

Table 12.8 Software Item PDR—Entry and Exit Criteria

SI PDR Entry Criteria	SI PDR Exit Criteria
Software Requirements Specification is baselined.	Preliminary software architectural level Software Design is defined and documented.
System/subsystem and Software Requirements are baselined.	Software/Systems Engineering Environment (S/SEE) configuration is defined and internally controlled.
Software Design implementation trade studies are available.	Software requirements baseline is approved and verified to satisfy system/subsystem functional and non-functional performance requirements.
Make/buy decisions are completed.	Preliminary Interface Control Documents and the Software Test Plan are documented.
Interchangeability and irreplaceability decisions baselined.	Software incremental builds are defined and requirements are allocated to them.
	Software risk management process is defined and implemented.
	Initial evaluation of Commercial Off-the-Shelf and reusable software is completed.
	Software development estimates and support requirements updated.
	Compliance with relevant documentation requirements in the documentation production matrix (Table13.1) and the Software Review Exit Criteria listed in Appendix G.
	Conduct and conclude a successful SI PDR.
	EXIT EVENT: The SI is approved and software Detailed Design is ready to start.

provide effective customer access and visibility of these activities. Technically knowledgeable members from the acquisition and supporting organizations should be invited to attend the System PDR. Almost all systems involve a significant amount of software, so the System PDR should include the CSwE and Senior Software Engineers working as integral members of the Systems Engineering Acquisition Team.

For complex systems, the system-level PDR activities should be concluded only after completion of all software—or other lower level reviews—and before Detailed Design begins. Each program must determine the best sequence of system-level and software-level PDRs for their program. The SwIPT and SE should review changes made to the SRS. The Software Design should also be reviewed to determine if it fully implements the higher-level functional and design architecture and specified requirements. Changes to requirements should be analyzed for impact to the design. The analysis should cover both internal and external interfaces.

The SwIPT and SE should analyze the allocation of functional requirements to elements of the software architecture for consistency and completeness. Where incremental development is being used, allocation of requirements to software builds should also be reviewed for consistency and completeness. System Engineers and the SwIPT must collaborate to

ensure that the processes used to arrive at functional and performance requirements, for each SI and Configuration Item (CI), complies with the following:

- Required trades and allocations are completed.
- A balanced and integrated approach is demonstrated.
- An audit trail from functional baseline to customer requirements, substantiating changes as necessary, is established.
- The selected design approaches are confirmed through evaluation of risk mitigation and resolution.
- The physical architecture and design is assessed to determine adequacy, completeness, and realism of proposed development functional requirements.
- The reviews affirm that the planned integrated system meets functional baseline requirements and that the System Preliminary Design meets customer requirements.

At the conclusion of the System PDR, a configuration controlled development specification or description must be complete and ready for use in Detailed Design. A basic requirement for a successful System PDR is to identify and establish relevant software exit criteria. Recommended entry and exit criteria for the System PDR is shown in Table 12.9.

Table 12.9 System Preliminary Design Review—Entry and Exit Criteria

System PDR Entry Criteria	System PDR Exit Criteria
System/subsystem functional and performance requirements are baselined, including the SRS with needed changes incorporated after the SAR.	Preliminary SI and interface designs are established, documented, reviewed, and determined to be ready for Detailed Design.
Preliminary system software and SI architecture is established and accommodated within the system architecture.	Confirmation that the software specified requirements are satisfied by the design architecture and approach.
Software specified requirements and preliminary design are satisfied by the Subsystem/System Design architecture and approach.	System/Software Development Environment requirements and configuration are defined and internally controlled.
Preliminary Software Designs and interface requirements are complete and documented.	Software increments (builds) are defined and allocated.
Software Design implementation trade studies and make/buy decisions are completed.	COTS and reusable software are identified and verified to meet requirements.
Compatibility is established between the SIs and all configuration items.	Software test plans (as required) are complete.
Scheduled SI and subsystem PDRs have been successfully completed.	Software development estimates, support requirements and progress metrics are updated.
Engineering Change Proposals are up-to-date including software impacts.	Software development files are established and maintained current.
	EXIT EVENT: The formal System PDR is concluded, Preliminary Designs are complete and approved, and Detail Design is ready to begin.

12.6 Collaboration during Acquisition Phase B: Engineering and Manufacturing Development

The collaborative SE and SwE activities performed during *Engineering and Manufacturing Development* are shown in Table 12.10 along with the reviews related to each activity.

12.6.1 Collaboration for the System Critical Design Review

The *System Critical Design Review* (System CDR) assesses the System Design as captured in product specifications for each configuration item (hardware and software) in the system, and ensures that each product in the product baseline has been captured in the Detailed Design documentation. It is a multi-disciplined technical review to ensure that the system under review can proceed into development (fabrication for hardware), demonstration, and test, and that it can, or should be able to, meet performance requirements within the documented cost, schedule, risk, and other system constraints.

> *SwE has an important supporting role during the System CDR.*

A software build is a version of the software comprising a specified subset of the requirements for the full system during the time the software build is being developed. Suggested entry and exit criteria for the System CDR from a software perspective are shown in Table 12.11.

System Detailed Design should address the Software Design as part of the subsystem and system-level Critical Design Reviews. The SRS should be reviewed for changes and traceability to the Detailed Design. The Software Design should be reviewed to ensure it fully implements the higher-level functional architecture and specified requirements. Impact of changes in requirements should be analyzed for impact to the design.

If necessary, requirements should be reallocated, and designs adjusted, to be consistent and complete. Analysis of the System Detailed Design should cover internal and external interfaces, and the following allocations analyzed for consistency and completeness:

- Allocation of functional requirements to software configuration items, components, and units
- Allocation of software requirements to incremental builds or blocks

12.6.2 Collaboration for the SI Critical Design Review

The objective of Software Item Detailed Design and the SI Critical Design Review (SI CDR) is to identify implementation details for each Software Unit comprising the SI. System and Software Engineers must collaborate to define the specifics of the algorithms or processes an SU is to perform and determine details of the data structures used by the SU internally—and for interactions with other SUs. The resulting SU detail design descriptions are normally sufficient for code

Table 12.10 Collaboration during Milestone B: Engineering and Manufacturing Development

Acquisition Milestone	Key Software and System Collaborative Activities	Related Reviews
	Support to the IBR (*)	Integrated Baseline Review (IBR)
MILESTONE B Engineering and Manufacturing Development (EMD)	System Detailed Design	System Critical Design Review—see 12.6.1
	Software Item Detailed Design (**)	SI Critical Design Review (SI CDR)—see 12.6.2
	Software Unit Integration Testing (**)	SI Test Readiness Review (SI TRR)—see 12.6.3
	Software Item Qualification Testing (**)	Post-Test Review—see 12.6.3
	Subsystems Integration & Testing (**)	Subsystems Acceptance Test (SAT)—see 12.6.4
	Sw/Hw Integration and Testing	System Test Readiness Review (System TRR)—see 12.6.4
	Support to the FCA	Functional Configuration Audit—see 12.6.5
	System Qualification Review	System Verification Review—see 12.6.6 Production Readiness Review

(*) IBRs can take place at the start of the program, at the start of each milestone (A, B, or C), or as needed.
(**) These software activities are iterative.

Table 12.11 System CDR—Entry and Exit Criteria

System CDR Entry Criteria	System CDR Exit Criteria
Software architectural level and Detailed Design are established.	Software Detailed Designs and interfaces have been reviewed and are ready for implementation.
Software requirements baseline(s) satisfy system/subsystem requirements baselines.	The Software/Systems Engineering Environment (S/SEE) is implemented and ready to support implementation and test.
Software incremental builds or blocks are defined with allocated requirements.	Software test descriptions and draft Software Test Procedures are complete.
Specific requirements are satisfied by the design architecture and approach.	Software development progress metrics are being, collected, used and updated.
Scheduled software and subsystem CDRs have been successfully completed.	Software Development Files are established and maintained.
Software specifications, including changes since PDR, are complete.	Updated evaluation of Commercial Off-the-Shelf and reusable software is completed.
Engineering Change Proposals, including software impacts, have been reviewed.	Software development estimates and support requirements are updated.
	Applicable system software and interface control documents are baselined.
	Software risks are identified and mitigation plans have been developed.
	EXIT EVENT: The System CDR is concluded, the SIs are approved and ready to start the SI CDR.

developers to implement the design into code. The main *objectives* of the SI Detailed Design are to:

- Decompose the Software Item into Software Units
- Complete a description of the Detailed Design for each SU and record all results in the SDF
- Develop test cases, test procedures and test data for SU integration
- Baseline the applicable documentation that may include the: *Software Architecture Description*, *Software Design Description*, *Software Test Plan*, *Interface Design Description*, Software Master Build Plan, the *Data Base Design Document* and the *System/Subsystem Design Description*.

SI Detailed Design Tasks. A major activity of Detailed Design is decomposing the SI Architectural Design into the lowest level SUs. The design must be developed in sufficient detail to map the design to the features of the selected programming language, the target hardware, operating system, and network architecture. Major tasks performed during SI Detailed Design should include:

- *Refining the Design Model*: Adding additional details to the design model to accommodate detailed decisions and constructs necessary for implementation.

- *Defining Implementation Details*: Refining internal design to add data structures, attribute types, visibility, interfaces and usage mechanisms. Factors to consider include execution time, memory usage, development time, complexity, maintainability, reusable software and hardware resource utilization. Analysis and modeling may be necessary to determine the best design approach.
- *Generating Class Stubs*: Generate code header files and class stubs based on the object model definitions. Design complex class algorithms or logic using a Program Design Language (PDL).
- *Prototyping and Simulations*: Performing prototyping and simulation to validate critical processing areas, mitigate implementation risk, or to identify optimizations.
- *Developing Software Increment Test Cases*: Revising the software build/test plans to add test cases and to ensure all test support needs and plans are in place.
- *Generating and Reviewing Products*: Holding Peer Reviews on Detailed Design products, and adding the Detailed Design information to the SDD, IDD and STP.

Combined System and SI CDRs. The SI CDR may be held as part of the *System CDR*. Each program must determine the best sequence of the system-level CDR and software-level CDR for their program. Suggested entry and exit criteria for the SI CDR are shown in Table 12.12.

Table 12.12 Software Item CDR—Entry and Exit Criteria

SI CDR Entry Criteria	SI CDR Exit Criteria
SI requirements developmental baselines are confirmed to satisfy the system and subsystem requirements baselines.	Updated confirmation that the Software Item requirements, as specified in the contractor developmental baseline specifications, are fully satisfied by the Detailed Design description.
Software requirements and increments are planned, defined and allocated.	Detailed level Software Item Designs are established and reviewed, and are determined to be complete and ready for implementation.
Software System Architectural Design level is established.	Software Item test descriptions are complete.
Software specifications, including changes since PDR, are updated.	Draft Software Item test procedures are complete.
Preliminary Software Design is completed.	Detailed SI Design interface descriptions are complete.
Confirmation that the Software specified requirements are satisfied by the design architecture and approach.	Software Item development progress metrics are updated to reflect current development and design status.
SI metrics are tracking current development status and progress.	Software development estimates and support requirements are updated.
Software impacts from ECPs are available.	SI development files are established and maintained current.
	Compliance with relevant documentation requirements in the documentation production matrix (Table 13.1) and the Software Review Exit Criteria in Appendix G.
	EXIT EVENT: The SI CDR is concluded and the SIs are ready to start Coding and Unit Testing (see 11.5).

12.6.3 Collaboration for the SI Test Readiness Reviews

Software Item Test Readiness Reviews (SI TRR) are performed to ensure the software is ready to enter formal Software Item Qualification Testing (SIQT) and to ensure that resources are not wasted by commencing a SIQT event that will quickly fail due to lack of product readiness.

The objective of performing SIQT is to execute the test procedures as documented in the *Software Test Description* (STD), using products under Software Configuration Management (SCM) control, and in a witnessed test environment ensuring the software meets its requirements. The SIQT should always be preceded with the SI TRR. *This review ensures that all necessary test documentation, equipment, materials, and personnel are ready, and that coordinated test schedules are in place.* The Master Test Plan (MTP), sometimes called a Test and Evaluation Master Plan (TEMP), or its functional equivalent, is the basis for the software test planning and readiness activity leading to the SI TRR. Essential elements of TRR planning include:

- Identification of critical system characteristics, key features and subsystems
- Interfaces with existing or planned systems required for mission accomplishment

- Technical performance measurement results against user-defined requirements
- Developmental test schedules

The material presented at the SI TRR should include: Software Unit test results; SI dry run results; formal test environment description (hardware, test tools, and associated software); requirements verification at a lower level; test schedules; and SIQT tasks as described in the SI *Software Test Plan* and the SI STD. Typical documentation that may be reviewed includes the SRS, IRS, *Software User's Manual* (SUM), and the *Computer Operators Manual* (COM) as described in 11.9.3.

Software and System Engineers should collaborate in analyzing and recording SIQT results to finalize the SIQT activity by:

- Documenting test results in the *Software Test Report* (STR) or, for Support Software, in the SDF
- Performing a review of the STR, or verifying the capture of the test results in the SDF
- Conducting an optional Post-Test Review (PTR). The PTR may be called either a Build Turnover Review (BTR), or a Test Exit Review (TER), after completion of the SIQTs and the integration of Software Items

Table 12.13 Software Item Test Readiness Review—Entry and Exit Criteria

SI TRR Entry Criteria	SI TRR Exit Criteria
The requirements being tested (applicable SRS and IRS, or subsets) are identified.	All test facilities and resources are ready and available to support software testing within the defined schedule.
Traceability of test requirements to the SRS and IRSs is established.	Planned testing is consistent with the defined approach including Regression Testing.
All SI level test procedures are complete.	Software test descriptions and procedures are defined, verified and baselined.
Objectives of each test are identified.	The software being tested and the entire test environment is configuration controlled as applicable.
All applicable documentation is complete and controlled.	All lower level software testing has been successfully completed and documented.
The method for documenting and dispositioning test anomalies is acceptable.	Software metrics show readiness for Software Item Qualification Testing (SIQT).
	Software problem report system is defined and implemented.
	Software test baseline is established and controlled.
	Software development estimates are updated.
	Requirements that cannot be adequately tested at the SI level are identified and included in the system-level test.
	The Build Turnover Review is completed (optional).
	EXIT EVENT: The SI TRR has been completed, the exit criteria for the SI TRR has been satisfied, and the SI is ready for SIQT.

The STR must be analyzed for completeness and is subject to a document review. Suggested entry and exit criteria for the SI TRR are shown in Table 12.13.

The STR must be baselined once all review modifications have been incorporated and, when approved, must be maintained under SCM control. The SI TRR is successful when it is determined that the Software Test Procedures and the lower level test results form a satisfactory basis for proceeding to Software Item Qualification Testing (SIQT). A full discussion of SIQT is covered in Section 11.7 and is not repeated here.

12.6.4 Collaboration for the System Test Readiness Reviews

The *System Test Readiness Review* (System TRR) ensures that the system, or subsystem under review, is ready to proceed into formal System Test. The System TRR assesses test objectives, methods, procedures, and scope and confirms that required test resources have been properly identified and coordinated to support the planned tests. The System TRR verifies the traceability of planned tests to system requirements and user needs. The completeness of the test

procedures and their compliance with test plans and descriptions is also determined.

Although the collaboration of System and Software Engineering continues throughout system testing, the Systems Engineering Integration and Test Team defines the requirements for System Qualification Testing. There are no *formal* Software Developer responsibilities during System Test, except for the important tasks of analyzing software discrepancies, generating SCRs/SDRs as needed, and implementing software code changes resulting from the new SCRs/SDRs.

During the System TRR, test procedures are evaluated for compliance with the applicable specifications and the Master Test Plan. Open problem reports against the product being tested, the process used to develop the product, and the environment being used in the test are reviewed and assured to be acceptable. Suggested entry and exit criteria for the System TRR are shown in Table 12.14.

System Qualification Testing. During the System TRR, the system is assessed for development maturity, cost/schedule effectiveness, and risk to determine readiness to proceed to the formal System Qualification Testing (SQT) discussed in 10.4 and not repeated here. Testing that cannot

Table 12.14 System Test Readiness Review—Entry and Exit Criteria

System TRR Entry Criteria	System TRR Exit Criteria
The system requirements being tested are identified.	System test facilities and resources are ready and available to support system testing within the defined schedule.
Traceability of test requirements to the system or subsystem specifications is established.	System test descriptions and procedures are defined, verified and baselined.
All lower level tests are complete including subsystems integration and testing and the Subsystem Acceptance Test is successfully completed.	Planned testing is consistent with the defined approach.
Objectives of each test are identified.	The entire test environment is configuration controlled as applicable.
All applicable documentation is complete and controlled.	The System Test baseline is established and controlled.
The method for documenting and dispositioning test anomalies is acceptable.	Cost, schedule and performance risks have been identified and mitigation plans have been developed.
	The problem reporting system is defined and implemented.
	All metrics show readiness for system testing.
	EXIT EVENT: The system (or subsystem) being tested is approved and the System TRR is completed and ready to proceed to formal system testing.

be adequately verified at the SI level must be included in the test plan and should be reviewed during the System TRR.

Additional Testing Procedures. As shown in Figures 3.4, 3.5, 12.1 and 12.2, there are three additional optional testing audits following the SQT that may be required by your contract: Functional Configuration Audits (FCAs) described in 12.6.5; System Verification Review (SVR) described in 12.6.6; and the System Physical Configuration Audit (System PCA) described in 12.7.1.

12.6.5 Collaboration for the Functional Configuration Audit

Functional Configuration Audits are conducted, *when required by the contract*, to verify that Hardware Items (HIs) and Software Items have achieved the requirements specified in their functional and performance requirements specifications. Hardware and software FCAs may be held independently or concurrently. Normally, a series of FCAs is held to cover each relevant HI and SI in a new development. The entry and exit criteria for the System FCA should be included in the *Integrated Management Plan*. FCAs may be an incremental part of the System Verification Review as discussed in Subsection 12.6.6.

Software FCAs are intended to confirm that each SI is verified and tested relative to the allocated requirements in the Software Requirements Specification, Interface Requirements Specifications, and relevant higher-level specifications. System

Engineers play an important role in this collaborative process. Each SI in the system should be part of an FCA verification process.

System and Software Engineers should be part of the FCA Team and should focus on confirming the adequacy and completeness of the verification of the Hardware Configuration Items, and the SIs. The FCA must verify that the specified software requirements are satisfied as demonstrated and recorded in the verification records (test results). System and Software Engineers must reach a technical agreement on the validity and degree of completeness of the operations and support documentation including, *as applicable*: the Software Test Reports; Software User's Manual; Computer Operations Manual; Computer Programming Manual (CPM); and the Firmware Support Manual (FSM).

Internal software FCAs may also be performed as part of the Developer's baseline process, i.e., internal to the Developer's processes. Software FCAs also verify that the tested SIs were designed, coded, and tested following defined program processes in the planning documents such as the *Software Development Plan*, the *Configuration Management Plan*, and the *Software Test Plan*. Table 12.15 provides the typical entry and exit criteria for the FCAs.

Software FCAs may be conducted on a single SI, a group of SIs, or incrementally on developed software builds. When software is developed incrementally, when more than one FCA is conducted, a final Software FCA must be conducted to ensure all specified requirements have been satisfied. In

Table 12.15 Functional Configuration Audit—Entry and Exit Criteria

FCA Entry Criteria	FCA Exit Criteria
System and subsystem functional and development specifications are completed.	Allocated baselines for software functions are defined.
Functional/development specifications, (e.g., SRS and IRS) are completed.	Verification of functional and performance requirements are complete.
Draft product design specifications are completed.	Readiness confirmed for the next level of FCA or readiness for production.
Software-related test deficiency reports that can be resolved are resolved.	FCA minutes, identifying open discrepancies and actions for resolution, are completed.
Software test plans, descriptions, and procedures are completed.	Relevant metrics are being collected and appropriate Corrective Actions are being taken.
	EXIT EVENT: The FCA has been successfully concluded and the system is ready for the next higher-level FCA or SVR as required.

cases where SI verification can only be completely determined after System Integration and testing, the final Software FCA would be conducted using the results of the System Integration tests.

12.6.6 Collaboration for the System Verification Review

The *System Verification Review* confirms the completeness of the system development and readiness for deployment and sustainment. The Hardware Items, Software Items, subsystems and the full system are verified to demonstrate that they satisfy the performance and functional requirements contained in the functional and allocated baselines. The SwIPT provides support in confirming that software verifications are complete and the software is successfully integrated with the total system. SVR is the culmination of incremental reviews which confirms completeness of the products (i.e., HIs, SIs, and the subsystems which comprise the system).

Concurrency. The SVR establishes and verifies final product performance. The SVR can be conducted concurrently with a Production Readiness Review (PRR) if that review is required in your contract. The System Functional Configuration Audit (System FCA) may also be conducted concurrently with the SVR as discussed in 12.6.5. Combining the SVR with the PRR and FCA into one event is obviously an efficient approach if allowed by your contract.

The System FCA can also audit the system's adequacy and compliance with the verification process, as well as the completeness and sufficiency of the recorded verification/test results relative to the specified requirements. In this case, the System FCA may preclude the need for the SVR.

Verification testing should be accomplished prior to a formal SVR and this testing documented as planned events in the IMP and *the Software Development Plan*. Traceability between the system, SI requirements, design, and code should also be verified. Table 12.16 provides the suggested entry and exit criteria for the SVR.

12.7 Collaboration during Phases C and D: Production, Deployment and Sustainment

In order to assure complete coverage, System Engineers and Software Engineers collaborate during acquisition Milestone C: Production and Deployment. The collaboration during Milestone C involves the *Physical Configuration Audit*, and it may also involve the *System Test Readiness Reviews* (System TRR), discussed in Subsection 12.6.4, depending on which phase you contract places the System TRR. The Software PCA and the Hardware PCA may be independent events or they may be held concurrently. Likewise, the System PCA may be a separate event or may be combined with the software and hardware PCAs. The key collaborative tasks during Milestone C and D are shown in Table 12.17.

12.7.1 Collaboration for the Physical Configuration Audit

The *System Physical Configuration Audit* involves both hardware and software. It is a formal *examination* of the "as built" Hardware Items and Software Items against their technical documentation to establish or verify the HI/SI's product baseline. A PCA should be held, if required by your contract,

Table 12.16   System Verification Review—Entry and Exit Criteria

SVR Entry Criteria	SVR Exit Criteria
Incremental FCAs for each HI, SI and subsystem are completed.	HI and SI allocated baselines, product specifications and functional/performance requirements are verified.
System and subsystem functional/development specifications are completed.	Software requirements, design and code traceability are established.
HI and SI functional and development specifications, (e.g., SRS and IRS) are completed.	SIs are verified through integration and test of hardware, software and subsystems.
HI and SI product design specifications are completed.	Software implementation is verified against specified system requirements.
Compliance with specified requirements is confirmed with verification and test results.	FCA minutes, identifying open discrepancies and actions for resolution, are completed.
	Required operational and support documentation, including the Software Development Files, are complete and available.
	EXIT EVENT: Readiness for deployment of the system and the software is verified.

Table 12.17   Collaborative Activities and Reviews during Milestones C and D

Acquisition Milestone	Key Software and System Collaborative Activities	Related Reviews
MILESTONE C PRODUCTION & DEPLOYMENT	Transition to Operations and Sustainment	Physical Configuration Audit
	Finalization of User and Support Manuals	Operational Test Readiness Review
MILESTONE D SUSTAINMENT	Update System Requirements	Full Deployment Decision Review

for each HI/SI after completion of the acceptance testing of the first system designated for operational deployment. This review should be conducted in accordance with established Configuration Management guidelines.

The objective of the System PCA is to confirm that all HI/SI PCAs have been satisfactorily completed; items that can be baselined only at the system-level have been baselined; and required changes to previously completed baselines have been implemented (e.g., deficiencies discovered during testing have been resolved and implemented). The System PCA should not be started unless its related *System Functional Configuration Audit* has been completed or is being accomplished concurrently. Entry and exit criteria for the System PCA should be included in the IMP if the PCA is required by your contract.

The Software Developer should identify specific differences, if any, between the physical configuration of the SI and the configuration that was used during the FCA. A list delineating both approved and outstanding changes against the SI should be provided and should identify approved *Software Discrepancy*

Reports and approved deviations/waivers to the software and interface requirements' specifications. Table 12.18 provides an example of the PCA entry and exit criteria.

The *Software PCA* should include a detailed audit of design documentation, and operations/support documents. Documentation to be audited may include the *Software Product Specification* (SPS), *Interface Design Description* and *Software Version Description* (SVD), or their equivalents. Satisfactory completion of a Software PCA, and approval of the product specification and the Interface Design Document, or their equivalents, are necessary in order to establish the SI's product baseline.

12.7.2   Collaboration during Sustainment

Purpose of Software Sustainment. Software Sustainment, often referred to as Operations and Maintenance (O&M), or just "Maintenance," *consumes about two-thirds of the total life cycle cost* and may last for decades. A key software development objective is to produce a software product that is easy

Table 12.18 Physical Configuration Audit Example—Entry and Exit Criteria

PCA Entry Criteria	PCA Exit Criteria
All relevant lower level HI/SI PCAs have been completed in a satisfactory manner.	PCA discrepancies have been documented as action items with responsible organizations and suspense dates established for resolution of the discrepancies.
System, subsystem, and HI/SI product specifications, including interface specifications, are completed.	Software product SI specifications have been verified against the as-coded software.
System, subsystem and HI/SI SVRs and FCAs, including test reports, showing that all tests have been performed as required, and are completed.	HI specifications have been verified against the as built product.
Information defining the exact design of the HI and SI is identified, internally controlled, and ready for PCA.	Software support requirements and operational needs are completed and verified for accuracy.
Required Software information, whether electronic or hard copy, has been confirmed ready for the PCA.	Version description documents are completed and verified for accuracy.
	EXIT EVENT: Software is approved as ready for transition to sustainment.

to maintain and is operationally cost-effective. The purpose of the Software Sustainment activity is to:

■ Operate the system in accordance with its intended functions and adapt the software to changes in the hardware or operational environment so that it continues to perform its intended functions.
■ Enhance the operational software when required so that it, and the system, accomplishes new functions or provides improved performance.
■ Correct known problems in the software not resolved during development and testing.
■ Correct residual problems identified after deployment and correcting errors generated during sustainment after implementing changes, enhancements and improvements.
■ Provide modifications to improve system maintainability and reliability.

During the sustainment activity, the collaboration of System and Software Engineers is as critical, if not more so, than during the original development and testing. Fixing problems and making changes and enhancements to the software and hardware during sustainment always has an impact on the software, so the collaboration of software and system engineering is essential. System Engineers should be aware of, and involved with, all of the Software Sustainment activities. Chapter 16 is entirely devoted to managing Software Sustainment.

12.7.3 Software Maintenance Plan

Adequate facilities, Support Software, personnel and documentation must be available after the development activities

are completed so that the software can be maintained in an operational and sustainable condition. The best, and possibly the only, way to ensure all of this will be available when needed is to require the preparation of a realistic *Software Maintenance Plan* (SMP) relatively early during software development and to obtain a commitment to follow it. The customer or acquisition agency must spearhead this task. System engineers should be involved with the development of the SMP and help to ensure compliance.

Even if all the details of the SMP cannot be completed, a draft should be prepared as early as possible to ensure the software development contract contains the essential provisions. The SMP should be revised during the development effort—especially after key development milestones. The plan should not be allowed to become out-of-date or the team responsible for maintaining the plan to become dormant.

SMP Participants. In addition to Software and System Engineering participation, the SMP participants should include the team involved with the acquisition, maintenance and use of the software. This team, often called the Computer Resources Working Group (CRWG), should be composed of applicable representatives from the acquisition, maintenance and user organizations and possibly logistics personnel. One of the first decisions by the CRWG, and a key element of the SMP, is the selection of the proposed organization that will be responsible for the operational sustainment activity.

12.7.4 The Sustainment Organization

The selected sustainment organization may be an organization affiliated with the customer, an outside contractor or within the system development organization. If your system is a government-funded program, maintaining the system *within the affected government community* is often the best option especially if the system has a long life, has strict operational security requirements, or involves the need for an extensive geographically distributed maintenance capability.

The designated sustainment organization must be provided the same support software that was used during software development. There will be significant cost savings if the maintenance contract specifies that the needed support software, and other related facilities and services, are included in the sustainment contract rather than adding them to the contract later. Personnel with the specific software, system and application knowledge are a key element in establishing the project's long-term sustainment capability. A substantial portion of this knowledge can be obtained if key maintenance personnel can be selected early and involved with reviews, evaluations and test activities during the software development effort.

12.7.5 Transition to Operations

As discussed in Section 11.9, this activity is concerned with the preparation, installation and checkout of the executable software, on the target system, at a customer or user site. Upon successful completion of the System Qualification Test for the final build, and all problem reports that can be closed are closed, the software development cycle is completed and the software is ready for transfer to the customer for system acceptance testing. System Engineers should be involved with this process.

Prior to actually releasing the software for use, there remains software and documentation preparation work that must be completed. The tasks necessary to prepare the software and software-related products necessary for a user to run the software includes preparing: the executable software, version descriptions, databases, user manuals and installation at the user site. Preparing software for use, at the user's site, ensures that there is a smooth transition of software into the actual operational system. These tasks should begin well before completion of the SQT.

System Engineers must ensure that the tasks and products of this activity are in compliance with the *System Verification Plan* (written by the SEIT Team). This planning must be coordinated with hardware installation schedules and environment support including the establishment of schedules, plus the identification of resources and personnel required for the installation. This activity may also include the planning, preparation, and presentation of required user training.

12.7.6 Transition to Sustainment

As discussed in Section 11.10, the software and documentation preparation work that must be completed in order to transition the application and support software to System Sustainment includes preparation of the:

- Executable software and source files
- Version descriptions for the maintenance site
- "As built" Software Item Design and related information
- Support and maintenance manuals
- Updating system requirements
- Transition to the designated sites.

The sustainment site may be one or many satellite sites where System Sustainment is performed. Software transition involves considerable advance planning and preparation that must start early in the life cycle to ensure a smooth transition to the sustainment organization. The tasks and products of this activity must be in compliance with the *Software Transition Plan* (STrP) that defines the plans for transitioning the software, test beds and tools to the maintenance center's facilities.

The preparation of required maintenance manuals, such as the *Computer Programming Manual* and *Firmware Support Manual*, should begin early in the Software Design activity and continue into subsequent activities as pertinent information becomes available. The final updated versions of the system and software requirements and design descriptions should also be included as needed. The preparation may include the planning, preparation and presentation of maintenance training if required.

12.7.7 Maintenance of COTS-Intensive Systems

Sustainment becomes more complicated with the increased use of third-party software such as *Commercial Off-the-Shelf* software products. A COTS-intensive system may be a suite of multiple components—including reuse code, legacy code and COTS from multiple vendors—plus custom components to achieve the desired functionality. Maintaining such a composite of software elements presents a significant Software Maintenance challenge. System and Software Engineers must coordinate and collaborate during the sustainment activity to address (at least) the following key issues:

- *Parallel Testing Capability*: Incremental updates and development must take place without affecting ongoing operations.
- *Planning for Upgrades and Obsolescence*: Most COTS software products undergo new releases about every year or so; old releases will eventually be unsupported.
- *Data Rights*: Rights to the source code is essential in case the vendor goes out of business or if major changes are needed.

- *Technology Advancement*: Sustainment includes maintaining a technology refresh plan.
- *Vendor Licenses*: Transition of license management tasks needs to be jointly planned.
- *Information Assurance Testing*: System Regression Testing must be performed on all upgrades and patches related to security support and must be continuous during sustainment.

- *Design for Interchangeability*: The system architecture group should isolate the COTS products with minimal interfaces, rather than intermingling the COTS with the developed software, to facilitate replacement.
- *Software Risk Management*: Sustainment organization must have the resources and capability to identify and analyze risks and either remove the risks or implement workaround procedures.

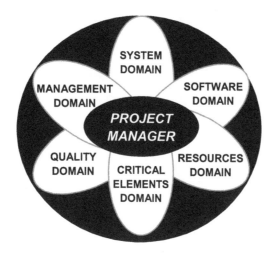

CRITICAL SOFTWARE ELEMENTS DOMAIN

5

Chapter 13

Software Documentation and Work Products

The right document, with the right content, available when needed at a critical decision point in the project, can be as close to a silver bullet as you can get.

During the software development process, various documents are needed at different phases of the life cycle. An overview of the plans for production of software documentation must be included in the Software Development Plan (SDP). Not all of the documentation and work products described will be needed by every program; in fact, it is almost certain that *no project, regardless of size, will need all of them*. Every project needs to decide what is needed, or contractually required and what makes sense for their situation and adds value to the project. This chapter describes specific software documents, plus other *software work products* (that are not documents per se, such as diagrams, databases, etc.), in the following sections:

- Purpose and Value of Software Documentation (13.1)
- Software Documentation Production (13.2)
- Core Software and System Documents (13.3)
- Software Management and Quality Control Plans (13.4)
- Software Work Instructions and Procedures (13.5)
- Non-Document Work Products (13.6)

13.1 Purpose and Value of Software Documentation

Software documentation supports the basic requirement to provide a well-defined and consistent *software baseline* during the Software Acquisition Life Cycle. Good documentation

provides the critical requirements and design information to Software Developers, provides management and customer visibility into the development process and progress, and helps to resolve the training problem created by personnel turnover.

Customer Visibility, Insight and Involvement. Customer technical insight and visibility into the development process is necessary to enhance communication, and helps to provide control and allows redirection if necessary. This insight and visibility is provided by day-to-day oversight of the development process, effective interface with the Developers, definition and monitoring of metrics and *Technical Performance Measures* (TPMs) and review of selected software development documentation in order to help ensure that the software product under development satisfies technical, cost, schedule and supportability requirements.

Software documentation (i.e., specifications, plans and reports) and software reviews provide the customer with evidence of the project's progress. Normally, your customer does not have the time or the required technical knowledge to become intimately familiar with everything about the system software. Good documentation, frequent reviews and modular development will provide your customer with sufficient visibility into what is being developed and a measure of the progress achieved. Customer visibility should occur as soon as possible in the development process to ensure that their requirements are being built into the system rather than added later, at much greater expense.

Document Traceability. As discussed in Chapter 2, software specification, design and test documents do not stand by themselves but must maintain consistent bi-directional traceability between each other. Each software requirement in the *Software Requirements Specification* (SRS) must link or trace up and back to one or more requirement in the *System/Subsystem Specification* (SSS). Similarly, each function or module in the *Software Design Document* (SDD) must trace up to one or more requirements to the SRS.

Each test defined in the *Software Test Procedures* must trace through the code and design that are tested back to the pertinent requirements in the SRS and SSS. After approval, the SRS and SDD serve as baselines for subsequent software development activities. The test procedures, Software Test Description (STD) and Software Test Report (STR) also should require approval, but they do not normally become baselines since they do not define mandatory characteristics of the final software product.

Staff Turnover. Almost every system development will have some turnover of personnel during development so software source code should be written so that it will be easily understood by new personnel. In addition, documentation can be used as an introductory and quick reference tool, providing a valuable aid for new project personnel. Also, documentation used and needed during the Software Sustainment period must be kept up-to-date. If the documentation is not current, it will impede and increase the cost of the maintenance effort.

13.2 Software Documentation Production

The example *Software Documentation Production Matrix* shown in Table 13.1 is an important guide as it summarizes the preparation of the normally required documentation during the Software Development and Test Life Cycle for large programs. It identifies when the normal preparation of a draft (D), preliminary (P) and baselined (B) versions

Table 13.1 Software Documentation Production Matrix Example

Software Development Activities	SRS	IFCD	SMBP	SDD SAD	IDD DBDD	STP	STD	STR	SVD SPS	SUM STrP	FSM CPM
Software Requirements Analysis	Pa	B	D								
SI Architectural Design	B	U	P	P	D/P	P					
Software Item Detailed Design[b]	U		B	B	B	B				D	
Software Implementation and Unit Testing[b]	U				U	U					D
Unit Integration and Testing[b]						U	D/P		D		
SI Qualification Testing[c]						U	P/B	B	P	P	P
SI/HI Integration and Testing[c]							U		B		
System Qualification Testing									U		
Preparing for SW Transition to Operations									U	B[d]	
Preparing for SW Transition to Maintenance									U		B

MATURITY LEGEND: D = Draft In Process; P = Preliminary Baseline Completed; B = Baselined; U = Updated Baseline (as needed); SOFTWARE DOCUMENTATION: SRS = Software Requirements Specification; IFCD = Interface Control Document; SMBP = Software Master Build Plan; SAD = Software Architecture Description; SDD = Software Design Description; IDD = Interface Design Description; DBDD = Data Base Design Document; STP = Software Test Plan; STD = Software Test Description; STR = Software Test Report; SVD = Software Version Description; SPS = Software Product Specification; SUM = Software User's Manual; FSM = Firmware Support Manual; CPM = Computer Programming Manual; STrP = Software Transition Plan; SI/HI = Software Item / Hardware Item.

[a] In this example, the SRS contains the Interface Requirements Specification (IRS), Software Requirements Traceability Matrix (SRTM) and Requirements Test Verification Matrix (RTVM).
[b] Iterative for each build.
[c] This activity may be iterative, in reverse order, or concurrent.
[d] Other optional user manuals include: Computer Operation Manual (COM); Software Center Operations Manual (SCSOM); and the Software Input / Output Manual (SIOM).

of documents are prepared, as well as when baselined documents are updated (U).

Some documents may not be contractually deliverable but are prepared because they are needed (such as unit test plans, test descriptions and special reports). In the Table 13.1 example, the SRS contains the *Interface Requirements Specification* (IRS), *Software Requirements Traceability Matrix* (SRTM) and *Requirements Test Verification Matrix* (RTVM). Table 5.7, in Chapter 5, contains a similar table but it is a breakdown of software documentation mapped to formal reviews.

It is also recommended to include in your SDP, or in an appendix to it, a master index of *all* software documentation. Software work products may include documentation, test results, non-document work products and the source code itself. Minimum work products vary according to software category and, of course, the program's *Contract Data Requirement List* (CDRL). Chapter 11 described the detailed software development activities and its subsections contain a list of software work products produced during each activity of the life cycle.

13.3 Core Software and System Documentation

Your contract should identify the required work products to be delivered. If the contract does not specifically identify deliverable work products, you must identify the documents and work products *needed* by the project. The selection of documentation that provides value to your project may be considered a matter of judgment. However, as a guide, the following tables provide a general set of core software documents, plus a set of system-level documents, that are can be generally considered a minimum set. Tables 13.2 and 13.3 provide a brief statement of the primary purpose of each document. Regardless, if the document does not add value to your project, don't produce it.

First Impressions. How documents look are important. It will not take the customer very long to examine or browse the documents they receive to get a first impression and first impressions are hard to reverse. Obviously, a document that has a beautiful cover but is basically "content

Table 13.2 Core Software Development Documents

Document Name	Document Describes
Software Development Plan (SDP)	The Software Development Life Cycle process
Software Requirements Specification (SRS)	What the software must do
Software Design Document (SDD)	How the software will do it
Software Test Plan (STP)	How it will be shown that the software fulfills its requirements
Software Test Description (STD)	The specific inputs, scenario, and acceptance criteria for each test
Software Test Report (STR)	The results obtained when the test procedures were conducted
Software Product Specification (SPS)	What was coded, tested and delivered
Software Version Description (SVD) (can be an updated version of the SPS)	What is included in the specific version of the software that is delivered
Software Users Manual (SUM) Software User's Manual (SUM) (essential if a human operator has direct access to and control over the software operation)	How a human operator uses the software and interprets its results

Table 13.3 Core System Development Documents

Document Name	Document Describes
System/Subsystem Specification (SSS)	What the total system must do
Interface Requirements Specification (IRS)	What requirements are imposed on the software-to-software and hardware-to-software interfaces
Interface Design Document (IDD)	How the Software Items communicate with each other and with the hardware

free" is unacceptable. It has to *look good on both the inside and the outside* as I realized once in the following scenario:

> **Lessons Learned.** As a college co-op student, I spent 18 months (alternating quarters) working at McDonnell Aircraft Corp. in St. Louis where I worked as a trainee rotating through 50 departments. A new model carrier-based aircraft was being prepared for final approval by the Navy. It was rolled out to the tarmac that morning and the Navy brass were to arrive that afternoon. The original Mr. McDonnell came out to inspect it and told the engineers, according to first-hand reports, to roll it back in because it was not ready. The Chief Engineer said "Mr. Mac this plane has been checked out many times; everything works perfectly. Why do you believe it is not ready?" Mr. Mac said, "It's not shiny enough!"

What has this story got to do with documentation? The answer is *perception*. If your customer reviews your documentation and it looks poor, there will be a perception that the entire system you have developed is poor. If you write

good documents, and they "look good," and have substantive content, the first impression will be good, and this can have a major psychological positive impact on the System Test and customer approval process.

13.4 Software Management and Quality Control Plans

Software management and quality control plans represent important adjuncts to the SDP that document specific implementation details not covered in the SDP. These plans may be an SDP Addenda or bound separately. In either case, Table 13.4 contains an example list, and brief descriptions, of the purpose for typical management and quality control plans that should be produced—*if they provide value added to your program*. The production matrix, shown in Table 13.1 above, does not include *software management and quality control plans* that may be required or needed such as the following plans:

■ Software Development Plan (or equivalent, is always needed)
■ Software Measurement Plan (needed on medium to large programs)

Table 13.4 Candidate Software Management and Quality Control Plans

Name of Plan	Purpose of Plan
Software Measurement Plan or Guidebook	Describes the approach, guidelines and "how to" instructions for establishing a standard software metrics program across the software development effort. It contains specific user instructions as to what measurements to make, when to make them, calculations needed to translate the measurements into useful management data, analysis techniques and report format examples.
Software Subcontract Management Plan	This plan may be included in the program's Subcontract Management Plan. It describes what software is subcontracted, and to whom, responsibilities of the Subcontract Management Team, identification of its members, responsibilities of the Software Subcontract Technical Manager, subcontract tracking and oversight of software activities and references to contractual commitments.
Software Risk Management (or Mitigation) Plan	The plan for determining and mitigating software-related risks. It describes the approach to identification and management of risks inherent in the development effort including reliability, design, cost and schedule risks. It should assign a risk severity level to each identified risk, define risk handling plans where needed, the process for assuring implementation, and provide plans for maintaining and improving maturity levels of team members. It may be included in the program's Risk Management Plan.
Software Data Management Plan	Provides details on the scheduling, formatting, delivery, storage and control for program deliverable and non-deliverable software documentation and media. It describes how the program provides: current program information; expedient interchange and access of controlled data to program personnel; timely delivery of contract deliverables; the repository and central access point for software documentation; the data accession list; document storage media control; and the focal point for software-related information. It *must* also include the mechanism for electronic access to the data by the customer.

(Continued)

Table 13.4 (Continued)   Candidate Software Management and Quality Control Plans

Name of Plan	Purpose of Plan
Software Reviews Plan	Provides software management with the controls necessary to oversee software development review activities and provides Software Engineers with the standards and practices required to conduct software development reviews. It describes what reviews will be held, the objectives of each review, when the reviews will be held, products reviewed, and establishes the entry and exit criteria for each review.
Software COTS/Reuse Plan	Covers COTS/Reuse software evaluation, selection, procurement, development environment requirements, special COTS/Reuse Configuration Management procedures, acceptance procedures, integration and implementation, maintenance, evolution and vendor monitoring and management.
Software Resource Estimation Plan	It should describe the derivation of software resources needed including software size; staffing; development effort; schedules and milestones; costs; and critical computer resources needed. It should also describe the processes for making estimates; periodic refinements of the type of measurements collected; documenting results; and using parametric estimating models.
Software Roles and Responsibilities	Summarizes the roles and responsibilities for each Software Engineering skill group (usually in tabular form) including: SwIPT Lead, Chief Software Engineer, Subsystem Chief Software Engineer, Software Process Lead, SwIPT Software Lead, SwIPT Software Integration and Test Lead, Software Item Lead, Software Engineers, software test engineers and Software Configuration Management (see Appendix C).
Software Safety Plan	Describes the Safety Critical safeguards that must be built into the software when human safety is involved. It may be incorporated into other documents such as the "System Safety Program Plan" or the "Risk Management Plan."
Software Configuration Management Plan	Establishes a plan for creating and maintaining a uniform system of configuration identification, control, status accounting, and audit for software and work products throughout the software development process including the Corrective Action Process.
Software Quality Assurance Plan (or (Software Quality Program Plan)	Establishes a planned and systematic software quality process to ensure that the software products and software processes comply with program contractual requirements as well as program process and product standards. It identifies the activities performed by the SQA organization in the development of all SIs and describes the SQA policies, procedures and activities to be used by all Software Development Team members. It is normally produced by the SQA Engineer assigned to your project.
Software Quantitative Management Plan	A high-level plan for establishing quantitative management on a program including quality goals, customer goals, other goals to supplement the IMP, priorities and metric limits. It may be part of the Software Measurement Plan.
Software Process Improvement Plan	Describes how process improvement is integrated into the management culture, and the plans for implementing a managed, iterative and disciplined process for improving software quality; increasing productivity; reducing cost and schedule; and eliminating activities with little or marginal value. It should describe the controls, coordination and information feedback needed from the software development process; defect detection, removal and prevention process; the quality improvement process; and software metrics.
Software Peer Review Plan	Defines the procedures, data collection, responsibilities, and reporting guidelines for inspections and evaluations of software products.

- Software Subcontractor Management Plan (needed if you have subcontractors)
- Risk Management Plan (needed on most programs)
- Data Management Plan (usually needed on large programs)
- Software Reviews Plan (usually needed on most programs)
- Software COTS/Reuse Plan (needed if you utilize reused software)
- Software Resource Estimation Plan (usually needed on large programs)
- Software Roles and Responsibilities (needed on large programs)
- Software Safety Plan (needed if human lives are involved)
- Software Configuration Management Plan (needed on medium to large programs)
- Software Quality Assurance Plan (almost always needed)
- Software Quantitative Management Plan (usually needed on large programs)
- Software Process Improvement Plan (needed if you are at a low maturity level)
- Software Peer Review Plan (needed on medium to large programs)

If your project is part of a large program, some of the management and quality control plans may be produced at the program level and used by all projects. If applicable, *Data Item Descriptions* (DIDs) should be listed on a *CDRL* in order to ensure the software work products are delivered as required under the contract. Each DID should provide a full description of the contents of each deliverable software document. Note that the program-level *Master Test Plan* is not a software document so it may be necessary to develop a more detailed project-level *Software Test Plan* and include it as an addendum to the SDP.

13.5 Software Work Instructions and Procedures

The defined software development process, as captured in the SDP at a relatively high level, should be elaborated through detailed Work Instructions and/or Procedures. These instructions or procedures should contain detailed directions for the day-to-day implementation of the software process. In other words, *the SDP describes what needs to be done and the procedures describe how to do it*. For large organizations, the work procedures provide standardized methods for performing specific tasks.

A list of the Work Instructions and Procedures can be included in an SDP Appendix, but the Work Instructions and Procedures themselves should be bound separately as

they are typically voluminous (sometimes three linear feet of three-ring binders). Table 13.5 contains a partial example list of Work Instructions and Operational Procedures that should be included in the SDP.

Work Instructions and Operational Procedures are seldom developed for, or by, each project. Detailed procedures are usually based on heritage or parent organizational sources for similar activities and then customized for your program's use if necessary. For activities shared across each program, common procedures may be developed. The Software Engineering Process Group (SEPG) should maintain an inventory of approved software procedures for your project.

Lessons Learned. Section 13.5 discusses detailed "how to" software procedures. There is another type of procedure that can be called "operational procedures" typically developed by the company you work for and are mandated for all projects and programs. Clearly, larger companies need

Table 13.5 Work Instructions and Operational Procedures—Example

Work Instruction Number	Work Instruction Name
#########	Action Item Processing
#########	Change Control Process
#########	Coding and Software Unit Test Planning and Reporting
#########	Configuration Management—Audits and Reports
#########	COTS Baseline Management and Product Evaluations
#########	Handling of Deviations and Waivers
#########	Electronic Data Management
#########	Formal Reviews
#########	Schedule Development, Approval and Maintenance
#########	Software Configuration Control Board
#########	Software Development File (SDF) Format
#########	Software Disaster Recovery and Backup Storage
#########	Software Integration and Test
#########	Software Peer Reviews
#########	*Etc.*

more standardization for control purposes; however, sometimes the "procedure gurus" get out of control. For example, early in my career, I worked for an organization that required a series of approvals for *all* purchases. I ordered a $6 wall clock. After reviewing the time it took for all the required approval signatures, I calculated it cost the organization $75 to buy the $6 clock. You may not have any influence on corporate procedures, but if it results in an unreasonable adverse impact on your budget, at least complain loud enough for the procedure gurus to hear it.

13.6 Non-Document Software Work Products

Non-document software work products are *not* included in the Documentation Production Matrix in any of the above tables. It is important to note that not all of the software work products are documents. The following are examples of software work products that may be produced during software development and may or may not be contractually required to be delivered:

- Software Requirements Database
- Software architecture, data flow and interface design diagrams
- Engineering memos
- Status and productivity reports
- Software use cases and scenarios
- Simulation Models and design captured in Object-Oriented (OO) Models
- State Transition, Software Hierarchy and Functional Block diagrams
- Design review packages
- Formal briefings (e.g., from JMRs or JTRs)

Chapter 14

Software Estimating Methods

A critical element of the project planning process is developing a realistic understanding of how much the software project is likely to cost, and how long it is likely to take, before the project starts.

Estimating is a challenging and difficult task. Managing costs and schedules was discussed in Chapter 8; this chapter addresses the methodologies used to *estimate cost and time*, and its management implications, in the following four Sections:

- Software Estimating Fundamentals (14.1)
- Software Size Estimation Methods (14.2)
- Software Code Growth (14.3)
- Software Size Estimation Tools (14.4)

Software tools should be used extensively to assist in cost and time estimation. They range from spreadsheets and project management software to specialized simulation and cost estimating tools such as the tools briefly described in Section 14.4. Using these tools reduces the incidence of calculation errors, speeds up the estimation and updating process, and allows consideration of multiple costing alternatives.

14.1 Software Estimating Fundamentals

Software estimates of effort, time, and costs are *living* estimates based on the best information available at the time of the estimate. *Estimates are not a commitment carved in stone.* They are a *snapshot* based on what is known about the system *at the point in time when the estimate is made.* The initial estimate and all subsequent updates should be prepared utilizing a standardized process, based on changes to the project requirements, interfaces and other factors, throughout the life of the project.

Lessons Learned. You may have learned in kindergarten that "haste makes waste" and that old adage is absolutely applicable to developing software-intensive systems but it is paramount in *estimating*; you must take the time necessary to do it right to the best of your ability.

14.1.1 Estimation Accuracy

Software Managers are often measured against how well they carry out their budgets, so you should do whatever you can do to improve your estimating abilities. The accuracy of software cost and time estimates depends primarily on *when* the estimate is made. The improvement in accuracy as the project moves through the Development Life Cycle is discussed in Section 14.3, but producing a reasonably accurate estimate *before* the project begins is the real challenge. Sometimes, non-sophisticated methods can work as described in the following method I once used for smaller systems:

Lessons Learned. Many years ago I was responsible for preparing *fixed price* software development contracts (a dangerous way to make a living) for a software programming and consulting company. I took the following approach to estimate the project development cost and schedule. Since these automation projects were relatively small in scope, I was extremely careful to make sure I had compiled a *detailed and documented specification* of exactly what the customer wanted. Then I gave the specification to the assigned Programmers to review, along with my best effort to define the end products, and asked them to give me their best estimate of the time it will take them to implement.

If the Programmer was a neophyte, with little or no experience in the subject matter, I multiplied his/her estimate by *four*. If the Programmer was experienced and had some background in the subject matter, I multiplied the estimate by *three*. If the Programmer was an expert, with a lot of experience in the subject matter, I multiplied the estimate by *two*. You may think this was a "tongue-in-cheek" method to conduct serious business (actually it was), but I must tell you that *my estimates were usually close to the final numbers*. I am not going to recommend this approach but it worked for me very well back then for relatively small projects, so if there is no other way, remember this as a rule-of-thumb.

The larger your project is, the more essential it is to have a good estimate, but a large project is more difficult to estimate. There are no global solutions and no silver bullets. It is a difficult and normally a time-consuming task but you, as the Software Project Manager, must roll up your sleeves and get it done. Depending on the chosen estimation method, or methods, you could do some of it alone, but it is recommended to get advice during the process from your Senior Developers, Designers and Testers as well as from experts in using the estimation tools selected for your project.

14.1.2 Effort Distribution over the Life Cycle

If you organize software-intensive systems into three group sizes, small, medium and large, you will find that the *percentage distribution* of the development cost and time for *each activity* varies. For example, for a small system, the percentage of total effort devoted to Coding and Unit Testing may be 50% or more. In a very large system, Coding and Unit Testing may be as low as 10% of the total effort. Figure 14.1 is my notional representation of the variability of the effort distribution of each major development activity for small, medium and large projects.

14.1.3 Calculating Equivalent Source Lines of Code (ESLOC) for Reused Software

When Software Design and/or code is *reused*, the estimation of its real cost is usually based on an approach called the *ESLOC* count. The premise to this approach is that you are not building the reused software from the ground up and that *some portion* of the design, code and/or testing does not require development effort and should not be considered a cost item. The method to be used for calculating ESLOC must be described in your *Software Development Plan*.

A common approach to calculating ESLOC is to set the proportionate weighting factors for *designing* (40%), *coding*

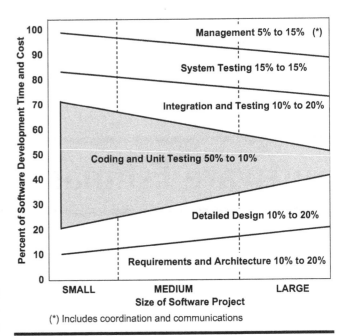

Figure 14.1 Effort allocation over the Development Life Cycle.

(30%) and *testing* (30%) for the reused software product. Programs may deviate from these standard proportions. A weighting breakdown of 40%, 20%, 40% is also often used so you must decide what is appropriate for your project. The ESLOC count is calculated by estimating the percentage of *new* design, coding and testing needed for the deliverable product and multiplying the sum of these weightings by the lines of code in the reused product. The general formula is:

$$ESLOC = \text{Lines of Code} \times \big[\text{percent design changes}$$
$$+ \text{percent code changes} + \text{percent retesting} \big]$$

For example, assume an existing software product, or COTS product, with 1000 Source Lines of Code was selected for reuse by a module in your program which has very similar functionality. Upon examination of the reused product, an estimate is made that only 10% of the design needs to be changed, 30% of the code must be rewritten, and 60% of the reused software module needs to be retested. In this example, using the 0.40, 0.30 and 0.30 weighting rule, the ESLOC is 310 and is calculated as follows:

$$1000\big[(0.10 \times 0.40) + (0.30 \times 0.30) + (0.60 \times 0.30)\big]$$
$$= 1000\big[0.04 + 0.09 + 0.18\big] = 1000\big[0.31\big] = 310$$

Another approach is having experienced Programmers examine the reuse product and, through the Delphi technique (described in Subsection 14.2.4 below), the ESLOC can be arrived at by a consensus.

14.2 Software Size Estimation Methods

There are three generic types of software estimation: expert *judgment*; formal mathematical-based *models*; and some *combination* of them. Table 14.1 is a summary of the most popular estimation approaches. There are strong proponents of each approach and even stronger proponents of the tools that support each estimation approach.

> **Lessons Learned.** My experience has demonstrated, without question, using the results of any two approaches to the same estimate is far superior to relying solely on one approach (or tool). *Corroboration* is the key to accuracy as the confidence level of the estimate increases dramatically when two independent estimation approaches arrive at the same, or close to the same, result. It could be two expert judgment estimates, one expert judgment and one formal model, or (my best choice) two formal parametric models. The best use of tools is to derive *ranges of estimates* and gain an understanding of the *sensitivities* of those ranges to changes in various input parameters.

14.2.1 Bottom-Up versus Top-Down Software Estimating

Bottom-Up Estimating. This type of estimating requires each task to be broken down into small and understandable components and is described in this Guidebook as the Work Breakdown Structure (WBS). As discussed in Subsection 1.5.2 and Section 8.2, the WBS is an enumeration of *all* work activities to be performed by the contract, structured into a hierarchy that organizes the work activities into short, understandable, manageable, and trackable tasks. Individual estimates are developed to determine what specifically is needed to meet the requirements of each of these smaller components of the project.

The estimates for the smaller individual components are then aggregated to develop a larger estimate for the entire task. In doing this, the estimate for the task as a whole is typically far more accurate than making one large estimate which typically will not as thoroughly consider all of the individual components of a task. In general, as the size of the task being estimated becomes smaller, the accuracy increases. Also, some estimates will be too high, and some will be too low, but if you have enough of them, the law of averages will offset the highs and the lows thus promoting a more accurate estimate.

> **Lessons Learned.** I prefer the advantages of bottom-up estimating because lower-level employees take a personal interest in the estimating process and that can potentially improve their motivation and morale. Team members work together and their estimates are taken to the next higher level until reaching you or your senior management level for approval. Though lower-level team members help to develop and implement the planned estimate, it is primarily the Project Manager's responsibility to see that the project is completed within budget and on time.

Top-Down Estimating. Senior Managers may be reluctant to accept advice or guidance from lower-level employees, so top-down planning gives them control of the decision-making process. However, senior management must be specific with their expectations if they want those who are not part of the planning process to follow the plan. Often this type of planning, which may rely on incentives, creates problems with motivation and moral. Using top-down planning in project management does not take full advantage of your talented employees who could (and do) have much wisdom to offer the project in a bottom-up estimating approach. On the positive side, top-down planning allows for the division

Table 14.1 Software Estimation Approaches

Estimation Approach	Type	Tools to Support Implementation
Bottom-up (based on WBS)	Expert Judgment	Project management software
Parametric Model	Formal Models	COCOMO II; SEER–SEM; SLIM; PRICE–S
Group Estimation	Expert Judgment	Wideband Delphi; Planning Poker (Agile)
Size-based Model	Formal Analysis	Function Points; Use Cases; Story Points (Agile)
Analogy/Comparison	Expert Judgment	Weighted Micro Function Points; ANGEL (UK)
Combination	Expert Judgment	Expert judgment and a parametric model
Dual Parametric Models	Formal Models	Comparison of two parametric models

of a project into strategic steps which can be studied and the tasks properly planned and assigned.

To create a top-down estimate, you need to identify similar projects and consider what made them take as long as they did. Then combine the aspects of those past projects with what you know about your current project to come up with an overall estimate. A top-down estimate is *based on analogies rather than on task breakdowns*. In my opinion, it's more of an *educated guess* but not a wild guess. The objective is *not* to be precise but to produce a *Rough Order of Magnitude* (ROM). The typical steps are:

■ Identify 3–5 similar projects that are completed. These are projects with similar characteristics such as similar architecture and size that started with about the same amount of information.

■ Determine how long each project took in man months. Hopefully, this data is either documented or you have access to the Managers or Developers of those projects.

■ Identify the factors or cost drivers that were the major influencers on each project's duration. This is the hard part because such data may not be documented, but you should seek out those who were involved to extract as much information as you can.

■ Apply what you learned to your project and discuss with your team the comparison of the new project versus the completed projects.

The top-down estimation approach may be able to be performed in a shorter time period than the bottom-up approach, however, utilizing the WBS process is worth the time it takes because it provides a more accurate estimate. In any case, large systems must have a WBS, so bottom-up estimating prevails.

14.2.2 Parametric Estimating Models

This type of model refers to an estimation technique which utilizes the statistical relationship that exists between a series of historical data and a delineated list of other applicable variables. Some examples of these variables include square footage in a construction project, the number of lines of code that exist in a specific software application, and other similar variables. One valuable aspect of parametric estimating is the higher levels of accuracy that can be built into it (depending on the accuracy of input data). Some of the common parametric models are described in the software estimation tools Section 14.4. Parametric estimating helps acquisition-side (customer/user) estimators to:

■ Provide an independent cost estimate as a means of comparing budgets and Developer estimates, and weigh the impacts of potential change orders quickly, before requesting the changes.

■ Build product breakdown structures to simplify performing an assessment of alternatives and estimates to complete (ETCs).

■ Provide a framework for collaboration and sharing results throughout the program.

■ Improve program affordability management through a life cycle approach, minimizing risk and cost and delivering better value by improving the efficiency and effectiveness of the project.

Supply-side (Developer) estimators can leverage parametric estimating to:

■ Accelerate the evaluation of bid/no-bid scenarios.

■ Increase estimate credibility while reducing proposal cycle time and costs to develop estimates.

■ Simplify the ability to respond to, or recommend, alternative design and program execution options.

14.2.3 Analogy/Comparison Estimation

A pure Analogy/Comparison Estimation approach involves finding similar projects in historical data, retrieved from completed software projects, which may have been developed by different organizations, to assist Project Managers in performing an appropriate cost estimation of their software project or subsystem. Analogous estimating is a form of *expert judgment* and is also referred to as Case-Based Reasoning (CBR).

> **Lessons Learned.** At Lockheed Martin, I sometimes used a "Software Size Estimator" database developed by the Space Systems Division. That division saved and compiled statistics from *completed* software-intensive programs over many years and developed an effective database-driven software sizing methodology that supported the cost estimating process. The cost data was collected at the Software Item (SI) level and contained *functional characteristics* of the developed software, processing modes, languages, etc. that assisted in preparing sizing estimates by *Analogy and Comparison to your program*. If this approach is applicable to your line of business, you should start an initiative to develop such a database or get corporate to initiate the task company-wide.

14.2.4 Delphi Techniques

The Delphi technique was conceived at The Rand Corporation in the late 1940s to make predictions about future events; it was named after the Greek oracle of antiquity, located at Delphi. It is a consensus technique. This is generally how it works. Three or more members of your team are given all the details you have at that time regarding a proposed software development and they are asked to make an estimate regarding the time needed to implement it. Each member makes their estimate in a preliminary round without consulting the other participants. The first round results are then collected, tabulated and provided to each participant for review. In the second round, the participants are again asked to make the same estimate, but this time with full knowledge of the other participants' estimates in the first round.

The second round results in a narrowing of the range of the estimates, providing a reasonable consensus by the group. Estimates that are much higher, or lower, than the consensus are discussed to evaluate the rationale used by those estimators. The original Delphi technique avoided group discussions; the *Wideband Delphi* technique (developed by Barry Boehm in 1981) added group discussions between assessment rounds. Wideband Delphi is a useful technique for coming to a reasonable estimate when you need "expert opinion" and do not have comprehensive system requirements. Stellman and Greene (2006, pp. 39–48) describe the Delphi estimation technique in depth.

14.3 Software Code Growth

Estimates of code size made early in the life cycle of large complex systems are often not even close to the final code size. As the program progresses through the development activities, and the degree of clarity of the deliverable product increases, code size estimates become much more accurate. Figure 14.2 is my notional graphical representation of this inherent attribute of software code growth. The X-axis shows a multiplier percentage, at each development activity, to more accurately predict what the final code size will be depending on when the estimate is made. Figure 14.2 also shows a likely upper and lower range and a most likely plot.

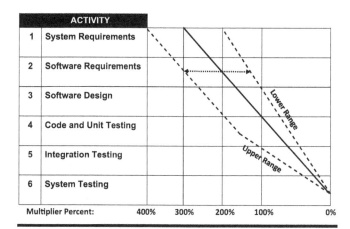

Figure 14.2   Software code growth multiplier (notional).

For example, if an estimate is made during software requirements (activity two in the figure), the chart indicates that a final code size estimate made during that activity predicts it is likely to grow from 150% to 300% with a 200% growth being the most likely final code size. For smaller and less complex systems, the three lines in Figure 14.2 would shift to the right.

The range of potential growth decreases more rapidly after Coding and Unit Testing (CUT). Programmers are notoriously optimistic about the time it will take them to accomplish a software task. Okay, so I really should not pick on Programmers since it is human nature to be that way. Some readers of this Guidebook may think this multiplier chart (Figure 14.2) is too drastic; well, keep records, and you will probably be surprised at the results. It may be a worst case scenario for small projects, but the larger the project, the more accurate this notional graph becomes.

Lessons Learned. A rule-of-thumb, with more real-world applications than you might imagine, is called the 90–90 rule of project estimation. It goes like this: *the first 90% of the task takes 90% of the time, and the last 10% also takes 90% of the time!* Don't laugh because I have been tracking this for a long time and am amazed at how often it is true. One main reason this happens is due to *deferring* the hardest parts of the project to the end. Another big reason is that the estimator is afraid to tell the funding organization what they really think it will cost because that is likely to result in a cancellation of the project. That "hill at the end" is always there; even the Boston marathon has "heartbreak hill" at the end so *plan for the heavy effort that is always present during the last 10% of almost any project* (it also applied to preparing this Guidebook).

If you do not have a good method of estimating how long a *new* project will take, or how much it will cost, use the 90–90 rule and multiply the best estimate by 1.8 (or just double it) and in the end you should be a lot closer than the original estimate.

There are many psychological factors potentially explaining the strong tendency toward *over-optimistic effort estimates* that need to be dealt with to increase the accuracy of effort estimates. These factors are present even when using formal estimation models because much of the input to these models is judgment-based. It is human nature to be optimistic (for most of us).

Factors that have been demonstrated to be important causes of over-optimism are: *wishful thinking, anchoring* (the common human tendency to rely too heavily on the first piece of information offered when making decisions), *planning fallacy* (a built-in optimistic bias) and *cognitive dissonance* (where an individual experiencing inconsistent, contradictory beliefs, ideas or values at the same time, tends to become psychologically uncomfortable, and is motivated to try to reduce this discomfort by over-optimism).

14.4 Software Size Estimation Tools

Some of the most popular software size estimation tools used in the past are briefly profiled below and include COCOMO II, SEER–SIM, Programmed Review of Information for Costing and Evaluation (PRICE–S®), Function Points and Software Lifecycle Management (SLIM®) (*the use of registered trade names in this material is not intended to infringe on the rights of the trademark holder*). These commercial cost models are important tools for estimating software cost and resources required. However, their accuracy during very *early phases*, well before development begins, is usually limited and questionable because their key input—realistic estimates of total software lines of code—are seldom available at that time.

Unfortunately, some of these cost models can produce widely different results when you make very small, and sometimes subjective, changes to their key input parameters. Furthermore, cost models do not always include all of the related software development costs such as software support to System Engineering tasks; hardware/software integration tasks; Software Configuration Management; Software Quality Assurance; database population, and other considerations.

> **Lessons Learned**. If possible, I *strongly* recommend you *use two estimation tools as a sanity check* of one against the other. If you get widely divergent results, then one (or both) are likely inaccurate. If the results of both are close, the level of confidence of the estimate rises dramatically.

14.4.1 Constructive Cost Model II (COCOMO II®)

COCOMO is an algorithmic software cost estimation model originally developed by Boehm (1981). *COCOMO II* (Boehm, 2000) is the successor of COCOMO 81 and is better suited for estimating modern software development projects. It provides more support for modern software development processes and includes an updated project database. The model uses a basic *regression formula* with parameters that are derived from historical project data plus current as well as future project characteristics. COCOMO consists of a hierarchy of three increasingly detailed and accurate formats:

■ Basic COCOMO: best for quick, early, Rough Order of Magnitude estimates of software costs

■ Intermediate COCOMO: takes into account the differences in project attributes
■ Detailed COCOMO: adds the influence of individual project activities (analysis, design, etc.)

The program size is expressed in estimated Thousands of Source Lines of Code (KSLOC). COCOMO identifies three *classes of software projects*:

■ *Organic Projects*: *Small* teams with *good* experience working with *loose* requirements
■ *Semi-Detached Projects*: *Medium*-sized teams with *mixed levels* of experience working with a mix of *rigid* as well as *less than rigid* requirements
■ *Embedded Projects*: Developed within a set of "tight" constraints. It also includes a combination of organic and semi-detached projects (hardware, software, operational, etc.)

Basic COCOMO computes software development effort (and cost) as a function of program size The Basic COCOMO version has limited accuracy because it does not account for differences in project attributes, hardware constraints, personnel quality and experience, modern tools and techniques, etc. Intermediate COCOMO provides more accurate results.

Intermediate COCOMO computes software development effort as a function of program size and a set of "cost drivers" that include subjective assessments of product, hardware, personnel and project attributes. This COCOMO extension includes a set of four "cost drivers," each with a number of subsidiary attributes resulting in a total of 15 attributes. The principal *cost drivers* are:

■ *Software attributes* including required software reliability; size of the application database; and complexity of the product
■ *Hardware attributes* including run-time performance constraints; memory constraints; volatility of the virtual machine environment; and required turnabout time
■ *Personnel attributes* including analyst capability; Software Engineering capability; applications experience; virtual machine experience; and programming language experience
■ *Project attributes* including use of software tools; application of Software Engineering methods; and the required development schedule

Each attribute receives a rating on a six-point scale that range in importance from "very low" to "extra high." An effort multiplier from a standard COCOMO table applies to the rating. The product of all effort multipliers results in an *Effort Adjustment Factor* (EAF)—the typical range is 0.9 to 1.4.

Detailed COCOMO incorporates all characteristics of the intermediate versions plus an assessment of the cost driver's impact on each phase of the Software Engineering process. The detailed model uses different "effort multipliers" for each cost driver attribute. These *Phase-Sensitive* effort multipliers determine the amount of effort required to complete each phase. All of the software is divided into modules, COCOMO is applied to each module to estimate effort, and the sum is total effort.

14.4.2 System Estimation & Evaluation of Resources/Software Estimation Model (SEER–SEM®)

Also called *SEER for Software* (Galorath, 2010), this cost estimation model is an algorithmic project management software application designed specifically to estimate, plan and monitor the effort and resources required for any type of software development and/or maintenance project. SEER (having a double meaning to one who has the ability to foresee the future) relies on parametric algorithms, knowledge bases, simulation-based probability and historical precedents to allow Project Managers, Engineers, and Cost Analysts to estimate a project's cost schedule, risk and effort before the project is started.

Since the initial release in 1988 by Galorath Inc., SEER–SEM has undergone vast upgrades including major enhancements to the core math behind the model that handles the realities of projects rather than just a Rayleigh curve approximation, plus many more knowledge bases. In 2003, SEER–SEM added significant new features such as Goal Setting (allowing projects to be managed) and Risk Analysis (allowing Project Managers to make changes to estimates). Version 6 of SEER for Software was the first to allow SEER to both input and output through various Microsoft products, such as Excel. Version 7 of SEER included better handling of projects that stretch beyond their optimal effort.

SEER Version 10.3 represents perhaps the first time that SEER could be integrated to support *all* phases of a software project's lifecycle. The size of SEER for Software has grown to over 200,000 Source Lines of Code and shifted from simply a means to generate work estimates through parametric modeling, to a system that enhances those results with simulation-based probability and over 20,000 historical cases to draw conclusions from. Capabilities of this model will certainly change over time.

SEER–SEM has also branched into cost estimation products for hardware and electronics (SEER–H); Manufacturing (SEER–MFG); and Information Technology (SEER–IT) a version of SEER created to aid IT professionals in estimating the design, build and maintenance of information technology infrastructures and service management projects.

SEER for Software is composed of a group of models working together to provide estimates of effort, duration, staffing, and defects. These models can be briefly described by the questions they answer:

- How large is the estimated software project (Lines of Code, Function Points, Use Cases, etc.)?
- What is the likely productivity of the Developers?
- What amount of effort and time are required to complete the project?
- How does the project outcome change when schedule and staffing constraints are applied?
- How should activities and labor be allocated into the estimate?
- Given expected effort, duration and labor allocation, how much should the project cost?
- Given the information provided, what is the expected quality of the delivered software?
- How much effort will be required to adequately maintain and upgrade a fielded software system?
- How is the project schedule progressing and when will it end?
- Is this development achievable based on the technology involved?

14.4.3 Programmed Review of Information for Costing and Evaluation

Also referred to as "PRICE–S" by the software community, it was developed by RCA in the 1960s for internal hardware estimates for aerospace applications. PRICE—S was a pioneer in the science of parametric modeling and has grown into a family of models supporting estimates for hardware, software, microcircuitry, life cycle costs, etc. The primary input to the model is *Source Lines of Code* combined with organizational performance factors to produce a *top-down estimate* of effort and development time plus a number of useful outputs including progress forecasts.

PRICE® systems expertise in cost estimating software provides benefits for cost-related functions, including the ability to improve coordination and value throughout the organization by virtue of the Estimating System Integration Framework. The PRICE True Planning's life cycle approach to program affordability management helps improve the efficiency of project selection, control and delivery over the long-term while minimizing risk and cost. It also provides a frame of reference for collaboration and sharing results throughout the program and throughout the supply chain.

> **Lessons Learned.** I have used PRICE–S successfully in the past. If you use it, I recommend hiring a trained expert to help because I found that small changes in some of the parameters can

make a huge change in the results. PRICE–S is not the only parametric tool to have this issue. These tools are powerful, but you need to know how to use them to obtain accurate results. Also, as discussed earlier, using two tools as a sanity check for verification of results is recommended.

14.4.4 Function Points

While Source Lines of Code is an accepted way of measuring the absolute *size* of the code from the Developer's perspective, metrics such as *Function Points* capture software size functionally from the user's perspective. The Function-Based Sizing (FBS) metric extends Function Points so that hidden parts of software such as complex algorithms can be sized more readily. The cost (dollars or hours) of a unit is calculated from past projects. The recognized standards for sizing software based on Function Points is defined by the International Function Point User Group (IFPUG): ISO/IEC 20926:2009 (ISO/IEC, 2009); and COSMIC (ISO/IEC 19761, 2011).

The process begins with the identification of *functional user requirements* of the software and each one is categorized into one of five types: outputs, inquiries, inputs, internal files and external interfaces. Once the function is identified and categorized into a type, it is then assessed for complexity and assigned a number of Function *Points*. Each of these functional user requirements maps to an end-user business function, such as a data entry for an input or a user query for an inquiry.

This distinction is important because it tends to make the functions measured in Function Points map more easily into user-oriented requirements, but it also tends to hide internal functions (e.g., algorithms), which also require resources to implement. Recently there have been different approaches proposed to deal with this perceived weakness, implemented in several commercial software products. Variations of the IFPUG method designed to make up for this (and other weaknesses) include:

- *Early Function Points*: Adjusts for problem and data complexity with questions that yield a somewhat subjective complexity measurement; it eliminates the need to count data elements.
- *Engineering Function Points*: Elements (variable names) and operators (e.g., arithmetic, equality/inequality, Boolean) are counted.
- *Bang Measure*: Defines a function metric based on 12 primitive (simple) counts that affect or show Bang, defined as "the measure of true function to be delivered as perceived by the user." Bang measure may be helpful in evaluating a Software Unit's value in terms of how much useful functionality it provides.

- *Feature Points*: This adds changes to improve applicability to systems with significant internal processing (e.g., operating systems, communications systems).
- *Weighted Micro Function Points*: Adjusts Function Points using weights derived from program flow complexity, vocabulary, object usage, and algorithmic intricacy.

As the Software Project Manager, you need to evaluate the value of Function Points to your project especially if a good measuring technique, namely estimating lines of code, is available.

14.4.5 Software Lifecycle Management

SLIM, also called the *Putnam Model* (Putnam, 2003), is an empirical model that works by collecting software project data (for example, effort and size) and *fitting a Rayleigh curve to the data*. Estimates of future effort are made by providing size and calculating the associated effort using the equation which best fits the original data. Created by Lawrence Putnam, Sr., in 1978, it was a pioneering work in the field of software process modeling and is still widely used. SLIM describes the time and effort required to finish a software project of a specified size.

SLIM is part of the suite of proprietary tools Putnam's company, Quantitative Software Management (QSM), has developed based on his model. Putnam found that software staffing profiles over time followed the mathematical "Rayleigh distribution curve" and derived an equation that plots *effort* as a function of *time* yielding a *Time-Effort Curve*. The points along the curve represent the estimated total effort to complete the project at that *time*. One of the distinguishing features of the Putnam Model is that total effort usually decreases as the time to complete the project is extended. This is normally handled in other parametric models by a schedule relaxation parameter.

14.4.6 Web-based Tools

A newer concept in software estimation is the advent of web-based tools. These tools can be used from anywhere with a web browser by anyone that can capture basic requirements. It involves a server side and a client side. All calculations and data are performed and stored in one central (public) server/database. The client communicates using standard web services. The client side code is available as open source to download. The client is typically built with PHP; all code is in one single PHP file. This makes the web-based tool easy to share and integrate into corporate management or web systems enables software development effort estimation using different methods including: COCOMO II, WBS, Delphi, Analogy/Comparison Estimation and Custom modular estimation for web and mobile applications.

14.4.7 Custom Tools

Custom tools, for internal use only, are often built by a variety of organizations to address their specific software estimation efforts. A personal example of this took place when I was a member of a team that developed a ground software cost estimation model for use during a very early phase well before the project started and well before anyone had any idea of code size (Bowers et al., 2008). It was a desktop tool, not requiring the need to estimate lines of code, but instead took *basic characteristics* of the proposed ground software system to estimate software size and staff months.

In this personal example, the tool user goes through three steps during which 4–6 inputs are required. Initially, the user chooses the type of modification (small, similar or major new capability). Next, the user chooses the primary function of the new software selected from a list of functional characteristics derived from the ground software Work Breakdown Structure. Finally, the user selects the key program parameters (such as the number of payloads, downlink bandwidth, orbital tracking, etc.) for the program being estimated. Using a fixed estimate of productivity, the model produces three estimates of effort—low, nominal and high.

> **Lessons Learned.** Building custom tools are not easy to create, and they are typically expensive. Unless you can cost-justify an effort of this nature, I recommend going with a commercial product and, if it is not a perfect fit, tailor it to your needs. There are risks to this approach but they are usually less, and less expensive, than building your own custom tools.

14.4.8 Good, Fast and Cheap

It seems that the same "cosmic force" discussed earlier prevents most project developments (especially software) from *simultaneously* having high quality, completed on time, and inexpensive to produce. This well-known rule is sometimes called the "Quality Triangle Dilemma," and it is embedded in stone as the proverb: *Good, fast and cheap–you can only have two* as shown in Table 14.2.

The purpose of discussing this is to make sure you don't find yourself in this dilemma by trying to achieve all three elements simultaneously. You can always try to make your project *better, faster and cheaper*, but no matter how hard you try, it cannot be of high quality, completed quickly and inexpensive at the same time. You need to decide which of the two are more important to your project.

The Quality Triangle Dilemma in Table 14.2 is generally true; however, it omits an important fourth element: *Scope*. My version of this dilemma, that may be more valuable to a Software Project Manager, includes four elements: cost/budget (equivalent to cheap), schedule (equivalent to fast), quality (equivalent to good), plus scope (functionality). These four elements are defined as:

- Meeting *cost and budget* means the project can be completed within the allocated funding.
- Meeting *schedule* means the project can complete all tasks within the planned time period.
- Meeting *quality* means that the work products (primarily the code and documentation) meet the highest standards for correctness and completeness.
- Meeting *scope* means the project fulfills the customer's functional requirements.

The concept here is *if three of these elements are defined by the customer, the Project Manager must be allowed to determine the fourth element*. With this set of four elements, that I am calling the *Quality Quaternion*, the customer should not be allowed to dictate all four elements. The Quality Quaternion Decision Table is shown in Table 14.3.

If you should be so unfortunate to become the Manager of a project where the customer has indeed mandated all four elements, your task is going to be tough. You, and your team, are going to work a lot of overtime *trying* to improve performance and make the system better, faster and cheaper. Unless you are producing software that can endanger human lives, there is some slack in the quality of the product that you can manipulate to your advantage because quality is not always observable on the surface. Most of the time, the customer will want changes during development. This may give you a great opportunity to improve the situation by requoting

Table 14.2 Quality Triangle Dilemma

If you want the Development to be:	The Product will NOT be:
Fast *and* Cheap	Good (High Quality)
Good *and* Fast	Cheap
Cheap *and* Good	Fast (to produce)

Table 14.3 Quality Quaternion Decision Table

If the Customer specifies:	The Project Manager determines:
Cost/Budget; Schedule; Quality	Scope
Scope; Schedule; Quality	Cost/Budget
Cost/Budget; Quality; Scope	Schedule
Cost/Budget; Scope; Schedule	Quality

the cost and time for making these changes based on a worst case scenario.

Lessons Learned. When I was a consultant in the Washington, DC, area working for a relatively small consulting company, we were frequently bidding for government work against a division of a very large computer company. They always won when the *cost* of the system was the predominant award criterion. They won because they severely underbid the cost of development. They knowingly took this approach in anticipation of *changes* that almost always came during development and then they heavily overcharged the customer to make those changes thus recouping funding well beyond their underbid. As a small company, we could not risk being able to recover an underbid, but the big company could take such a risk. This is a tricky game but it may be an option when the competition is stiff, and you can afford to take that risk.

Another tack you might take is by implementing the easy functions early in the Development Life Cycle and postponing the more difficult and more time-consuming functions for a later time. When it becomes clear that your project is not going to make the cost or schedule milestones, your customer may be persuaded to replan the project and increase the cost, move out the schedule, or reduce the functionality.

If you are faced with a customer demanding an undoable task, you must mount an aggressive offense to figure out how to get the job done without losing your customer, your shirt, your reputation, or all three. The situation you are in may not be the fault of the customer but rather an overzealous attempt by your marketing organization making unrealistic commitments to get the contract.

Lessons Learned. I have been in this predicament and have observed that when a customer proposes an unrealistic cost or schedule, in the long run it turns out to cost them more, and take longer, than it would have been with a realistic cost and schedule estimate up front. If you believe the contract specifics are undoable, try to convince the customer of that *before* the contract is signed.

Whatever you decide to do, at the beginning of an unrealistic project you must discuss strategy with your senior management and be honest about the inevitability of not being able to meet all of your customer's demands for cost, schedule, quality and scope. If you want to protect your reputation, you should have an open discourse with your Manager so you will not be the scapegoat when the project gets into trouble. Closed door strategy meetings with your management should also be held periodically throughout the Development Life Cycle to review options and ideas for recovery of setbacks.

Under promising and over delivering is far better than the reverse.

Chapter 15

Managing Software Performance with Measurements

Managing the development of software-intensive projects is a knowledge-intensive process, and much of that knowledge comes from taking measurements. If you cannot measure it, you cannot control it.

The depth of *knowledge* available to the Project Manager concerning his/her project is directly proportional to the success of the project. So, how do Project Managers get this knowledge? The answer is it comes from several sources. However, the primary source is from *measurements* (also referred to as software *metrics*). Measurements and its related *Management Indicators* are critical to the software management process because *effective management controls are dependent on timely and accurate measurements and appropriate Corrective Actions* taken as a result of analyzing those measurements.

The plans to manage software development on your project using software measurements must be included in either the *Software Development Plan* (SDP), a *Software Measurement Guidebook* (SMG) typically an addendum to the SDP, a *Quantitative Management Plan* (QMP) that may be an appendix to the *Integrated Management Plan* (IMP)—or in all three at appropriate levels of detail.

The SMG and QMP must describe how the project sets and uses metrics including objectives, thresholds, plans, actuals, historical data in managing development and sustainment and how the metrics are used to influence program decisions. Software measurement initiatives should be in accordance with the software measurement standard(s) imposed on the program which may include:

- Existing organizational and contractually imposed software measurement policies, standards and procedures
- The international standard: ISO/IEC 15939, "Software Engineering—Software Measurement Process" (ISO/IEC, 2010)
- "Practical Software Measurement: Objective Information for Decision Makers," (McGary, 2001)

The software measurement initiative must be tailored to each individual project. The Project Manager, and the Software Development Team, must use *Software Management Indicators* to aid in managing the software development process and communicating the status of the software development effort to the customer, program management, and stakeholders. A brief description of *managing software performance with measurements* is covered in this chapter in the following ten sections:

- Principal Objectives of Software Measurement (15.1)
- Continuous Feedback Control Systems (15.2)
- Approach to Software Management Measurements (15.3)
- Types of Software Measurements (15.4)
- Software Measurement Categories and Indicators (15.5)
- Software Measurement Set (15.6)
- Measurement Data Collection Time Frames (15.7)
- Software Measurement Information Specification (15.8)
- Analysis and Reporting of Software Management Indicators (15.9)
- Software Indicator Thresholds and Red Flags (15.10)

15.1 Principal Objectives of Software Measurement

Typical objectives of a software measurement initiative are to provide:

- Relevant and timely information to help Software Managers, Leads, Developers, and Engineers perform their responsibilities correctly, on-time, within budgets, and at a high quality
- Vital tracking information for Project Managers to facilitate the reduction of the project's software cost, schedule, and technical risks by taking timely Corrective Actions
- A practical, efficient and up-to-date management methodology and basis for quantitative software development control, status determination and activity re-planning
- Historical records of performance for trend analysis and other value-added information, to support continuous Software Process Improvement

15.2 Continuous Feedback Control Systems

A fundamental aspect of the typical software measurement initiative is a continuous improvement approach through a closed-loop feedback control system as depicted by the three examples in Figure 15.1. It shows the *principles of feedback*

control with a generic example, an example for the common thermostat, and a (high level) feedback control loop as it might be applied to software development.

Lessons Learned. The simplicity of the concept of a continuous closed-loop feedback control system in Figure 15.1 hides its terrific usefulness as an *initial model* in depicting the relationship of the elements in such a system. I have used this simple approach many times to *begin the* graphical depiction of the feedback control system planned for my project.

Management Indicators. Figure 15.2 shows an example of a *Closed-Loop Software Management Control Process*. The Management Indicators (resources, process, product and quality) provide feedback data so that Corrective Actions can be applied to improve the development process, maximize resource utilization and predict and adjust the quality of the products. The resources and process produce the products. Figure 15.2 essentially depicts the essence of the SEI Software Process Maturity Level 4. It is an important model that should be used as a goal. However, the relative simplicity of Figure 15.2 can also be misleading as it *does not convey the inherent complexity of achieving a smoothly running, unencumbered, skillfully managed closed-loop software development control system.*

Corrective Action. Timely Corrective Actions must be taken because that will result in realizing maximum benefits from the software measurement program. Without Corrective

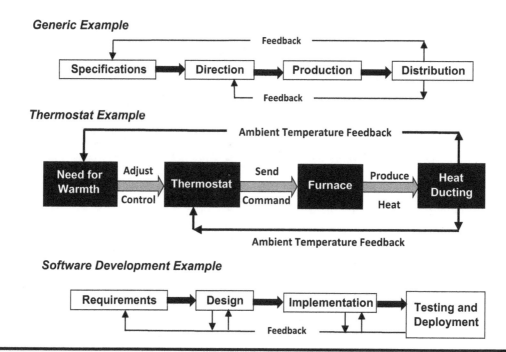

Figure 15.1 Continuous feedback control examples.

Actions, the measurement program would become merely a historical archive, and the cost of maintaining it would be difficult to justify. However, historical information can and should be collected and used for trend analysis, productivity calculations and continuous process improvement.

Reporting Formats. Software metrics can be represented in graphical or tabular formats consistent with your organizational guidelines, or your program requirements, plus your personal preferences and management style. Plotted data are

strongly recommended since trends are vital. The typical reporting period for formal measurement reports is monthly, however, under certain dynamic conditions informal reporting (such as weekly) might be needed. Significant changes should be *reported whenever they occur*, and formally reported in the next reporting period.

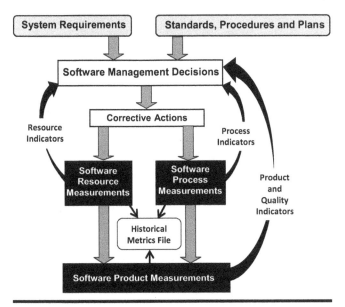

Figure 15.2 Closed-loop software management control process.

15.3 Approach to Software Management Measurements

An example of a top-down management measurement approach is depicted in the framework shown in Figure 15.3. It shows a tailorable hierarchy of four groups. Multiple users may have similar objectives but each group has a different perspective and different information needs. *The effort to collect, analyze and document metrics must be consistent with their value to the users and to the project.*

Goal-Question-Metric (GQM) Paradigm. The example of a top-down approach in Figure 15.3 is generally based on the *GQM* paradigm originally proposed by Basili (1984). At the top of Figure 15.3, software management establishes program/project goals (Group 1) resulting in a set of Measurement Categories (Group 2) to determine progress in meeting the goals. The categories identify a set of Measurement Indicators (Group 3) that provides information needed by the Group 2 categories. The measurement indicators are produced from the detailed software base measurements, as well as Derived Measurements (Group 4), that

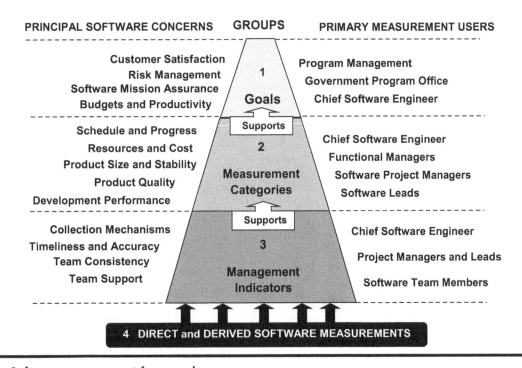

Figure 15.3 Software measurement framework.

must be collected to provide the data needed by the indicators. Following is a brief description of the four groups.

- *Goals (Group 1)*: Top-level Group 1 program/project goals are of primary interest to Senior Managers, the customer and the CSwE. They provide an effective means to appraise and track software milestones and overall project trends. Data to support the goals can be derived from a combination of measurements collected and calculated from lower groups.
- *Software Measurement Categories (Group 2)*: The Group 2 software categories address the question: What information do Software Managers, Leads and Developers need in order to manage their task in a timely and effective manner and be responsive to program/project goals? As shown in Figure 15.3, the categories are schedule and progress; resources and cost; product size and stability; product quality; and development performance.
- *Software Management Indicators (Group 3)*: Group 3 focuses on specific Software Management Indicators that must be collected to support the Group 2 Measurement Categories (see 15.4).
- *Software Direct and Derived Measurements (Group 4)*: *Direct* measurements (raw data) must be collected, plus the needed *Derived* measurements must be calculated, in order to provide the data needed for the Management Indicators in Group 3.

The hierarchy and relationship between Measurement Categories, Management Indicators and the direct and derived measurements (also called base measurements) is discussed in Section 15.6.

15.4 Types of Software Measurements

There are two basic types of software measurements (metrics) that are collected: direct and derived:

- **Direct Measurements.** These are measurements at the base, or primitive, level and include software product or process *attributes* for each measurement period such as lines of code produced, labor hours expended, number of defects found, etc. Direct measurements are functionally independent of other measures. Whenever possible, automated tools should be used for the data collection task.
- **Derived Measurements**. Derived measurements involve the need to collect more than one direct measurement and it requires a calculation. For example, to collect the monthly "requirements volatility" management indicator, it is necessary to collect measurements of the number of requirements added, deleted and modified this month divided by the total number of active requirements last month. Another example is the combination of code produced and labor hours to derive (calculate) productivity metrics.

15.5 Software Measurement Categories and Indicators

Figure 15.4 depicts the Measurement Categories and Indicators supporting the five *key software measurement questions* that Software Managers must have periodic responses to in order to effectively manage the software development effort.

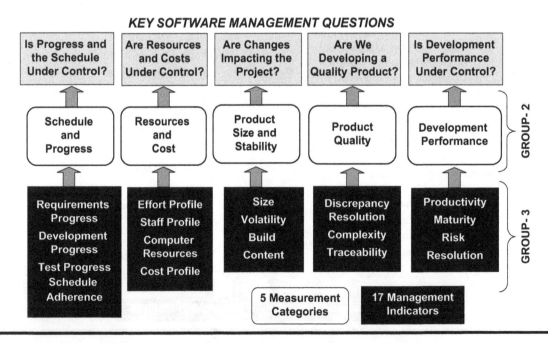

Figure 15.4   Categories and indicators support key management questions.

There is no restriction to adding more key questions if needed but these five questions should constitute a minimum set. The five *Measurement Categories* (Group 2) shown in Figure 15.4 directly provides answers to the five key questions and the five categories are supported by 17 *Management Indicators* (Group 3).

15.6 Software Measurement Set

Table 15.1 contains an example of a *software measurement set* summarized from numerous metrics programs I have managed. It includes the three levels (Measurement Categories, Management Indicators, and Direct/Derived Measurements)

Table 15.1 Software Measurement Set Categories, Indicators, Measures and Insights

Measurement Categories	Management Indicators	Measurements	Insights Provided
Schedule and Progress	Requirements Progress	Requirements Defined Requirements Verified and closed	Tracks status of requirements definition and effectiveness of the requirements definition process
	Development Progress	Components and Units Defined Units Designed, Coded & Tested (*)	Visibility of SU design, coding, testing and integration plus effectiveness of Peer Reviews
	Testing Progress	Test Cases Developed Test Cases Dry Run Test Cases Performed & Passed (*)	Tracks the ability to maintain and control software integration testing progress
	Schedule Adherence	Project Milestones (*) Scheduled Activities	Tracks the ability to maintain the overall software development schedule and schedule realism
Resources and Cost	Effort Profile	Labor Hours by Activity & Trends (*) Rework Hours by Activity	How well the cost of software development is being controlled
	Staff Profile	Staffing Level—actual versus planned Staff Experience Levels Staff Turnover	Potential schedule impacts, indication of timely task completion, and ability to maintain staff competency
	Target Computer Resources	CPU / Input-Output / Memory Utilization Response time	Identifies changes in computer resource limitations and provides time to rectify shortfalls
	Cost Profile	Earned Value Performance (*) Schedule—Cost Performance Index Schedule—Cost Variance	Tracks earned value and budget versus actual work performed versus work scheduled
Product Size and Stability	Size	Requirements Size and Growth Lines of Code Size/Origin/ Growth (*)	Status of estimated scope of work; indicates potential schedule, cost and staffing problems
	Volatility	Requirements Volatility Lines of Code Volatility	Indicates stabilization of software development and potential cost and schedule impacts
	Build Content	Requirements Implemented per Build	Indicates the extent of deferred functionality to later builds causing potential schedule impacts

(Continued)

Table 15.1 (Continued)   Software Measurement Set Categories, Indicators, Measures and Insights

Measurement Categories	Management Indicators	Measurements	Insights Provided
Product Quality	Design Complexity	Measures, quantifies and evaluates the structure of SUs with Cyclomatic Complexity	Indicates complexity and quality of the code, expected testing effort and ease of maintenance
	Action Item Resolution	Action Item Status and Aging Action Item Report Type and Source	Visibility and management of open action items and the aging of action items in-process
	Traceability	Requirements Traceability (*)	Tracks requirements traceability to design and to Test Cases
Development Performance	Productivity	Ratio of Code Size versus Labor hours.	Tracks the effectiveness of development performance
	Maturity	SDRs Opened versus Closed & Aging (*) Development Defect Density Path Testing Success	Tracks effectiveness of the Corrective Action and testing processes; provides an indication of software reliability
	Risk Resolution	The Number of Risk Mitigation actions opened versus closed.	Indicates the effectiveness of the risk mitigation process

(*) Can be considered a minimum set of base measurements.

consistent with Groups 2, 3 and 4 as shown above in Figure 15.3. The last column in Table 15.1 contains a brief description of the insights provided by each measurement. A table of this nature should be included in the SDP or in a *Software Measurement Guidebook* that can be an addendum to the SDP.

There are 50 measurements listed in Table 15.1. If applicable, all of them should be collected when needed and used in the management of large software-intensive systems. Smaller systems can justify using a reduced set. The asterisks shown in Table 15.1 identify eight basic measurements that may be considered a minimum set. Selection of a minimum set is judgmental. You must select the measurements needed for tracking and control of your project.

15.7 Measurement Data Collection Time Frame

Not all measurements are collected and reported at the same time. For example, testing measurements are not needed (or available) during requirements generation. The duration of time when Management Indicators are collected depends on the life cycle model and your project's needs. Figure 15.5 is a generic example of the typical data collection time durations for common software development activities. Please note that measurements are equally useful and valuable during the sustainment phase.

Management indicators should be tailored to the system being developed. Additional indicators and supporting base measurements can be added to address critical or unique needs of each program. The measurements collected during system requirements, design and qualification testing, as shown in Figure 15.5, cover the Software Engineering tasks performed in collaboration with Systems Engineering. Some anomalies are expected, for example, the instability of requirements is expected during SI requirements analysis but would present a serious situation if that was occurring during SIQT. Keep in mind that for software developed in builds, some of the activities are repeated in each build.

15.8 Software Measurement Information Specification

In order to use software measurements effectively, you should have and use a mechanism that specifies exactly what will be measured and how the resulting data will be analyzed to produce results that satisfy your information needs. That mechanism has been called a *Measurement Information Specification* (MIS) that defines the data that will be collected, the computations that will be performed on that data, and how the resulting data will be reported and analyzed.

One of the most comprehensive approaches to defining an MIS was published by the *Practical Software Measurement* (PSM) group (McGarry, 2001). The following description is

Measurement Categories	Management Indicators	SRD	SI Req	SI AD	SI DD	CUT/ UI&T	SIQT	SQT
Schedule and Progress	Requirements Progress							
	Development Progress							
	Testing Progress							
	Schedule Adherence							
Resources and Cost	Effort Profile							
	Staff Profile							
	Target Computer Resources							
	Cost Profile							
Product Size and Stability	Size							
	Volatility							
	Build Content							
Product Quality	Discrepancy Resolution							
	Design Complexity							
	Traceability							
Development Performance	Productivity							
	Maturity							
	Risk Resolution							

SRD = System Requirements and Design   SI Req = Software Item Requirements
SI AD = Software Item Architectural Design   SI DD = Software Item Detailed Design
CUT/UI&T = Code and Unit Test / Unit Integration and Test   SIQT = Software Item Qualification Test
SQT = System Qualification Test   ▪▪▪▪▪▪▪ = Possible data collection period

Figure 15.5   Software measurement data collection time frames—example.

a summary adaptation from the PSM. Remember that tailoring a procedure or process to the needs of your project is always recommended and the PSM is no exception. The MIS is defined here as composed of three specifications:

- *Information Need Specification:* What information you need to know and collect
- *Base Measurement Specification:* What you need to measure to gain that knowledge
- *Derived Measurement Specification:* Calculations also needed to gain that knowledge

The basic elements of the MIS are shown in Figure 15.6. This is a simplified view of the specifications identified by the PSM. The data required by these specifications should be included in a *Software Measurement Guidebook* that is normally an addendum to the SDP. The three specifications are briefly described below.

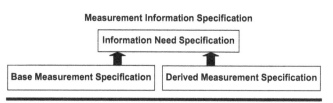

Figure 15.6   Basic elements of the Measurement Information Specification.

15.8.1 Information Need Specification

Table 15.2 is an example format for the specification used to describe the *information need* and the *measurable concepts* to address that need. The format in Table 15.2 contains fields for identifying relevant entities as well as the base and derived measures that are responsive to the information need.

To help clarify how the Measurement Information Specification is used, Table 15.3 is an example derived from

Table 15.2 Format of the Information Need Specification

Name	Measurement Description
Information Need	What the measurement user (e.g., Project Manager or Project Team member) needs to know in order to make informed decisions and take Corrective Actions.
Information Measurement Category	A logical grouping of information needs provided to give structure to the measurements. The five recommended Measurement Categories are: schedule and progress, resources and costs, product size and stability, product quality, and development performance as shown in Table 15.1
Measurable Concept	Satisfying the information need by defining the types of data and objects that must be measured.
Relevant Entities	Relevant Entities may include process or product elements of a project such as project tasks, plans/estimates, resources, and deliverables.
Base Measure	The single property or root characteristic of the data that is quantifiable.
Derived Measure	A measure that is *calculated* as a function of two or more base measures or a formula used to calculate the derived measure.

Table 15.3 Example of the MIS Information Need Specification for Staff Profile

Measurement	Staff Profile
Information Need	Evaluate staffing requirements to see if staffing assumptions are being realized.
Information Category	Resources and Cost.
Measurable Concept	■ Compare planned staffing requirements to actual staffing provided to determine staffing status ■ Compare planned experience to actual experience to identify staffing competency shortfalls ■ Compare staffing gained and lost to plan to identify staffing turnover and trends
Relevant Entities	Planned Headcount Actual Headcount
Base Measures	■ Planned Head Count—Total ■ Planned Head Count—Experience Level Category 1 ■ Planned Head Count—Experience Level Category 2 ■ Actual Head Count—Total ■ Actual Head Count—Lost ■ Actual Head Count—Gained ■ Actual Head Count—Experience Level Category 1 ■ Actual Head Count—Experience Level Category 2
Derived Measure	The following derived measure is used to graph these indicators: Staffing Volatility Index.

(Abelson et al., 2011) that shows how it can be applied to the *Staff Profile* Management Indicator.

15.8.2 Direct and Derived Measure Specifications

Table 15.4 is an example format for the Direct (also called Base) and Derived Measurement Specifications supporting the Information Need Specification. Some measures are used to define multiple measurement constructs. Not all measures require the specification of a derived measure.

The appendixes in Abelson et al. (2011) contains detailed descriptions of the base and derived measurements including where and how they are obtained, how often reported, the scale used, and unit of measurement.

Lessons Learned. I have authored many software metrics guidebooks at the project, division and even at the corporate level (Gechman and Stratton, 1994), and have always been able to produce useful metrics guidebooks less complex than the full PSM approach. Your contract may require

Table 15.4   Direct and Derived Measure Specifications

Measurement	Direct Measure Specifications
Direct Measures	A measure of a single attribute defined by a specified measurement method (e.g., planned number of lines of code).
Measurement Methods	The logical sequence of operations defining the counting rules for each base measure.
Type of Method	The method used to quantify an attribute as either (a) subjective, involving human judgment or (b) objective, using established rules to determine numerical values.
Scale	The ordered set of values or categories that are used in the base measure.
Type of Scale	The type of the relationship between values on the scale, either: ■ Nominal—the measurement values are categorical, as in defects by their type. ■ Ordinal—measurement rankings, as in assignment of defects for severity levels. ■ Interval—measurement values having equal increments. ■ Range—a range of real numbers for equal quantities of the attribute.
Unit of Measurement	The standardized quantitative amount that will be counted to assign value to the base measure, such as an hour or a line of code.
Measurement	**Derived Measure Specifications**
Derived Measures	A measure that is calculated as a function of two or more base measures.
Measurement Function	The formula that is used to calculate the derived measure.

use of the PSM approach; if not you should decide if your project needs to use the full PSM approach, a derivation of it, or use it as the foundation for a software measurement program tailor-made for your project. In any case, *an effective measurement program is your secret management weapon.*

15.9 Analysis and Reporting of Software Management Indicators

Understanding the root cause of a problem is the most effective path to preventing it.

As the Software Project Manager, you must make sure software metrics data is reported to program management and the customer at least monthly to provide frequent status checks of the development effort. When potential problems are identified, the affected Software Lead must analyze the problem indicators to determine if the data accurately reflects a real or developing problem and if timely management action is needed to correct the problem. Coordination with the Integration and Test Team may be necessary for support in analyzing the potential problem or developing and implementing a correction to the problem.

The SEPG should review the Software Management Indicators to determine their effectiveness. Software indicators

should be added and deleted as their utility and cost/benefit is determined. The following questions should be considered when deciding when to add or delete a management indicator:

■ Is the metric providing needed information to SwIPTs in sufficient time to implement management actions to minimize cost or schedule impacts?
■ Does the metric accurately measure the software development process activity it is intended to measure and does it provide useful and meaningful status information?

Lessons Learned. I highly recommended for medium-to-large developments that you spearhead the establishment, and conduct, a joint Customer—Developer *Software Measurement Working Group*. It is a best practice as it will ensure the information needs of all participants are met. I have done this and the working groups have all been useful and highly cost-effective. If your software is being developed by geographically separated organizations, a working group like this is indispensable.

The SEPG at the program or subsystem level must perform analysis of the Software Management Indicators in accordance with their scope of control. The SEPG, or the responsible working group, should meet at least monthly to review the results of the metrics analysis and report their findings to the Subsystem SwIPT, the CSwE and the SPM. Findings should also be distributed to the Software Leads for their use in managing their daily tasks. The SPM or CSwE should periodically report status results to program management.

Quantitative Management Plan. The *Quantitative Management Plan*, if such a plan is prepared, should define the establishment of program goals and the *methods* used for collection, analyzing, controlling and reporting performance data in terms of the goals. In addition, it should present *strategies* for achieving the goals, performing causal analysis and determining Corrective Actions.

15.10 Software Indicator Thresholds and Red Flags

Software Indicator Thresholds. Management decisions based upon the analysis of Software Management Indicators should use thresholds to highlight or flag non-nominal conditions. Thresholds should be established by either the responsible working group for the Subsystem SwIPT, by the CSwE, by the Project Manager, or some combination of those. When a value is outside the nominal conditions, the Subsystem SwIPT should determine how the problem can be fixed to bring the values within the stated limits and if the out-of-limit condition is an acceptable design-related decision. Table 15.5 is a *cut-off portion* of an example *table of allowable thresholds* for a few of the software indicators. Each program and project must decide on what is an allowable threshold.

Software Indicator Red Flags. Potential problems with software management effectiveness can be anticipated at the time the RFP, or contract, is issued if specific requirements for the measurement program are not identified. For example,

the Software Developer and the Program Office should be aware of red flags such as the examples listed in Table 15.6.

Lessons Learned. Red flags are also evident when "metrics gurus" want to measure everything! For example, measuring software productivity strictly by measuring the lines of code a Programmer produced is essentially meaningless. A clever Programmer could work hard at writing 100 lines of creative code that replaces 300 lines of poorly written code; rather than ding that Programmer for only writing 100 lines of code he or she should be rewarded for making the software more efficient.

Use common sense for identifying potential red flags. For example, look for metrics reported at a level too high to detect a serious item that can be lost in the summing-up process. The top number may look good, but that hidden serious item may be a harbinger of a mission failure.

A red flag is obvious if there are numbers, like the numbers in Figure 15.7, that are... hard to believe.

Table 15.6   Software Indicator Program Red Flags

RED FLAG if the contract, or RFP, does NOT require:
Formal delivery of periodic measurement reports and analysis of software management indicator data.
A Software Measurement Guidebook, or Quantitative Measurement Plan, delivered with the proposal.
Measurement data to be presented in a useful, unambiguous, and easy to read graphical form.
Definition of measurement collection durations.
Measurement commitments to be flowed down to subcontractors.
Documented allowable indicator threshold deviations and a plan to determine root cause for deviations.
A clear plan for Corrective Action when deviations are identified.
Sufficient measurements to effectively monitor program status (excessive indicators may also be a red flag).

Table 15.5   Software Indicator Thresholds—Example

Software Information Categories	Software Management Indicators	Threshold
SCHEDULE AND PROGRESS	Requirements Progress	+10% from Plan
	Development Progress	+10% from Plan
	Test Progress	+10% from Plan
	Schedule Adherence	+10% from Plan
RESOURCES AND COST	Effort Profile	+10% from Plan
	Staff Profile	+10% from Plan
	Computer Resources	+10% from Plan

Figure 15.7   An amazing coincidence?

Chapter 16

Managing Software System Sustainment

Maintaining a deployed system, responsive to changing customer needs over a long time frame (sometimes decades), is as important, and can be just as challenging, as the original implementation.

Software Sustainment (often called Software Maintenance) appears as the last chapter in this Guidebook, but not because it is unimportant—it is actually *the longest in duration and most expensive activity* as it can consume about *two-thirds of the total Software Life Cycle cost*. From a customer usage perspective, some might argue it is also the most important activity of the Software System Lifecycle—assuming you delivered a successful system that was used until it is replaced or retired.

Software sustainment includes the processes, procedures, people, materiel, funding and information required to fully support, maintain and operate the software portions of a system. *Sustainment can be viewed as a superset of the Software Maintenance activity.* For example, Software Sustainment addresses issues that may not be included in normal maintenance activities such as day-to-day operations; updating documentation; expansion of deployment locations; security and data access activities; Configuration Management; user support; and hiring and training of users, operators and administrators.

Preparing for software transition to operations and then transition to sustainment is discussed in Sections 11.9 and 11.10 and will not be repeated here. Chapter 16 is focused on the management issues encountered *during sustainment*. The Software Development Plan (SDP), or other Software Maintenance plans discussed below, must address the Software Sustainment issues and the process for managing and resolving these issues. The management of Software Sustainment is covered in the following six sections:

- Software Sustainment Objectives (16.1)
- Planning for Software Sustainment (16.2)
- The Software Sustainment Plan (16.3)
- Software Sustainment Organization (16.4)
- Key Software Sustainment Issues (16.5)
- Contract Closure and System Retirement (16.6)

16.1 Software Sustainment Objectives

As the Software Project Manager, your original software development objectives were to produce a software product that provides the required functionality, is easy to access and use, and is cost-effective to maintain. After delivery, the Software Sustainment tasks are to make sure all the needed elements work effectively to satisfy all of those objectives. Generally, there are four types of Software Sustainment tasks:

- *Corrective*: Fixing errors involves *correcting known problems* in the software not resolved during development plus correcting problems and errors identified and generated after deployment. An important part of corrective fixes is to update the documentation to reflect those changes.
- *Adaptive*: Software must be periodically modified for compatibility with inevitable *changes* in the system hardware or operational environment so that the software continues to perform its intended functions. Adaptive maintenance requires identifying the requirements that are affected by the environmental changes, identifying the design changes needed, and implementing and testing those changes in the code.
- *Perfective*: Periodically, the software needs to be enhanced so that it, or the system, accomplishes *new or*

improved functionality. In some cases, this could involve a major or even a complete redevelopment effort.

■ *Preventive*: Preventive maintenance involves modifications to *improve performance*, activities needed to maintain the current system, or changes to implement improvements in maintainability and reliability.

During corrective and adaptive maintenance, the users know there is a problem. When modifications are made to make the software "better," users do not perceive there may be new problems created, so great care must be taken to avoid introducing defects and transforming a good working system into one that does not work well.

Retesting and Configuration Management play an important role in the process regardless of the type of sustainment performed. Even if the modifications are small, retesting is always required after changes are made to the software. Depending on the extent of the changes, retesting and Regression Testing may be more time-consuming, and more complex, than it was for the original system. Retesting must ensure that the original parts still work, that the modifications work as required, and that the functionality of the original system is not inhibited by the changes. The retesting process during sustainment should be documented in detail as this is not a simple task.

Configuration Management (CM), discussed in Section 6.3, is just as important during sustainment as it was during development. CM manages the changes, determines when the updates are released and ensures that all users in the field are running the proper version of the software.

16.2 Planning for Software Sustainment

Your customer, or their *Program Office*, should be (and usually are) directly involved with your maintenance organization regarding the planning and management of Software Sustainment.

16.2.1 Sustainment Planning Considerations

During the original development process, Software Sustainment *must not* be treated as an afterthought. Decisions made during early development activities, especially during Software Design, can have a significant impact on the cost of sustaining software and its related maintenance tasks. Many systems, especially DoD systems, are operational for a long time (sometimes decades), so *planning for Software Sustainment* requires careful consideration during three key activities during the original system development:

■ *Architecture and Design*: Efficient and effective sustainment systems must be well architected and designed. Effective sustainability is a direct result of quality architecture and design. A good design should be modular, highly cohesive (i.e., each module performs a distinct task), and loosely coupled (so that each module depends as little as possible on other modules for its functionality). Design guidelines such as these should be part of, or referenced by, the SDP. The architecture and design must also plan for Commercial Off-the-Shelf (COTS) evolutionary changes.

■ *Coding*: Coding is another important area where improved maintainability can be achieved. Good coding standards are key to this achievement along with an enforcement process. Frequent Peer Reviews of the code, during the modifications, also helps to improve maintainability.

■ *Documentation*: This is probably the most important activity to improve maintainability. When modifications to the software are needed long after delivery, Software Engineers familiar with the system are usually not available. The sustainment personnel will have to try to uncover errors and implement fixes based on existing code and documentation they did not produce.

It is vitally important that the software documentation accurately reflects the current state of all parts of the system including the requirements, design rationale, architecture and design models (e.g., UML diagrams), coding, test cases, and test results. Documentation may be hard copy or in electronic format and frequently are embedded in tools. If the documentation ever becomes worthless, the inevitable result is a maintenance nightmare and a costly rework.

16.2.2 Sustainment Planning Issues

There are significant issues that must be addressed when upgrading the operational software to new software releases. For example, answers to the following upgrading issues need to be addressed when planning for Software Sustainment:

■ Does the system architecture support *isolating* a subset of equipment for installing and testing new software releases without affecting ongoing operations?
■ Can the same networks be *simultaneously* used for test data and ongoing operational data?
■ Can a test database be installed *with* an operational database?
■ Can operations be transferred to an off-site *backup facility* while the new software release is installed and tested?
■ Is the backup facility a full copy and *faithful representation* of the operational environment?
■ Are *procedures* in place to simultaneously maintain an earlier version of the software while the new version is being developed? In that context, how will changes

made in the version under maintenance be incorporated in the new version being developed?

The ability to upgrade current operational systems to new software releases, without affecting ongoing operations, must be planned for and incorporated into system requirements from the beginning of the project.

16.3 The Software Sustainment Plan

After the original system development activities are successfully concluded, adequate facilities, support software, support personnel and needed documentation must be available so that the software can be maintained in an operational and sustainable condition. A good way to ensure all of this will be available when needed is to require preparation of a *Software Sustainment Plan*, often referred to as the *Software Maintenance Plan*. Draft versions of the SSP should be prepared relatively early during the original software development. Periodic status reporting should be in place to ensure the plan is followed to accomplish the desired sustainment environment. Table 16.1 is an example outline of the SSP.

Even if the details of the SSP cannot be completed during early development activities, a preliminary draft should be prepared early, *so you don't forget about it*, and to ensure the software development process contains the essential sustainment considerations and provisions. The SSP should be revised during the development effort—especially after key

Table 16.1   Example Outline of the Software Sustainment Plan

1. Introduction (Purpose, goals, and scope of the Software Sustainment effort)
2. References (Documents that constrain or support the Software Sustainment effort)
3. Definitions (Defines or references all terms required to understand the plan)
4. Software Maintenance Overview (the organization, scheduling priorities, tools, techniques, resources, responsibilities, and methods used in the sustainment process)
4.1 Organization
4.2 Scheduling Priorities
4.3 Resource Summary
4.4 Responsibilities
4.5 Tools, Techniques and Methods
5. Software Sustainment Process (actions performed for each phase of the sustainment process defined in terms of input, output, process and control)
5.1 Problem/modification identification/classification and prioritization
5.2 Analysis
5.3 Design
5.4 Implementation
5.5 System Testing
5.6 Acceptance Testing
5.7 Delivery
5.8 Risk Management
5.9 Configuration Management
6. Software Sustainment Reporting Requirements (how information will be collected and provided to members of the sustainment organization)
7. Software Sustainment Administrative Requirements (The standards, practices and rules for anomaly resolution and reporting)

(Continued)

Table 16.1 (Continued)   Example Outline of the Software Sustainment Plan

7.1 Anomaly Resolution and Reporting
7.2 Deviation Policy
7.3 Control Procedures
7.4 Standards, Practices and Conventions
7.5 Performance Tracking
7.6 Quality Control of Plan
8. Software Sustainment Procedures (the procedures to be followed in recording and presenting the outputs of the maintenance process)

development milestones. The plan should not be allowed to become out-of-date or the team responsible for the plan to become dormant. An overview of the SSP can be contained in the project-level SDP.

Life Cycle Management Plan (LCMP). The LCMP is a potential option for containing the software sustainment planning information. However, the LCMP may not be required by your program, if it is required, it may not be produced by the software organization. If that is the case, the LCMP typically would not have sufficient software-related details needed, and it would need to be augmented by an addendum to the LCMP, or an addendum added to either the Software Management Plan (SMP) or the SDP.

If neither the LCMP nor the SMP are produced by your program, and the required sustainment information does not exist in an equivalent document, then *all* of the needed Software Sustainment details must be included in an addendum to the SDP, or in the Software Development Files (SDF), or both.

16.4 Software Sustainment Organization

The selected Software Sustainment organization may be the customer, a customer-related organization, a contractor, or the original Developer organization who is awarded a separate contract for operating and maintaining the system. In any case, the selected sustainment organization must be able to acquire the knowledge, documentation, support software, and *data rights* to the software.

The software data rights issue can be a *show stopper*; it should have been addressed before awarding the development contract and should not wait until the sustainment contract. The Software Sustainment organization must have access to the *same* support software that was used during software development.

Personnel with the specific software and application knowledge for the system are a key element in establishing the project's sustainment capability. A substantial portion of this knowledge can be retained if the selected sustainment project personnel are partially involved with reviews, evaluations, and test activities during the original software development effort. However, in the real world, the sustainment project personnel are not likely known or selected early in the original development process.

16.5 Key Software Sustainment Issues

Sustainment becomes more complicated with the increased use of COTS software products. A COTS-intensive system may be a suite of multiple Off-the-Shelf components—including reuse code, legacy code, and COTS from multiple vendors—plus custom components to achieve desired functionality. Maintaining such a composite of software elements presents a significant Software Maintenance challenge. COTS software is also discussed in Subsection 1.8.3, Section 9.7, and Appendix D that contains criteria for evaluating COTS and reusable software products.

Table 16.2 is an example summary of key Software Sustainment issues and pitfalls that are typically encountered; the table pertains to both COTS-intensive systems and systems that are not highly COTS dependent. Your project SDP should address the planned approach for handling these sustainment issues and the approach to avoiding the pitfalls you may encounter during the extended use of COTS software.

16.6 Development Project and Contract Closure

Properly ending a project, and its contract, may be as important as initiating it.

There are two contracts of concern here: the contract for the original system development; and the separate contract

Table 16.2   Typical Key Software Sustainment Issues

Issue	Resolution
Parallel Testing Capability	Incremental updates and related development tasks must take place without affecting ongoing operations.
Planning for Upgrades and Obsolescence	Funding for these upgrades must be planned for. Most COTS software products undergo new releases every two years; old releases will eventually be unsupported.
Software Data Rights	Data rights to the source code and documentation is essential; this issue should be resolved before awarding the development contract.
Technology Advancement	The sustainment group must create and maintain a thorough technology refresh plan.
Vendor Licenses	Transition of license management tasks needs to be jointly planned in advance.
Information Assurance Testing	System Regression Testing must be performed on all upgrades and patches and Information Assurance requirements must be satisfied.
Design for COTS Interchangeability	Sustainment may be impossible if the COTS vendor goes out of business. The system architecture should *isolate the COTS products* with minimal interfaces, rather than intermingling COTS with developed software; this approach will facilitate replacement of the COTS product. Obtaining the COTS source code (e.g., having it put in escrow) is usually not a good idea as it is almost never adequately documented and is typically so huge that it may be impossible to maintain.
Software Risk Management	The sustainment organization must have the resources and capability to identify and analyze risks and perform effective risk mitigation.
Ability to test software updates	Adequate tools, and trained expertise, must be available at the sustainment site and the maintenance environment must support running multiple software versions.
Supporting processes	Implement an effective and sufficient software fault management process, and other supporting tasks during sustainment.
Qualified sustainment personnel	Provide formal training for Software Sustainment personnel. Quickly replace key software development staff that leave. Hire staff members who are experienced in managing COTS-intensive systems.
Incomplete Software Sustainment documentation	Keep software documentation up-to-date and usable. Specify the delivery of a complete set of software documentation, in sufficient detail, in the development contract. Documentation includes text, code, models, diagrams, etc.
Ineffective Configuration Management	An effective Configuration Management process is essential to a successful sustainment effort. The CM system must be able to handle multiple versions in support concurrently and must maintain the exact version and configuration of the software at each operational site, and in each computer at the operational sites. That means the CM system must track the software version installed in each workstation and server. This is especially important during upgrades.
Version synchronization	Maintain synchronization of development versions by feeding back into the development versions the changes made to the version being supported. If this is not performed effectively, the result will be multiple diverging baselines of the same system with a large cost downstream to get everything in synchronization.

for sustainment. That means there are two different closure events: *closing* the development contract after the system development effort is completed, and then *retiring* the system and closing that contract after its mission is over. It is possible the system development contract could include an extension for conducting sustainment.

System *retirement* may also be called system *decommissioning* or system *sun-setting*. The goal is to remove the system from production as seamlessly as possible. Operational systems are removed from production for several reasons, including:

■ The system is being completely replaced because it has become obsolete
■ The contract supporting System Sustainment has ended

- The system is no longer needed to support a new business model
- Continuation of the system is no longer cost-effective
- The system has become redundant during consolidation of operations after a corporate merger

Retirement of software-intensive systems can be a complex issue faced by many organizations as legacy systems are removed and replaced by new systems. This task must be performed with minimal impact on business operations. The retirement task can be difficult to execute easily, especially if it involves regulatory or environmental considerations or the formal return or disposal of government furnished equipment.

The deployment of a new release usually includes instructions to remove the previous release, and it is usually a relatively simple exercise. There are times, however, when you may not retire a release when a new version is available. This may happen if you cannot require users to migrate to the new release or if you must maintain an older system for backward compatibility.

Project Closure Events. The original system development contract and the long-term sustainment (maintenance) contract are usually large, and may be complex, contracts. Sometimes, closing or retiring a large contract may be so involved that the process of doing so becomes a project of its own. To be more specific, there is an important difference between ending a development *contract* and closing the development *project;* the difference between these closure events can be summarized as follows:

- *Contract Closure* (also called Procurement Closure) procedures are specific to the *contract* with your customer. This involves verifying that the terms and conditions of the contract were completed and all work was performed satisfactorily, making sure that all payments were made, updating of contract records to reflect final results, confirming completion of exit criteria for contract closure, and formally closing out all subcontracts associated with the project. Most of this is normally performed by your contracts and legal groups, but as the SPM you should be involved in the contract closure process.
- *Project Closure* (sometimes called Administrative Closure) involves doing everything your organization must do to close out the *project* including collecting the project's records, writing down lessons learned, and formally releasing all Project Team members. You also need to verify how well the work was done, because you need to assess your project's failures or successes and document and archive *lessons learned* for future use. As the SPM you will likely lead the project closure process.

The Project Closure Process. The key activities involved in the project closure process are:

- *Formal Sign-Off:* After the products, services, or results are delivered, you must obtain a formal sign-off from the customer to formally declare the project completed.
- *Lessons learned analysis and documentation:* Once you receive a formal sign-off from your customer, you must: analyze the positives and negatives of your project; and compile and document your project's lessons learned for use in future projects.
- *Release Resources:* After receiving formal sign-off from your customer, formally release and return the resources (human and material) to their source departments.
- *Subcontractor Closures:* All vendor/supplier contracts must also be formally closed with a formal closure process. Once they have delivered the products to you and you have delivered the same to your end customer, the contract paper should be closed.
- *Indexing and Archiving the Project Files:* When all the required deliverables are accepted by the customer, you should compile, index and archive the project files for future use. There should be a central data bank where such files are stored.
- *Celebration:* When all of the above is completed, and the project is declared a success, all of your team members should share that success. A celebration party is a great morale builder and team motivator that you can capitalize on in future projects. In addition, it is a good idea to hold milestone completion parties on long-term projects. Try to involve your end customer and third-party vendors in the celebrations.

A Smooth Project Closure Process. The importance of the project closure process should be conveyed to your Development Team. Your Project Team may relax after the sustainment task is complete, so they may not be fully prepared for the important system closure process. From the beginning, a major objective of the Project Life Cycle should be an efficient and smooth project closure process, including final sign-off by the customer. Project closure delays can be prevented by the involvement of the customer throughout the Project Life Cycle. Budget overruns and project delays affect the organization adversely, and a smooth project closure process will be cost-effective for your business.

If your home organization has a lessons learned database, it may help you recognize the difficulties encountered during previous project closure processes, and help you develop a suitable strategy for a smooth closing. Make a list of all the activities and steps required to close your project, review it regularly, and then when the time comes, execute the closure process efficiently and professionally.

Final Thoughts

Software Engineering has been around for well over 50 years; yet, we are still on the steep part of the growth curve and nowhere near full maturity. Fasten your seatbelts, you can't imagine what is yet to come! The ultimate goal of this Guidebook is to help guide future software developments toward achieving success in the delivery of software-intensive systems—especially large systems.

Francis Bacon (1561–1626) wrote, "Some books are to be tasted, others to be swallowed, and some few to be chewed and digested; that is, some books are to be read-only in parts; others to be read but not curiously; and some few to be read wholly, and with diligence and attention."

It is my desire to have this Guidebook "read wholly, and with diligence and attention." I hope you use it that way.

Appendix A: Software Acronyms

Table A.1   Definition of Software Acronyms

Acronym	Definition
ADM	Acquisition Decision Memorandum
AFOTEC	Air Force Operational Test and Evaluation Center
ANSI	American National Standards Institute
AoA	Analysis of Alternatives
AOTR	Assessment of Operational Test Readiness
AOTR	Assessment of Operational Test Readiness
API	Application Programming Interface
APO	Acquisition Program Office (or PMO)
ASP	Acquisition Strategy Panel
BAR	Build Architecture Review
BOE	Basis of Estimate
BTR	Build Turnover Review
C/R	COTS / Reuse (a software class)
C4ISR	Command, Control, Communications, Computers, Intelligence, Surveillance and Reconnaissance
CAIV	Cost As An Independent Variable
CAP	Corrective Action Plan
CARD	Cost Analysis Requirements Document
CASE	Computer Aided Software Engineering
CBA-IPI	CMM^SM-Based Assessment–Internal Process Improvement
CCB	Configuration Control Board
CCE	Component Cost Estimate
CCR	Critical Computer Resource

Acronym	Definition
CDD	Capabilities Development Document
CDR	Critical Design Review
CDRL	Contract Data Requirements List
CI	Configuration Item
CIP	Contract Implementation Plan
CM	Configuration Management
CMMI^SM	Capability Maturity Model Integrated
CMP	Configuration Management Plan —or— Cloud Management Platform
COCOMO	Constructive Cost Model
COM	Computer Operation Manual
CONOPS	Concept of Operations (same as OCD)
COTS	Commercial Off-the-Shelf
CPI	Cost Performance Index
CPM	Computer Programming Manual
CPU	Central Processor Unit
CR	Change Request (system level)
CRF	Change Request Form
CRWG	Computer Resources Working Group
CSCI	Computer Software Configuration Item (current acronym is SI)
CSwE	Chief Software Engineer
CTE	Critical Technology Elements
CUT	Coding and Unit Testing
CWBS	Contract Work Breakdown Structure
CV	Cost Variance
DAB	Defense Acquisition Board

Acronym	Definition
DAG	Defense Acquisition Guidebook
DBDD	Data Base Design Description —or— Data Base Design Document
DID	Data Item Description
DM	Data Management
DMS	Data Management System
DoDAF	Department of Defense (DoD) Architecture Framework
DOORS	Dynamic Object-Oriented Requirements System (tool)
DR	Discrepancy Report (system level)
DT&E	Development Test and Evaluation
ECP	Engineering Change Proposals
ECR	Engineering Change Request
EDIN	Electronic Data Interchange Network
EIA	Electronic Industries Association
EM	Engineering Memorandum
EMD	Engineering and Manufacturing Development
EQT	Element Qualification Test
ERB	Engineering Review Board
ESLOC	Equivalent Source Line of Code
ETC	Estimate to Complete
EVMS	Earned Value Management System
FAA	Federal Aviation Administration
FCA	Functional Configuration Audit
FMECA	Failure Modes, Effects and Criticality Analysis
FOC	Full Operation Capability
FQT	Formal Qualification Test
FSM	Firmware Support Manual
FSW	Flight Software
GFE	Government Furnished Equipment
GFS	Government Furnished Software
GG/GP	Generic Goals/Generic Practices
GIG	Global Information Grid

Acronym	Definition
GOTS	Government Off-the-Shelf
GPS	Global Positioning Satellite/System
GSE	Ground Support Equipment
GUI	Graphical User Interface
HAR	Hardware Acceptance Review
HI	Hardware Item
HIQT	Hardware Item Qualification Test
HSI	Human System Integration
I&T	Integration and Test
IA	Information Assurance (Security)
IAS	Information Assurance Strategy
IBR	Integrated Baseline Review
ICA	Independent Cost Assessment
ICAT	Independent Cost Analysis Team
ICD	Initial Capabilities Document
IDD	Interface Design Description
IDR	Interim Design Review
IEEE	Institute of Electrical & Electronics Engineers
IFCD	Interface Control Document
ILS	Integrated Logistics Support
IMF	Integrated Management Framework
IMP	Integrated Master Plan
IMS	Integrated Master Schedule
INCOSE	International Council of Systems Engineering
IPPD	Integrated Process & Product Development
IPT	Integrated Product Team
IRD	Interface Requirements Document
IRR	Integration Readiness Review
IRS	Interface Requirements Specification
ISO	International Standards Organization
IT&V	Integration Test and Verification
ITAR	International Traffic in Arms Regulation

Acronym	Definition
ITR	Initial Technical Review
IV&V	Independent Verification and Validation
JTR	Joint Technical Review
KDP	Key Decision Point
KPP	Key Performance Parameter
KSLOC	Thousand SLOC
LCC	Lifecycle Cost
LCMP	Life Cycle Management Plan
LOE	Level Of Effort
MC	Mission Control
MDA	Milestone Decision Authority
MDD	Material Development Decision
MIL STD	Military Standard
MIS	Measurement Indicator Specification
MMC	Mission Management Center
MOSA	Modular Open System Architecture
MSDL	Master Software Development Library
MTP	Master Test Plan (see STEP)
NDI	Non-Developmental Item
NSA	National Security Agency
OCD	Operational Concept Description
O&M	Operations and Maintenance
OOA/OOD	Object-Oriented Analysis/Object-Oriented Design
OT&E	Operational Test and Evaluation
OTRR	Operational Test Readiness Review
PA	Product Assurance —or— Process Area
PCA	Physical Configuration Audit
PDR	Preliminary Design Review
PDRR	Program Definition and Risk Reduction
PIP	Process Improvement Program
PM	Program Manager
PMO	Project Management Office (or APO)
PMP	Program Management Plan

Acronym	Definition
POE	Program Office Estimate
PP	Planning Package
PRR	Production Readiness Review
PSP	Personal Software Process
PTR	Post-Test Review
PUI	Program Unique Identifier
QMP	Quantitative Management Plan
QTRR	Qualification Test Readiness Review
RAN	Resource Allocation Notice
RE	Responsible Engineer
RFP / RFI	Request for Proposal/Information
RMA	Reliability, Maintainability & Availability
RMP	Risk Management Plan
ROI	Return On Investment
ROM	Rough Order of Magnitude
RTVM	Requirements Test Verification Matrix
S/SEE	System/Software Engineering Environment
SA/SD	System Analysis/System Design
SAD	Software Architecture Description
SAR	Software Requirements & Architecture Review
SAT	Subsystem Acceptance Test
SBDR	Software Build Design Review
SBRAR	Software Build Requirements and Architecture Review
SCAMPI	Standard CMMI Approval Method for Process Improvement
SCCB	Software Configuration Control Board
SCM	Software Configuration Management
SCMP	Software Configuration Management Plan
SCMT	Subcontractor Management Team
SCR/SDR	Software Change Request/Software Discrepancy Report
SCS	System Critical Software (software class)

Acronym	Definition
SCSOM	Software Center Operations Manual
SDCE	Software Development Capability Evaluation
SDD	Software Design Description
SDE	Software Development Environment
SDF	Software Development Folder (or File)
SDL	Software Development Library
SDLC	Software Development Life Cycle
SDP	Software Development Plan
SDR	System Design Review (same as SFR) —or— Software Discrepancy Report
SDRL	Subcontractor Data Requirements List
SE	Systems Engineering
SE&I	Systems Engineering and Integration
SEE	Software Engineering Environment
SEI	Software Engineering Institute
SEIT	Systems Engineering Integration & Test
SEMP	Systems Engineering Management Plan
SEN	Software Engineering Notebook
SEP	Systems Engineering Plan
SEPG	Software Engineering Process Group
SETA	Systems Engineering & Technical Assistance
SFR	System Functional Review (also SDR)
SG/SP	Specific Goals/Specific Practices
SI	Software Item (formerly called CSCSI)
SI&T	Software Integration and Testing
SIOM	Software Input/Output Manual
SIP	Software Installation Plan —or— Software Implementation Process
SIQT	Software Item Qualification Testing
SLATE	System-Level Automation Tool for Engineers
SLC	System Life Cycle
SLOC	Source Lines Of Code
SMBP	Software Master Build Plan

Acronym	Definition
SMP	Subcontract Management Plan or Software Maintenance Plan
SoS	Scrum of Scrums, or System of Systems
SOA	Service Oriented Architecture
SOO	Statement of Objectives
SOW	Statement of Work
SPAR	Software Process Assets Repository
SPCR	Software Problem Change Report
SPE	Software Process Engineer (see SPL)
SPI	Software Process Improvement —or— Schedule Performance Index
SPICE	Software Process Improvement and Capability Determination
SPL	Software Process Lead (see SPE)
SPR	Software Peer Review
SPS	Software Product Specification
SQA	Software Quality Assurance
SQAP	Software Quality Assurance Plan
SQPP	Software Quality Program Plan
SQT	System Qualification Testing Subsystem Qualification Testing
SRD	System Requirements and Design
SRHP	Software Risk Handling Plan
SRMP	System Risk Management Plan
SRR	System Requirements Review
SRS	Software Requirements Specification
SRTM	Software Requirements Traceability Matrix
SS	Support Software (a software class) —or— System Specification
SSDD	System/Subsystem Design Description
SSME	Software Subject Matter Experts
SSP	Standard Software Process
SSPM	Software Standards & Practices Manual
SSR	Software Specification Review
SSS	System/Subsystem Specification

Acronym	Definition
ST&E	System Test and Evaluation
STD	Software Test Description
STE	Software Test Environment
STEP	System Test & Evaluation Plan or MTP
STP	Software Test Plan
STR	Software Test Report (or Results)
STrP	Software Transition Plan
SU	Software Unit
SUM	Software User's Manual
SVD	Software Version Description
SVR	System Verification Review
Sw	Software (also SW or S/W)
SWAMP	Software Acquisition Management Plan
SWAPI	Software Acquisition Process Improvement
SwCCB	Software Change Control Board
SwE	Software Engineering
SWEBOK	Software Engineering Body of Knowledge
SWED	Software Entity Database
SwIPT	Software Integrated Product Team
SIS	Software Intensive System
T&E	Test and Evaluation
TBX	To Be Reviewed, or To Be Determined, or To Be Supplied, etc.

Acronym	Definition
TD	Technology Development (Milestone A)
TDS	Technology Development Strategy
TEMP	Test and Evaluation Master Plan
TER	Test Exit Review
TES	Test and Evaluation Strategy
TIM	Technical Interchange Meeting
TMP	Technology Maturation Plan
TPM	Technical Performance Measurement
TRD	Technical Requirements Document
TRL	Technology Readiness Level
TRM	Test Requirements Matrix
TRR	Test Readiness Review
TSP	Team Software Process
TT&C	Telemetry, Tracking & Control (or Command)
UI&T	Unit Integration and Testing (also UIT)
UML	Unified Modeling Language
V&V	Verification and Validation
VCRM	Verification Cross Reference Matrix
WBS	Work Breakdown Structure
WI	Work Instruction
WP	Work Package

Appendix B: Software Definitions

Acceptance: An action by an authorized representative of the customer by which the customer assumes ownership of software products as a partial or complete performance of a contract.

Action Item: A documented required task tracked to completion.

Activity: An activity is an action that takes place over a period of time; it has a beginning and an end. Activities are composed of tasks needed to accomplish the objectives of an activity.

Agile Development Methodology: An Agile software project is a generic name covering a number of methods and iterative approaches for software development. Agile methods are based on the principles of human interaction, where software evolves through collaboration of self-organizing, cross-functional teams. As the name implies, to be "agile" is conceived to be flexible and more quickly able to respond to changes (see Section 4.3).

Algorithm: A process or set of well-defined rules to be followed in calculations, or other problem-solving operations for the solution of a problem, especially by a computer in a finite number of steps.

Algorithm Software: Algorithm code that has been verified to meet all functional and performance requirements for data quality, timeliness, and execution within the architecture.

Approval: Written notification by an authorized representative of the customer that a Developer's plans, requirements, designs, or other aspects of the project appear to be sound and can be used as the basis for further work. Such approval does not shift responsibility from the Developer to meet contract requirements.

Architecture: The organizational structure of a system or Software Item, identifying its components, their interfaces and a concept of interaction among them.

Associate Developer: An organization that is neither a prime contractor nor subcontractor to the Developer, but who has a development role on the same level as the prime contractor.

Availability: The probability that a developed system, prior to the start of a mission, will be mission capable when a mission is needed. Availability accounts for system faults, and the time it takes to restore the system to a mission capable state following failure (maintainability).

Baseline: An initial standard, or measurement, against which future status, progress, and changes are compared and measured. The budget can serve as one baseline, the schedule as another, etc. Software baselines describe a particular version of the software (e.g., increment, document, build or release) and consists of a set of internally consistent requirements, design, code, build files and user documentation.

Base Measure: A specific single measurement (raw data). See Direct Measurement.

Behavioral Design: The design of how an overall system or a Software Item will behave, from a user's point of view, in meeting its requirements, ignoring the internal implementation of the system or the Software Item. This design contrasts with Architectural Design, which identifies the internal components of the system or the Software Item, and with the Detailed Design of those components.

Build: (1) A version of the software that meets a specified subset of the requirements that the completed software will meet. (2) The period of time during which such a version is developed. Note: The meaning of the terms "build" and "version" is up to the Developer; for example, it may take several versions to reach a build, a build may be released in several versions, or the terms may be used as synonyms.

Commercial Off-the-Shelf Software: COTS software is a generic term often referred to as "third-party software." See Reusable Software Product.

Communications Plan: Document that defines the content, method, and frequency of communications between the Project Manager, members of the Project Team, stakeholders, and management.

Computer Hardware: Devices capable of accepting and storing computer data, executing a systematic sequence of operations on computer data, or producing control outputs. Such devices can perform

301

substantial interpretation, computation, communication, control, or other logical functions.

Computer Program: A combination of computer instructions and data definitions that enable computer hardware to perform computational or control functions.

Computer Software Configuration Item: The CSCI was the former name for a Software Item (SI).

Computer Software Unit: See Software Unit (SU).

Configuration Control Boards: A hierarchy of control groups with different levels of control and responsibilities at the system and subsystem levels. Includes the Engineering Review Board (ERB).

Corrective Action Process: The CAP handles problems or issues detected in software work products.

Critical Path: The sequence or chain of interdependent activities in the project that takes the longest time to complete. This sequence determines the shortest schedule for the project. Any delay in a Critical Path activity increases the project schedule.

Customer: The acquiring organization that procures and manages the contract for a software-intensive system, software product or software service from a supplier for itself or another organization. The customer is also known as the buyer.

Data Item Description: The DID contains the format, content, and preparation instructions for a data product generated by the specific activities and tasks. Examples include DIDs for the *Software Development Plan* (SDP), *Software Design Description* (SDD), *Software Test Report* (STR) and *Software User Manual* (SUM).

Database Management System: An integrated set of computer programs that provide the capabilities needed to establish, modify, make available, and maintain the integrity of a database.

Database: A collection of related data stored in one or more computerized files in a manner that can be accessed by users or computer programs via a database management system (DBMS).

Decision Criteria: Numerical thresholds, targets or patterns used to determine the need for action or further investigation, or to describe the level of confidence in a given result. Decision criteria help to interpret the results of measurement. Decision criteria may be calculated or based on a conceptual understanding of expected behavior.

Deliverable Software Product: A software product that is required by the contract to be delivered to the customer or other system recipient.

Dependability: Defines the probability that the system will be able to complete the mission even if there is a failure given the system was available at the start of the mission. Dependability accounts for system faults and the time it takes to restore the system to a mission capable state following a failure.

Derived Measurement: A derived measurement is a function, or arithmetic calculation, of two or more values of direct (or base) measurement. For example, lines of code produced by your team, divided by effort expended during the same time period, can provide an indicator of productivity.

Design: Those characteristics of a system or Software Item that are selected by the Developer in response to the customer requirements. Some will be elaborations or derived requirements, such as definitions of all error messages in response to a requirement to display error messages; others may be implementation-related, such as decisions about what Software Units and logic are needed to satisfy the requirements.

Developer: The organization that develops software products or performs development activities (including requirements analysis, design, coding and testing through acceptance) during the Software Life Cycle Process. It may include new development, modification, reuse, re-engineering, maintenance or any other activity that results in software products needed to be responsive to the system requirements.

Direct Measurement: Also called a "base" measure, it is a measure of a single attribute defined by a specified measurement method. A direct measure is functionally independent of other measures.

Document/Documentation: A collection of data, regardless of the medium on which it is recorded, that generally has permanence and can be read by humans or machines.

Earned Value Measurement System: The EVMS is a measurement of the actual work accomplished.

Electronic Data Interchange Network: Secure remote access available to all authorized users.

Element (or Factory) Acceptance Test: A set of formal criteria, such as a procedure, whose execution satisfies a set of requirements agreed to by an authorizing agency. EAT (or FAT) is performed at the contractor's software development facility. This test may also be called the "Element Qualification Test."

Efficient: Having a high ratio of output-to-input; working or producing with a minimum of waste or time.

Element: A configuration item within a subsystem, consisting of integrated hardware and software.

Evaluation: The process of determining whether an item or activity meets specified criteria.

Firmware: Firmware is a type of software that is dedicated to a hardware device, and resides as read-only software *on* the hardware device.

Hardware Item: HIs are an aggregation of hardware that satisfies an end-use function and is designated for purposes of specification, interfacing, qualification testing, Configuration Management, or other purposes. It is also referred to as a Configuration Item (CI).

Heritage: A previous baseline or baselines that the current software version may be based on.

Human System Integration: A disciplined, unified and interactive Systems Engineering approach for integrating human considerations into System Design, development and life cycle management.

Increment: A defined pass through the Program Life Cycle, including a sequential set of Software Life Cycle activities; may include multiple planned software builds.

Incremental Model: The Incremental Life Cycle Model is a multi-build model. Software requirements analysis and Architectural Design, plus initial Detailed Design, code, integration, and test are completed in the first build. Additional capabilities are added in subsequent builds through detail design, code, integration, and test activities. This model supports delivery of an interim capability of the final product.

Independent Verification and Validation: IV&V is a systematic evaluation of software products and activities by a third-party not responsible for developing the product being evaluated.

Indicator: A measure that provides an estimate or evaluation of specified attributes. See Management Indicators.

Information Need: The insight necessary to manage objectives, goals, risks and problems.

Information Product: One or more indicators, and their associated interpretations, that are responsive to an information need.

Integrated Management Plan: The IMP is a system-level master program plan that must be augmented with more detailed software-specific plans.

Integrated Master Schedule: The IMS is a system-level master program schedule that must be augmented with more detailed software-specific schedules.

Integrated Product Team: See Software Integrated Product Team (SwIPT).

Integration and Test: I&T combines tested entities into the next higher entity (e.g., lower level SUs into higher-level SUs), then tests the interactions between entities to verify the entities work correctly with each other, in accordance with test plans. Also verifies processes (e.g., tasks) synchronize correctly with processes in other components.

Interface: A relationship among two or more entities in which the entities share, provide, or exchange data, such as Software Item to Software Item (SI/SI), Software Item to hardware item (SI/HI), Software Unit to Software Unit (SU/SU), Software Item to user, etc. An interface is not a SI, SU, or other system component; the interface is the relationship among them.

Joint Review: A process or meeting involving representatives of both the customer and the Developer, during which project status, software products, and/or project issues are examined and discussed.

Key Performance Parameter: KPPs are the minimum attributes or characteristics considered essential for an effective capability. Failure to meet a KPP threshold can be cause for the system concept to be reevaluated or the program to be reassessed or terminated.

Legacy: The extent to which software may impact future programs or other software by nature of functionality or problems that may be introduced.

Life Cycle: The complete set of phases a system goes through, beginning with its conception and ending with its replacement or retirement from service.

Maintainability: Defines the probability that, when a failure occurs, the system can be retained in or returned to a mission capable state within a specified period of time. Pre-mission maintainability supports availability. During a mission, this defines the probability that the system can be retained in or returned to an operational state within a specified period of time when a failure occurs.

Maintainer: An organization that performs sustainment (also called maintenance) activities for a system after it is developed; it may be an integral part of the customer's organization.

Maintenance (of Software): Software maintenance is a component of Sustainment and involves the activities needed to keep the operational system working to satisfy the user's needs. See Sustainment.

Maintenance Organization: The organization that is responsible for modifying and otherwise sustaining the software after transition from the development organization to the operational environment.

Management Indicators: Management Indicators are the basis for analysis and decision-making. Measurement is usually based on imperfect information, so quantifying the uncertainty, accuracy, or importance of indicators is an essential component of understanding the actual value of the indicator.

Measurement: A set of operations having the objective of determining the value of a measure.

Measurement Process: The process for establishing, planning, performing and evaluating measurements within an overall project, enterprise or organization.

Metrics: See Measurement as the terms are often used interchangeably.

Milestone: A point in time representing the completion of an activity, phase or stage.

Model: A model is a calculation combining one or more direct and/or derived measures with associated decision criteria. It is based on an understanding of, or assumptions about, the expected relationship between the component measures and/or their behavior over time. Models produce estimates or evaluations relevant to defined information needs.

Modified Code: Code previously constructed and reused with modifications. See Reused Software.

Module: A text file containing Source Lines Of Code (SLOC). In C++ this is generally a single class.

Non-Deliverable Software Product: A software product that is not required by the contract to be delivered to the customer or other designated recipient.

Process Improvement: See Software Process Improvement.

Process: An organized set of interrelated or interacting activities, involving the practical application of accepted principles, conducted to transform inputs into outputs and achieve specific results.

Project: A project is a temporary endeavor to produce a specific product, service, upgrade or result within a defined period of time and is normally constrained by funding.

Project Charter: A document that announces the project, gives it a name, states its purpose, identifies the Project Manager, and announces his or her level of authority. May be called Project Vision or Scope Plan.

Prototyping: Building an early "experimental" product or portion of the product that enables customers and Developers to examine some aspects of the system to decide if it is suitable or modifiable for the final product. It provides a better understanding of the requirements and the interfaces, an ability to test throughput speeds, develop environmental testing, etc. (See Subsection 4.1.4)

Qualification Testing: Testing performed to demonstrate to the customer that a system or Software Item meets its specified requirements. Software Item Qualification Testing (SIQT) is performed in an environment *functionally equivalent to the target environment* and is intended to verify and validate all software requirements. SIQT is performed prior to acceptance testing.

Quality: The quality of a product is the degree to which the product attributes, such as capability, performance or reliability, meet the needs of the customer or mission, as specified through the requirements definition and allocation process. Also, see Software Quality.

Rapid Prototyping: See Prototyping.

Re-engineering: The process of examining and altering an existing system to reconstitute it in a new form. Re-engineering may include: *reverse engineering* (analyzing a system and producing a representation at a higher level of abstraction, such as the design *from* the code); *restructuring* (transforming a system from one representation to another); *re-documentation* (analyzing a system and producing user or support documentation); *forward engineering* (using software products derived from an existing system, with new requirements, to produce a new system); *retargeting* (transforming a system to install it on a different target system); and *translation* (transforming source code from one language or version to another).

Release: The distribution of a new product, or new function, or fixes for an existing product. There are typically three types: (1) *Document Release*—after approval of a document by the CCB, the Data Center releases the document and posts it as a replacement of previous versions; (2) *COTS or Reuse Release*—an update to a product from the vendor. If, after review of the update, the updated product is accepted for further development, it is released to development; (3) *Software Release*—distribution of new versions of software to integration that include approved and tested changes.

Reliability: The probability that a system will be able to complete its mission without failure assuming the system was available for use at the start of the mission. Reliability does not account for returning the system to a "mission capable" state (maintainability).

Requirement: (1) A functional characteristic that a system or Software Item must possess in order to be acceptable to the customer. (2) A mandatory statement in a standard or contract specification. Specification statements must be: unambiguous; correct; complete; realistic; testable; quantitative if possible; traceable; with only one requirement per statement.

Reusable Software Product: A software product previously developed for one use but having the potential for other uses. It may have been developed specifically to be usable on multiple projects or in multiple roles on one project.

Risk Management Plan: The RMP is a program level document for identifying the process for risk planning, identification, assessment, prioritization, handling and monitoring.

Software: Computer code, programs, procedures, documentation and data pertaining to the operation of a software-intensive system. Data may include information in databases, rule bases, or configuration data. Procedures may include, for example, interpreted scripts

Software Change Request: SCRs are documents used to enhance or improve the software product or make changes to commitments, plans or baselines.

Software Development File: Also called a *Software Development Folder*, the SDF is a repository for material pertinent to the development of a software-intensive system. The SDF is typically electronic and may include (either directly or by pointer) rationale and constraints related to requirements analysis, design, and implementation; Developer-internal test information; and schedule and status information.

Software Development Library: The SDL is a controlled collection (archive) of software, documentation, other intermediate and final software products, and associated tools and procedures used to facilitate the orderly development and subsequent maintenance of software.

Software Development (or Implementation) Process: An organized set of activities performed to translate user needs into software products See Chapter 11.

Software Development: A set of activities, with related procedures, that results in software products. Software development may include new development, modification, reuse, re-engineering, maintenance, or any other activities that result in developing software products.

Software Development Environment: The SDE covers the facilities, hardware, software, firmware, procedures, and documentation needed to develop the system and perform qualification, and other, testing of software. SDE elements may include computer aided software engineering (CASE) tools, compilers, assemblers, linkers, loaders, operating systems, debuggers, simulators, emulators, documentation tools, and database management systems.

Software Discrepancy Reports: SDRs are used to document unexpected error conditions or anomalies.

Software Engineering Environment: Another name for the Software Development Environment.

Software Engineering Notebook: SWENs are repositories of program unique information that provide an organized method of communicating technical information and a repository for historical data.

Software Engineering: The application of a systematic, disciplined, quantifiable approach to the development, operation and sustainment of software, i.e., the application of the engineering discipline to software. It is sometimes used as a synonym for software development.

Software Engineering Process Group: The SEPG is focused on: evaluating the implementation and progress of the defined software development process; identifying areas for potential process improvement; and performing evaluations to determine if the observed deficiencies are the result of non-compliance with, or inadequacy of, your current policies and procedures (see Subsection 6.1.1).

Software Implementation Process: The activities responsible for developing and testing the software products and includes the following activities: Software build planning and estimating; Software build requirements; Software architectural design; Software Detailed Design; Coding and Unit Testing; Software Unit integration and testing; and Software Item Qualification Testing. See Chapters 3 and 11.

Software Integrated Product Team: The SwIPT is the key team of Developers, assembled from applicable engineering disciplines, to design, develop, produce and support an entire product or system.

Software Item: SIs are an aggregation of Software Units, or higher-level software components, that satisfy specific end-use functional requirements. Software Items are selected based on trade-offs among software function, size, host or target computers, support strategies, plans for reuse, criticality, interface considerations, need for separate documentation, and other factors. The SI was formerly called a Computer Software Configuration Item(CSCI).

Software Master Build Plan: The SMBP maps the incremental functionality, capabilities and requirements allocated to all software builds.

Software Process Improvement: The sequential steps for the improvement of the software development process and the changes needed to implement those improvements.

Software Process Lead: The SPL is a trained change agent responsible for facilitating all software process tasks for the program. Sometimes the SPL is called the Software Process Engineer.

Software Product: Software or associated information created, modified, or incorporated to satisfy a contract. Examples include plans, requirements, design, code, databases, test information, and manuals.

Software Project Manager: The SPM is the key individual with responsibility and authority for directing the software project or portion assigned, and the Project Team.

Software Quality: The ability of delivered software to satisfy its specified functional, performance and interface requirements, including the required dependability, reliability, maintainability, availability, security, supportability, and usability.

Software-Intensive System: A SIS is where the Software Component is the predominant factor in accomplishing the common system objectives. The SIS consists of software code, the computer equipment on which the software operates, and interfaces with other software and Hardware Items. It may also include the human interactions as well as the applicable policies, procedures and regulations.

Software Test Environment: The STE includes the facilities, hardware, software, firmware, procedures and documentation needed to perform qualification and other testing of the system being developed. Elements may include but are not limited to simulators, code analyzers, test case generators, and path analyzers, and may also include elements used in the Software Development Environment.

Software Transition: The set of activities that enables responsibility for a software system, or a portion of it, to pass from one organization to another.

Software Unit: The SU is the basic element in the design of a Software Item. It may be a major subdivision of a Software Item, a component of that subdivision, a class, object, module, function, routine, or database. Software Units may occur at different levels of a hierarchy and may consist of other Software Units. Software Units in the design may or may not have a one-to-one relationship with the code and data entities (routines, procedures, databases, data files, etc.) that implement them or with the computer files containing those entities. The SU was previously called a *Computer Software Unit* (CSU).

Stakeholders: Those persons and organizations that have an interest in, and impact resulting from, the performance and completion of the project. The customer or user of a product created through a project is usually a primary stakeholder.

Statement of Work: The SOW is a contractual document that defines the work to be performed for a specific project under contract. It defines the goals, scope, and constraints of the project in terms of *what* needs to be done, *not how* to do it.

Subsystem: A subsystem (may also be called a segment) is the next level down from the entire system; subsystems are typically a physical or functional subset of the entire system.

Supplier: A supplier is an organization that enters into a contract with the customer for the development and implementation of a system, software product or software service under the terms of a contract.

Supportability: The degree to which System Design characteristics and planned logistics resources meet pre-mission readiness and operational utilization requirements. The primary performance requirements that characterize supportability are availability, dependability/reliability, and maintainability.

Sustainment (of Software): The set of activities that takes place to ensure that software installed for operational use continues to perform as intended and fulfill its intended role in system operation. Software sustainment includes maintenance, technical support, aid to users and related activities.

System Design: The process of defining, selecting, and describing solutions to requirements in terms of products and processes. It is also the product of the design activities that describes the solution.

System: A system is a combination of interacting elements organized to achieve one or more stated purposes. It may be a product or the services it provides. It is a composite of hardware, software, human skills, and techniques capable of performing or supporting an operational role. It includes all operational equipment, related facilities, materials, software, services and personnel required for its operation.

Systems Engineering Integration and Test: The SEIT Team is part of the Systems Engineering organization and has primary responsibility for all of the system-level tasks described in Chapter 10.

Tailoring: A manual process of reviewing software standards mandated by the home organization to ensure the project's common software development process is compatible with those standards. The tailoring process also includes developing waivers against specific company process standards in favor of those needed for the project's common process.

Technical Performance Measures: TPMs are a measurement that indicates progress toward meeting critical system characteristics (technical parameters) that are specified in requirements or constrained by System Design. Technical parameters tracked with TPMs have clearly measurable target and threshold values. TPMs are used to track computer system resource margins (CPU usage and I/O memory margins).

Test: "Test" and "testing," refer to the activities of verifying that the software implementation meets the design (unit and integration testing) and verifies that the requirements are satisfied (qualification testing).

Testbed: A system testing environment containing the target hardware, operational software, instrumentation, simulations, software tools and any other supporting element needed to conduct a test of the full or partial system on the ground.

Testability: A design characteristic of hardware and software, which allows the status of an item to be confidently determined or verified in a timely fashion.

Thread: An end-to-end functional capability traced through the system and verified by creating an input and observing the intended output.

Tiger Team: A small special purpose group set up for a short time to investigate a specific matter.

Turnover: The process of providing a software product from one team member to another, or from the Developer to the customer, in accordance with pre-established quality criteria.

User: Refers to the end-users of the software-intensive system being developed.

Validation: Ensures that the final product incorporates all of the system requirements.

Verification: Ensures that each developed function works correctly.

Version: See Build.

Work Breakdown Structure: The WBS is a basic planning document that breaks down the project into its constituent tasks or activities. It lists the specific work needed to complete *all* aspects of the project.

Work Instructions: Documentation containing detailed directions for the day-to-day implementation of the software process. It is also referred to as a work procedure.

Work Package: The smallest element of work with allocated budget and schedule against which progress is reported and tracked.

Work Product: Any artifact produced by a development process that may or may not be deliverable to your customer.

Appendix C: Software Roles and Responsibilities for Skill Groups

Tables C.1–C.11, summarize the principal roles and responsibilities for 11 Software Engineering skill groups. These tables are examples and should be tailored to make them specific to your program or project. These tables are a partial coverage of all software-related skill groups.

- Chief Software Engineer (Table C.1)
- Subsystem Chief Software Engineer (Table C.2)
- Software Process Lead (Table C.3)
- Software Integrated Product Team (SwIPT) Lead (Table C.4)
- Software Integration and Test Lead (Table C.5)
- Software Item Team Lead (Table C.6)
- Software Engineer/Programmer (Table C.7)
- Software Test Engineer (Table C.8)
- Software Configuration Manager (Table C.9)
- Software Quality Assurance Manager (Table C.10)
- Software Subcontract Manager (Table C.11)

The information in these tables is presented as an example of software roles and responsibilities—they are not intended to limit an individual's responsibilities or to limit the type of skills you may need for your program. The intent is to define a minimum set of *responsibilities* and how various individuals *interact to facilitate consistency across a program*. Tables like these could be included in, or referenced by, the various SDP sections where the listed Software Engineering roles and responsibilities are performed.

NOTE: Parts 2–5 of this Guidebook also contain summary lists of roles and responsibilities.

Table C.1   Roles and Responsibilities of the Chief Software Engineer

Roles	Responsibility
Chief Software Engineer	Provide software oversight and insight across the program Lead and coordinate system software activities Report overall software status to program management Empowered to work software problems at any level Run and prepare (if applicable) Software Management Reviews Recommend award fee score for software subcontractors
SEPG	SEPG Chair
SDP	Key contributor, reviewer and approver
Software Appraisal	Prepare for and support appraisal
Work Instructions and Procedures	Review all work instructions, procedures and local processes
Planning	System level software planning Review all software schedules for consistency across program
Cost/Schedule Reporting	Review SwIPT cost/schedule information
Requirements	Review and assess software requirements across the program
Architecture Design	Review/assess software architecture design across the program Contributor, reviewer and approver of system architecture documents
Risk Management	Represents software at the Risk Management Board
Software Test	Review and assess software test plans, procedure and reports across the program
Software Metrics	Consolidate metrics from SwIPTs into program-level metrics Analyze metrics for trends across program Facilitate program and SwIPT level action in response to metrics Coordinate definitions of metrics and measurements
Problem Reports	Address problems that span SwIPTs and system interfaces Address problems that affect functional performance
Review Boards	Engineering Review Board (ERB) member or moderator Subsystem SwIPT Software Configuration Control Board as needed Program Configuration Control Board (CCB)

Also, see Section 1.6.4 for a description of responsibilities for the Chief Software Engineer's team.
SDP = Software Development Plan; SEPG = Software Engineering Process Group.

Table C.2 Roles and Responsibilities of the Subsystem Chief Software Engineer

Roles	Responsibility
Subsystem Chief Software Engineer	Provide oversight and insight into the subsystem (as applicable) software activities Lead and coordinate the subsystem's software activities Report the subsystem's software status to Chief Software Engineer and program management Empowered to work software problems at any level within the team Prepare for and support Software Management Reviews
SEPG	Member
SDP	Key contributor and reviewer
Software Appraisal	Prepare for and support appraisal
Work Instructions and Procedures	Review work instructions, procedures and local processes
Planning	Perform subsystem level software planning Review software schedules for consistency across program
Cost/Schedule Reporting	Review the subsystem's software cost and schedule information
Requirements	Review and assess software requirements across the subsystem
Architecture Design	Contribute, review and assess software architecture/design across the subsystem
Risk Management	Represents subsystem software at the Risk Management Board as required Identify, assess, mitigate, monitor and close software risks
Software Test	Review and assess software test plans, procedure and reports across the subsystem
Software Metrics	Consolidate metrics from the subsystem's SwIPTs into program-level metrics Analyze metrics for trends across the subsystem's Software Items Facilitate program and SwIPT level action in response to metrics Coordinate metric definitions
Problem Reports	Address problems that span the subsystem's SwIPTs Address problems that span system interfaces Address problems that affect functional performance
Review Boards	Lead the subsystem's SwIPT SCSCBs Member of the program SCCB

SDP = Software Development Plan; SEPG = Software Engineering Process Group; SCCB = Software Configuration Control Board.

Table C.3 Roles and Responsibilities of the Software Process Lead

Roles	Responsibility
Software Process Lead or Software Process Engineer	Software process owner; responsible for defining and maintaining the program's software process captured in the SDP Represents the project with the program and/or corporate SEPG Plan and coordinate software training
SEPG Administrative Chair	Chair forum for review and concurrence of software work instructions/procedures Chair horizontal coordination of software work instructions/procedures across SwIPTs
SDP	SDP owner
Software Appraisal	Lead software appraisal preparation effort and customer interface (customer may perform the assessments)
Work Instructions and Procedures	Plan and coordinate procedure development Owner of program-common software procedures
Planning	Review and concur with work instructions/procedures that implement the processes called out in the SDP
Cost/Schedule Reporting	Content and format Review software process cost schedule tracking
Requirements	Reviewer
Architecture Design	Reviewer
Risk Management	Maintains the program software risk mitigation plan
Software Test	Reviewer
Software Metrics	Content and format of measurement reports Reviewer of measurement reports
Problem Reports	Reviewer Elevates problem trends to corporate SEPG
Training	Ensure development and maintenance of the Program Training Plan

SDP = Software Development Plan; SEPG = Software Engineering Process Group.

Table C.4 Roles and Responsibilities of the Software Integrated Product Team Lead

Roles	Responsibility
SwIPT Lead	Lead and coordinate subsystem software activities Report software status to the Software Project Manager Advise, coach, and resolve conflicts within the SwIPT Software Team Prepare for and support Software Management and Technical Reviews
SEPG	Member
SDP	Key contributor, reviewer and stakeholder
Software Appraisal	Prepare for and support appraisal
Work Instructions and Procedures	Ensure work instructions/procedures are followed
Planning	SwIPT level software planning Responsible for subsystem software bidding information
Cost/Schedule Reporting	Manage software baselined schedule Manage software budget (if not allocated to the SI level) Capture earned value (if budget held at this level) Consolidate SI earned value into SwIPT earned value data Note: Budget must be portioned to SIs and earned value collected at the SI level
Requirements	Support allocation of requirements to SIs
Architecture Design	Provide technical guidance on trade studies, Systems Engineering, design, and vendor selection
Risk Management	Identify, assess, mitigate, monitor and close software risks
Software Test	Provide technical guidance on integration and testing
Software Metrics	Consolidate metrics from SIs into SwIPT level metrics Collect SwIPT level metrics Analyze metrics for trends across SwIPT Facilitate Software IPT level action in response to metrics
Problem Reports	Track software problems and ensure problems are resolved by their due date
Review Boards	Subsystem SwIPT SCCB chair

SDP = Software Development Plan; SEPG = Software Engineering Process Group; IPT = Integrated Product Team SI = Software Item.

Table C.5 Roles and Responsibilities of the Software Integration and Test Lead

Roles	Responsibility
Software Integration and Test Lead	Ensure software meets the requirements defined in the SRSs Ensure Software IPT SIs collectively meet the subsystem specification (as appropriate) Perform SI integration
SEPG	Member
SDP	Reviewer
Software Appraisal	Prepare for and support appraisal (if conducted)
Work Instructions and Procedures	Prepare test specific work instructions/procedures
Planning	Software IPT level software integration and test planning
Cost/Schedule Reporting	SI earned value review; software earned value, cost, schedule
Risk Management	Identify, assess, mitigate, monitor and close software risks
Software Test	Coordinate integration and testing between Software Teams
	Owner of subsystem software integration sequence, integration test plans, threads, test cases and test procedures
	Integration test Lead
	Review and approve integration test results
Software Metrics	Collect and report software integration and test metrics
	Analyze metrics and take appropriate action
Problem Reports	Record and track problems
Review Boards	Subsystem Software IPT SCB member

SDP = Software Development Plan; SEPG = Software Engineering Process Group; IPT= Integrated Product Team; SCCB = Software Configuration Control Board.

Table C.6 Roles and Responsibilities of the Software Item Team Lead

Roles	Responsibility
Software Item Lead	Lead and coordinate the SI software activities
SEPG	Member
SDP	Reviewer
Software Appraisal	Prepare for and support appraisal
Work Instructions and Procedures	Implement work instructions/procedures
Planning	Perform SI level software planning
Cost/Schedule Reporting	Manage SI Software Team based on baselined schedule Manage SI budget (if not held by the SwIPT Lead) Capture earned value (if budget held at this level) Report schedule status to SwIPT Lead
Requirements	Allocate requirements to lower level and higher level SUs Assign SUs to Software Engineers
Architecture Design	Oversee SI level architecture design Coordinate inter-SI interface design Verify SU design meets SI requirements
Risk Management	Identify, assess, mitigate, monitor and close software risks
Software Test	Review SU testing for SI integration Develop SI test plans/procedures for SI integration
Software Metrics	Collect and report SI metrics Analyze metrics and takes appropriate action
Problem Reports	Assign problems to Software Engineers on the team

SDP = Software Development Plan; SEPG = Software Engineering Process Group; SI = Software Item SU = Software Unit.

Table C.7   Roles and Responsibilities of the Software Engineer/Programmer

Roles	Responsibility
Software Engineer/Programmer	Development and test code for individual SUs
SEPG	Participate as needed
SDP	Read, understand, follow and recommend improvements
Software Appraisal	Prepare for and support appraisal as needed
Work Instructions and Procedures	Implement work instructions/procedures
Planning	SU level software planning
Cost/Schedule Reporting	Schedule/report SU activities
Requirements	Define/derive SI requirements and interfaces for assigned products
Architecture Design	Design and develop assigned SU architecture
Risk Management	Identify, assess, mitigate, monitor and help close software risks
Work Products	Code and Unit Test assigned SUs
Software Test	Perform SI integration and test
Software Metrics	Collect SU level information and provides to SI lead
Problem Reports	Work assigned problems

SDP = Software Development Plan; SEPG = Software Engineering Process Group.

Table C.8 Roles and Responsibilities of the Software Test Engineer

Roles	Responsibility
Software Test Engineer	Perform verification of software requirements
SEPG	Participate as needed
SDP	Understand and follow
Software Appraisal	Support appraisal as needed
Work Instructions and Procedures	Implement work instructions/procedures
Planning	SI qualification test planning
Cost/Schedule Reporting	Schedule/report SI qualification test activities
Requirements	Test software to meet requirements
Risk Management	Identify, assess, mitigate, monitor and help close software risks
Software Test	Develop, dry run and execute test procedures, test cases, test data, databases, test drivers, scripts, etc. and report results
Software Metrics	Collect SI qualification test information and provide to SI test lead
Problem Reports	Work assigned problems

SDP = Software Development Plan; SEPG = Software Engineering Process Group.

Table C.9 Roles and Responsibilities of the Software Configuration Manager

Roles	Responsibility
Software Configuration Manager	Establish software baselines; identify items to be placed under software CM control Manage changes to items under software CM control Perform baseline status accounting Perform subcontractor software baseline library audits Manage COTS software and changes to COTS software in the development environment and in operational software
SEPG	Member
SDP	Contributor, reviewer and approver
Software Appraisal	Prepare for and support appraisal
Work Instructions and Procedures	Author SCM specific work instructions/procedures
Requirements	Establish requirements baseline
Architecture Design	Establish architecture design baselines
Software Test	Build test software from source code and provide configured software for testing
Software Metrics	Collect and report problem report metrics
Problem Reports	Track problems
Review Boards	Administer local Subsystem SwIPT SCCB

SDP = Software Development Plan; SEPG = Software Engineering Process Group; SCCB = Software Configuration Control Board

Table C.10 Roles And Responsibilities of the Software Quality Assurance Manager

Roles	Responsibility
Software Quality Assurance Manager	Monitor compliance with program and corporate processes and standards Report to the program Product Assurance Manager Report status and findings to SI Lead, SwIPT Lead, Chief Software Engineer, Software Process Lead, SwIPT leads and Project Manager Perform subcontractor Software Quality Assurance system audits
SEPG	Member
SDP	Contributor and reviewer; SQA Lead is an approver
Software Appraisal	Prepare for and support appraisal
Work Instructions and Procedures	Author SQA specific work instructions/procedures
Requirements	Audits requirement baseline
Architecture Design	Audit design baseline
Risk Management	Identify, assess, mitigate, monitor and help close software risks
Software Test	Audit configured software Witness testing where requirements are verified
Software Metrics	Collect and report audit metrics and SQA non-compliance metrics
Problem Reports	Close the SQA non-compliance reports
Review Boards	Subsystem SwIPT SCCB member

SDP = Software Development Plan; SEPG = Software Engineering Process Group; SCCB = Software Configuration Control Board.

Table C.11 Roles And Responsibilities of the Software Subcontract Manager

Roles	Responsibility
Software Subcontract Manager	Technical subcontract aspects of a software subcontract Function similar to an SwIPT Lead over the subcontract Participate in Peer Reviews Perform software subcontractor oversight Approve all SDRLs as called out in the contract
SEPG	Member
SDP	Contributor, reviewer and stakeholder
Software Appraisal	Prepare for and support appraisal
Work Instructions and Procedures	Review and concur with software work instructions or procedures from subcontractor
Planning	Review and concur with subcontractor's SDP Annex and other software-related plans
Cost/Schedule Reporting	Review software subcontractor cost and schedule performance against baseline
Requirements	Responsible for requirement flow down to software subcontractor
	Review and concur with software subcontractor SRS(s) and IRS(s),
Architecture Design	Review and concur software subcontractor's Software Architecture Description, SDD(s), IDD(s) and DBDD(s)
Software Test	Review and concur software subcontractor test approach, plans, and results
Software Metrics	Review all software metrics and ensure subcontractor takes Corrective Action when indicated by the metrics
Problem Reports	Address problems that affect functional performance
Review Boards	Participate in software subcontractor SCCB(s)

SDRL = Subcontractor Data Requirements List; SDP = Software Development Plan; SCCB = Software Configuration Control Board; SEPG = Software Engineering Process Group.

Appendix D: Criteria for Evaluating Commercial Off-the-Shelf (COTS) and Reusable Software Products

The following list describes example criteria for evaluating COTS and other reusable software products. Evidence for complying with the criteria must be provided by prototype testing and/or stress testing within your system environment.

Does the COTS/Reuse Product you are considering have:

- The ability to provide the required functional capabilities; operate with contract constraints; achieve necessary performance goals; and provide required protection (safety, security and privacy).
- High reliability and maturity as evidenced by the product's established track record.
- Inherent product Supplier/Vendor viability including:
  - Compatibility of the COTS/Reuse Product supplier's future direction with your projected system needs (including both software and platform emphasis).
  - Supplier long-term business prospects and commitment to the COTS/Reuse Product.
  - Type and quality of supplier support available—current and planned.
- Suitability and operability for incorporation into the new system architecture including:
  - Compatible software architecture and design features and product limitations identified.
  - Absence of obsolete technologies.
  - Quality of the product's design, code and documentation.
  - Need for re-engineering and/or additional code development (e.g., wraps, "glue" code).
  - Compatibility with the set of reusable software products.
- Ability to remove or disable features and capabilities not required by the new system and the impact if those features cannot be removed or disabled or are not removed or disabled.

- Interoperability with other system and system-external elements including:
  - Compatibility with system interfaces and ability to interface with legacy systems.
  - Adherence to standards (e.g., open systems interface standards).
- Availability of personnel knowledgeable about the reusable software product plus training required, additional staff required, and vendor or third-party support required.
- Availability, completeness, accuracy and quality of documentation and source files.
- Testability as evidenced by the ability to identify and isolate faults.
- Acceptability of reusable software product licensing and data rights including:
  - Restrictions on copying and distributing the reusable software product or documentation.
  - Warranties available and license or other fees applicable to each copy.
  - Customer's usage and ownership rights, especially to the source code.
  - Ability to place source code in escrow against the possibility of the Vendor/Developer going out of business.
  - Absence of unacceptable restrictions in the standard license (e.g., export restrictions).
- Supportability and ability to make changes to the COTS/Reuse Product, including:
  - Likelihood the COTS/Reuse Product will need to be changed and the feasibility/difficulty of accomplishing that change, when changes are not made by the Vendor or Product Developer.
  - Feasibility and difficulty of accomplishing needed changes.
  - Priority of changes required by this system versus other changes being made.

- Likelihood that the changed version will continue to be maintained by the Vendor.
- Likelihood of being able to modify future versions to include changes.
- Impact of changes on life cycle costs.
- Impact if the current version is not maintained by the Vendor/Developer or if changes are not able to be incorporated into future versions.

■ Impacts of expected upgrades to the COTS/Reuse product including:
- Frequency of COTS/Reuse Product upgrades and modifications being made by the vendor.
- Feasibility and difficulty of incorporating the new version of the COTS/Reuse Product into your system and the impacts if the new version is not incorporated.
- Impact on development costs and schedule to incorporate upgrades.

■ Compatibility of planned upgrades of the COTS/Reuse product with system and software development plans and schedules including:

- Compatibility of planned upgrades with build content and schedules.
- Dependencies among COTS/Reuse Products.
- Potential for an incompatible set of COTS and/or Reusable Software Products.
- Potential for schedule delays until all dependent COTS/Reuse Products are upgraded.

■ Criticality of the specific functionality provided by the COTS/Reuse Product and the availability of alternate source(s) for that functionality.

■ Short and long-term cost impacts and trade-offs of using the COTS/Reuse Product including:
- Amount of management reserve needed to handle COTS/Reuse uncertainties.
- Ability to tolerate COTS/Reuse Product problems beyond the program's control at any point in the System Life Cycle.
- Ability to incorporate continuous evolution of the product by the supplier during system development and sustainment.

Appendix E: Pre-Milestone A Plans and Strategies

Addendum to Chapter 12: Pre-Milestone A Plans

Cost Analysis Requirements Description (CARD): The Program Office is responsible for preparation of the draft CARD during Pre-Milestone A. It is a description of the salient technical and programmatic features of a program whose costs are to be estimated. The CARD provides analysts with the basic information they need to *estimate the program's cost* to the detail identified in the life cycle cost model. Software Engineering is an essential participant in development of the CARD.

Configuration Management Plan (CMP): Describes the configuration identification, naming, control, change management and auditing of the hardware and software systems to be developed and delivered to the customer. An initial "working" version is needed early in Pre-Milestone A to manage the data being developed as part of the Concept Analysis (Also see Section 6.3).

Human Systems Integration (HSI) Plan: The HSI Plan describes how the overall system performance will be optimized by integrating manpower, personnel, training, human factors, safety and occupational health, personnel survivability, and habitability considerations in the acquisition process. Software Engineering and the SSMEs generate this plan.

Modular Open Systems Approach (MOSA) Plan: The MOSA Plan describes the integrated business and technical strategy for assessment and implementation of open systems in the DoD. Software Engineering and the SSMEs are responsible for generating this plan.

Risk Management Plan (RMP): The Acquisition Team is required to develop an RMP during Pre-Milestone A. This involves developing a comprehensive list of hardware and software risks that might impact the performance, cost or schedule of the space system acquisition. The RMP describes the means by which risks will be identified and managed. The activity of analyzing risks starts before MDD but becomes formal in Pre-Milestone A. If System Engineers assume certain hardware risks can be solved or mitigated via software,

SwE must evaluate and confirm that assumption. Software engineers also participate in audits and periodic reviews of the risk assessments and the RMP.

Software Acquisition Management Plan (SWAMP): Program Office planning is a key activity during Pre-Milestone A. The SWAMP provides the foundation for the other data products developed during Pre-Milestone A as well as the plans to be executed during later milestones. The SWAMP describes the software acquisition plans, processes and activities for the Program Office as well as the organizational roles and responsibilities in executing them. It assists the CSwE and SwE in maintaining technical insight and involvement with the contractors' software development effort. The SWAMP is prepared by SwE, and it is tightly coupled with the Systems Engineering Plan (SEP).

Software Acquisition Process Improvement (SWAPI) Implementation Plan: The SWAPI describes the methods that will improve the efficiency and effectiveness of SMC acquisition processes and software management. The SWAPI plan is tightly coupled with the SWAMP and is usually developed after the SWAMP is a mature document.

Addendum to Chapter 12: Pre-Milestone A Strategies

Data Management Strategy (DMS): The DMS describes the life cycle management of all data, software, and intellectual property. The DMS includes the disciplined processes and systems that will be used to plan for, acquire, and/or access, manage, and use data throughout the life cycle. Development of the DMS is an important adjunct to the *CMP*, which describes how control of the data is achieved (see Section 9.5).

For data that may be transmitted over the Global Information Grid (GIG), the DMS must be developed in coordination with the Net-Centric Data Strategy. The DMS also needs to take into consideration the development of the Information Assurance Strategy which describes the mechanisms for protecting the data and assuring its authenticity. The DMS is usually produced by SwE, with support from SE and the DM group, and it is essentially developed concurrently with the other Pre-Milestone A strategies.

Information Assurance (IA) Strategy (IAS): The IAS describes the analysis of potential security issues as uncovered during the risk analysis and DoD Architecture Framework (DoDAF) activities. A discussion of DoDAF is beyond the scope of this Guidebook. The IAS must describe the approach for obtaining Certification and Accreditation from the National Security Agency (NSA) for handling classified information including the management of keying material and cryptographic equipment.

The IAS must also identify all long-lead items and must be coordinated with the IMS. The IAS is developed in concert with the Test and Evaluation Strategy (TES), discussed below, with respect to obtaining the 'Interim Authority to Test' and the 'Interim Approval to Operate' when classified information is to be transmitted over the air or over the network. The IAS should be produced by SwE with support from SE and NSA personnel.

Technology Development Strategy (TDS): Early during Pre-Milestone A, SwE and SE should collaborate to produce a preliminary analysis to identify the set of Critical Technology Elements (CTEs) for the program and corresponding strategies for attaining at least Technology Readiness Level (TRL) level 6 prior to acquisition Milestone B. Hardware, software, and other elements that need to mature the CTEs to TRL level 6 should be identified. A discussion of TRLs is beyond the scope of this Guidebook.

The capabilities identified in the ICD, and later investigated in the Analysis of Alternatives and requirements development activities, should be supported by the content of the TDS. The TDS must include specific cost, schedule and performance goals, as well as exit criteria for technology demonstration(s).

The *Technology Maturation Plan* (TMP) is a key contributor to the TDS and is closely coordinated with the SEP. The TMP describes the approach for assuring CTEs achieve the required maturity during the life cycle of the program. The TMP developed during Pre-Milestone A becomes an important driver for the technology readiness assessment performed in Milestone B.

Net-Centric (NC) Data Strategy: The NC Data Strategy describes the approach for assuring seamless interoperation with the GIG. This is where the Net-Ready Key Performance Parameter (KPP) is usually discussed. (See the Defense Acquisition Guidebook (DAG) for additional requirements applicable to developing the NC Data Strategy.)

Test and Evaluation (T&E) Strategy (TES): The purpose of T&E is to verify the system performs as expected by the users. The Program Manager, in coordination with the customer, and others as appropriate, should develop an integrated TES for the program during Pre-Milestone A. Software Engineering should be fully involved with this process. The TES is developed in concert with the TRD and the SEP. In Milestone A, the TES is used to start the development of the *Master Test Plan* (MTP), also called the *Test & Evaluation Master Plan* (TEMP).

Software Engineering and SE should collaborate to establish realistic schedules for accomplishing the evaluation of analyses performed, planned test events, report preparation, and estimates of the level of support they will require. They should ensure that the exit criteria (the required test results or analyses) of the integrated test events will satisfy the objectives of Systems Engineering, Software Engineering, Specialty Engineering, test and quality communities participating in the event.

The T&E Working-Level Integrated Product Team, and the Program Manager, must ensure that the Development T&E (DT&E) program is robust in order to achieve a successful Operational T&E (OT&E) outcome—what happens here in the earliest milestone of the acquisition can have far-reaching effects. The T&E Working-Level Integrated Product Team should include representatives from SwE and other agencies as needed.

Appendix F: Annotated Outline of the Software Development Plan (SDP)

The following is an example outline of the *Software Development Plan*. It is derived from and is consistent with the standards listed in Appendix I.

1.0 Scope. The scope section of the SDP can be divided into the following four paragraphs:

1.1 Identification. This paragraph can contain a full identification of the system and the software to which this document applies, including, as applicable, identification number(s), title(s), abbreviation(s), version number(s), and release number(s).

1.2 System Overview. This paragraph can briefly state the purpose of the system and the software to which this document applies. It can describe the general nature of the system and software; summarize the history of system development, operation, and maintenance; identify the project sponsor, customer, user, Developer, and support agencies; identify current and planned operating sites; and list other relevant documents.

1.3 Document Overview. This paragraph can summarize the purpose and contents of the SDP and can describe any security or privacy considerations associated with its use.

1.4 Relationship to Other Plans. This paragraph can describe the relationship of the SDP to other project management plans.

2. Referenced Documents. This section can list the number, title, revision, and date of all documents referenced in the SDP. This section can also identify the source for all documents.

3. Overview of Required Work. This overview section of the SDP can be divided into paragraphs as needed to establish a context for the planning described in later sections. It can include an overview of:

- Requirements and constraints on the system and software to be developed
- Requirements and constraints on project documentation
- Position of the project in the System Life Cycle
- The selected program/acquisition strategy or any requirements or constraints on it

- Requirements and constraints on project schedules and resources
- Other requirements and constraints, such as on project security, privacy, methods, standards, interdependencies in hardware and software development, etc.

4. General Requirements. This section of the SDP can be divided into paragraphs 4.1 and 4.2. If different builds or different software on the project require different planning, these differences can be noted in the paragraphs. In addition to the content specified below, each paragraph can identify applicable risks or uncertainties and plans for dealing with them. The planning in each subparagraph of SDP Section 4 can cover all contractual clauses regarding the identified topic.

4.1 Software Development Process. This paragraph can describe the software development process to be used. The planning can cover all contractual clauses concerning this topic and identification of the Software Developmental Life Cycle model(s) to be used, including: planned builds, if applicable, their build objectives, and the software development activities to be performed in each build.

4.2 General Requirements for Software Development. This paragraph can be divided into the following subparagraphs:

4.2.1 Software Development Methods. This paragraph can describe or reference the software development methods to be used. Included can be descriptions of the manual and automated tools and procedures to be used in support of these methods. The methods can cover all contractual clauses concerning this topic. Reference may be made to other paragraphs in this plan if the methods are better described in context with the activities to which they will be applied.

4.2.2 Standards for Software Products. This paragraph can describe or reference the standards to be followed for representing requirements, design, code, test cases, test procedures and test results. The standards can cover all contractual clauses concerning this topic. Reference may be made to other paragraphs in this plan if the standards are better described in context with the activities to which they will be applied. Standards for code can be provided for each programming language to be used. They can include:

- Format Standards (such as indentation, spacing, capitalization, and information sequence)
- Standards for header comments (for example, name/identifier of the code; version identification; modification history; purpose; requirements and design decisions implemented; notes on the processing (such as algorithms used, assumptions, constraints, limitations, and side effects); and data notes (inputs, outputs, variables, data structures, etc.)
- Standards for other comments (such as required number and content expectations)
- Naming conventions for variables, parameters, packages, procedures, files, etc.
- Restrictions, if any, on the use of programming language constructs or features
- Restrictions, if any, on the complexity of code aggregates

4.2.3 Traceability. This paragraph can describe the approach to be followed for establishing and maintaining bi-directional traceability between levels of requirements, between requirements and design, between design and the software that implements it, between requirements and qualification test information, and between computer hardware resource utilization requirements and measured computer hardware resource utilization.

4.2.4 Reusable Software Products. This paragraph can be divided into the following subparagraphs:

4.2.4.1 Incorporating Reusable Software Products. This paragraph can describe the approach to be followed for identifying, evaluating, and incorporating reusable software products, including the scope of the search for such products and the criteria to be used for their evaluation. It can cover all contractual clauses concerning this topic. The candidate or selected reusable software products, known at the time this plan is prepared or updated, can be identified and described, together with benefits, drawbacks, and restrictions, as applicable, associated with their use.

4.2.4.2 Developing Reusable Software Products. This paragraph can describe the approach to be followed for identifying, evaluating and reporting opportunities for developing reusable software products. It can cover all contractual clauses concerning this topic.

4.2.5 Assurance of Critical Requirements. This paragraph can be divided into the following subparagraphs to describe the approach to be followed for handling requirements designated critical:

- 4.2.5.1 Safety
- 4.2.5.2 Security

- 4.2.5.3 Privacy Protection
- 4.2.5.4 Dependability, Reliability, Maintainability and Availability
- 4.2.5.5 Assurance of other Mission Critical Requirements

4.2.6 Computer Hardware Resource Utilization. This paragraph can describe the approach to be followed for allocating computer hardware resources and monitoring their utilization.

4.2.7 Recording Rationale. This paragraph can describe the approach to be followed for recording rationale that will be useful to understand the key decisions made on the project. It can interpret the term "key decisions" for the project and state where the rationale are to be recorded.

4.2.8 Access for Customer Review. This paragraph can describe the approach to be followed for providing the customer or its authorized representative access to Developer and Subcontractor facilities for review of software products and activities.

5. Plans for Performing Detailed Software Development Activities. This section of the SDP covers the details of the developmental activities. It is focused primarily on "what" needs to be done, "when" it needs to be done, and "who" should do it. In general, it is not focused on "how" to do it as that level of detail is typically handled by procedures that back up the SDP.

Section 5 can be divided into the paragraphs described below. Subsections corresponding to non-required activities for a project may be satisfied by the words "Not applicable," but the subsection should be included with those words. If different builds or different software on the project require different planning, these differences should be noted. The discussion of each activity can include the approach (plans, processes, methods, procedures, tools, roles and responsibilities) to be applied to the:

- Analysis or other technical tasks involved
- Recording of results
- Preparation of associated deliverables, if applicable

For each activity, include entrance criteria, inputs, tasks to be accomplished and products to be produced, verifications to be used (to ensure tasks are performed according to their defined processes and products meet their requirements), outputs and exit criteria.

The discussion can also identify applicable risks or uncertainties and plans for dealing with them. Reference may be made to SDP Section 4.2.1 if applicable methods are described there. The planning in each subparagraph of SDP Section 5 can cover all contractual clauses regarding the identified topic.

5.1 Project Planning and Oversight. This paragraph can be divided into the following subparagraphs to describe the approach to be followed for project planning and oversight:

- 5.1.1 Software Development Planning
- 5.1.2 Software Item Test Planning
- 5.1.3 System Test Planning
- 5.1.4 Planning for Software Transition to Operations
- 5.1.5 Planning for Software Transition to Maintenance
- 5.1.6 Adherence to and schedule for updating the plans

5.2 Establishing a Software Development Environment. This paragraph can be divided into the following subparagraphs to describe the approach to be followed for establishing, controlling, and maintaining a Software Development Environment:

- 5.2.1 Software Development Environment
- 5.2.2 Software Integration and Test Environment
- 5.2.3 Software Development Library
- 5.2.4 Software Development Files
- 5.2.5 Non-Deliverable Software

5.3 System Requirements Analysis. This paragraph can be divided into the following subparagraphs to describe the approach to be followed for participating in System Requirements Analysis:

- 5.3.1 Analysis of User Input
- 5.3.2 Operational Concept
- 5.3.3 System Requirements

5.4 System Design. This paragraph can be divided into the following subparagraphs to describe the approach to be followed for participating in System Design:

- 5.4.1 System-Wide Design Decisions
- 5.4.2 System Architectural Design

5.5 Software Requirements Analysis. This paragraph can describe the approach to be followed for software requirements analysis.

5.6 Software Design. This paragraph can be divided into the following subparagraphs to describe the approach to be followed for Software Design:

- 5.6.1 Software Item-Wide Design Decisions
- 5.6.2 Software Item Architectural Design
- 5.6.3 Software Item Detailed Design

5.7 Software Implementation and Unit Testing. This paragraph can be divided into the following subparagraphs to describe the approach to be followed for Software Implementation and Unit Testing:

- 5.7.1 Software Implementation
- 5.7.2 Preparing for Unit Testing
- 5.7.3 Performing Unit Testing
- 5.7.4 Revision and Retesting
- 5.7.5 Analyzing and Recording Unit Test Results

5.8 Unit Integration and Testing. This paragraph can be divided into the following subparagraphs to describe the approach to be followed for Unit Integration and Testing:

- 5.8.1 Preparing for Unit Integration and Testing
- 5.8.2 Performing Unit Integration and Testing
- 5.8.3 Revision and Retesting
- 5.8.4 Analyzing and Recording Unit Integration and Test Results

5.9 Software Item Qualification Testing. This paragraph can be divided into the following subparagraphs to describe the approach to be followed for Software Item Qualification Testing:

- 5.9.1 Independence in Software Item Qualification Testing
- 5.9.2 Testing on the Target Computer System
- 5.9.3 Preparing for Software Item Qualification Testing
- 5.9.4 Dry run of Software Item Qualification Testing
- 5.9.5 Performing Software Item Qualification Testing
- 5.9.6 Revision and Retesting
- 5.9.7 Analyzing and Recording Software Item Qualification Test Results

5.10 Software/Hardware Item Integration and Testing. This paragraph can be divided into the following subparagraphs to describe the approach to be followed for participating in software/hardware item integration and testing:

- 5.10.1 Preparing for Software/Hardware Item Integration and Testing
- 5.10.2 Performing Software/Hardware Item Integration and Testing
- 5.10.3 Revision and Retesting
- 5.10.4 Analyzing and Recording Software/Hardware Item Integration and Test Results

5.11 System Qualification Testing. This paragraph can be divided into the following subparagraphs to describe the approach to be followed for participating in System Qualification Testing:

- 5.11.1 Independence in System Qualification Testing
- 5.11.2 Testing on the Target Computer System

- 5.11.3 Preparing for System Qualification Testing
- 5.11.4 Dry run of System Qualification Testing
- 5.11.5 Performing System Qualification Testing
- 5.11.6 Revision and Retesting
- 5.11.7 Analyzing and Recording System Qualification Test Results

5.12 Preparing for Software Transition to Operations. This paragraph can be divided into the following subparagraphs to describe the approach to be followed for preparing for software use:

- 5.12.1 Preparing the Executable Software
- 5.12.2 Preparing Version Descriptions for User Sites
- 5.12.3 Preparing User Manuals
- 5.12.3.1 Software User Manuals
- 5.12.3.2 Computer Operations Manuals
- 5.12.4 Installation at User Sites

5.13 Preparing for Software Transition to Sustainment. This paragraph can be divided into the following subparagraphs to describe the approach to be followed for preparing for software transition to the Sustainment Phase:

- 5.13.1 Preparing the Executable Software
- 5.13.2 Preparing Source Files
- 5.13.3 Preparing Version Descriptions for the Maintenance Site
- 5.13.4 Preparing the "As Built" Software Item Design and Other Related Information
- 5.13.5 Updating the System/Subsystem Design Description
- 5.13.6 Updating the Software Requirements
- 5.13.7 Updating the System Requirements
- 5.13.8 Preparing Maintenance Manuals
  - 5.13.8.1 Computer Programming Manuals
  - 5.13.8.2 Firmware Support Manuals
- 5.13.9 Transition to the Designated Maintenance Site

5.14 Software Configuration Management. This paragraph can be divided into the following subparagraphs to describe the approach to be followed for Software Configuration Management:

- 5.14.1 Configuration Identification
- 5.14.2 Configuration Control
- 5.14.3 Configuration Status Accounting
- 5.14.4 Configuration Audits
- 5.14.5 Packaging, Storage, Handling, and Delivery

5.15 Software Peer Reviews and Product Evaluations. This paragraph can be divided into the following subparagraphs to describe the approach to be followed for software peer reviews and product evaluations:

- 5.15.1 Software Peer Reviews
- 5.15.1.1 Prepare for Software Peer Reviews
- 5.15.1.2 Conduct Peer Reviews
- 5.15.1.3 Analyze Peer Review Data
- 5.15.2 Software Product Evaluations
- 5.15.2.1 In-Process and Final Software Product Evaluations
- 5.15.2.2 Software Product Evaluation Records
- 5.15.2.3 Independence in Software Product Evaluations

5.16 Software Quality Assurance. This paragraph can be divided into the following subparagraphs to describe the approach to be followed for Software Quality Assurance:

- 5.16.1 Software Quality Assurance Evaluations
- 5.16.2 Software Quality Assurance Records
- 5.16.3 Independence in Software Quality Assurance
- 5.16.4 Software Quality Assurance Non-Compliance Issues

5.17 Corrective Action. This paragraph can be divided into the following subparagraphs to describe the approach to be followed for the Corrective Action Process:

- 5.17.1 Problem/change reports—These reports can include items to be recorded such as: project name, originator, problem number, problem name, software element or document affected, origination date, category and severity, description, analyst assigned to the problem, date assigned, date completed, analysis time, recommended solution, impacts, problem status, approval of solution, follow-up actions, corrector, correction date, problem type, version where corrected, correction time and description of solution implemented.
- 5.17.2 Corrective Action System

5.18 Joint Technical and Management Reviews. This paragraph can be divided into the following subparagraphs to describe the approach to be followed for joint technical and Management Reviews:

- 5.18.1 Joint Technical Reviews
- 5.18.2 Joint Management Reviews

5.19 Risk Management. This paragraph can describe the approach for performing risk management.

5.20 Software Management Indicators. This paragraph can describe the approach to be used for software measurement throughout the life cycle. It can also include the specific software measurements to be used (that is, collected, analyzed, reported and used for decision-making, Corrective Actions and reporting to the customer), including which measurements will be reported by life cycle activity (e.g., requirements, design, code, integration, test).

5.21 Security and Privacy. This paragraph can describe the approach for meeting the security and privacy requirements.

5.22 Subcontractor Management. This paragraph can describe the approach to performing subcontractor management.

5.23 Interface with Software Independent Verification and Validation (IV&V) agents. This paragraph can describe the approach for interfacing with the software IV&V agents.

5.24 Coordination with Associate Developers. This paragraph can describe the approach for performing the coordination with Associate Developers, working groups, and interface groups.

5.25 Improvement of Project Processes. This paragraph can describe the approach for performing improvements to the project processes.

6. Schedules and Activity Network. This section can provide:

- Schedule(s) identifying the activities in each build and showing initiation of each activity, availability of draft and final deliverables and other milestones and completion of each activity.
- An activity network, depicting sequential relationships and dependencies among activities and identifying those activities that impose the greatest time restrictions on the project.

Only a broad overview schedule is typically included within the SDP; detailed schedules and activity networks are usually not included within the SDP, but their locations are noted in the SDP in order to avoid updating the SDP every time a schedule changes.

7. Project Organization and Resources. This section can be divided into the following paragraphs to describe the project organization and resources to be applied in each build.

7.1 Project organization. This paragraph can describe the organizational structure to be used on the project, including the organizations involved, their relationships with one another and the authority and responsibility of each organization for carrying out required activities.

7.2 Project Resources. This paragraph can describe the resources to be applied to the project. It can include, as applicable:

- 7.2.1 Personnel resources, including:
  - The estimated staff loading for the project (number of personnel over time)
  - The breakdown of the staff-loading numbers by responsibility (for example, management, Software Engineering, software testing, Software Configuration Management, software product evaluation, Software Quality Assurance)
  - A breakdown of the skill levels, geographic locations, and security clearances of personnel performing each responsibility
  - The rationale for effort, staff loading and schedule estimates, including software cost
  - and schedule estimation techniques, the input to those techniques (e.g., software size and software cost driver parameters), and any assumptions made
- 7.2.2 Overview of Developer facilities to be used, including geographic locations in which the work will be performed, facilities to be used, and secure areas and other features of the facilities as applicable to the contracted effort.
- 7.2.3 Customer-furnished equipment, software, services, documentation, data and facilities required for the contracted effort. A schedule detailing when these items will be needed can also be included.
- 7.2.4 Other required resources, including a plan for obtaining the resources, dates needed and availability of each resource item.

8. Notes. This section can contain any general information that aids in understanding this document (e.g., background information, glossary, rationale). This section can include an alphabetical listing of all acronyms, abbreviations, and their meanings as used in this document and a list of any terms and definitions needed to understand this document.

Annexes. Annexes may be used to provide information published separately for convenience in document maintenance (e.g., charts, classified data). As applicable, each annex should be referenced in the main body of the SDP where the data would normally have been provided. Annexes may be bound as separate documents for ease in handling and should be lettered alphabetically (A, B, etc.).

Appendix G: Exit Criteria for Software Reviews

The software exit criteria must be defined to the level of completeness called for in the SDP, based on the Software Life Cycle Model used, and appropriate to the current development activity.

Table G.1   Exit Criteria for the SAR, PDR and CDR

EXIT CRITERIA for the Software Requirements and Architecture Review (SAR), Software Preliminary Design Review (PDR), and the Software Critical Design Review (CDR)
Software requirements, including software interface requirements and software architecture views: ■ Are correct, complete, consistent, feasible, verifiable and clear (including derived requirements) ■ Are traced to and fully implement their parent requirements ■ Have verification methods and verification levels specified
Non-developmental items (COTS and reuse software) are fully integrated into the software architecture.
The design of each Software Item (SI) is clear, correct, complete, consistent and elaborated to the detail of the Software Units (SU) comprising the SI.
Software Architecture and Design adequately addresses: ■ Use of all applicable standards and satisfies all applicable interoperability-related requirements ■ End-to-end processing across external and internal interfaces for operations, maintenance and training ■ Operational database management and control functions ■ Functional and performance requirements for states and modes and requirements for survivability ■ Supportability, fault management, diagnostics, dependability, reliability, maintainability and availability
Engineering analyses, models and/or simulations, demonstrate that the software architecture, Detailed Design and computer resources meet Key Performance Parameters (KPPs) with adequate margins.
Safety, Information Assurance and human systems integration analyses are consistent with the software architecture, design and resources.
Engineering analyses and trade studies demonstrate adequacy of the software algorithms, architecture, Detailed Design and computer resources.
Software qualification test plans and procedures are valid, complete and consistent with the software architecture, design, and test plans and all software requirements are fully allocated to tests described in the test plans.
The master software build plan is complete, feasible, executable and consistent with Software Requirements, architecture, test plans, and schedules.
Software risk assessments and the software Risk Management Plan (RMP) are integrated with the system RMP.
Effective software risk handling plans are in place, and activities are being performed in accordance with the plans.

(Continued)

Table G.1 (Continued)   Exit Criteria for the SAR, PDR and CDR

EXIT CRITERIA for the Software Requirements and Architecture Review (SAR), Software Preliminary Design Review (PDR), and the Software Critical Design Review (CDR)
Software size estimates are supportable and software cost models have been calibrated with real data.
Software cost and schedule estimates have enough margin to cover estimated risks and support.
Software schedules are consistent with higher-level schedules.
The SDP is consistent with the updated IMP, SEMP and other plans.
The SDP describes a set of processes, methodologies, tools, environments and life cycle models that are feasible, appropriate for program scope and complexity, and used consistently by all team members.
The software development and test environments are integrated and consistent with Systems Engineering environments across all team members.
Definitions for software metrics are documented, clear, and correct and include reasonable thresholds for triggering defined Corrective Action.
Software measurements (metrics) and Technical Performance Measures (TPM) are being collected, analyzed, reported and used for decision-making.
The software problem/deficiency report status indicates that adequate progress is being made resolving problems.

Appendix H: Chapter Highlights

Chapter 1 Highlights

Software Project Management Introduction

- The Guidebook is a "shopping list" of what *must* be done, what *should* be done, what *can* be done, and what *may* be done if applicable to your project (1.1).
- Acronyms are an inherent part of the daily life of a Software Engineer (1.1).
- To clarify the difference between programs and projects, view the overall hierarchy as programs consisting of projects, projects consisting of activities and activities consisting of tasks (1.2).
- A project is a temporary endeavor to produce a specific product, service, upgrade or result within a defined period of time and is usually constrained by funding (1.2).
- Magic is not a participant in the delivery of successful software-intensive systems; they must be planned, well-executed and managed (1.3).
- Performing project management functions, in an organized framework of activities and processes, is the job of the Software Project Manager (SPM) (1.3).
- As the SPM on medium-to-large software-intensive projects, the essential level of technical expertise means your ability to understand *what* is being done by your team, not necessarily a detailed understanding as to exactly *how* it is being performed (1.3).
- Develop an ability to recognize the type of problems that have a propensity for self-healing and when you should not be prematurely over-reactive (1.3).
- A Software Project Manager must be flexible and open-minded because, when dealing with large complex systems, frequent changes are a certainty (1.3).
- A software-intensive system is an organized collection of hardware, software, people, policies, methods, procedures, regulations, constraints and related resources interacting with each other, within the system's set boundaries, followed to accomplish common system objectives (1.4.1).
- Systems Engineering is a formal interdisciplinary field of engineering focused on how complex projects should be designed, implemented, tested and managed (1.4.2).

- On many programs and projects, Software Engineering is located organizationally within the Systems Engineering group (1.4.2).
- Software Engineering and Systems Engineering are tightly coupled in today's software-intensive systems—and that includes essentially all major systems. Software considerations need to be involved in every system decision (1.4.3).
- Early involvement of Software Engineering in the System Life Cycle is a major help in mitigating the risks of excessive cost, effort and rework (1.4.3).
- The System Life Cycle (SLC) is a cradle to grave perspective starting with an initial conceptual analysis of the proposed system and ending with an ultimate replacement or retirement of the system (1.4.4).
- A working product in the hands of the customer is just the beginning of the longest and most costly portion of the System Life Cycle–System Sustainment (1.4.4).
- As the SPM, you must provide the context for your project by defining the boundaries and scope of your contractual responsibilities (1.4.5).
- The software-intensive system you are developing can incorporate one, or many, other systems within the boundaries of your system (1.4.6).
- A one-page big picture flowchart overview is a key mechanism for understanding and identifying the required system functionality (1.4.7).
- Strategic planning up front usually makes the difference between success and failure (1.5).
- Every software development program should have a Software Development Plan (SDP), but for large programs it is a critical element for success. An SDP, or similar document, is required by essentially all software development standards (1.5.1).
- The SDP is focused on *what* will be done and should be backed up with detailed operational procedures that describe *how* to do it. An incomplete, poorly written, unorganized or inadequate SDP is a clear red flag (1.5.1).
- The Work Breakdown Structure (WBS) is a key software budgetary planning and control activity that organizes and decomposes, in a hierarchical structure,

all the project tasks into smaller, more manageable, and controllable components (1.5.2).

■ Finding, recruiting, nurturing and retaining a competent staff is the single most important function of a Software Project Manager (1.6).

■ For a software-intensive project, the organization chart should always be structured to facilitate software management visibility and software technical oversight.

■ Development of very large and complex software-intensive systems may require hundreds of personnel involved in software management, development and support tasks (1.6.2).

■ Software Project Managers and Lead Programmers have different responsibilities (1.6.3).

■ The quality of a software system is directly related to the process used to create it—and the Chief Software Engineer is the core of the software process (1.6.4).

■ The development of software products is performed by Software Integrated Product Teams (SwIPT); your cross-disciplinary team of Developers (1.6.5).

■ Each WBS element should be the primary responsibility of a single SwIPT (1.6.5).

■ Effective interpersonal communication with your team, in an intellectually honest fashion, along with effective team conflict management and resolution, are the keys to successful Software Project Management (1.6.6).

■ SPMs must figure out how to please the customer—even if they are *wrong* (1.7).

■ There must be a definition of the *category* assigned to each software entity because not every software entity needs to have the full set of documentation, the full set of reviews, the full set of metrics, and the same level of testing (1.8).

■ System Critical Software (SCS) is physically part of, dedicated to, and/or *essential to full performance* of the system (1.8.1).

■ Support Software (SS) aids in system hardware and software development, test, integration, qualification and maintenance (1.8.2).

■ Commercial Off-the-Shelf (COTS)/Reused (C/R) software is third-party *non-developmental* software (1.8.3).

■ Applicable software standards must provide *value added* to your project (1.9).

Chapter 2 Highlights

Software Project Management Activities

■ A major objective facing all SPMs is the task of *balancing scope and resources in order to achieve effective project integration and control* (2.2).

■ A software-intensive system, coded to perfection by world-class Programmers, will fail if it does not address the needs of your customer. There is nothing, except for a capable staff, that will contribute more to the success of your project than having a valid set of requirements (2.3.1).

■ Requirements are often complex and usually not fully defined up front; in that case, a life cycle methodology that allows for *frequent refinement* of requirements during development, such as evolutionary or spiral development, must be used (2.3.1).

■ An automated requirements management and traceability database should be used by every large software-intensive project (2.3.1).

■ An important aspect of your SPM requirements management responsibilities is to control what is called requirements *scope creep* (2.3.1).

■ Software risk management is the process of identifying, measuring and assessing the risk and then developing strategies to manage its mitigation or elimination (2.3.2).

■ A fundamental requirement for effective project management is the ability to *measure and track performance in a timely manner to determine the status of your project and allow you time to take Corrective Actions* (2.3.3).

■ Management of cost and schedules is not a *singular* activity because it involves multiple activities that *interact, overlap and support* each other (2.3.4).

■ Essentially everything an SPM does is directly or indirectly related to quality (2.3.5).

■ As the SPM, you are responsible for managing the project stakeholders (2.3.6).

■ Excellent communication with your stakeholders is an essential foundation of controlling expectation management (2.3.6).

■ Software *work products* are essential artifacts of the software development process (2.3.7).

■ Software integration and testing activities are pervasive (2.3.9).

■ Regression Testing is performed to make sure that recent changes have not impacted the software that was previously tested and caused unintended and undesirable consequences (2.3.9).

■ Estimates given at the start of a project may haunt you for the whole project, so it is very important to have as good an estimate as you can make from the beginning (2.3.10).

■ Process improvement assumes you have a process that can be improved!(2.4).

■ The Software Engineering Process Group (SEPG) plays a key role—especially when development activities span organizational, administrative, geographic or functional boundaries (2.4.1).

- The Software Quality Assurance (SQA) group is not responsible for software quality; as the SPM, you and your software Development Team are responsible for building quality into the product, and you are accountable for delivering a quality product (2.4.2).
- A *system* involves the combination of all hardware, software, firmware, and often people, working synergistically together to produce a desired operational result (2.4.3).
- The Software Configuration Management (SCM) process is an *essential development control activity* that begins during requirements definition (2.4.3).
- Software Peer Reviews (SPRs) are *structured methodical examinations* of software work products by the Developers, with their peers, to identify existing defects, to recommend needed changes, and to identify potential problems (2.4.5).
- The Capability Maturity Model Integrated (CMMI) is a *process improvement, training and process maturity appraisal* program (2.4.6).
- Six Sigma has been successful at many software organizations because it invokes a formal, structured process to improve software quality compared to the haphazard approach those organizations used in the past (2.4.7).
- System Sustainment typically consumes about two-thirds (or more) of the total System Life Cycle cost. Maintaining a deployed system, responsive to changing customer needs over its lifetime (sometimes decades), can be just as challenging, as the original system development task (2.5).

Chapter 3 Highlights

System and Software Life Cycle Processes

- The processes followed by a project are the secret to its success (3.0).
- Following defined and structured processes, applicable to your project, will dramatically increase the probability of a successful software-intensive system implementation (3.0).
- A process is a series of activities, involving tasks, procedures, constraints, and applied resources, that produces one or more planned outputs (3.1.1).
- During the latter activities of the Software Development Life Cycle, the cost of not following the process is significantly greater than following a process (3.1.2).
- The System Life Cycle covers the *entire system* starting with an initial conceptual analysis of the proposed system and ending with the eventual replacement or retirement of the system.

- The Software Development Life Cycle (SDLC) starts during System Definition and ends at the start of System Sustainment. Embedded within the SDLC are part of the System Definition task; the entire Software Implementation Process (SIP); and the System Integration, Testing and Verification (IT&V) process (3.4).
- The SIP is repeated for each build or spiral and may be repeated multiple times during System Sustainment to implement new functions and make needed modifications (3.5).
- Naming conventions for each build must be established up front by assigning unique alphanumeric designations (3.5).
- The System Integration, Testing and Verification (System IT&V) process involves activities where individual software modules are combined and then tested and verified as a group (3.6).
- The generic software IT&V process normally involves five testing stages (3.6).
- Sustainment is the longest and most expensive activity of the System Life Cycle. From the user's standpoint, it may also be the most important activity (3.7).
- Significant savings can be realized if software entities can be placed into a class that does not require the same level of attention required by System Critical Software that is critical to full functionality and performance of the system (3.8).
- Testing is not optional; it is an inherent and critical element of every software-intensive system. Managing the test process basically involves: estimating the cost of testing; tracking and monitoring the actual cost of testing; and making corrections to stay on track (3.9).
- As the SPM, especially on large programs, you should not be directly performing the actual testing. However, you should be directly involved with deciding the appropriate testing strategy planned for your project (3.9.2).
- Larger systems are usually tested using a mixture of strategies rather than any single approach. Whatever testing strategy is adopted, it is always sensible to adopt an incremental approach to subsystem and system testing (3.9.2).
- If *any* system element changes, including both software and hardware changes, *software testing needs to be repeated* (3.9.3).
- Test *risk analysis* is appropriate for most software development projects (3.9.4).
- A critical characteristic of a *good* requirement is that it must be *testable* (i.e., verifiable) (3.9.5).
- If your software-intensive system in fully compliant with your customer's requirements, and it passes all of the qualification testing, *it still might fail* if you failed

to test it in its intended real operational (flight-like) conditions and expected environment (3.9.6).

■ For large projects, or ongoing long-term projects, automated testing can be cost-effective (3.9.7).

■ There are many variables involved, and techniques available, in deciding *how much testing is needed* to produce a dependable software product (3.9.8).

Chapter 4 Highlights

Software Development Methodologies

■ Software development methodologies, also called process models, are frameworks used to plan and control the process of developing software-intensive systems (4.1).

■ A Software Development Life Cycle model should almost always be used to describe, organize, monitor and control software development activities but there is considerable confusion regarding the differences between some of them (4.1).

■ More than one Software Development Life Cycle Model may be needed for different types of software applications (4.1).

■ A brief description of common *Software Development Process Models* in alphabetical order (4.1):

– Agile: The term *Agile* covers a number of methods and approaches for software development. It is focused on frequent delivery and testing of small portions of the full system (4.3).

– Evolutionary: The software is developed in a series of builds with increasing functionality; requirements are defined for each evolutionary build as each build is developed (4.1.1).

– Incremental: This model requires that all of the requirements are defined up front; the software product is then developed in a series of builds with cumulative increasing functionality (4.1.2).

– Iterative: Not really a software development model but more of a *quality improvement approach* where a fully developed and delivered system is periodically updated and improved with each new release of the product (4.1.3).

– Prototyping: Building an early experimental portion of a system to better understand interfaces and requirements, to test throughput speeds, develop environment testing, etc. (4.1.4).

– Spiral: A risk-driven development process that is: (1) a cyclic approach that grows a system's functionality incrementally focused on decreasing its degree of risk; and (2) a set of anchor point milestones for ensuring stakeholder commitment to acceptable system solutions (4.1.5).

– Unified: A variation of the Spiral Model exemplified by the IBM Rational Unified Process® (RUP®). It is not a single prescriptive process but an *adaptable process framework* intended to be tailored using the Object-Oriented Unified Modeling Language (UML) (4.1.6).

– Waterfall: A linear sequential software development model that requires all functionality and design requirements to be defined up front and each development activity to be completed before the next activity begins, although some overlap is allowed (4.1.7).

■ The Agile movement is as a collection of *lightweight* software development methods in reaction to the more *heavyweight process-oriented* development methods perceived by the critics to be too structured, too regulated, too much documentation, and often micro-managed (4.3).

■ A common pitfall for *enterprise-wide Agile adoption* is a lack of defining exactly *what does Agile mean* in your organization (4.3.1).

■ "Agile" implies flexibility as well as rapidity with these core principles: self-organizing teams, self-directed teams, daily stand-up meetings, minimal processes and principles, frequent releases, continuous testing, supporting infrastructure, and customer collaboration (4.3.2).

■ There are *three* sides to every story. You can call the three sides *the positive, the negative and the neutral* or call them *the optimistic, the pessimistic, and the realistic* (4.3.3).

■ Goals for a complex *mega system* are much more than just "code that works" (4.3.4).

■ Agile is not a silver bullet; regardless, if applied properly in the right environment, even on smaller entities of large systems, Agile can be an effective methodology (4.3.5).

■ The principal roles performed by an Agile/Scrum Team may be called by different names depending on the Agile methodology being used, but the basic concepts are very similar: Scrum Master, the Scrum Team members, Product Owner, and the customer (4.3.6).

■ Agile has definite value and applicability to specific situations, but it *should not be forced* into environments where it does not seamlessly meet the tenants of the Agile methodology. Agile is orders-of-magnitude better than the all too common chaotic ad-hoc approach which is completely void of any process (4.3.8 and 4.3.9).

■ Every large system, and most mid-sized projects, must have, *and follow*, a formal program management methodology by maintaining an approved *Integrated Master Plan* (IMP) coupled with a related *Integrated Master*

Schedule (IMS), or equivalent, to provide a complete schedule and activity network for all program and system activities (4.4).

■ The SDP should include, or reference, an *activity network* depicting sequential relationships and dependencies among all software activities, and identify those activities that impose the greatest restrictions on the project (4.4).

■ As the SPM, you must identify the standards to be used or modified for your project (4.5).

■ Whatever hierarchy is chosen for your program, for product levels and their directly related testing levels, you must ensure it is followed to avoid confusion and to significantly enhance the probability that all the pieces of the software puzzle will fit together properly (4.5.2).

Chapter 5 Highlights

Software Management Domain

■ A goal is a destination with a planned path and a deadline; planning is an ongoing task (5.1).

■ Planning is critical at the start of the Development Life Cycle as it is the foundation for initially producing the software plans required to implement and perform the software development process and for the formation of Software Teams required to execute those plans (5.1).

■ Preparation of the *Software Configuration Management Plan* (SCMP), and the *Software Quality Program Plan* (SQPP) should be assigned to your SCM and SQA Teams (5.1).

■ The Software Development Plan, or an equivalent document whatever you may call it, is your *key* software planning document (5.1.1).

■ Site-specific SDPs (also called SDP Annexes) can be produced containing specific and/or unique policies and procedures applicable to them only that expands on, but *does not conflict with*, the policies and procedures defined in the program-level SDP—except for approved waivers (5.1.1).

■ The quality and attention to detail in the SDP is a major source selection *evaluation criteria* when bidding for government work (5.1.1).

■ The first step in planning is to review software requirements since the scope of the software task is established by identifying system requirements to be satisfied by the software products (5.1.2).

■ A database should be produced and periodically updated to provide a mechanism for identifying, profiling and tracking *all Software Items* (SIs) on the project (5.1.2).

■ A software *build* is a portion of a system that satisfies, in part or completely, an identifiable subset of the total end-item or system requirements and functions (5.1.4).

■ A comprehensive Software Master Build Plan (SMBP) must be provided to map the incremental functionality, capabilities and requirements allocated to each build (5.1.4).

■ Software measurement data must be used to compare actual software size, cost, schedule and progress against the established plan so that you can take timely Corrective Actions (5.1.5).

■ The status of software should be reviewed (usually weekly) at the subsystem-level meetings and at monthly program status meetings (5.1.5).

■ Tasks performed that are *not* called out in the contract as a system requirement can be considered "gold plating" unless the task is a derived requirement essential to system functionality. A passion for excellence and efficiency is much less expensive to attain than a passion for perfection (5.1.6).

■ Software Test Engineers, at the subsystem level, are responsible for documenting their *Software Test Plan* (STP), *Software Test Descriptions* (STD), and *Software Test Reports* (STR) to verify that the SIs meet their allocated requirements (5.1.7).

■ Software Developers and Test Engineers have a *support role* in the planning and execution of system testing (5.1.8).

■ Planning and preparation, for transition to operations and maintenance, should start relatively early in the life cycle to ensure a smooth transition to the users' site (5.1.9).

■ Risk management is one of the most important responsibilities of a Software Project Manager: it must be a proactive continuous process throughout the Software Development Life Cycle (5.2).

■ The *Software Risk Handling Plan* (SRHP), typically an Addendum to the SDP, is the project's principal plan for identifying and mitigating software risks (5.2.2).

■ Periodic Technical and Management Reviews of software products and status must be conducted at the following levels: system, subsystem, Software Item level and Software Unit (5.3).

■ Joint Technical Reviews (JTRs) and Joint Management Reviews (JMRs) must be conducted to ensure product correctness and completeness and to elevate visibility into the status of evolving products. Attendees must have the knowledge and the authority to make technical, cost and schedule decisions (5.3.1 and 5.3.2).

■ Safety and security concerns can significantly impact your software architecture (5.4.1 and 5.4.2).

■ Applying Human-Computer Interface (HSI) techniques improves total system performance and can reduce both human errors and the cost of operations across the system's life cycle (5.4.3).

■ System Critical Software are software elements required for full system functionality, that if not performed, performed out-of-sequence, or performed incorrectly, will directly or indirectly cause the system to fail (5.4.5).

■ The Software Development Environment (SDE) must be sized to include the capacity to support post-deployment Software Support requirements, thus promoting long-term maintainability (5.4.5).

■ The management of software subcontractor teams is almost always a significant challenge (5.5).

■ On large projects, a Subcontractor Management Team (SCMT) must be established (5.5.1).

■ Software Verification and Validation (V&V) are two related but separate procedures (5.6).

■ Planning for the disposal of the system while you are working so hard to develop it may seem to be counterproductive; however, system disposal can be a big headache if not properly planned for in advance (5.7).

■ The definition of "done" should not be elusive; you are responsible for making sure that every staff member precisely understands what is being built and what is meant by "*done*" (5.8).

Chapter 6 Highlights

Software Quality Domain

■ Empowering everyone on your team, to act on the basis of understanding and following the processes in which they are involved, will result in better software products and services (6.0).

■ All work is part of some type of process; a process is the transformation of input into output through work that adds value (6.0).

■ The contemporary approach to continuous improvement now deals with building quality into every process and product from the beginning (6.0).

■ *Software Process Improvement* (SPI) involves identifying process areas needing improvement and developing new processes and procedures to implement those improvements (6.1).

■ A *Software Engineering Process Group*, or an equivalent organization, must be established as it is the heart of your SPI effort (6.1.1).

■ The SEPG membership typically includes the Chief Software Engineer (the Chair), Software Process Lead (SPL), SCM and SQA Leads, and organization representatives. The SEPG is also responsible for producing the program- or project-level SDP (6.1.1).

■ Recommendation for SPI must be presented to the SEPG, for review and evaluation of the cost and schedule impacts, and then to the CSwE for approval (6.1.2).

■ Be aware that there are *unintended consequences* to every decision (6.1.2).

■ The Software Process Lead is a trained change agent responsible for facilitating all software process tasks and is critical to an effective Software Process Improvement program (6.1.4).

■ The collection and sharing of *lessons learned* between programs and projects can, in the long run, provide substantial benefits to the organization (6.1.5).

■ The SQA group is *not* responsible for software quality; the Software Development Team is responsible for building quality into the product and the Software Project Manager is accountable for delivering a quality product (6.2).

■ Depending on the size of your project, at least one Software Quality Engineer should be assigned to your project by the SQA organization (6.2.1).

■ Evaluating the quality of software products requires a "big picture" perspective (6.2.2).

■ The SQA group must maintain records for all evaluations performed in order to provide objective *evidence* that the evaluations were conducted (6.2.3).

■ Software Quality Engineers (SQE) support software development as an active member of the subsystem they are supporting. However, SQEs must maintain a direct reporting line to their SQA organization and not be in a direct reporting line to the program they are supporting (6.2.4).

■ *Software Configuration Management* is an essential development control activity that begins during requirements definition (6.3).

■ SCM is responsible for all tasks necessary to control baselined software products and to maintain the current status of the baselined products throughout the Development Life Cycle (6.3).

■ Typically, SCM has a *three-tiered* Configuration Management scheme: the development site, the subsystems, and the program level. Software libraries are maintained at each level (6.3.1).

■ *Baselines* are standards or product versions against which future status, progress and changes are compared and measured (6.3.2).

■ *Version Control* involves the management of changes to source code, documentation and other collections of information (6.3.2).

■ The *Software Configuration Management Plan* documents the policies and procedures for conducting required SCM for all Software Items and establishes the plan for creating and maintaining a uniform system of configuration *identification*, *control*, *status accounting* and *audit* for software work products throughout the software development process (6.3.4).

■ Changes must be controlled through the change control process to avoid "scope creep" resulting in massive disruption and potential project cancellation (6.3.4).

■ The *Corrective Action Process* (CAP), often called *change management*, is triggered when performance deviates significantly from the plan, when defects are identified in the software work products, or when enhancements and improvements are proposed (6.4).

■ To report problems or changes with baselined software products, *Software Discrepancy Reports* (SDRs) and *Software Change Requests* (SCRs)—or similar names—must be used as part of the Corrective Action Process (6.4.1).

■ On large programs, there is typically a hierarchy of *Configuration Control Boards* (CCB) with different levels of control and responsibilities (6.4.3).

■ If a decision you made is wrong, and the problem it creates does not become apparent until well into the project, admit your mistake and take Corrective Action as soon as you can (6.4.4).

■ A Software Peer Review is a structured, methodical examination of software work products by the Developers, with their peers, to identify defects, to recommend needed changes, and to identify potential problems. SPRs can provide immeasurable value to your program (6.5).

■ Defects should be removed as early as possible because the later in the life cycle the defect is found the more expensive it is to fix—a lot more expensive (6.5.1).

■ The number of defects expected from a Peer Review depends on when in the life cycle the review is conducted. You will (should) find more during the early stages of development (6.5.5).

■ The key control gate to provide effective in-process quality checking is not allowing products that do not meet readiness criteria to pass to the next implementation process (6.5.6).

■ Action items from Software Product Evaluations must be tracked to closure and monitored to make sure the number of action items does not severely impact your cost and schedule (6.5.6).

■ The Capability Maturity Model Integrated is a *framework* for building process improvement systems, related training and process maturity appraisal programs (6.6).

■ The principal purpose of the CMMI is *to improve operational performance* by improving the efficiency of production, delivery, and outsourcing thereby lowering the cost and raising the quality. CMMI is not a development standard or a Development Life Cycle (6.6.1).

■ The CMMI structure involves three *Areas of Interest* (constellations), supported by *Process Areas* (PA) that can be carried out using either the *Staged or Continuous Representation* Models. (6.6.2).

■ CMMI *Process Areas* are a cluster of related practices that, when implemented collectively, satisfy a set of *goals* considered important for making improvements in that area (6.6.2).

■ The *Continuous Representation* is a model that provides the flexibility to focus only on the Process Areas that are closely aligned with your project or organization's business objectives (6.6.3).

■ The *Staged Representation* is a model structure where attaining the goals of a set of process areas establishes a maturity level and each level builds a foundation for subsequent levels (6.6.3).

■ The Staged Representation is a popular approach because it provides an easy migration from the CMM to CMMI and has been around for a lot longer than the Continuous Representation (6.6.3).

■ Organizations seeking a maturity rating are *not certified in CMMI*; they are *appraised*. You get appraised by an External Appraisal Team lead by a certified Lead Appraiser (6.6.5).

■ A smooth way to gain maximum improvement of your processes is by following CMMI as a *guide* to building a systemic process improvement infrastructure tailored to your project needs (6.6.5).

Chapter 7 Highlights

Managing the Software Project Team

■ People are your most important resource. Finding, recruiting, nurturing and retaining a competent staff is the single most important function of a Software Project Manager (7.0).

■ Applying extra funding and resources to *finding the right people* for your staff, will provide a high rate of return—exceeded only by your acumen in actually *picking* a compatible team from the candidates you found (7.1).

■ Job descriptions are a valuable marketing tool for attracting qualified candidates so you should give it the attention it deserves (7.1.2).

■ You can speed up the resume review process if you have a streamlined procedure of correlating each resume to the job descriptions (7.1.3).

■ Staffing is a *two-way process* for each of you to learn more about the other in order to decide if there is a good match; involve candidates in decisions that affect them (7.1.3).

- Unless you are absolutely certain you can manage a bad fit, err of the side of caution and don't make a job offer to a candidate rejected by your team (7.1.3).
- Make very sure the Software Engineers you hire have the experience, background and *mindset* for the job they are being hired for—in other words, put the right pegs in the right holes (7.1.4).
- Social media and the internet have had a profound impact on the hiring process. They provide Hiring Managers with a new cost-effective lens through which to evaluate job seekers (7.3.1).
- Generations have different personalities that change over time. The approach to hiring and managing your staff must take into account their generational background (7.3.2).
- According to a recent study, the top three employment *expectations* of the college class of 2015 were: Personal growth opportunities; Job security; and Good benefits package (7.3.3).
- Software Managers should understand the key factors that motivate and de-motivate the performance of their staff (7.4).
- People will have the incentive to work much harder, and to produce a higher quality product, if they believe their efforts really matter (7.4.1).
- One of the easiest and most effective ways a Software Manager can motivate their staff is to provide them with the latest and greatest "toys" and technology (7.4.3).
- Giving recognition and praise to members of your staff when their work is performed in an exemplary manner, essentially costs you nothing, takes very little time to do, and has a profound impact (7.4.4).
- It is unreasonable to expect your staff to work hard *all the time*. Outlets are forms of "play" that can pay large dividends (7.4.5).
- Make sure that the salaries received by your staff are fair and adequate (7.4.6).
- Avoid toxic people who are cynical and abrasive as their negativity can be very disruptive (7.4.7).
- One approach is not to promote someone until they have successfully performed at the level to which they are being promoted (7.4.8).
- In addition to technical respect, you need to earn personal respect and the best way to do that is to show respect to your staff. Always act in an ethical and professional manner (7.4.10 and 7.4.11).
- Communications is one of the most fundamental skills of life and is a prerequisite to problem-solving. Effective communication is a cornerstone to successful project management (7.5).
- The lack of good communications is often the *root cause* of many management problems (7.5.1).

- The SPM must ensure that *needed* information is flowing to the team regardless of location (7.5.4).
- If possible, place Programmers into the type of programming disciplines that he/she is most experienced with and most comfortable working in (7.6).
- The success and quality of the software product is directly related to the *process* used to create it. For large complex projects, a structured software development process, tailored to the needs of the project, must be followed (7.6.1).
- Before you hire the programming staff for your project, you need to determine what *types* of Programmers your project needs (7.6.2).
- Understanding the typical personalities, you will find in Programmers can be a major asset to you in managing them (7.6.3).
- A major theme of this Guidebook is that there are software development *processes* that must be followed, after tailoring, to the needs of your large software-intensive system (7.6.4).
- As the SPM, you have the important responsibility of evaluating the performance of your programming staff and helping them *improve their performance* (7.6.5).
- Harness the ideas and wisdom of your employees; create a *continuous feedback culture* (7.7).
- Strive to develop what can be called a *healthy organizational culture* to increase productivity, growth, and efficiency and reduce both counterproductive behavior and turnover (7.7.1).
- An *environment of excellence* for a Software Development Team involves three major components: the work *environment*, the work *atmosphere* and the work *infrastructure* (7.7.2–7.7.4).
- No world-class software development methodology or process improvement strategy can overcome serious problems that can result from *mismanagement of interpersonal conflicts* (7.7.5).
- *Compromise* is my strong preference for resolving conflicts, but reality checking also works by making staff members who have conflicts realize there are *three sides to every story* (7.7.5).
- The best tools are of little value, and a waste of resources, if no one knows how to use them (7.8).

Chapter 8 Highlights
Managing Software Costs and Schedules

- The process of controlling software costs and schedules is a critical SPM function (8.0).
- Every major program must have, *and follow*, a formal program management methodology by maintaining

an approved program-level *Integrated Master Plan* directly mapped to a program-level *Integrated Master Schedule* (8.1).

■ An overall master schedule may be included in the initial SDP. The detailed software schedules are typically updated so frequently that they should *not* be part of the updated SDP, but their location should be referenced in the SDP (8.1).

■ A Work Breakdown Structure is a decomposition of *all* the project tasks into smaller, manageable, and controllable components or tasks; there should be only one WBS per contract (8.2).

■ The WBS typically is prepared in two versions: The *product* hierarchy indicating how components are organized and the *activity* hierarchy indicating the activities pertinent to each component (8.2).

■ Budgets are formal statements of financial resources allocated to specific activities over a specified time period. Project Managers are allocated a budget for their project, and they *must manage it* (8.3).

■ Software development is a very labor-intensive effort; salaries and related employee overhead expenses are typically the largest and most important software budget item (8.3.1).

■ Acquiring specific expertise for a relatively short duration, not available from your full-time staff, may be accessible only through the consulting route (8.3.2).

■ Software Managers must work closely with their financial controls group to determine, if, when and how to capitalize equipment over several years (8.3.3).

■ The training budget should be given the attention, and funding, it needs and deserves (8.3.5).

■ The software development schedules must show the details of the proposed builds and how they relate to overall program milestones (8.4).

■ PERT and CPM charts are very similar in their approach, but there are definite differences (8.4).

■ As the Software Project Manager, you should do everything in your power to *start activities at the earliest possible start date* (8.4.2).

■ Actual cost does not represent actual work accomplished; the actual cost curve shows only the cost incurred and usually does *not reflect the amount of real work performed* (8.5).

■ To obtain a true status of your project, the *earned value management control system* (EVMS) identifies *actual* work accomplished or "what you got for what you paid for" (8.5).

■ A formal *Cost/Schedule Control System* (C/SCS) has been proven to be a very effective management tool for control of large systems (8.5).

■ The *Cost Account Manager* (CAM) is the most significant contributor to the successful operation of the C/SCS as well as successful completion of internal audits and customer project reviews (8.5.1).

■ Software performance measurement in the C/SCS consists of providing status, evaluating performance and forecasting future activities at the cost account level (8.5.3).

■ The analysis of earned value cost and schedule performance can be displayed graphically (8.5.5).

■ The Integrated Process and Product Development (IPPD) technique is a management approach that *simultaneously* integrates *all* essential development activities (not just software) through the use of multidisciplinary teams to optimize the design, manufacturing, and supportability processes (8.6).

Chapter 9 Highlights

Managing Software Facilities, Reuse and Tools

■ The adequacy of your facilities has a significant impact on the productivity of your team (9.0).

■ Software reuse is not a trivial issue. Projects that plan to use a significant amount of COTS software, or any reusable software product, must address COTS/Reuse in considerable detail in their SDP (9.0).

■ The *Software Development Environment* consists of *the hardware*, *software*, *procedures* and *documentation* necessary to support the software development effort (9.1).

■ Each software development subsystem must have a Software Test Environment (STE) that supports integration and testing of its SIs as part of its integrated SDE (9.2).

■ Two generic levels of Software Development Libraries are normally used to implement Software Configuration Management: the Master Software Development Library (MSDL) and a Software Development Library (SDL) at each site (9.3).

■ Poor record keeping is an easy pitfall for SPMs. *Software Development Files* (SDF), also called *Software Project Workbooks*, are essential organized project data repositories (9.4).

■ SDFs should be inspected and audited throughout the program, to determine compliance with the SDP, with at least one inspection performed during each build and prior to major reviews (9.4.2).

■ Non-deliverable software consists of non-operational software developed, purchased or used during software development but *not required by the contract to be delivered* (9.4.3).

■ Data Management (DM) provides the *interchange and access of controlled project data* to program personnel and the customer, supports timely delivery of contract deliverables and addresses key issues such as disaster recovery and data rights (9.5).

■ The Developer's parent organization should have in place an effective and comprehensive electronic data management system for the storage, retrieval and distribution of program related software documentation and work products (9.5.1).

■ With the emergence of *cloud computing*, the entire approach to the storage and retrieval of data is changing, but there are some current cloud-related challenges to overcome (9.5.1).

■ Plans for disaster recovery should be included in the SDP or in an external plan (9.5.2).

■ Proprietary concerns regarding vendor products can be a major issue in source selection (9.5.3).

■ The physical working environment, and other aspects of the environment for Software Developers, can have a major impact on their productivity, morale, retention and even recruiting (9.6).

■ Development Teams should consider using reusable software products wherever possible (9.7).

■ The approach to be followed for identifying, evaluating and incorporating reusable software products must be described in the SDP (9.7.1).

■ Commercial Off-the-Shelf software can have a major impact on the reduction of schedule risks and cost risks, but the process of including COTS components may be difficult (9.7.2).

■ The method of integrating selected COTS components may impose additional constraints on the software architecture (9.7.4).

■ Requesting the vendor to modify their COTS product to meet the needs of your program is generally considered high risk and is definitely *not* recommended (9.7.5).

■ New tools will not make an ineffective process more effective; new tools are not a panacea for fixing problems—but they can make an effective process more efficient (9.8).

■ In general, the use of *Computer Assisted Software Engineering* (CASE) tools can be classified into three types: tools, workbenches and life cycle environments but these distinctions are flexible (9.8.1).

■ Despite the significant potential advantages for your project or your organizations adopting CASE technology, there are some risk factors that you must be aware of including unrealistic expectations and inadequate training, standardization and process control (9.8.2).

Chapter 10 Highlights

Systems Engineering Domain

■ The success or failure of a large software-intensive system is based upon a clear understanding by the customer and the Developers as to what is being built even though the precision of that understanding will change during development; this is called Requirements Development (10.0).

■ In this Guidebook, the Systems Engineering Domain includes four principal activities directly involving Software Engineering; two are early in the life cycle and two are late in the life cycle:
  - System Concept and Requirements Development.
  - System Design.
  - Software Item and Hardware Items (SI/HI) Integration and Testing (I&T).
  - System Qualification Testing (10.0).

■ There are various types of requirements that an SPM may be involved with including functional (or design); performance (or operational); interface; organizational; and acceptance (10.0).

■ The way a requirement is phrased is just as important as the requirement itself (10.0).

■ The major objective of System Concept and Requirements Development is the specification and documentation of *system and subsystem-level* requirements (10.1).

■ The major output documents resulting from the requirements analysis activity are typically preliminary versions of the *System/Subsystem Specifications* (SSS), the *Operational Concept Description* (OCD) and a top-level *Interface Specification* (10.1).

■ All system, subsystem-to-subsystem, and subsystem-to-external requirements and interfaces should be maintained in a *Requirements Database* (10.1.2).

■ During the System Design activity, system characteristics should be refined through trade studies, analyses, simulation, prototyping and *functional decomposition* to define subsystem *Software Items* and *Hardware Items* as well as SI interfaces to other SIs and HIs (10.2).

■ The SDP is a process document, not a design document; however, it is recommended that your SDP contain high-level Architectural Design overviews of the system and the software (10.2.1).

■ System-wide Software Design decisions and their rationale should be documented by the Systems Engineering Integration and Test (SEIT) in *Engineering Memoranda* and the *System/Subsystem Design Description* (SSDD) (10.2.3).

■ The software and hardware *Integration and Testing* process involves integrating SIs with other interfacing SIs, integrating SIs with Hardware Items, testing the groupings, and continuing this process until all interfacing SIs and HIs in the system are integrated and tested successfully (10.3).

■ The following products are developed during the SI/HI I&T activity:

– A baselined *Software Product Specification* (SPS)
– A baselined *Software Version Descriptions* (SVD) supporting the current software release
– An updated *Software Test Description* (10.3)

▪ The Software Developers and Software System Engineers must develop an *integration strategy* that defines a systematic *approach for integrating the SIs into the complete software release* (10.3).

▪ System Qualification Testing (SQT), often called System Acceptance Testing, involves verifying that *all* system requirements have been met—including the system interface requirements (10.4).

▪ SQT must be performed by "independent" Test Engineers—*not the Developers* (10.4.1).

▪ *Test-Like-You-Fly* (TLYF), also called *Test-Like-You-Operate*, means that even if your system is fully compliant with the customer's requirements, it still might fail if you failed to test it in its intended real operational (flight-like) condition and environment—*before* the system goes operational. TLYF is *not* a replacement for *any* required testing function (10.4.9).

▪ Ignorance, anywhere, is also a flaw! The bottom line is, if you change anything in a complex system that has previously been tested—*retest!*

Chapter 11 Highlights

Software Engineering Domain

▪ The Software Engineering Domain is the core of your software-intensive system. If you are going to manage the software development process, you must first understand it (11.0).

▪ Software *requirements* are the foundation of successful software-intensive systems (11.1).

▪ *Allocated* software requirements are flowed down (*decomposed*) from the Systems Engineering requirements allocation process. *Derived* software requirements are not specifically flowed down but are arrived at through logic and reasoning as needed to perform needed functionality (11.1.1).

▪ Requirements must also be specifically evaluated for safety, security, privacy protection, dependability, reliability, maintainability and availability. A single project Requirements Database should be used to capture all requirements (11.1.1).

▪ Software requirements analysis work products include Software Requirements Specification (SRS); Software Requirements Database; Traceability Matrixes; Software Development File; Software Master Build Plan (SMBP draft); and the Interface Control Document (IFCD) (11.1.2).

▪ Uncontrolled changes or continuous growth in the planned scope of a project is called requirements creep and is a major challenge for Project Managers (11.1.3).

▪ There is no requirement for the software requirements analysis activity to be entirely completed prior to the start of the Software Design activity. When following iterative life cycle models, the software requirements analysis activity may be repeated iteratively for each build (11.1.6).

▪ Software Item Design includes: *SI Architectural Design; and SI Detailed Design* (11.2).

▪ The Software Design activity tasks are intended to be performed as consecutive steps of increasing levels of design specificity (11.2.2).

▪ Documentation produced during the Software Design activity for each SI includes the: Software Architecture and Design and interface descriptions; test plan, models and diagrams; traceability products in the Requirements Database; and an updated SMBP (11.2.3).

▪ The objective of *SI Architectural Design* is to describe the high-level organization of the SIs in terms of Software Units (SUs) and their relationships (11.3.1).

▪ The Detailed Design activity involves decomposing the SUs from the SI Architectural Design into the lowest level Software Units (11.4.1).

▪ Software Design must include understanding and consideration of the human aspect of the "man-machine" interface, also referred to as the Human System Integration (HSI) (11.4.2).

▪ The objective of software *Coding and Unit Testing* (CUT) is to: convert the SU Detailed Design into computer code and databases that are inspected and unit tested; confirm successful completion; generate test descriptions; review source code; and execute test cases; (11.5).

▪ The objective of the Software *Unit Integration and Testing* (UI&T) activity is to perform a systematic and iterative series of integration builds on SUs that have successfully completed Code and Unit Test, and build them up to a higher level SU, or SI, for the current build (11.6.1).

▪ The objective of Software Item Qualification Testing (SIQT) is to demonstrate that the SI meets the system, performance and interface requirements allocated to the SI being tested (11.7.1).

▪ Documentation products normally produced during the SIQT activity for each SI include *Software Test Description*; *Software Test Report(s)*; an updated *Software Test Plan*; and traceability products from the *Requirements Database* (11.7.2).

▪ During hardware and software integration, the Software Team is usually in a *support role* to the SEIT group. SI/HI I&T involves integrating Software Items with interfacing SIs and HIs and testing to confirm they work as intended (11.8).

■ Although ensuring a smooth transition to systems operations takes place at the end of the Development Life Cycle, consideration of these tasks should occur much earlier and preferably concurrently with design, development, and testing throughout the life cycle (11.9).

■ The preparation for software transition to sustainment includes preparation of the documentation and software products required by maintenance personnel at the maintenance center to perform their maintenance tasks (11.10).

Chapter 12 Highlights
Collaboration of Software and Systems Engineering

■ The Software Engineering (SwE) and Systems Engineering (SE) domains are tightly coupled because that collaboration is essential to the project's success and, as the SPM, you must make sure that collaboration takes place (12.0).

■ The collaboration of SwE and SE has two perspectives:
  – There are system development activities that are the responsibility of SE, but these activities are *supported by SwE*.
  – There are software development activities that are the responsibility of SwE, but these activities are *supported by SE*.

■ System and Software Engineers support each other during *five software acquisition periods* that include four logical *management approval points* during the System Life Cycle: *Pre-Milestone A* Concept Studies; *Milestone A* Concept Development; *Milestone B* Engineering and Manufacturing Development (EMD); and *Milestones C and D* Production, Deployment and Sustainment (12.0).

■ *Milestone A:* After the concepts of the system are defined and refined, management approval is needed to either proceed to the planning Phase A, revise the concept, or cancel the proposed program (12.2).

■ *Milestone B:* After the concept is developed to a point where overall system requirements and plans are prepared, management approval is needed to advance to the execution Phase B (EMD) involving the development of the Preliminary Design and design review (12.2).

■ *Milestone C:* At the end of Phase B, management approval is needed to embark on full-scale development and production of the hardware and software system.

The *Software Implementation Process* takes place during the production Phase C (12.2).

■ *Milestone D:* When it is confirmed that the system developed meets the requirements, approval is needed to deploy the system and move it from the development lab to operations and then into long-term sustainment (maintenance) until the system is eventually replaced or retired (12.2).

■ As the system evolves during the *Pre-Milestone A* Concept Studies, the *Initial Capabilities Document* (ICD) can be replaced by a draft *Capability Development Document* (CDD) that will ultimately support the Milestone A decision to proceed (12.4.3).

■ The challenge is the early recognition of system capabilities that *will require software contributions* in their development or implementation, as well as assessing the feasibility of proposed software solutions (12.4.3).

■ During Pre-Milestone A, the customer's Program Office is usually responsible for the preparation of the *Technical Requirements Document* (TRD) although they will evolve during the life cycle (12.4.3).

■ The *System Requirements Analysis* activity is led by SE but support from the Software Integrated Product Team is essential. The key SE–SwE collaborative activities during system requirements analysis are: *system* requirements analysis; *System* Functional Design; *software* requirements analysis; and *Software* Item Architectural Design (12.5).

■ The collaborative SE and SwE activities performed during Acquisition Phase B—Engineering and Manufacturing Development are: Support of Integrated Baseline Reviews; System Detailed Design; Software Item Detailed Design; support to the Integration and Testing of Software Units, Software Items and hardware; the FCA; and the System Qualification Review (12.6).

■ The collaborative SE and SwE activities during Acquisition Phases C and D—Production, Deployment and Sustainment are Transition to Operations and Sustainment; Finalization of User and Support Manuals; and updated system requirements as required (12.7).

■ Software transition involves considerable advance planning and preparation that must start early in the life cycle to ensure a smooth transition to the sustainment organization (12.7.6).

Chapter 13 Highlights
Software Documentation and Work Products

■ The right document, with the right content, available when needed at a critical decision point in the project, can be as close to a silver bullet as you can get (13.0).

■ Not all of the potential documentation or work products will be needed by every program; in fact, it is almost certain that no project, regardless of size, will need *all* of them (13.0).

■ Software documentation supports the basic requirement to provide a well-defined, consistent and documented *software baseline* during the Software Acquisition Life Cycle (13.1).

■ Complete and good documentation: provides critical requirements and design information to Software Developers; provides management and customer visibility and involvement into the development process and progress; provides consistent bi-directional requirements traceability; and helps to resolve the training problem created by personnel turnover (13.1).

■ If the contract does not specifically identify deliverable work products, as the SPM, you must identify the documents and work products *needed* by your project. The following may be considered a minimum set of core system and software documents (13.3):
  – System/Subsystem Specification
  – Interface Requirements Specification (IRS)
  – Interface Design Document (IDD)
  – Software Development Plan (SDP)
  – Software Requirements Specification
  – Software Design Document (SDD)
  – Software Test Plan
  – Software Test Description
  – Software Test Report
  – Software Product Specification (SPS)
  – Software Version Description (can be an updated version of the SPS)
  – Software Users' Manual (SUM)

■ How documents look are important. It will not take the customer very long to examine or browse documents they receive to get a first impression and first impressions are hard to reverse (13.3).

■ Software management and quality control plans represent important adjuncts to the SDP that document specific implementation details not covered in the SDP (13.4).

■ The defined software development process, captured in the SDP at a relatively high level, should be elaborated through detailed Work Instructions and/or Procedures (13.5).

■ Not all of the software work products are documents (13.6).

Chapter 14 Highlights
Software Estimating Methods

■ A critical element of the project planning process is developing an understanding of how much the software project is likely to cost, and how long it is likely to take, before the project starts (14.0).

■ Software tools are and should be used extensively to assist in cost and time estimation (14.0).

■ Software estimates of effort, time and costs are *living* estimates based on the best information available at the time of the estimate. *Estimates are not a commitment carved in stone* (14.1).

■ Software Managers are often measured against how well they carry out their budgets, so you should do whatever you can do to improve your estimating abilities (14.1.1).

■ When Software Design and/or code is reused, the estimation of its real cost is usually based on an approach called the *Equivalent Source Lines of Code* (ESLOC) count (14.1.3).

■ There are three generic types of software estimation: expert *judgment*; formal mathematical-based *models*; and some combination of them. Using the results of any *two* approaches as a sanity check for the same estimate is far superior to relying solely on one approach (or one tool) (14.2).

■ When considering bottom-up versus top-down estimating, I prefer the advantages of bottom-up estimating because lower-level employees take a personal interest in the estimating process and that can potentially improve their motivation and moral (14.2.1).

■ Parametric estimating models are an estimation technique utilizing statistical relationships that exist between a series of historical data and a particular delineated list of other variables (14.2.2).

■ Analogous estimating is an analogy approach that involves finding similar projects in historical data that can be compared to your project to help estimate your cost (14.2.3).

■ The Delphi technique is a commonly used team consensus estimating approach (14.2.4).

■ Early estimates of code size for large systems are typically not even close to the final code size. As the program progresses through the development activities, and the degree of clarity of the deliverable product increases, code size estimates become much more accurate (14.3).

■ If you do not have a good method of estimating how long a *new* project will take, or how much it will cost, use the 90–90 rule and multiply the best estimate by 1.8 (or just double it) and in the end you should be a lot closer than the original estimate (14.3).

■ The most popular software size estimation tools used in the past include COCOMO II, SEER–SIM, PRICE–S, Function Points and SLIM (14.4.1–4.4.7).

■ The "Quality Triangle Dilemma" is embedded in stone as the proverb: "Good, fast and cheap—you can only have two." However, my *Quality Quaternion* is more valuable to a Software Manager because it adds a fourth important element: Scope (14.4.8).

■ Under promising and over delivering is far better than the reverse (14.4.8).

Chapter 15 Highlights
Managing Software Performance with Measurements

■ Managing the development of software-intensive projects is a knowledge-intensive process, and much of that knowledge comes from taking measurements. The depth of *knowledge* available to the Project Manager is directly proportional to the success of the project (15.0).

■ Effective management controls are dependent on timely and accurate measurements and appropriate Corrective Actions taken as a result of analyzing those measurements (15.0).

■ A fundamental aspect of a typical software measurement initiative is a continuous improvement approach through a closed-loop feedback control system (15.2).

■ The effort to collect, analyze and document metrics must be consistent with their value to the needs of the project (15.3).

■ A typical top-down management measurement approach is a tailorable hierarchy of four groups. Software management establishes *program/project goals* (Group 1) resulting in a set of *Measurement Categories* (Group 2) to determine progress in meeting those goals. The categories identify a set of *Measurement Indicators* (Group 3) that provides information to support the Group 2 categories. The indicators are produced from detailed software *Base Measurements* (Group 4) that must be collected to provide the data needed by the indicators (15.3).

■ The Measurement Categories and indicators support a minimum of five *key software measurement questions* that SPMs must have periodic responses to in order to effectively manage the software development effort. The five *key questions* are (15.4):
  - Is the project's progress and its schedule under control?
  - Are resources and cost under control?
  - Are changes impacting the project's cost and schedule?
  - Are we developing a quality product?
  - Is development performance under control?

■ The five *Measurement Categories* (Group 2) that provide answers to the above questions are Schedule and Progress; Resources and Cost; Product Size and Stability; Product Quality; and Development Performance. There are 17 *Management Indicators* (Group 3) that provide support to the five categories (15.4).

■ Not all measurements are collected or needed at the same time (15.5).

■ In order to use software measurements effectively, you should have a mechanism that specifies exactly what will be measured and how the resulting data will be analyzed to produce results that satisfy your information needs. That mechanism can be called a *Measurement Information Specification* (MIS) that defines the data that will be collected, the computations that will be performed on that data, and how the resulting data will be reported and analyzed (15.7).

■ Understanding the *root cause* of a problem is the most effective path to preventing it (15.8).

■ For medium-to-large developments, you should spearhead the establishment and conduct of a joint Customer–Developer *Software Measurement Working Group* (15.8).

■ Management decisions based upon the analysis of software Management Indicators must use *thresholds* to flag non-nominal conditions (15.9).

Chapter 16 Highlights
Managing Software System Sustainment

■ Maintaining a deployed system, responsive to changing customer needs over a long time frame is as important, and can be just as challenging, as the original implementation (16.0).

■ Software Sustainment can consume about two-thirds of the total Software Life Cycle cost. From a customer perspective, it may be the most important activity of the Software System Life Cycle (16.0).

■ Software sustainment includes the processes, procedures, people, materiel, funding and information required to fully support, maintain, and operate the software portions of a system. Sustainment can be viewed as a *superset* of the Software Maintenance activity (16.0).

■ Generally, there are four types of Software Sustainment tasks: Corrective; Adaptive; Perfective; and Preventive. Retesting and Configuration Management play an important role in the process regardless of the type of sustainment performed (16.1).

- Your customer, or their *Program Office*, should be (and usually are) directly involved with the planning and management of Software Sustainment (16.2.1–16.2.3).
- The ability to upgrade current operational systems to new software releases, *without affecting ongoing operations*, must be planned for and incorporated into system requirements from the beginning of the project (16.2.2).
- Draft versions of the *Software Sustainment Plan* (SSP), also called the *Software Maintenance Plan*, should be prepared relatively early during the software development process (16.3).
- The Software Sustainment organization must have access to the same *Support Software* that was used during software development (16.4).
- Table 16.2 is an example summary of key software issues and pitfalls that are typically encountered during sustainment (16.5).
- Properly ending a project may be considered as important as initiating it. There are two different closing issues: *closing* the development contract after the system development effort is completed, and *retiring* the system after its mission is over (16.6).
- The key activities involved in the closure process of the developmental contract (16.6):
  - Formal sign-off
  - Lessons learned analysis and documentation
  - Release resources
  - Subcontractor closures
  - Indexing and archiving the project files
  - Celebration

Final Thoughts: Francis Bacon (1561–1626) wrote "Some books are to be tasted, others to be swallowed, and some few to be chewed and digested; that is, some books are to be read-only in parts; others to be read but not curiously; and some few to be read wholly, and with diligence and attention." *It is my desire to have this Guidebook "read wholly, and with diligence and attention."* I hope you use it that way.

Appendix I: References by Category

Acquisition Management

Adams, R., S. Eslinger, K. Owens and M. Rich. 2006. *Reducing Software Acquisition Risk: Best Practices for the Early Acquisition Phases*. The Aerospace Corporation Report, TR-2006(8550)-1.

DoD. 1991. *The Program Manager's Guide to Software Acquisition Best Practices*. Version 2.0. Department of Defense Software Acquisition Best Practices Initiative.

Eslinger, S., M. Gechman, D. Harralson, L. Holloway and F. Sisti. 2006. *Software Acquisition Management Plan (SWAMP) Preparation Guide*. The Aerospace Corporation TOR-2006(1455)-5743.

GAO. 2008. *Defense Acquisitions: Assessment of Selected Major Weapons Programs*. U.S. Government Accountability Office, GAO-08-467SP.

Harrell, J. et al. 2009. *Software Acquisition Guidebook for Space System: Pre-Milestone A*. The Aerospace Corporation, TOR-2009(8506)-9.

USAF. 2003. *Guidelines for Successful Acquisition and Management of Software-Intensive Programs*. Condensed Version. Department of the Air Force, Software Technology Support Center.

Agile/Scrum Management

Boehm, B. and R. Turner. 2004. *Balancing Agility and Discipline: A Guide for the Perplexed*. Pearson Education.

CMMI Institute. 2016. Agile Performance: Agility Depends Upon Capability. *CrossTalk: Journal of Defense Software Engineering*, November/December 2016.

deSouza, B. 2015. *Agility Comes With Maturity*. CIO New Zealand.

DoD AT&L. 2013. *Agile in the DoD*. Department of Defense. January–February 2013. DoD.

Kennaley, M. 2010. *SDLC 3.0: Beyond a Tacit Understanding of Agile*. Fourth Medium Press.

Nir, M. 2016. *The Agile PMO: Leading the Effective, Value Driven, Project Management Office*. Sapir Publishing.

Schiel, J. 2017. *Enterprise-Scale Agile Software Development*. CRC Press

Schwaber, K. 2004. *Agile Project Management with Scrum*. Microsoft Press.

Stern, T. 2016. *Lean and Agile Project Management: How to Manage Projects Better, Faster and More Effective*. Productivity Press.

Stuart, J. 2010. *10 Pitfalls to Enterprise Agile Adaption*. Best Practices White Paper, Version-1. Construx Software.

Stuart, J. et al. 2012. *Five Things Every Software Executive Should Know about Scrum*. Construx Software.

Cost and Estimation Management

ANSI/EIA. 2010. *ANSI/EIA-748-B: Earned Value Management Systems*. American National Standards Institute/Electronics Industries Association.

Boehm, B. et al. 2000. *Software Cost Estimation with COCOMO II*. Prentice Hall.

Boehm, B. 1981. *Software Engineering Economics*. Prentice-Hall.

Bowers, J., M. Gechman, W. Macaulay, A. Unell and M. Brodner. 2008. *Development of an Early-Phase Ground Software Cost Estimation Model*. The Aerospace Corporation, ATR-2009(8217)-1.

DeMarco, A. 2010. *The PRICE True Planning® Estimating Suite*. White Paper. Price Systems.

Fleming, Q., J. Koppelman. 2010. *Earned Value Project Management*. Project Management Institute.

Galorath. 2010. *SEER for Software Workshop: Estimate Training Manual*. Galorath Incorporated.

LMSC. 1994. *Software Estimating Guide*. Lockheed Missiles & Space Company, Inc. LMSC/SSD.

Stutzke, R. 2005. *Estimating Software-Intensive Systems: Projects, Products and Processes*. SEI Series in Software Engineering.

Management Related

Archibald, R. and S. Archibald. 2015. *Leading and Managing Innovation: What Every Executive Team Must Know About Project, Program and Portfolio Management*, 2nd Edition. Auerbach Publications.

Atesman, M. 2017. *Engineering Management in a Global Environment: Guidelines and Procedures*. CRC Press.

Ball, J. 2008. *Professionalism is for Everyone*. The Goals Institute.

Belker, L., J. McCormick and G. Topchik. 1981. *The First Time Manager*, 6th Edition. AMACOM.

Boehm, B. 2000. *The Art of Expectation Management*. IEEE Computer Society.

Covey, S. 1991. *Principle-Centered Leadership*. Simon and Schuster.

Ewing, D. 1964. *The Managerial Mind*. The Free Press of Glencoe/ The Macmillan Company.

Guber, P. 2011. *Tell to Win: Connect, Persuade and Triumph with the Hidden Power of Story*. Crown.

McCarthy, C. 2015. *Program Management in Defense and High Technology Environments*. Auerbach Publications.

Pew. 2010. *Millennials: A Portrait of Generation Next*. Pew Research Center.

Republic Media. 2015. *Top Companies to Work for in Arizona—2015*. CareerBuilder® Republic Media, AZcentral. com. The Arizona Republic, 28 June 2015.

Templar, R. 2011. *The Rules of Management*. Pearson Education/FT Press.

Measurement (Metrics) Management

Abelson, L., S. Eslinger, M. Gechman, C. LeDoux, M. Lieu and K. Korzec. 2011. *Software Measurement Standard for Space Systems*. The Aerospace Corporation, TOR-2009(8506)-6.

Basili, V. and D. Weiss. 1984. A Methodology for Collecting Valid Software Engineering Data. *IEEE Transactions on Software Engineering*, Vol. SE-10.

Gechman, M. and D. Stratton. 1994. *Guidelines for the Lockheed Corporate Software Metrics Program*. Lockheed Corporation Joint Software Task Force, Calabasas, CA.

Gechman, M., K. Kao and R. Armstrong. 1995. Implementation and Operation of a Comprehensive Software Metrics Program. *Proceedings of the 7th Annual Software Technology Conference*, Salt Lake City, Utah.

Goodman, P. 1993. *Practical Implementation of Software Metrics*. McGraw-Hill.

ISO/IEC. 2010. *ISO/IEC 15939: Software Engineering—Software Measurement Process*. International Organization for Standardization/International Electro-technical Commission.

Jones, C. 2017. *A Guide to Selecting Software Measures and Metrics*. CRC Press/Auerbach Publications.

Kan, S. 2003. *Metrics and Models in Software Quality Engineering*, 2nd Edition. Addison-Wesley.

McGary, J. et al. 2001. *Practical Software Measurement: Objective Information for Decision Makers*. Addison-Wesley.

Putnam, L. and W. Myers. 2003. *Five Core Metrics: The Intelligence Behind Successful Software Management*. Dorset House.

Process/Methodology Management

Aldridge, E., Jr. 2002. *Evolutionary Acquisition and Spiral Development*. Memorandum, Washington, DC: Office of the Under Secretary of Defense for Acquisition, Technology and Logistics.

Cockburn, A. 2008. Using Both Incremental and Iterative Development. *CrossTalk: The Journal of Defense Software Engineering* 21(5).

Davis, W. 1992. *Tools and Techniques for Structured Systems Analysis and Design*. Addison-Wesley.

DAU. 2009. *Integrated Defense Acquisition, Technology and Logistics Life Cycle Management System—Process Chart*. Version 5.3.4. DoD: Defense Acquisition University.

DeMarco, T. and P. Plauger. 1979. *Structured Analysis and System Specification*. Prentice-Hall.

Fowler, M. 2000. *Refactoring*. Addison-Wesley.

Gechman, M. and S. Eslinger. 2011. *The Elements of an Effective Software Development Plan—Software Development Process Guidebook*. The Aerospace Corporation, ATR-2011(8404)-11.

Gechman, M. 2003. *NPOESS Software Development Plan*. Operational SDP for the National Polar-Orbiting Environmental Satellite System. Document D31417-01. Northrop Grumman Space Technology.

Humphrey, W. 1991. *Introduction to the Personal Software Process*. Software Engineering Institute Series. Addison-Wesley.

Humphrey, W. 2000. *Introduction to the Team Software Process*. Software Engineering Institute Series. Addison-Wesley.

Humphrey, W. 1990. *Managing the Software Process*. Software Engineering Institute Series. Addison-Wesley.

Kruchten, P. 2000. *The Rational Unified Process: An Introduction*. Addison-Wesley.

McConnell, S. 1996. *Rapid Development*. Microsoft Press.

Ward, P. and W. Mellor. 1985. *Structured Development for Real-Time Systems*. Yourdon Press.

Whitten, J., V. Barlow and L. Bentley. 1991. *System Analysis and Design Methods*, 3rd Edition. McGraw-Hill.

Product and Process Improvement

Adams, R., S. Eslinger, K. Owens and M. Rich. 2004. *Software Acquisition Best Practices: Experiences from the Space Systems Domain*. The Aerospace Corporation, TR-2004(8550)-1.

Ahern, D., A. Clouse and R. Turner. 2001. *CMMI Distilled: A Practical Introduction to Integrated Process Improvement*. Addison-Wesley.

Chrissis, M. M. Konrad and S. Shrum. 2010. *CMMI: Guidelines for Process Integration and Product Improvement*, 2nd Edition. Addison-Wesley.

Cox, I. et al. 2016. *Visual Six Sigma: Making Data Analysis Lean*, 2nd Edition. John Wiley.

Florac, W. and Carleton, A. 1999. *Measuring the Software Process: Statistical Process Control for Software Process Improvement*. Addison-Wesley.

ISO/IEC. 1993. *ISO/IEC 15504: Software Process Improvement & Capability Determination (SPICE)*. Joint International Organization for Standardization/International Electro-technical Commission.

Kolawa, A. and D. Huizinga. 2010. *Automated Defect Prevention: Best Practices in Software Management*. John Wiley—IEEE Computer Society Press.

Lockheed Corp. 1994. *Guidelines and Tools for Continuous Improvement*. Lockheed Corp., LC-944.

SEI. 2006. *Standard CMMI® Appraisal Method for Process Improvement (SCAMPI)*. Ver. 1.2: Method Definition. CMU/SEI-2006-HB-2. Software Engineering Institute/Carnegie-Mellon University.

SMC. 2006. *Software Acquisition Process Improvement*. Instruction 63-103. USAF: Space and Missile Systems Center.

Project Management

Andler, N. 2011. *Tools for Project Management, Workshops and Consulting: A Must-Have Compendium of Essential Tools and Techniques*, 2nd Edition. John Wiley.

Berkun, S. 2005. *The Art of Project Management*. O'Reilly.

Carstens, D. et al. 2017. *Project Management Tools and Techniques: A Practical Guide*. CRC Press.

Chemuturi, M. and T. Cagley. 2010. *Mastering Software Project Management: Best Practices, Tools and Techniques*. J. Ross Publishing.

DeMarco, T. and T. Lister. 2013. *Peopleware: Productive Projects and Teams*, 3rd Edition. Addison-Wesley.

Donaldson, H. 1978. *A Guide to the Successful Management of Computer Projects*. Halsted Press.

Ensworth, P. 2001. *The Accidental Project Manager: Surviving the Transition from Techie to Manager*. John Wiley.

Eslinger, S. 2006. *The Position of Software in the Work Breakdown Structure (WBS) for Space Systems*. The Aerospace Corporation, TOR-2006(9550)-3.

Gechman, M. and S. Eslinger. 2010. *An Overview of Software Issues During Operations and Maintenance*. The Aerospace Institute, Presentations to the Space System Operations (S2040 class).

Heagney, J. 2012. *Fundamental of Project Management*, 4th Edition. AMACOM.

Hill, G. 2013. *The Complete Project Management Office Handbook*, 3rd Edition. Taylor & Francis Group: CRC Press.

Hughes, B. and M. Cottrell. 2009. *Software Project Management*, 5th Edition. McGraw-Hill.

Jalote, P. 2002. *Software Project Management in Practice*. Addison-Wesley.

Jones, C. 2004. Software Project Management Practices: Failure versus Success. *CrossTalk: The Journal of Defense Software Engineering*, 17(10).

Karten, B. 2016. *Project Management Simplified: A Step-by-Step Process*. CRC Press.

Kerzner, H. 2013. *Project Management Metrics, KPIs and Dashboards*. John Wiley.

Kerzner, H. 2017. *Project Management: A Systems Approach to Planning, Scheduling and Controlling*, 12th Edition. John Wiley.

Kogon, K. et al. 2015. *Project Management for the Unofficial Project Manager*. Ben Bella Books.

Larson, E. and C. Gray. 2014. *Project Management—the Managerial Process*, 6th Edition. McGraw-Hill Series, Operations and Decision Sciences.

Letavec, C. 2006. *The Program Management Office*. J. Ross Publishing.

Luckey, T. and J. Phillips. 2006. *Software Project Management for Dummies*. John Wiley.

Lutchman, C. 2017. *Project Execution*. CRC Press.

Mantle, M. and R. Lichty. 2013. *Managing the Unmanageable: Rules, Tools and Insights for Managing Software People and Teams*. Pearson Addison-Wesley.

Marchewka, J. 2016. *Information Technology Project Management*, 5th Edition. John Wiley.

McConnell, S. 1998. *Software Project Survival Guide*. Microsoft Press.

Murray, A. 2016. *The Complete Software Project Manager: Mastering Technology from Planning to Launch and Beyond*. John Wiley.

McDonald, J. 2010. *Managing the Development of Software Intensive Systems*. John Wiley

Phillips, D. 2000. *The Software Project Manager's Handbook: Principles That Work at Work*. IEEE Computer Society.

Phillips, J. 2010. *IT Project Management: On Track From Start to Finish*, 3rd Edition. McGraw Hill.

PMI. 2013. *Guide to the Project Management Body of Knowledge (PMBOK® Guide)*, 5th Edition. Project Management Institute.

Royce, W. 1970. Managing the Development of Large Scale Software Systems. *Proceedings of IEEE WESCON*, Los Angeles, CA.

Stellman, A. and J. Greene. 2006. *Applied Software Project Management*. O'Reilly.

Thayer, R. 1991. *Software Engineering Project Management*. IEEE Computer Society Press.

Tinnivello, P. Ed. 2017. *Project Management*. CRC Press.

Villafiorita, A. 2014. *Introduction to Software Project Management*. CRC Press/Auerbach.

Yourdon, E. 1991. *Death March: The Complete Software Developers Guide to Surviving "Mission Impossible" Projects*, 2nd Edition. Yourdon Press.

Quality Management

Binder, R. 1991. *Can a Manufacturing Quality Model Work for Software?* IEEE Computer Society.

Davis, D. 2005. Experience with Capture-Recapture. *Proceedings of the 2005 SEPG*. Software Engineering Process Group. Seattle, WA.

Guarro, S. and W. Tosney. 2010. *Mission Assurance Guide*. The Aerospace Corporation, TOR-2007(8546)-6018.

Kan, S. 2003. *Metrics and Models in Software Quality Engineering*, 2nd Edition. Addison-Wesley.

Owens, K. and B. Troup. 2003. *A Practical Guidebook for Performing Software Capability Appraisals*. The Aerospace Corporation, TR-2003(8550)-1. ETG.

Owens, K. and M. Tagami. 2008. *Recommended Software-Related Contract Deliverables for National Security Space System Programs*. The Aerospace Corporation, TOR-2008(8506)-8108.

Peters, T. and R. Waterman, Jr. 2003. *In Search of Excellence: Lessons Learned from America's Best-Run Companies*. Harper Business Book.

Redwine, S. Jr. (Ed.). 2010. *Software Assurance: A Curriculum Guide to the Common Body of Knowledge to Produce, Acquire and Sustain Secure Software*. V-1.2. U.S. Department of Homeland Security.

Standish Group. 2016. *2016 Chaos Report*. The Standish Group International.

Related Documents and Articles

Adams, S. 1996. *The Dilbert Principle*. Harper Business.

Barker, J. 1998. *Discovering the Future: The Business of Paradigms*. ILI Press.

Copper, A. 2016. *More than a Quarter of Arizonans Say they are Leaving Religion Behind*. Mesa Republic (USA Today Network), Mesa, AZ., August 2016.

Eide, E. and M. Hilmer, 2016. Do Prestigious Colleges Pay Off? It Depends on the Major. *The Wall Street Journal*, Wealth Management Journal Report.

Glass, R. 1998. *Computing Calamities: Monumental Computing Disasters*. Prentice Hall Professional Technical Reference.

Herzberg, F., B. Mausner and B. Snyderman. 1959. *The Motivation to Work*. John Wiley.

Leveson, N. 2004. The Role of Software in Spacecraft Accidents. *AIAA Journal of Spacecraft and Rockets*, 41(4): 564–575.

Lucas, H. 1975. *Why Information Systems Fail*. Columbia University Press.

McGregor, D. 1960. *The Human Side of Enterprise*. McGraw-Hill.

Peter, L. and R. Hull. 2011. *The Peter Principle: Why things Always Go Wrong*. Harper Collins.

Tavani, H. 2003. *Ethics & Technology: Ethical Issues in an Age of Information and Communication Technology*. John Wiley.

Verzuh, E. 2015. *The Fast Forward MBA in Project Management*, 5th Edition. John Wiley.

Requirements Engineering

Alexander, I. and R. Stevens. 2002. *Writing Better Requirements*, Pearson Addison-Wesley

Endsley, M. 2015. Human Systems Integration Requirements Analysis, Chapter 5 of the *APA Handbook of Human-Systems Integration*. Boehm-Davis, D. and Lee, J. (Eds.). APA Books

IEEE 14143:00. 2000. *IEEE Recommended Practice for Software Requirements Specification*. IEEE Computer Society Press.

Kotonya, G. and I. Sommerville. 2000. *Requirements Engineering: Processes and Techniques*. John Wiley.

Kulak, D. and E. Guiney. 2000. *Use Cases: Requirements in Context*. Pearson Addison-Wesley

Rumbaugh, J. 1994. Getting Started: Using use Cases to Capture Requirements. *Journal of Object-Oriented Programming*, 7(5): 8–23.

Thayer, R. and M. Dorfman. 1997. *Software Requirements Engineering*. IEEE Computer Society Press.

Wiegers, K. 2003. *Software Requirements: Practical Techniques for Gathering and Managing Requirements throughout the Product Life Cycle*. Microsoft Press.

You, R. 2001. *Effective Requirements Practices*. Addison-Wesley.

Risk Management

DAU. 2003. *Risk Management Guide for DoD Acquisitions*. Ver. 2.0. Defense Acquisition University.

DeMarco, T. and T. Lister. 2003. *Waltzing with Bears: Managing Risks on Software Projects*. Dorset House.

Herrmann, D. 1999. *Software Safety and Reliability*. IEEE Computer Society.

Kendrick, T. 2015. *Identifying and Managing Project Risk*. AMACOM.

Software Engineering

Bott, F. et al. 2001. *Professional Issues in Software Engineering*, 3rd Edition. Taylor & Francis Group.

Bourque, P. and R. Fairley (Eds.). 2014. *Guide to the Software Engineering Body of Knowledge. SWEBOK*. Version 3.0, IEEE Computer Society.

Brooks, F. 1995. *The Mythical Man Month: Essays on Software Engineering*. 20th Anniversary Edition. Addison-Wesley.

Christensen, M., M. Dorfman and R. Thayer. 2002. *Software Engineering*. IEEE Computer Society Press.

Farkas, E. 2017. *Managing Web Projects*. CRC Press.

Gamma, E. et al. 2000. *Design Patterns: Elements of Reusable Object-Oriented Software*. Addison-Wesley.

Gechman, M., D. Houston and R. Wilkes. 2010. *Software Regression Testing Approaches for the Mission Planning Element of AEHF*. Prepared for the USAF Space and Missiles Systems Center, The Aerospace Corporation, TOR-2010(1475)-4.

Gechman, M. 2011. *The Collaboration of Software and Systems Engineering*. The Aerospace Corporation, TOR-2010(1475)-5.

Gilb, T. 1988. *Principles of Software Engineering Management*. Addison-Wesley.

LaFore, R. 2002. *Object-Oriented Programming in Microsoft C++*, 4th Edition. Pearson Education.

Lipman, S. et al., 2015. *C++ Primer*, 5th Edition. Addison-Wesley.

Pfleeger, S. 2001. *Software Engineering: Theory and Practice*, 2nd Edition. Prentice Hall.

Sommerville, I. 2010. *Software Engineering*, 8th Edition, Pearson Education.

Stroustrup, B. 2013. *The C++ Programming Language*, 4th Edition. Pearson Education.

Weisfeld, M. 2000. *The Object-Oriented thought Process*, 4th Edition, Addison-Wesley.

Witt, T. 2011. *IT Best Practices: Management, Teams, Quality, Performance and Projects*. CRC Press/Auerbach Publications.

Standards

Abelson, L., S. Eslinger, M. Gechman, C. LeDoux, M. Lieu and K. Korzec. 2011. *Software Measurement Standard for Space Systems*. The Aerospace Corporation, TOR-2009(8506)-6.

ACM/IEEE-CS. 1999. *Software Engineering Code of Ethics and Professional Practice*. Joint Task Force on Software Engineering Ethics and Professional Practices, Version 5.2.

Adams, R. et al. 2005. *Software Development Standard for Space System*. The Aerospace Corporation, TOR-2004(3909)-3537, Revision B.

Air Force Standard SMC-S-21. 2009. *Technical Reviews and Audits for System, Equipment and Computer Software*. U.S. Air Force Space and Missile System Center, Volume 1.

ANSI/AAMI. 2009. *HE75: Human Factors Engineering – Design of Medical Devices*. American National Standards Institute/Association for the Advancement of Medical Instrumentation.

ANSI/EIA. 2010. *ANSI/EIA-748-B: Earned Value Management Systems*. American National Standards Institute/Electronics Industries Association.

DoD. 1998. *MIL-STD-8820: System Safety Program Requirements*. U.S. Department of Defense.

DoD. 2012. *MIL-STD-1472: Design Criteria Standard—Human Engineering, Rev G*. U.S. Department of Defense.

EIA/IEEE. J-STD-016-1995. *Standard for Information Technology—Software Life Cycle Processes—Software Development—Customer-Supplier Agreement.* Issued for trial use by Electronics Industries Association /Institute of Electrical and Electronic Engineers.

EIA. 2002. *Engineering Bulletin HEB1: Human Engineering – Principles and Practices.* Electronics Industries Association.

Eslinger, S. 2010. *Space Systems Software Testing: The New Standards.* The Aerospace Corporation, ATR-2007(8365)-1.

FAA. 2003. *HF-STD-001: Human Factors Design Standard.* Federal Aviation Administration.

IEEE/EIA. 1998. *Industry Implementation of International Standard ISO/IEC 12207: Standard for Information Technology—Software Life Cycle Processes—Life Cycle Data.* Joint guide developed by the Institute of Electrical and Electronic Engineers/Electronics Industries Association.

ISO/IEC 15504: 1993. *Software Process Improvement & Capability Determination (SPICE).* Joint International Organization for Standardization/International Electro-technical Commission.

ISO 9001:2005. Quality Management Systems—Fundamentals and Vocabulary. International Organization for Standardization.

ISO/IEC 15939: 2010. *Software Engineering—Software Measurement Process.* Joint International Organization for Standardization/International Electro-technical Commission.

ISO/IEC 20926: 2009. *Software and Systems Engineering—Software Measurement—IFPUG Functional Size Measurement Method.* Joint International Organization for Standardization/International Electro-technical Commission.

ISO/IEC 1976: 2011. *Software Engineering—COSMIC: A Functional Size Measurement Method.* Joint International Organization for Standardization/International Electro-technical Commission.

ISO/IEC/IEEE 15288: 2015. *System and Software Engineering—System Life Cycle Processes.* International Organization for Standardization.

SEI. 2006. *Standard CMMI® Appraisal Method for Process Improvement (SCAMPI).* Ver. 1.2: Method Definition. CMU/SEI-2006-HB-2. Software Engineering Institute/ Carnegie-Mellon University.

Systems Engineering

Diernback, L. 2005. *Flight Software*, Chapter 15 of the *Space Vehicle Systems Engineering Handbook.* The Aerospace Corporation, TOR-2006(8506)-4494.

DoD/OSD. 2008. *Systems Engineering Guide for Systems of Systems.* Department of Defense, Office of the Secretary of Defense.

Endsley, M. 2016. *Building Resilient Systems Via Strong Human System Integration.* Defense AT&L Magazine. January-February 2016. U.S. Department of Defense.

Gechman, M. 2011. *The Collaboration of Software and Systems Engineering.* The Aerospace Corporation, TOR-2010(1475)-5.

INCOSE. 2004. *Guide to Systems Engineering Body of Knowledge.* International Council on Systems Engineering.

Teets, P. 2004. *Revitalizing the Software Aspects of Systems Engineering.* Under Secretary of the Air Force Memorandum, 04A-003.

USAF. 2010. *Life Cycle Systems Engineering.* USAF Instruction AFI 63-1201. U.S. Air Force.

Index